physica status solidi

VOLUME 8 · NUMBER 3 · 1965

AKADEMIE-VERLAG · BERLIN

THEORIA CUM PRAXI

Contents

(Continued on cover three)

physica status solidi

Volume 8 · Number 3 · Pages 637 to 936 and K 121 to K 212

March 1, 1965

AKADEMIE-VERLAG · BERLIN

Subscriptions and orders for single copies should be addressed to

AKADEMIE-VERLAG GmbH, 108 Berlin, Leipziger Straße 3—4 or to Buchhandlung KUNST UND WISSEN, Erich Bieber, 7 Stuttgart 1, Wilhelmstr. 4—6 or to Deutsche Buch-Export und -Import GmbH, 701 Leipzig, Postschließfach 160

Editorial Note:

"physica status solidi" undertakes that an original paper *accepted for publication* before the 8th of any month will be published within 50 days of this date unless the author requests a postponement. In special cases there may be some delay between *receipt* and *acceptance* of a paper due to the review and, if necessary, revision of the paper.

Schriftleiter und verantwortlich für den Inhalt: Professor Dr. Dr. h. c. P. Görlich, 102 Berlin, Neue Schönhauser Str. 20 bzw. 69 Jena, Humboldtstr. 26. Redaktionskollegium: Dr. S. Oberländer, Dr. E. Gutsche, Dr. W. Borchardt. Anschrift der Schriftleitung: 102 Berlin, Neue Schönhauser Str. 20, Fernruf: 426788. Verlag: Akademie-Verlag GmbH, 108 Berlin, Leipziger Str. 3—4, Fernruf: 220441, Telex-Nr. 011773. Postscheckkonto: Berlin 35021. — Die Zeitschrift „physica status solidi" erscheint jeweils am 1. des Monats. Bezugspreis eines Bandes MDN 60,-. Bestellnummer dieses Bandes 1068/8. Gesamtherstellung: VEB Druckerei „Thomas Müntzer" Bad Langensalza. — Veröffentlicht unter der Lizenznummer 1310 des Presseamtes beim Vorsitzenden des Ministerrates der Deutschen Demokratischen Republik.

Review Article

phys. stat. sol. 8, 639 (1965)

Institut für Theoretische Physik der Technischen Universität Dresden

Zum gegenwärtigen Stand der Theorie der Supraleitung

Von

H. WONN

1. Grundlegende Experimente und phänomenologische Beschreibung der Supraleitung

Im Jahre 1908 gelang H. K. Onnes die Verflüssigung des Heliums. Damit konnten Temperaturen bis zu 1 °K erreicht werden. Eines der ersten untersuchten Probleme war die Messung des elektrischen Widerstandes verschiedener Metalle bei tiefen Temperaturen. Dabei wurde 1911 von H. K. Onnes die Supraleitung entdeckt [1]. Das Verhalten normaler Metalle zeigt Fig. 1: Der elektrische Widerstand nähert sich bei $T \to 0$ dem Restwiderstand. Dieser wird durch Verunreinigungen und Gitterfehler bewirkt. Das Verhalten eines Supraleiters ist in Fig. 2 dargestellt: Der elektrische Widerstand verschwindet bei einer kritischen Temperatur T_c plötzlich.

Die kritischen Temperaturen reichen von Werten unterhalb von 0,1 °K bis etwa 18 °K. Wie mit Dauerstromversuchen in supraleitenden Metallringen bewiesen werden konnte, ist der elektrische Widerstand im supraleitenden Zustand exakt Null [2, 3].

Das in Fig. 2 gezeigte Verhalten ist nur richtig für den Gleichstromwiderstand idealer Supraleiter. Ein solcher idealer Supraleiter wird etwa durch einen Einkristall aus weißem Zinn verwirklicht. Bei vielen anderen Supraleitern, speziell bei Legierungen und Verbindungen, verschwindet der Widerstand nicht plötzlich, sondern allmählich.

Bei großen Frequenzen ändert sich der elektrische Widerstand beim Übergang normalleitend — supraleitend nicht plötzlich. Die optischen Eigenschaften im sichtbaren Gebiet ändern sich sogar überhaupt nicht [4]. Da diese aber von der elektrischen Leitfähigkeit σ abhängen (siehe z. B. [5]), bedeutet dies, daß σ in dem entsprechenden Frequenzbereich konstant bleibt. Bei kleineren Frequenzen, im fernen Ultrarot und im Mikrowellengebiet, ändert sich σ dagegen beim Übergang normalleitend — supraleitend. Einen Überblick über experimentelle Unter-

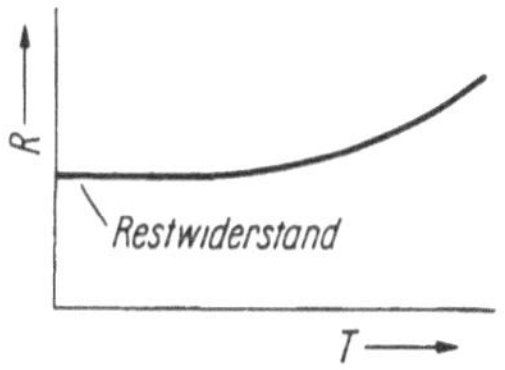

Fig. 1. Verhalten eines Normalleiters

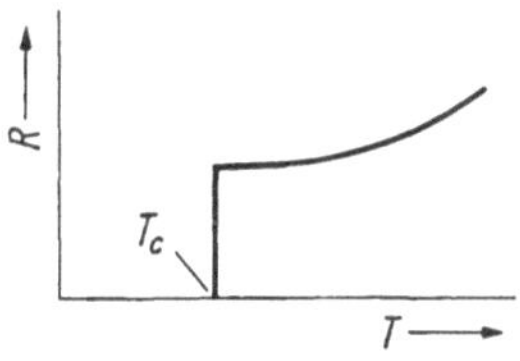

Fig. 2. Verhalten eines Supraleiters

suchungen der elektromagnetischen Eigenschaften der Supraleiter bei diesen Frequenzen geben Biondi et al. [6].

Die Supraleitfähigkeit kann durch genügend starke Magnetfelder wieder zerstört werden. Wird ein Supraleiter in ein Magnetfeld gebracht, so bleibt der elektrische Widerstand Null, solange das Magnetfeld H eine bestimmte kritische Feldstärke H_c nicht überschreitet. Ist H jedoch größer als das kritische Magnetfeld, kann kehrt der Widerstand zu dem normalen Wert zurück [7]. Das kritische Magnetfeld ist eine Funktion der Temperatur (vgl. Fig. 3). Es gilt näherungsweise

$$H_c(T) \approx H_0 [1 - (T/T_c)^2] \,. \tag{1}$$

Die Werte von $H_0 = H_c(0)$ liegen für metallische Elemente zwischen 30 Oe und 2000 Oe, für Legierungen und Verbindungen liegen sie zum Teil wesentlich höher. Auch wenn durch einen Supraleiter ein elektrischer Strom fließt, kann die Supraleitfähigkeit zerstört werden. Es gibt für jedes $T < T_c$ eine kritische Stromstärke, oberhalb der keine Supraleitung mehr möglich ist.

Die magnetischen Eigenschaften der Supraleiter wurden von Meissner und Ochsenfeld untersucht [8]. Diese Eigenschaften zeigen, daß die Supraleitfähigkeit nicht schlechthin unendlich große elektrische Leitfähigkeit ist. In einem Leiter mit unendlich großer Leitfähigkeit ist stets $\boldsymbol{E} = 0$. Bringt man einen solchen idealen Leiter in ein äußeres Magnetfeld $\boldsymbol{H}_a$, dann kann sich wegen

$$\boldsymbol{E} = 0\,, \qquad \frac{\partial}{\partial \boldsymbol{r}} \times \boldsymbol{E} = -\frac{\partial}{\partial t} \boldsymbol{B}_i = 0 \tag{2}$$

das Magnetfeld $\boldsymbol{B}_i$ in seinem Inneren nicht ändern; es bleibt Null. Genauso verhalten sich auch die Supraleiter. Ist das äußere Magnetfeld $H_a \equiv |\boldsymbol{H}_a|$ aber größer als das kritische Magnetfeld H_c, so wird beim Supraleiter die Supraleitfähigkeit zerstört, und das Magnetfeld durchdringt ihn. Die Supraleitfähigkeit läßt sich wieder herstellen, wenn das äußere Magnetfeld H_a bis $H_a < H_c$ verkleinert wird. Wäre der Übergang Normalleitfähigkeit — Supraleitfähigkeit lediglich ein Übergang von endlicher zu unendlich großer Leitfähigkeit, dann würde das Magnetfeld den Supraleiter auch nach dem Wiedereintritt der Supraleitfähigkeit durchdringen. Eine weitere Verkleinerung von H_a ließe das „eingefrorene" Magnetfeld im Inneren $B_i \equiv |\boldsymbol{B}_i|$ ungeändert (Fig. 4a). Die Experimente von Meissner und Ochsenfeld zeigten dagegen, daß das Magnetfeld beim Wiedereintritt der Supraleitfähigkeit aus einem Supraleiter (bis auf eine dünne Randschicht) herausgedrängt wird. Das Magnetfeld B_i im Inneren eines genügend dicken Supraleiters ist also stets Null (Fig. 4b). Diese Erscheinung heißt Meissnereffekt. Der Meissnereffekt ist von grundlegender Bedeutung für die Theorie. Bei einem idealen Leiter hinge B_i, und

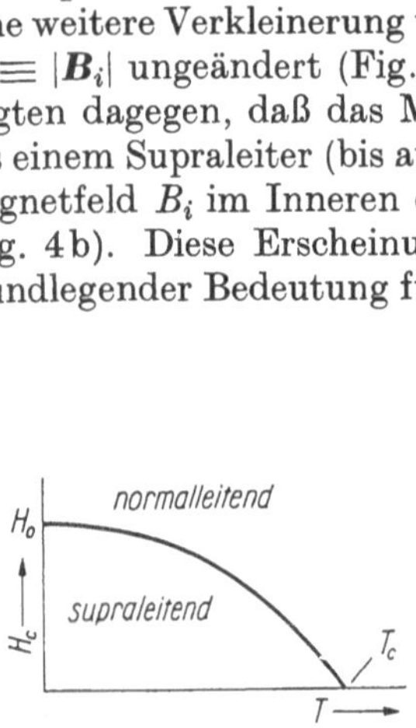

Fig. 3. Kritisches Magnetfeld in Abhängigkeit von der Temperatur

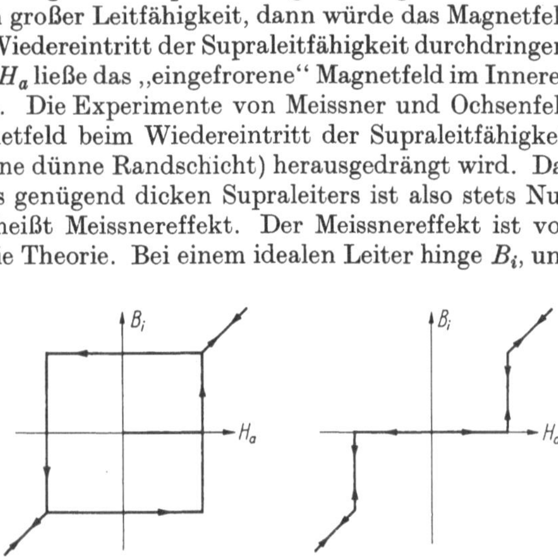

a b

Fig. 4. Unterschied des magnetischen Verhaltens. a) eines idealen Leiters und b) eines Supraleiters

Fig. 5. Temperaturabhängigkeit des Elektronenanteils der Wärmeapazität

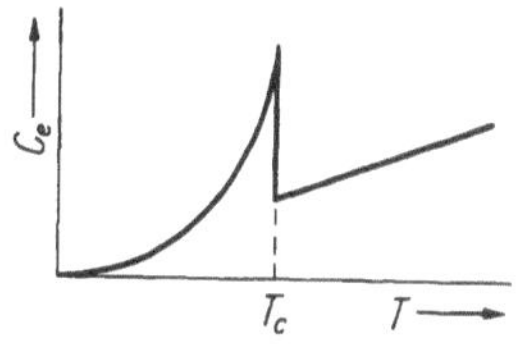

damit der innere Zustand, nicht allein von der Stärke des äußeren Feldes im gegebenen Zeitpunkt ab, sondern auch von den Feldstärkewerten früherer Zeiten. Wie der Meissnereffekt zeigt, ist B_i bei einem Supraleiter eindeutig durch das äußere Feld H_a festgelegt und von der Vorgeschichte unabhängig. Daher ist es möglich, den supraleitenden Zustand als thermodynamischen Gleichgewichtszustand und entsprechend den normalleitenden und den supraleitenden Zustand als zwei Phasen derselben „Substanz" aufzufassen.

Eine bemerkenswerte magnetische Eigenschaft der Supraleiter wurde erst vor wenigen Jahren entdeckt. Deaver und Fairbank [9] sowie Doll und Näbauer [10] konnten experimentell nachweisen, daß der magnetische Fluß durch einen supraleitenden Ring gequantelt ist. Die Größe des „Flußquants" ist $h/2\,e \approx 2 \cdot 10^{-15}$ Vs.

Von den thermodynamischen Eigenschaften der Supraleiter ist besonders die Wärmekapazität und deren Temperaturabhängigkeit interessant. Die Wärmekapazität besteht aus einem Anteil der Leitungselektronen C_e und einem Gitteranteil C_g:

$$C = C_e + C_g\,. \tag{3}$$

Der Gitteranteil C_g ändert sich beim Übergang normalleitend — supraleitend nicht. Daher wird nur der Elektronenanteil C_e betrachtet. Experimentelle Untersuchungen von Keesom und Van Laer [11] sowie von Corak und Shatterthwaite [12] lieferten die in Fig. 5 gezeigte Temperaturabhängigkeit von C_e.

Quantitativ gilt für die Temperaturabhängigkeit des Elektronenanteils im supraleitenden Zustand

$$C_e \approx A\, e^{-B/T}. \tag{4}$$

Diese Temperaturabhängigkeit deutet auf die Existenz eines Energiesprunges im Anregungsspektrum der Leitungselektronen hin. Wenn nämlich zwischen der Grundzustandsenergie und den Energien der angeregten Elektronenzustände eine Energielücke Δ liegt, dann ist entsprechend dem Boltzmanntheorem die Besetzungswahrscheinlichkeit der angeregten Zustände etwa proportional zu $\exp(-\Delta/k\,T)$. Dies führt dazu, daß die Wärmekapazität wie in (4) exponentiell von B/T abhängt. Nach der mikroskopischen Theorie entsteht ein solcher Energiesprung dadurch, daß sich die Elektronen in einem Bindungszustand befinden.

Eine weitere interessante Eigenschaft ist der Tunneleffekt bei Supraleitern. Zwischen zwei Metallen möge sich eine sehr dünne isolierende Schicht befinden. Wenn zwischen den beiden Metallen eine Spannung anliegt, dann fließt infolge des quantenmechanischen Tunneleffektes ein elektrischer Strom durch die Isolierschicht (= Potentialwall). Die zugehörige Strom-Spannungscharakteristik wird nichtlinear, wenn eines der Metalle supraleitend ist (oder beide supraleitend sind). Ursache hierfür ist das Auftreten der bereits erwähnten Energielücke im Anregungsspektrum der Elektronen. Einen Überblick über den Tunneleffekt bei Supraleitern findet man in einer Arbeit von Schmid [13].

Zur Zeit ist die Supraleitung bei 24 Elementen sowie bei einigen hundert Legierungen und Verbindungen bekannt. Viele dieser Verbindungen enthalten nur solche Elemente, bei denen bislang keine Supraleitfähigkeit festgestellt werden konnte. Die supraleitenden Verbindungen enthalten zum Teil auch nichtmetallische oder auch ferromagnetische Elemente [14]. Kürzlich wurde von R. A. Hein, R. Mazelsky et al. entdeckt, daß Supraleitfähigkeit auch in einem Halbleiter (Germaniumtellurid) auftreten kann [15].

Nach der Entdeckung des Meissnereffektes wurden die thermodynamischen Eigenschaften der Supraleiter von Gorter und Casimir theoretisch untersucht. Es gelang ihnen, das thermodynamische Verhalten der Supraleiter und speziell den Phasenübergang normalleitend — supraleitend aus einem phänomenologischen „Zweiflüssigkeitsmodell“ abzuleiten [16]. In diesem wie auch in den später untersuchten Zweiflüssigkeitsmodellen wird angenommen, daß im supraleitenden Zustand sich ein Teil der Elektronen in einem besonderen Zustand befindet und für die typischen Eigenschaften der Supraleiter verantwortlich ist (supraleitende Elektronen), während der übrige Teil sich weiterhin wie im normalleitenden Zustand verhält.

Eine phänomenologische Theorie, die die elektromagnetischen Eigenschaften der Supraleiter beschreibt, wurde von F. und H. London vorgeschlagen [17]. In dieser Theorie wurde das Ohmsche Gesetz der normalen Leitfähigkeit durch einen anderen Zusammenhang ersetzt. Entsprechend den Vorstellungen des Zweiflüssigkeitsmodells besteht die Gesamtstromdichte aus zwei Anteilen

$$\boldsymbol{j} = \boldsymbol{j}_n + \boldsymbol{j}_s \,. \tag{5}$$

Für den normalen Strom gilt weiterhin das Ohmsche Gesetz

$$\boldsymbol{j}_n = \sigma \boldsymbol{E} \,. \tag{6}$$

Der Suprastrom $\boldsymbol{j}_s$ genügt den Londonschen Gleichungen

$$\frac{\partial}{\partial t} \Lambda \boldsymbol{j}_s = \boldsymbol{E}, \qquad \frac{\partial}{\partial r} \times \Lambda \boldsymbol{j}_s = -\boldsymbol{B} \,. \tag{7}$$

Λ ist eine empirische temperaturabhängige Materialkonstante.

Aus der Londontheorie folgt, daß ein äußeres Magnetfeld innerhalb eines Supraleiters mit wachsendem Abstand von der Oberfläche rasch abklingt. Speziell für einen Supraleiter mit einer unendlich großen ebenen Begrenzungsfläche ergibt sich, für das Magnetfeld im Abstand x von der Oberfläche (Fig. 6)

$$B(x) = B_0 \, e^{-x/\lambda} \qquad \text{mit} \qquad \lambda = \sqrt{\Lambda/\mu_0} \,. \tag{8}$$

Das Magnetfeld dringt also in eine dünne Oberflächenschicht ein, deren Dicke etwa durch die Eindringtiefe λ gegeben ist. Im Inneren des Supraleiters, für $x \gg \lambda$, ist das Magnetfeld praktisch Null. Größenordnungsmäßig gilt für die Eindringtiefe $\lambda \approx 10^{-7}$ m.

Im Rahmen eines Zweiflüssigkeitsmodells ist die Materialkonstante Λ durch die Dichte der supraleitenden Elektronen (n_s) bestimmt; es gilt $\Lambda \sim 1/n_s$. In der Londontheorie ist Λ bzw. n_s innerhalb des gesamten supraleitenden Gebietes räumlich konstant. Im Jahre 1950 wurde von Ginzburg und Landau eine allgemeinere phänomenologische Theorie angegeben, in der n_s nicht mehr räumlich

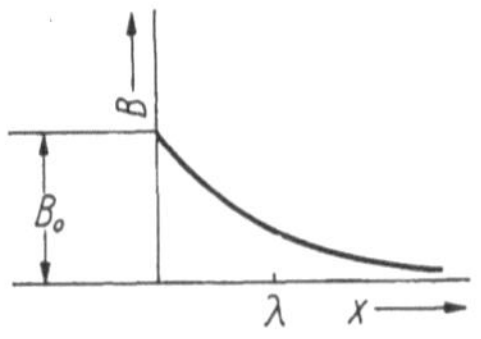

Fig. 6. Feldverlauf in einem supraleitenden Halbraum

konstant sein muß, sondern durch die physikalischen Bedingungen innerhalb des Supraleiters bestimmt ist [18]. Eine weitere Verallgemeinerung der Londontheorie stammt von Pippard [19]. In dieser nichtlokalen Theorie ist die Suprastromdichte $\boldsymbol{j}_s(\boldsymbol{r})$ am Ort $\boldsymbol{r}$ nicht mehr allein mit den Feldgrößen $\boldsymbol{E}(\boldsymbol{r})$ und $\boldsymbol{B}(\boldsymbol{r})$ des gleichen Ortes verknüpft wie in (7), sondern $\boldsymbol{j}_s(\boldsymbol{r})$ hängt auch von den Feldgrößen $\boldsymbol{E}(\boldsymbol{r}')$ und $\boldsymbol{B}(\boldsymbol{r}')$ benachbarter Orte $\boldsymbol{r}'$ ab.

2. Wechselwirkung und Isotopeneffekt

Das Ziel einer mikroskopischen Theorie der Supraleitung muß es sein, die Grundgleichungen der phänomenologischen Theorien zu erklären. Erste Versuche in dieser Richtung wurden von F. London unternommen. Er konnte, ausgehend von der quantenmechanischen Beschreibung des Supraleiters, die phänomenologischen Gleichungen (7) begründen. Hierzu war die Annahme nötig, daß sich die Wellenfunktionen der supraleitenden Elektronen nicht ändern, wenn der Supraleiter in ein Magnetfeld gebracht wird. Der Grund für eine derartige „Starrheit" der Elektronenwellenfunktion mußte irgendeine zwischen den Elektronen wirkende Wechselwirkung sein. Über die Natur dieser Wechselwirkung war zunächst nichts bekannt.

In einer Theorie von Welker [20] wurde angenommen, daß die Supraleitung durch eine magnetische Wechselwirkung zwischen den Elektronen zustande kommt. Nach den Theorien von Heisenberg [21] und von Born und Cheng [22] sollte die Supraleitung durch die Coulombwechselwirkung der Elektronen zustande kommen. Diese Bemühungen hatten aber nicht den gewünschten Erfolg. Einen entscheidenden Fortschritt brachte erst die Theorie von H. Fröhlich [23]. In dieser Theorie wurde angenommen, daß Supraleitung durch die Wechselwirkung der Elektronen über die Gitterschwingungen, die Elektronen–Phononen-Wechselwirkung, entsteht. Bei Supraleitern, die aus verschiedenen Isotopen bestehen, sind die Frequenzen der Gitterschwingungen von der Isotopenmasse M abhängig; ebenso wie für die Frequenz eines harmonischen Oszillators $\omega = \sqrt{a/m}$ (a Federkonstante, m Masse) gilt für die Frequenzen der Gitterschwingungen $\omega \sim \sqrt{1/M}$. Dies führt in der Theorie von Fröhlich zu einem Isotopeneffekt: Die kritische Temperatur T_c hängt von der Isotopenmasse M ab:

$$T_c \sim M^{-1/2} . \tag{9}$$

Ein solcher Isotopeneffekt konnte kurze Zeit später experimentell von Maxwell [24] und von Reynolds et al. [25] bestätigt werden.

Die Theorie von Fröhlich konnte zwar die Eigenschaften der Supraleiter nicht widerspruchsfrei erklären, aber die Grundidee, daß die Supraleitung durch die Elektronen–Phononen-Wechselwirkung bewirkt wird, schien richtig zu sein. Im Jahre 1957 wurde von Bardeen, Cooper und Schrieffer eine mikroskopische Theorie der Supraleitung angegeben (BCS-Theorie), die ebenfalls von dieser Grundidee ausgeht [26].

Auch die BCS-Theorie führt zu einem Isotopeneffekt. Sie liefert darüber hinaus viele Eigenschaften der Supraleiter in recht guter quantitativer Übereinstimmung mit den Experimenten. Man könnte daher denken, die Supraleitung sei schlechthin durch die Elektronen–Phononen-Wechselwirkung verursacht. Nun fanden aber Matthias et al., daß sich in den Übergangselementen Osmium und Ruthenium kein Isotopeneffekt nachweisen läßt [27]. Es wurde daher angenommen, daß in den Übergangselementen neben der Elektronen–Phononen-Wechselwirkung noch eine andere Wechselwirkung als Ursache für die Supraleitung mit berücksichtigt werden muß. Matthias vermutete, daß es sich hierbei um eine magnetische Spin-Austauschwechselwirkung handelt [13]. Andere Autoren nehmen an, daß bei den Übergangselementen eine Austauschwechselwirkung zwischen den Leitfähigkeitselektronen und den Elektronen innerer, nicht voll besetzter Schalen die Supraleitung bewirkt bzw. beeinflußt [28, 29, 30].

Mit Hilfe der Thermodynamik läßt sich aus dem kritischen Magnetfeld H_c die Differenz der Energien zwischen dem normalleitenden (E_n)- und dem supraleitenden (E_s)-Zustand berechnen:

$$E_n - E_s \approx \frac{\mu_0}{2} H_c^2 V . \tag{10}$$

V ist das Volumen des (kompakten) Supraleiters. Bei $T = 0$ entfällt dabei auf jedes Leitungselektron größenordnungsmäßig $(E_n - E_s)/N \approx 10^{-8}$ eV. Diese Energie ist sehr klein. Das wird deutlich, wenn man sie mit der Fermienergie (etwa 10 eV) oder mit der Korrelationsenergie (etwa 1 eV) vergleicht. Beim Übergang normalleitend — supraleitend ändert sich also die Energie nur wenig, aber die physikalischen Eigenschaften des Supraleiters verändern sich sehr auffallend. Beides wird verständlich, wenn man folgendes annimmt:

1. Nur die Elektronen einer dünnen Schicht unter der Fermioberfläche werden beim Übergang normalleitend — supraleitend beeinflußt. Die Dicke dieser Schicht ist etwa $\Delta\varepsilon \approx k\,T_c \approx 10^{-4}$ eV.

2. Nur diese Elektronen tragen im Mittel die Energie $\Delta\varepsilon \approx k\,T_c$ zu $E_n - E_s$ bei. Damit ist also nur der 10^{-4}te Teil der Leitungselektronen an der Supraleitung „beteiligt", und es ist klar, daß sich die Energie beim Übergang normalleitend — supraleitend nur wenig ändern wird. Dennoch können sich die physikalischen Eigenschaften des Supraleiters wesentlich ändern, weil die Beeinflussung der Elektronen an der Fermioberfläche stattfindet.

3. Die Cooper-Paare

In einem Supraleiter, wie auch allgemein in einem Metall, wirken zwischen den Leitungselektronen zwei verschiedene Wechselwirkungen: die abstoßende Coulombwechselwirkung und die Elektronen–Phononen-Wechselwirkung. Die Elektronen–Phononen-Wechselwirkung ist geschwindigkeitsabhängig. Sie wird anziehend, wenn die Impulse der Elektronen dicht bei der Fermioberfläche liegen. Ein Metall kann nun supraleitend werden, wenn die aus Coulombwechselwirkung und Elektronen–Phononen-Wechselwirkung resultierende Gesamtwechselwirkung anziehend ist. Aus dieser Forderung läßt sich ein Kriterium für das Auftreten der Supraleitung gewinnen [31]. Das eben Festgestellte gilt für Nichtübergangselemente. Für die Übergangselemente liegen die Verhältnisse nicht so klar. Aber auch bei diesen Elementen müßte für Supraleiter die resultierende Gesamtwechselwirkung, die zwischen den Leitungselektronen wirkt, anziehend sein. Denn nur bei einer effektiv anziehenden Wechselwirkung kann das Auftreten eines Energiesprunges im Anregungsspektrum erklärt werden.

Wegen dieser anziehenden mittleren Wechselwirkung muß das Elektronensystem des Leitfähigkeitsbandes gebundene Zustände besitzen. Um diese Zustände zu untersuchen, wurde von Cooper das folgende Problem behandelt [32]: Die Elektronen des Leitfähigkeitsbandes besetzen eine Fermikugel. Außerdem befinden sich außerhalb der Fermikugel zwei weitere Elektronen. Zwischen ihnen wirkt die resultierende anziehende Gesamtwechselwirkung. Dieses Modellproblem

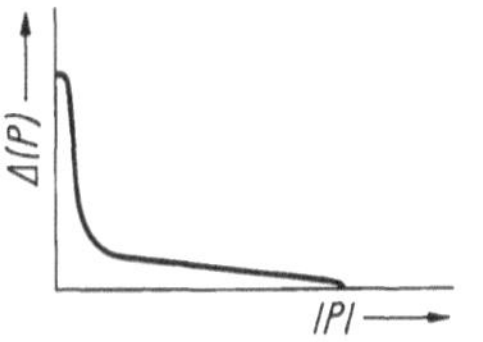

Fig. 7. Bindungsenergie eines Cooper-Paares in Abhängigkeit vom Gesamtimpuls

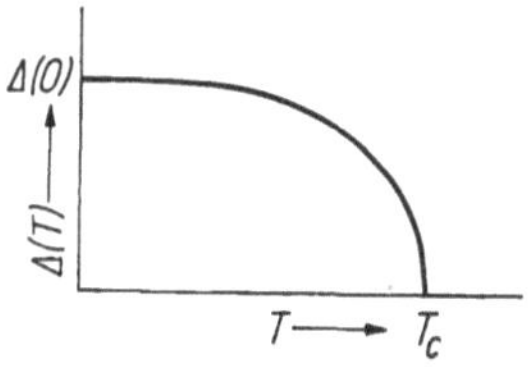

Fig. 8. Aus der BCS-Theorie folgende Temperaturabhängigkeit des Energiesprunges

ist exakt lösbar. Die beiden Elektronen besitzen einen gebundenen Zustand und ein Kontinuum freier Zustände. Der Bindungszustand tritt nur bei antiparallelen Spins auf. Er ist ein Singulettzustand. Die zugehörige Bindungsenergie Δ ist vom Gesamtimpuls $\boldsymbol{P} = \boldsymbol{p}_1 + \boldsymbol{p}_2$ der beiden Elektronen abhängig. Wie in Fig. 7 dargestellt, gibt es nur für $\boldsymbol{P} \approx 0$ eine merkliche Bindungsenergie. Für größere Gesamtimpulse geht $\Delta(\boldsymbol{P})$ rasch gegen Null. Die „festeste" Bindung tritt also auf, wenn der Gesamtimpuls der beiden Elektronen verschwindet. Zwei derartig gebundene Elektronen werden als Cooper-Paar bezeichnet.

Wie sich das gesamte System der Leitfähigkeitselektronen unter dem Einfluß der aus Coulombwechselwirkung und Elektronen–Phononen-Wechselwirkung resultierenden mittleren Wechselwirkung verhält, wurde in der BCS-Theorie untersucht [26]. Im normalleitenden Zustand besetzen die Elektronen des Leitfähigkeitsbandes bei $T = 0$ °K eine Fermikugel. Im supraleitenden Zustand sind die Elektronen in der Nähe der Fermioberfläche in Cooper-Paaren gebunden. Dies macht sich so bemerkbar, daß die Wellenfunktion des supraleitenden Grundzustandes eine Linearkombination von solchen Konfigurationen ist, in denen die Elektronenzustände $\boldsymbol{p}\uparrow$ und $-\boldsymbol{p}\downarrow$ entweder beide besetzt oder beide leer sind (siehe Fig. 11a). Die Wellenfunktionen für angeregte supraleitende Zustände sind ebenfalls Linearkombinationen von solchen Elektronenzuständen, wobei aber bestimmte „Paare" ausgelassen werden und die Elektronenzustände nur jeweils einzeln besetzt sind; z. B. sind die Zustände $\boldsymbol{p}'\uparrow$ und $-\boldsymbol{p}''\downarrow$ besetzt und die Zustände $-\boldsymbol{p}'\downarrow$ und $\boldsymbol{p}''\uparrow$ leer. Diese angeregten Zustände sind dadurch gekennzeichnet, daß ein bestimmtes Cooper-Paar „aufgebrochen" ist. Beim Aufbrechen eines Cooper-Paares muß die Bindungsenergie Δ aufgebracht werden. Dies führt dazu, daß im Anregungsspektrum der Elektronen ein Energiesprung vorhanden ist.

In der BCS-Theorie wird das Verhalten des Systems der Leitungselektronen nicht nur für $T = 0$ °K, sondern auch für endliche Temperaturen untersucht. Dabei zeigt sich, daß der Energiesprung (d. h. die mittlere Bindungsenergie der Cooper-Paare) von der Temperatur abhängt. $\Delta(T)$ wird mit zunehmender Temperatur kleiner und verschwindet bei der kritischen Temperatur T_c (siehe Fig. 8).

Die BCS-Theorie liefert weiter die Größe und die Temperaturabhängigkeit des kritischen Magnetfeldes H_c und der Wärmekapazität C_e. Die Übereinstimmung mit den experimentell gefundenen Beziehungen (1) bzw. (4) ist gut. Auch die elektrodynamischen Eigenschaften der Supraleiter lassen sich aus der BCS-Theorie erhalten. Bardeen, Cooper und Schrieffer konnten zeigen, daß zwischen dem Strom und den elektromagnetischen Feldgrößen ein ähnlicher Zusammenhang besteht wie in der phänomenologischen Theorie von Pippard [19].

Die mit Hilfe der BCS-Theorie erhaltenen Ergebnisse konnten auch mit anderen Methoden abgeleitet werden. Hier sind zu nennen: die Einführung von Quasiteilchenoperatoren durch Bogoljubov et al. [33] und durch Valatin [34], der Quasispinformalismus, der von Anderson [35] und von Koppe und Mühlschlegel [36] verwendet wurde, sowie die Methode der Greenschen Funktionen, die von Gorkov eingeführt wurde [37]. Speziell mit der letzten Methode konnten von Gorkov auch die Grundgleichungen der phänomenologischen Ginzburg-Landau-Theorie abgeleitet werden. Mit dieser Methode konnte ferner von Keller und Zumino [38] und von Weller [39] die Flußquantelung aus der mikroskopischen Theorie heraus begründet werden.

4. Quantitative Erklärung der Supraleitung

Die Tatsache, daß im supraleitenden Zustand alle Elektronen an der Fermioberfläche in Cooper-Paaren gebunden sind, hat zur Folge, daß in einem Supraleiter elektrische Ströme verlustfrei fließen können. Dies soll nun in dem folgenden vereinfachten Bild veranschaulicht werden.

In einem normalleitenden Metall besetzen die Elektronen des Leitfähigkeitsbandes bei $T = 0$ eine Fermikugel (Fig. 9a). Wird an dieses Metall eine elektrische Spannung gelegt, dann werden die Elektronen einerseits durch das entstehende elektrische Feld beschleunigt. Andererseits werden sie durch Stöße mit dem Gitter laufend „gebremst". Bei $T = 0$ handelt es sich vor allem um solche Stöße, die durch Verunreinigungen und Gitterfehler bewirkt werden (Restwiderstand). Im stationären Endzustand stellt sich die in Fig. 9b gezeigte Verteilung ein: Die Fermikugel ist als Ganzes verschoben. Ihr Mittelpunkt befindet sich beim Impuls $\boldsymbol{q}$. Die im Metall fließende Stromdichte ist proportional zu $|\boldsymbol{q}|$. (In Fig. 9 sind ebenso wie auch in Fig. 10, 11 und 12 die mechanischen Impulse dargestellt, und nicht die physikalisch unanschaulicheren kanonischen Impulse.)

Schaltet man nun die angelegte Spannung ab, dann wird die Fermikugel in ihre Ausgangslage zurückverschoben. Dies geschieht im einzelnen folgendermaßen: Aus dem rechten Kugelsegment von Fig. 10 gehen die Elektronen infolge der Stöße mit dem Gitter in das linke Kugelsegment über. Hierbei wird die Energie der Elektronen stets kleiner. Die den Elektronen entzogene Energie wird an das Gitter abgegeben und macht sich als Wärme bemerkbar. Es werden so lange Elektronen „umgesetzt", bis der Mittelpunkt der Fermikugel wieder im Ursprung liegt, d. h. bis kein Strom mehr fließt. Während des „Abklingprozesses" wird die Energie des Elektronensystems (kinetische Energie) kleiner und die Energie der Gitterschwingungen (Wärmeenergie) größer. Es liegt also ein irreversibler Prozeß vor.

Im supraleitenden Zustand besetzen die Elektronen des Leitfähigkeitsbandes ebenfalls näherungsweise eine Fermikugel, aber die Elektronen in der Nähe der

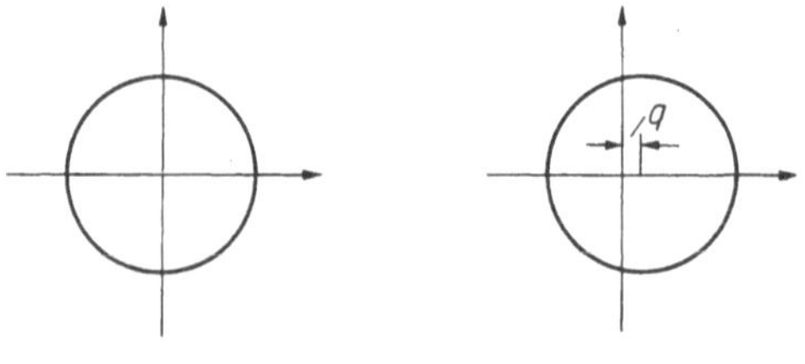

a b

Fig. 9. Verteilung der Elektronen des Leitfähigkeitsbandes im normalleitenden Zustand. a) Strom $\boldsymbol{j} = 0$; b) Strom $\boldsymbol{j} \neq 0$

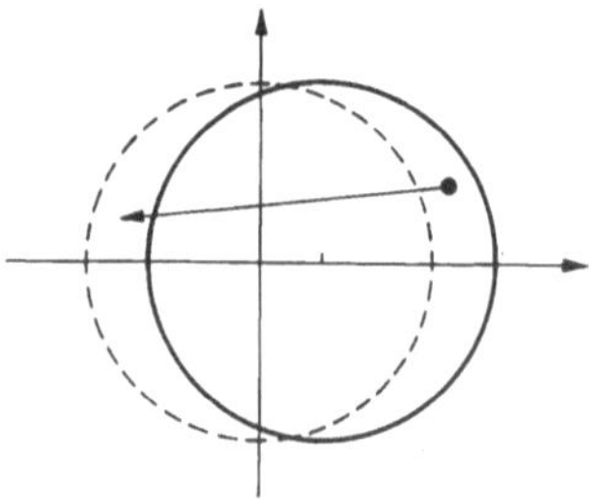

Fig. 10. Zur Diskussion des „Bremsprozesses" in einem Normalleiter

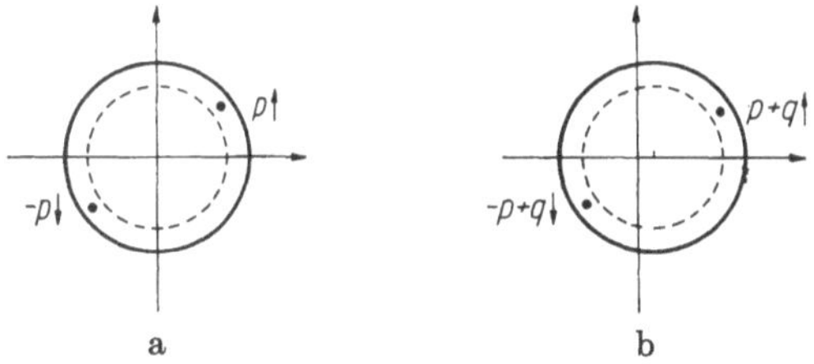

Fig. 11. Verteilung der Elektronen des Leitfähigkeitsbandes im supraleitenden Zustand. a) Strom $\boldsymbol{j} = 0$; b) Strom $\boldsymbol{j} \neq 0$

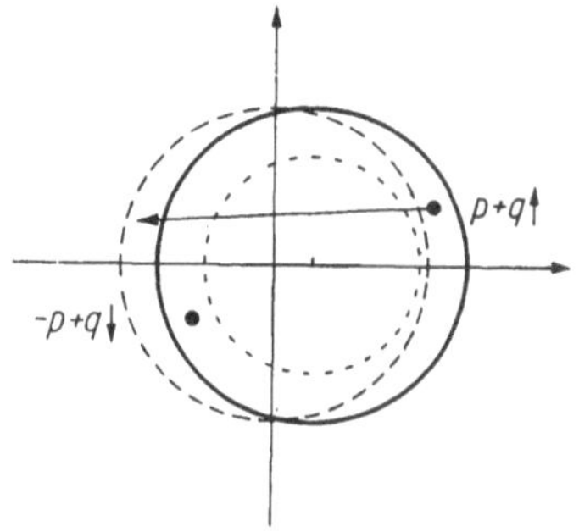

Fig. 12. Zur Diskussion des „Bremsprozesses" in einem Supraleiter

Fermioberfläche sind in Cooper-Paaren gebunden, d. h., die Wellenfunktion ist eine Linearkombination von Konfigurationen, bei denen die Elektronenzustände $\boldsymbol{p}\uparrow$ und $-\boldsymbol{p}\downarrow$ jeweils paarweise besetzt oder leer sind (Fig. 11a). Wenn im Supraleiter ein Strom fließt, dann liegt der Mittelpunkt der Fermikugel nicht mehr im Ursprung, sondern die Fermikugel ist als Ganzes verschoben. Dementsprechend ist die Wellenfunktion in diesem Zustand eine Linearkombinationen von Konfigurationen, bei denen diesmal die Elektronenzustände $\boldsymbol{p} + \boldsymbol{q}\uparrow$ und $-\boldsymbol{p} + \boldsymbol{q}\downarrow$ jeweils paarweise besetzt oder leer sind. Es ist wichtig, daß $\boldsymbol{q}$ für alle „Paare" gleich groß ist (Fig. 11b).

Im supraleitenden Zustand werden die Elektronen nicht „gebremst". Wenn nämlich ein Elektron infolge eines Stoßes mit dem Gitter aus dem Kugelsegment von Fig. 12 in das linke Kugelsegment übergeht, dann muß zuvor ein Cooper-Paar aufgebrochen werden. Daher muß zunächst die Bindungsenergie aufgebracht werden. Beim Übergang des Elektrons aus dem rechten Kugelsegment in das linke wird die Energie des Elektronensystems wieder kleiner, weil die kinetische Energie des übergegangenen Elektrons abnimmt. Bei kleinem Impuls $\boldsymbol{q}$ ist die durch den Übergang des Elektrons gewonnene kinetische Energie kleiner als die zum Aufbrechen des Cooper-Paars erforderliche Bindungsenergie. Das Elektronensystem kann daher an das Gitter keine Energie abgeben, sondern im Gegenteil, das Gitter müßte Energie an das Elektronensystem abgeben, damit das letztere „gebremst" werden könnte. Das ist aber bei $T < T_c$ aus thermodynamischen Gründen nicht möglich.

Bei genügend großen Impulsen $\boldsymbol{q}$ ($|\boldsymbol{q}| > q_{krit}$) ist dagegen die durch den Übergang des Elektrons freigewordene Energie größer als die benötigte Bindungsenergie. Dann ist eine „Bremsung" des Elektronensystems durch Stöße mit dem Gitter energetisch möglich, und die Supraleitung bricht zusammen. Da der Betrag der im Supraleiter fließenden Stromdichte $\boldsymbol{j}$ proportional zu $|\boldsymbol{q}|$ ist, bedeutet dies, daß die Supraleitung zerstört wird, wenn die Stromstärke einen bestimmten kritischen Wert überschreitet. Ebenso versteht man, daß die Supraleitung durch ein starkes äußeres Magnetfeld zerstört wird. Entsprechend dem Meissnereffekt kann ein Magnetfeld nicht in das Innere eines Supraleiters eindringen. Dies wird durch elektrische Ströme an der Oberfläche des Supraleiters bewirkt. Die Stärke dieser Ströme ist proportional zu $|\boldsymbol{H}_a|$. Überschreitet das äußere Magnetfeld $\boldsymbol{H}_a$ einen kritischen Wert, dann überschreiten auch die Ströme an der Oberfläche des Supraleiters ihre kritische Stromstärke, und die Supraleitung bricht zusammen.

Herrn Prof. Dr. W. Macke möchte ich für die Anregung zum Thema sowie für wertvolle Hinweise und helfende Diskussionen herzlich danken.

Literatur

[1] H. K. ONNES, Leiden Comm. **119b, 120b, 122b, 124c** (1911).
[2] H. K. ONNES, Leiden Comm. **140b, 141b** (1914).
[3] J. W. CROWE, IBM J. Res. Developm. **1**, 294 (1957).
[4] N. G. MCCRUM und C. A. SHIFFMAN, Proc. Phys. Soc. **A 67**, 368 (1954).
[5] W. MACKE, Wellen — Ein Lehrbuch der Theoretischen Physik, Akad. Verl.-Ges., Leipzig 1958.
[6] BIONDI, FORRESTER, GARFUNKEL und SHATTERTHWAITE, Rev. mod. Phys. **30**, 1109 (1958).
[7] H. K. ONNES, Leiden Comm., Suppl. **35** (1913).
[8] W. MEISSNER und R. OCHSENFELD, Naturwissenschaften **21**, 787 (1933).
[9] B. S. DEAVER und W. M. FAIRBANK, Phys. Rev. Letters **7**, 43 (1961).
[10] R. DOLL und M. NÄBAUER, Phys. Rev. Letters **7**, 51 (1961).
[11] W. H. KEESOM und P. H. VAN LAER, Physica **5**, 193 (1938).
[12] W. S. CORAK und C. B. SHATTERTHWAITE, Phys. Rev. **102**, 662 (1956).
[13] A. SCHMID, phys. stat. sol. **7**, 3 (1964).
[14] B. T. MATTHIAS, Phys. today **16**, 21 (1963).
[15] HEIN, GIBSON, MAZELSKY, MILLER und HUHN, Phys. Rev. Letters **12**, 320 (1964).
[16] C. J. GORTER und H. B. G. CASIMIR, Phys. Z. **35**, 963 (1934); Z. techn. Phys. **15**, 539 (1934).
[17] F. LONDON und H. LONDON, Proc. Roy. Soc. **A 149**, 71 (1935); Physica **2**, 341 (1935).
[18] V. L. GINZBURG und L. D. LANDAU, J. exper. theor. Phys. **20**, 1064 (1950).
[19] A. B. PIPPARD, Proc. Roy. Soc. **A 216**, 547 (1953).
[20] H. WELKER, Z. Phys. **114**, 525 (1939).
[21] W. HEISENBERG, Z. Naturf. **3a**, 65 (1948).
[22] M. BORN und K. C. CHENG, Nature **163**, 247 (1949).
[23] H. FRÖHLICH, Phys. Rev. **79**, 845 (1950).
[24] E. MAXWELL, Phys. Rev. **78**, 477 (1950).
[25] REYNOLDS, SERIN, WRIGHT und NESBITT, Phys. Rev. **78**, 487 (1950).
[26] J. BARDEEN, L. N. COOPER und J. R. SCHRIEFFER, Phys. Rev. **106**, 162 (1957); **108**, 1175 (1957).
[27] T. H. GEBALLE und B. T. MATTHIAS, IBM J. Res. Developm. **6**, 256 (1962).
[28] J. PERETTI, Phys. Letters (Netherlands) **2**, 275 (1962).
[29] K. ANDRES, J. L. OLSEN und H. ROHRER, IBM J. Res. Developm. **6**, 84 (1962).
[30] J. W. GARLAND, Phys. Rev. Letters **11**, 111 (1963).
[31] D. PINES, Phys. Rev. **109**, 280 (1958).
[32] L. N. COOPER, Phys. Rev. **104**, 1189 (1956).
[33] N. N. BOGOLJUBOV, V. V. TOLMACHOV und D. V. SCHIRKOV, Fortschr. Phys. **6**, 605 (1958).
[34] J. G. VALATIN, Nuovo Cimento **7**, 845 (1958).
[35] P. W. ANDERSON, Phys. Rev. **112**, 1900 (1958).
[36] H. KOPPE und B. MÜHLSCHLEGEL, Z. Phys. **151**, 613 (1958).
[37] L. P. GORKOV, J. exper. theor. Phys. **34**, 735 (1958); **36**, 1918 (1959).
[38] J. B. KELLER und B. ZUMINO, Phys. Rev. Letters **7**, 164 (1961).
[39] W. WELLER, phys. stat. sol. **2**, 1342 (1962).

(Received November 3, 1964)

Original Papers

phys. stat. sol. 8, 649 (1965)

Physics Department, University of Leicester

The Edge Emission in CdS Crystals

By

W. E. Spear and G. W. Bradberry

In this paper the mechanism of the green edge emission in pure, undoped CdS crystals is investigated. First, a systematic study of photoconductive, transport and luminescent properties and their dependence on heat treatment (350 °C) has been carried out. The photoconductive measurements between 77 and 130 °K indicate that a class II centre ($S_p \gg S_n$) lies 0.13 to 0.15 eV above the valence band. The experimental results are then compared with calculations based on a three centre model and carried out by electronic computer over a range of temperatures and centre densities. The observed luminescent properties can be predicted in agreement with experiment, but only if the radiative recombination takes place between a free electron and a hole trapped in the above centre (Schön-Klasens model). The effect of heat treatment and the nature of these centres are briefly discussed.

Es wird der Mechanismus der grünen Kantenlumineszenz von reinen, undotierten Kristallen untersucht. Die Photoleitungs-, Transport- und Lumineszenzeigenschaften und ihre Abhängigkeit von einer Wärmebehandlung (bei 350 °C) wurden systematisch untersucht. Die Photoleitungsmessungen zwischen 77 und 130 °K weisen auf ein Zentrum der Klasse II ($S_p \gg S_n$) mit einer energetischen Lage von 0,13 bis 0,15 eV über dem Valenzband hin. Die experimentellen Ergebnisse werden verglichen mit Rechnungen, die auf einem Drei-Zentren-Modell basieren und mit einer elektronischen Rechenmaschine für einen gewissen Temperatur- und Zentrendichtebereich ausgeführt wurden. Die beobachteten Lumineszenzeigenschaften sind nur dann in Übereinstimmung mit dem Experiment, wenn freie Elektronen mit Löchern, die an dem obengenannten Zentren angehaftet sind, strahlend rekombinieren (Schön-Klasens-Modell). Der Einfluß einer Wärmebehandlung und die Natur dieser Zentren werden kurz diskutiert.

1. Introduction

The term "edge emission" denotes a luminescent transition, observed in a number of II-VI compounds, in which the emitted photon energy lies close to that of the fundamental absorption edge. One of the most familiar examples is the green emission band which can be observed in CdS crystals cooled to liquid nitrogen temperature under excitation from a Hg arc.

Since the original observation of Kröger [1] the study of edge emission bands, both in doped and undoped CdS crystals, has formed the subject of a number of papers. The present work is concerned with an investigation of the green emission in "pure" CdS crystals, by which we mean that the specimens used did not contain any deliberately added impurities and were not treated in any way after growth.

The observation of the luminescent bands in pure specimens has led to much speculation as to their origin and at present the mechanism of the luminescent process still appears to be a controversial matter. The work of Lambe, Klick, and Dexter [2] has shown fairly convincingly that, contrary to the original interpretation of Kröger, the radiative recombination occurs between a free carrier and a carrier of opposite polarity trapped at an imperfection centre. However, the experimental work up to now does not seem to provide unambiguous evidence as to which carrier is trapped, the electron or the hole.

The two diametrically opposite models proposed in the literature are illustrated in Fig. 1. Model I, involving the recombination of a trapped hole and a free electron, was originally put forward by Schön [3] and by Klasens [4]. In model II, proposed by Lambe and Klick [5], the opposite process is envisaged.

A number of attempts to distinguish experimentally between the above models were based on a comparison between the decay rates of the luminescence and the photocurrent. Experiments of this kind were carried out on pure and doped crystals at 77 °K and 4 °K by Lambe [6], who regarded his results as evidence for model II. Broser et al. [7] on the other hand, observed a decay of the green emission at 4 °K in step with the photocurrent, and interpreted this in terms of model I. In another, somewhat different type of experiment, Lambe, Klick, and Dexter [2] used irradiation in the quenching band to increase the hole density. The crystal was first excited by U.V. at 4 °K and then kept in the dark. Subsequent illumination with 1 μm radiation produced a burst of green emission. A comparison of the luminescent and photocurrent decay was interpreted in favour of model II. Some doubt was thrown on the interpretation of the above experiments by the work of Pedrotti and Reynolds [8], which indicated that in the low temperature range used electrons may become bound at shallow levels below the conduction band.

As a general criticism Collins [9] pointed out that a comparison between the decay times is not sufficient to allow one to differentiate between the two models, since the decay times are related in a manner dependent on carrier concentration and recombination cross section.

In his own experiments Collins first investigated the spectral distribution of the blue and green emission; on the basis of his results, and using an exciton binding energy of 0.023 eV, he placed the centre responsible for green emission at 0.15 ± ± 0.01 eV from one of the band edges. Further experiments, with excitation by short 1 MeV electron pulses, were interpreted by Collins as evidence for model I.

A similar conclusion was reached by Niekisch [10] in his work on defect centres in CdS. The P.E.M. measurements of Auth [11] and Friedrich [12] are of interest in connection with model I as they clearly showed that a level of hole trapping centres exists at about 0.1 eV from the valence band; the above authors suggested that these centres were connected with the luminescent process.

Recently Halsted and Segall [13] interpreted edge emission in II-VI compounds in terms of a native double acceptor defect lying close to the conduction band (within 0.09 eV for CdS). This suggestion implies a recombination process according to model II. As the above defect was produced by heating the crystal to 900 °C in a Cd atmosphere and by subsequent quenching [14], it might be questioned whether this type of centre would generally be present in untreated crystals.

Razbirin suggested in a recent communication [15] that edge emission resulted from the annihilation of an exciton with the simultaneous ionisation of a defect

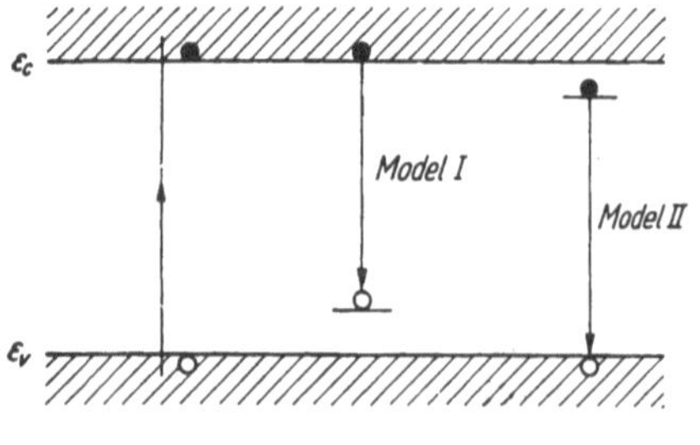

Fig. 1. The two proposed mechanisms for edge emission, model I (Schön-Klasens) and model II (Lambe-Klick)

centre. As this centre is assumed to be a neutral acceptor (0.12 eV above the valence band) the suggested luminescent transition is more closely related to model I than to model II.

In view of the diversity of opinion expressed in the literature as to the origin of the green edge emission, it appeared of interest to approach this problem in an essentially different way. The first step consisted of a systematic investigation on 12 CdS specimens of the following properties:

a) the luminescent intensity as a function of temperature and of excitation density,
b) the photoconductivity as a function of temperature and of excitation density,
c) the electron drift mobility,
d) infra-red quenching,
e) the effect of heat treatment on the above properties.

The results of experiments b), c) and e) were then fitted to a three centre model of a photoconductor and this was used to compute the rather sensitive luminescent properties according to both model I and model II. A comparison of the predicted and the experimental curves shows that within the scope of the three centre model only one of the proposed luminescent transitions is in agreement with experiment and fits into a consistent description of all the above properties.

2. Experimental Details

The measurements listed in the previous section were made in an attempt to establish a general pattern of behaviour in a number of undoped CdS crystals. For this reason, the specimens were obtained from three different sources (A. E. I. Rugby, E. M. I. Hayes, Birmingham University) to ensure that the results were not representative of specific properties determined by the growing technique or by impurities in a particular batch of starting materials.

Most of the experiments were carried out between 77 °K and 150 °K and the crystals were mounted in a vacuum specimen holder with automatic temperature control. Luminescence was excited with ultra-violet radiation, isolated from a high pressure mercury lamp by means of a Wratten 18B filter and an infra-red filter. The intensity of excitation was altered using neutral density filters which had previously been calibrated at 3650 Å. Absolute intensity measurements were made with a Hilger-Schwarz thermopile. In the calculation of the excitation density F, an approximate absorption depth of 1 μm has been assumed in all cases,

For the photoconductivity measurements under ultra-violet excitation crystals were fitted with evaporated indium electrodes in a gap-cell geometry. The $i - V$ characteristic of each cell was measured both in the dark and under the excitation levels used, as a check of the Ohmic behaviour of the electrodes. The effective depth of the region of current flow below the surface electrodes should be of the order of the ambipolar diffusion depth and, on the basis of previous work [16], a value of 1 μm has been used in the calculation of the conductivity and the free electron density n; values for the lattice mobility in the temperature range concerned were taken from the correlation graph shown in [16].

For the measurement of the spectral dependence of the luminescence a grating monochromator was used; quenching effects were generally studied by means of a small infra-red spectrometer. In an attempt to observe infra-red quenching in the 9 μm region, higher incident intensities were obtained by using a double filter of InSb and CaF_2.

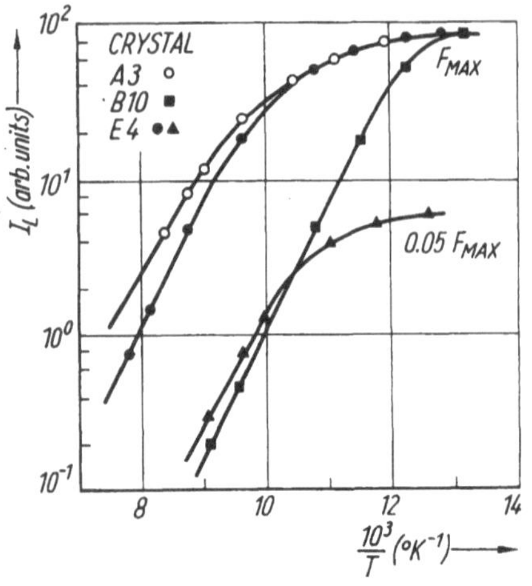

Fig. 2. Temperature dependence of the luminescent intensity at $F_{\max}$ and 0.05 $F_{\max}$

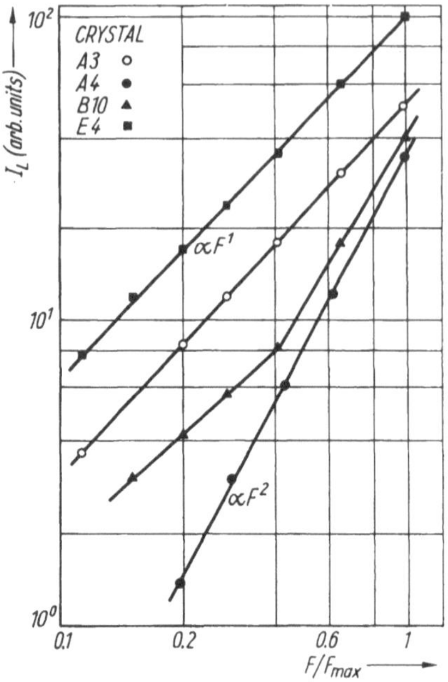

Fig. 3. Dependence of the luminescent intensity on excitation density normalised to $F_{\max}$ at 80 °K; the curves illustrate the three general patterns which were observed ▶

3. Experimental Results

3.1 Luminescent measurements

Fig. 2 shows typical examples of the temperature dependence of the luminescent intensity I_L between 77 °K and 150 °K. Except for one of the curves, the results were all obtained with the maximum excitation density $F_{\max}$ of about 4×10^{18} photons $cm^{-3}\, s^{-1}$. It can be seen that above about 95 °K, $L_L \propto \exp(\varepsilon_L/k\,T)$, and the experiments led to values of ε_L lying between 0.11 and 0.15 eV. This behaviour agrees with the results reported previously by Seiwert [17] who found values of ε_L between 0.09 and 0.12 eV and also interpreted ε_L as the activation

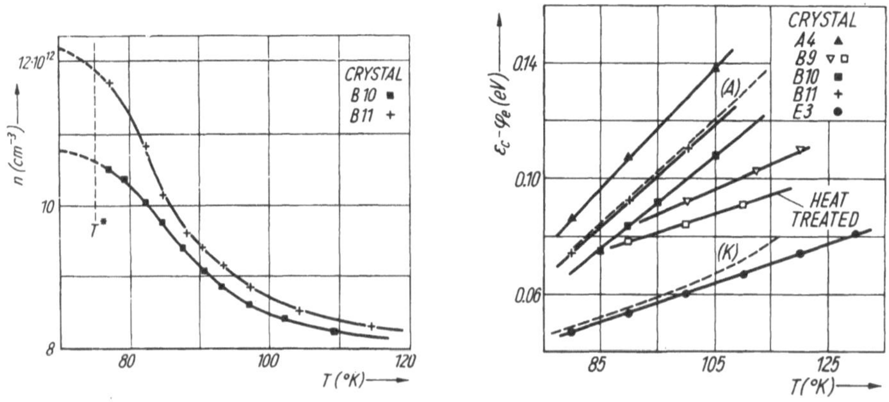

a b

Fig. 4. a) Electron density and b) position of quasi Fermi level below conduction band as a function of temperature. T^* denotes the estimated onset of thermal quenching, and the dotted lines in b) indicate the variation predicted by computations A and K. (In Fig. 4b read "ζ_e" instead of "φ_e")

energy of the luminescent site. The validity of this assignment will be discussed in Section 6. One of the curves in Fig. 2 was obtained at an excitation level of 0.05 F_{max} and indicates that the typical shape of the curves is independent of F.

Fig. 3 shows a logarithmic plot of the dependence of I_L at 80 °K on the excitation density F normalised to F_{max}. It illustrates the three general patterns which were observed: $I_L \propto F^2$, $I_L \propto F^1$, and $I_L \propto F^2$ at the higher excitation levels, tending towards a linear dependence at low F.

3.2 Photoconductivity measurements

The temperature dependence of the photoconductivity was investigated, using the excitation density F_{max}. At 80 °K, the electron density in the surface region, where the luminescence is excited, ranged from 10^{11} to 3×10^{14} cm^{-3} for the specimens investigated. Fig. 4a illustrates the pronounced decrease in n between 77 °K and 110 °K which was observed in most cases. In Fig. 4b the position of the quasi Fermi level for electrons, $\varepsilon_c - \zeta_e$, is plotted against T. In all specimens $\varepsilon_c - \zeta_e$ increased almost linearly with temperature, but it is significant that even at 115 °K the quasi Fermi level still lies closer than 0.15 eV to the conduction band in all cases.

Fig. 5 shows the dependence of the electron density n on the excitation density. The onset of visible luminescence has been indicated along the curves by an asterisk. It can be seen that for specimens A4, B9 and B12 supralinearity (i.e. $n \propto F^{>1}$) occurs in the luminescent region. Specimen A2, a non-luminescent crystal as grown, exhibits a high electron density with a half-power dependence on the excitation density.

3.3 Electron drift mobility measurements

The electron transport measurements on the crystals led to results in general agreement with those reported in two previous publications [16, 18]. Essentially two main groups can be distinguished. The first, (referred to as "high mobility crystals") pocesses an electron drift mobility μ_e at room temperature and above, which closely equals the lattice mobility. Below about 160 °K, μ_e begins to be controlled by shallow trapping centres, about 0.03 eV below ε_c, with a typical density of 10^{17} cm^{-3}. In the second group (the "low mobility crystals") μ_e is predominantly controlled by a deeper level, 0.15 to 0.16 eV below ε_c and having densities between 10^{15} and 10^{16} cm^{-3}. As has been shown, the presence of these centres reduced μ_e around room temperature to almost 1/10 of its previous value. The computations in Section 5 include examples of both the high and the low mobility cases.

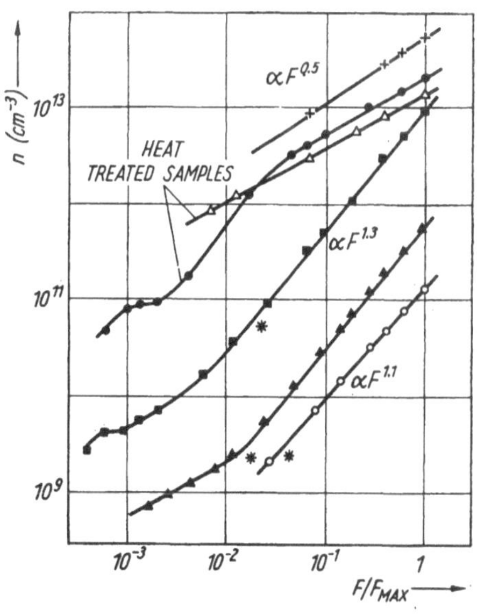

Fig. 5. Dependence of the electron density on excitation density, normalised to F_{max}. The onset of visible luminescence ist indicated by asterisks on the curves (+ A2 at 90 °K; ○ A4 at 78 °K; ▲ B9 at 80 °K and △ after heat treatment; ■ B12 at 90 °K; ● B12 at 104 °K after heat treatment)

3.4 Infra-red quenching

A number of infra-red quenching experiments were carried out to investigate centres lying close to the valence band and those in the vicinity of the middle of the forbidden band. Infra-red radiation in the wavelength range of 7 to 11 μm did not produce any measurable quenching effect. Quenching of the photoconductivity at $F = F_{max}$ with 1 μm radiation was observed, however, in several crystals both at room temperature and at 80 °K. The decrease Δn in the electron density, when the crystals were irradiated at room temperature with the same infra-red intensity, is given in Table 1.

Table 1

Results of infra-red quenching experiments at room temperature before and after heat treatment. The change in electron density, Δn, was produced by the same infra-red intensity in each case

Crystal	Δn before heat treatment	Δn after heat treatment
A 4	2×10^{10}	3×10^{13}
B18	1.5×10^{13}	4×10^{13}
A20	$< 10^{10}$	10^{13}
A 2*	2×10^{15}	—

* Non-Luminescent specimen.

3.5 Effect of heat treatment on above properties

Heat treatment of the specimens was carried out at 350 °C either in vacuum or in a nitrogen atmosphere, and led to the following observations:

1. A decrease of the luminescent intensity and, after heating for about two hours, a complete quenching of the emission.

2. A general increase in electron density as shown by specimens B9 and B12 in Fig. 5. The loss in luminescence is also accompanied by the loss of supralinearity; n now varies nearly as $F^{0.5}$ (cf. non-luminescent specimen A2).

3. In low mobility specimens, the heat treatment leads to a considerable increase in μ_e. Previous work [18] has shown that this is associated with a decrease in the density of the 0.16 eV centres.

4. Infra-red quenching experiments on heat treated crystals lead to the larger values of Δn given in Table 1.

The spectral distribution of the emission was checked for each specimen and agreed with that first reported by Kröger [1]. However, the red and orange emission bands reported by a number of authors [7, 10, 17] were not observed with the crystals used in this work.

4. Discussion of Experimental Results

The activation energy of the centres observed in transport measurements (0.16 eV) is almost identical to the activation energy of the luminescent centre as evaluated by Collins [9]. Although after heat treatment the simultaneous reduction in the luminescent intensity and the density of the centres might suggest a direct correlation between the two, the following considerations, similar to those put forward by Niekisch [10], throw doubt on such a correlation. The quasi Fermi level for electrons in the low temperature region always lies closer to the conduction band than the 0.16 eV centres, even in the most resistive specimen (A4). If it can be assumed that the occupation of these centres is controlled predominantly by the thermal release of electrons into the conduction band (see Section 5), it follows that throughout the luminescent temperature range the centres will remain almost completely filled with electrons. The observed exponential decrease of I_L with temperature (Fig. 2) can only be explained if the quasi Fermi level actually passes through the centres with increasing excitation. For this to occur in the temperature range from 80 °K to 120 °K, the luminescent sites would have to lie much closer to the conduction band; the emission should then occur at shorter wavelengths than those at which it is observed. In addition, a number of high mobility specimens, which from the analysis of the transport measurements contained fewer than 10^{14} cm^{-3} of the 0.16 eV centres, showed strong edge emission. This again would be difficult to explain if these centres were regarded as the luminescent site. The experimental evidence discussed in this section leads therefore to the conclusion that the 0.16 eV centres are not directly associated with the luminescent transition.

The observed supralinearity in the n vs. F curves, in conjunction with the temperature dependence of the electron density in the luminescent range, are regarded as significant factors in the interpretation of the results. An explanation of these effects, commonly observed in photoconductors, was put forward by Rose [19], Bube [20], and others in terms of two classes of recombination centres. Class I centres have equal cross-sections S_n, S_p for the capture of either carrier, whilst class II centres have a much larger cross-section for hole capture than for the subsequent capture of an electron ($S_p \gg S_n$). According to this model, supralinearity in n will occur when, with increasing excitation, the quasi Fermi level for holes (and the hole demarcation level) moves towards the valence band and closely approaches a level of class II centres. These begin to fill without contributing substantially to the recombination with free electrons and this results in an increased electron lifetime and leads to $n \propto F^{>1}$.

The extensive work of Bube [21] on CdS and other II-VI compounds has gone a long way towards establishing the essential validity of such a model; further evidence has recently been added by the direct determination of the capture cross-section of class II centres [22]. In view of this it would appear reasonable to associate the above experimental results with the presence of a class II centre. Unlike the well known sensitising centres, which in CdS are situated about 1 eV above the valence band, the observed thermal quenching of n around 90 °K suggests that the class II centre predominantly involved here should lie much closer to the valence band.

An attempt is made in the following section to fit the observed electrical properties of the specimens to a reasonably general model of a photoconductor. The above discussion indicates that this should include at least one class II centre.

5. General Model and Results of Computations

The model used in the following computations is essentially the three centre model analysed in detail by Bube [20]. Fig. 6 shows the centres and the transitions involved. Centre 1 is a hole trap lying ε_1 eV above the valence band and possessing cross-section S_{n_1}, S_{p_1} for electrons and holes, respectively. It is of the

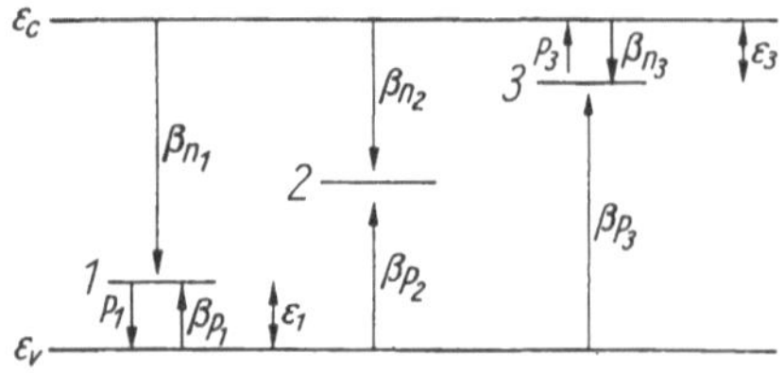

Fig. 6. Three centres model showing transitions

class II type, so that $S_{n_1} \ll S_{p_1}$. Centres 2, on the other hand, are typical class I centres with $S_{n_2} \simeq S_{p_2}$, lying sufficiently deep in the forbidden band that thermal release of trapped carriers may be neglected. Centres 3 are electron traps, empty in the absence of excitation. A quantum efficiency of unity is assumed in the generation of electron-hole pairs by optical excitation. The recombination constants β_{n_i} and β_{p_i}, given by the product of the thermal velocity and the corresponding capture cross-section (S_{n_i}, S_{p_i}), characterise the recombination between the ith level and each band respectively. Centre densities are denoted by N_i and the electron density in these states by n_i. $P_1 = N_v \beta_{p_1} \exp(-\varepsilon_1/kT)$ represents the probability per unit time of the thermal release of a hole in level 1 into the valence band. P_3 is defined in an analogous way.

From the rate equations in the steady state, the electron densities n_1, n_2, n_3 in the centres can be expressed in terms of the free carrier densities n and p and the above mentioned parameters (N, β, and P). Substitution of these equations into the condition of charge neutrality leads to a quartic equation between n and p. This was solved for p by electronic computer (Elliott 803) for $n = 10^6$, 10^7 to 10^{15} cm^{-3}, using an iterative method. Values of n and p were then fed into the rate equations and corresponding values of n_1, n_2, n_3 and F were printed out. The temperature dependence of these quantities was investigated by using the same programme with different values of P.

The parameters which characterise the centres were chosen according to the following considerations:

i) A reasonable value of 10^{-8} cm^3 s^{-1} (e.g. [10]) has been assumed for the recombination constant β_{p_1}, which implies that S_{p_1} is of the order of 10^{-15} cm^2. As suggested by the experimental results (Section 4) centre 1 must be of the class II type and, following Bube, β_{n_1} has been put at 10^{-12} cm^3 s^{-1}.

ii) Centre 2, which is a typical recombination centre, has a kinetically controlled occupation and both β_{n_2} and β_{p_2} were assumed to be 10^{-8} cm^3 s^{-1}.

iii) As suggested by the transport measurements, the predominant electron trap (centre 3) can have an activation energy ε_3 of either 0.15 eV or 0.03 eV. It has been assumed that these centres have the properties of a normal electron trap and that $\beta_{p_3} = \beta_{n_3} = 10^{-8}$ cm^3 s^{-1}.

iv) Bube has shown that the onset of supralinearity would be expected at an electron density n', where

$$n' \simeq \frac{(N_2 + N_3)\, P_1}{N_1\, \beta_{p_1}}\,.$$

The corresponding hole density p' must be such that $\beta_{p_1} p' < P_1$, so that the hole Fermi level lies above the centres, the occupation of which is controlled by the thermal release of holes into the valence band. The above condition for the onset of supralinearity allows us to estimate ε_1, the position of centre 1 with respect to the valence band. On the assumption that the centre densities are of the same order of magnitude, the n' values from Fig. 5 and from additional similar curves lead to ε_1 between 0.135 and 0.15 eV.

A further estimate of ε_1 should be obtainable from the onset of thermal quenching in the n vs. T curves. If this occurs at n^* and T^*, then

$$P_1 = N_v \beta_{p_1} \exp(-\varepsilon_1/k T^*) = \beta_{n_1} n^* .$$

Fig. 4a shows that T^* lies below the lowest available temperature, although for specimen B10 approximate values of $n^* \simeq 10^{13}$ cm^{-3} and $T^* \simeq 75$ °K can be obtained. This leads to $\varepsilon_1 = 0.138$ eV. These estimates show that, as far as the energy of the emission peak is concerned, centres 1 might be associated with the luminescent transition.

To avoid complicating the model any further, the deeper class II centres (about 1 eV above the valence band) have not been included. Because of their position far above the hole demarcation level and their extremely small electron capture cross-section [22], these centres will remain almost completely filled with holes in the temperature range of interest. Provided their density does not significantly exceed that of the other centres, it is though unlikely that this simplifying assumption would lead to very serious errors in the computations.

Table 2 lists the parameters used in the main group of computations. The β values have been kept the same throughout and the dependence on centre densities and activation energies is being investigated. Each computation has been performed for a number of temperatures between 77 °K and 135 °K. The results have generally been presented on a logarithmic plot of the densities n, p, n_1, n_2, n_3, and p_1 against the excitation density F, and Fig. 7a shows this for computation A at 77 °K.

The more general results of the calculations may be summarised in the following way:

1. The occurrence of supralinearity (see column 6 in Table 2) depends critically on the choice of the centre densities. In computation A (Fig. 7a) $n \propto F^{1.5}$ for

Table 2

Parameters used in computations and summary of main results. *: n_1 increases rapidly for F between 10^{16} and 5×10^{16} photons cm^{-3} s^{-1} and then varies linearly. The gradients listed in column 6 refer to excitation densities between 10^{16} and 10^{18} photons cm^{-3} s^{-1}

	1	2	3	4	5	6	7	8	9
Code	N_1 (cm^{-3})	N_2 (cm^{-3})	N_3 (cm^{-3})	ε_1 (eV)	ε_3 (eV)	n vs F (gradient)	ε_L (eV)	I_L vs F (gradient)	$(I_L)_{max}$ (cm^{-3} s^{-1})
A	10^{15}	10^{15}	10^{15}	0.135	0.15	1.5	0.22	1.6-2	$10^{16.5}$
B	10^{15}	10^{15}	10^{16}	0.135	0.15	1	0.16	1.5	10^{14}
C	10^{16}	10^{15}	10^{16}	0.135	0.15	1.7	0.16	1.6-2	$10^{15.5}$
D	10^{15}	10^{16}	10^{15}	0.135	0.15	1.6	0.25	1.9	$10^{15.5}$
E	10^{15}	10^{16}	2×10^{15}	0.135	0.15	1	0.22	—	10^{14}
F	5×10^{14}	10^{15}	10^{15}	0.135	0.15	1.1	0.19	1.9	10^{14}
G	5×10^{14}	10^{15}	5×10^{15}	0.135	0.15	1	0.13	1.7	10^{13}
H	10^{16}	10^{15}	10^{16}	0.145	0.15	1.6	0.17	1.6-2	10^{16}
J	10^{17}	10^{15}	10^{16}	0.145	0.15	*	—	1.3-2	10^{18}
K	10^{16}	10^{15}	10^{17}	0.145	0.03	0.5	0.11	1.0	10^{18}
L	10^{15}	10^{15}	10^{16}	0.145	0.03	0.5	0.14	1.0	10^{18}
M	10^{16}	10^{15}	10^{16}	0.145	0.03	0.5	0.14	1.0	10^{18}

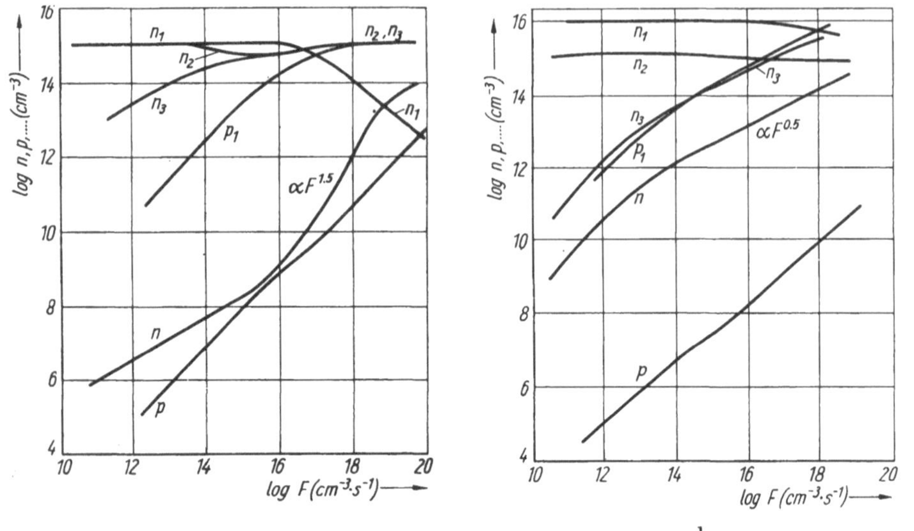

a b

Fig. 7. Results of computations at 77 °K. Carrier densities and electron occupation of centres are plotted against the excitation density: a) Computation A, b) computation K. For details see Table 2

values of F between 10^{16} and 10^{18} photons cm^{-3} s^{-1}. At the onset of supralinearity $\beta_{p_1} p' < 10$ so that the condition $\beta_{p_1} p' < P_1$ is satisfied.

If, however, $(N_2 + N_3)/N_1$ is increased as in computation B, then n' and also p' will increase. Supralinearity does not appear in B because with the larger p', $\beta_{p_1} p' \geqq P_1$. Computations D and E illustrate the very sensitive dependence of supralinearity on centre density, in this case N_3. Evidently, by proper choice of N_3, the range of gradients observed in the experiments may be accounted for.

2. As expected, an increase in N_3 decreases n over practically the whole range of F.

3. In all calculations p varies almost linearly over the range of F from 10^{10} to 10^{21} photons cm^{-3} s^{-1}, and is nearly temperature independent between 80 °K and 110 °K.

4. Centre 3 is practically filled throughout the luminescent range and remains so over the critical temperature range from 80 °K to 110 °K. This supports the assumption made in Section 4 that the kinetic recombination process (transition β_{p_3}) has no appreciable effect on the electron population in centres 3. The observed temperature dependence of edge emission cannot therefore be explained on this model by a decrease in either n_3 or p.

5. Alteration in the value of ε_1 from 0.135 eV (C) to 0.145 eV (H) increases the value of p_1, the other characteristics remaining sensibly constant. This indicates that the general conclusions are not critically dependent on the exact position of centre 1.

6. An increase in N_2 (D) reduces both p and p_1 by about the same factor. The effect on n and n_3 is negligible.

Computations K, L and M refer to the case $\varepsilon_3 = 0.03$ eV. A typical set of results (K) at 77 °K is shown in Fig. 7b. It is of particular interest that in the range of F leading to visible luminescence (10^{16} to 10^{18} photons cm^{-3} s^{-1}), n varies approximately as $F^{0.5}$, and reaches a value of about 10^{14} cm^{-3} near F_{max}. This agrees well with the general behaviour of high mobility specimens and of low mobility crystals after heat treatment.

6. Comparison between Computed and Experimental Behaviour

It is now possible to compare the experimental results of Section 3 with the predictions of the three centre model.

The luminescent results are of particular interest in this connection, especially the sensitive temperature dependence of I_L should be a critical test. Within the scope of the model there are two centres, 1 and 3, which according to their energetic positions should in principle be considered as the possible luminescent site, although the discussion in Section 4 makes centre 3 a doubtful choice. For model I, $I_L = \beta_{n_1}\, n\, p_1$, whilst model II gives $I_L = \beta_{p_3}\, p\, n_3$. The temperature dependence at F_{max}, predicted from a number of computations, is shown in Fig. 8. It is quite clear that model I predicts qualitatively and up to a point quantitatively the experimental temperature dependence of Fig. 2. On the other hand it can be seen that calculations based on model II lead to an almost temperature independent I_L, and the same applies to I_L curves obtained from all the other computations.

Column 7 of Table 2 lists the ε_L values which have been obtained from I_L curves calculated according to model I. Evidently ε_L is particularly sensitive to

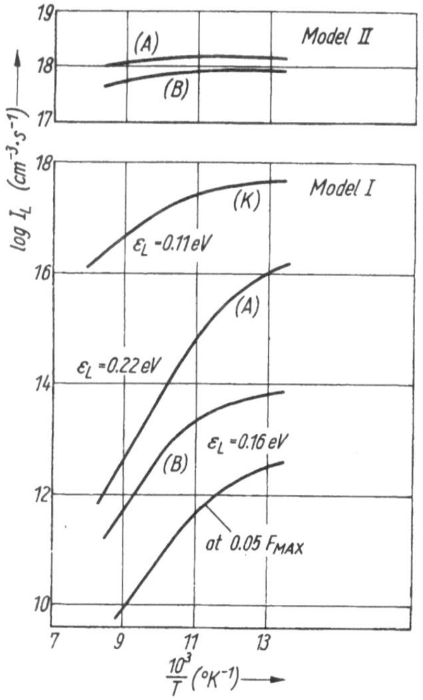

Fig. 8. Computed temperature dependence of I_L at F_{max} (except for lowest curve) for both models. Computation coding shown in brackets

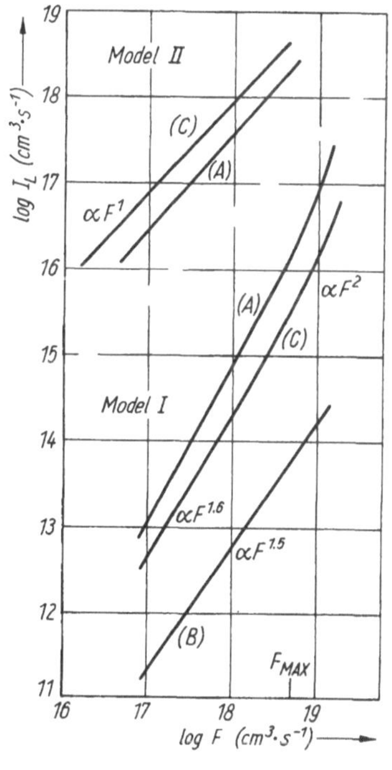

Fig. 9. Computed dependence of I_L on the excitation density at 77 °K. Computation coding shown in brackets. (The dimensions of I_L and F should read (cm^{-3} · s^{-1}) instead of (cm^3 · s^{-1}))

the density and position of centre 3. For instance an increase in N_3 from 10^{15} cm^{-3} (computation A) to 10^{16} cm^{-3} (B and C) reduces ε_L from 0.22 eV to 0.16 eV; a change in ε_3 from 0.135 eV to 0.03 eV results in a further decrease to 0.14 eV (L and M), whilst computation K gives $\varepsilon_L = 0.11$ eV. Variations in N_1 seem to have comparatively little effect. The calculations can therefore account satisfactorily for the range of observed ε_L, using reasonable centre densities, and ε_3 values determined from transport measurements. The results show that the simplifying assumptions made in Seiwert's analysis [17] cannot be entirely valid, so that in general the value of ε_L will be only a rough approximation to the activation energy of the luminescent site. A similar conclusion has been drawn by Klasens [23] in his discussion of model I.

The dependence of I_L on the excitation density, as predicted by both models, is shown in Fig. 9 for a number of computed cases. On the basis of model I, computations A, C, and H which led to supralinearity in n, show an F^2 dependence near F_{max}. With decreasing excitation the gradient decreases to about 1.6 (see column 8, Table 2). Calculation B, which did not lead to supralinearity, showed an $F^{1.5}$ dependence, whilst computations K, L, and M, characteristic of the high mobility specimens, predict a linear variation between I_L and F. Again, these results are in reasonable agreement with the observed pattern of behaviour (Fig. 3). Finally, in column 9 of Table 2 the calculated maximum intensity $(I_L)_{max}$ at F_{max} is listed. Experimental values range from 10^{16} to up to 10^{18} photons cm^{-3} s^{-1} for some high mobility specimens. This is in fact borne out by computations K, L, and M, although others with $\varepsilon_3 = 0.15$ eV give $(I_L)_{max}$ values which are far too low. An improvement in this respect can be obtained by increasing the value of β_{n_1} to 10^{-11} or 10^{-10} [11]. Subsidiary computations indicated that the general characteristics are unaffected by such a change, although the quantitative agreement with experimentally determined values of I_L is improved.

To test the calculated temperature dependence of n, the predicted position of the quasi Fermi level from computations A and K has been included in Fig. 4b. It can be seen that the calculated curves satisfactorily reproduce the trend of the experimental results.

From these comparisons it is evident that both the electrical and luminescent properties of the specimens can be explained in a consistent manner by the three centre model. It involves the basic assumption that the luminescent site has the property of a class II centre. In principle this is not unlikely, particularly as the recombination between a free electron and the neutral class II centre (after hole capture) is almost certainly a radiative process [22].

The class II centre introduces a significant asymmetry into the model, but it is not possible to reverse the roles of centres 1 and 3 because this would lead to disagreement with the observed photoconductive properties discussed in Section 4. Thus within the framework of the three centre model only a luminescent transition according to Schön-Klasens (model I) will fit into a consistent interpretation of all the observed facts. On this evidence it is suggested that model I describes the correct luminescent transition in pure CdS crystals.

7. Discussion

In view of the rather crucial importance of centres 1 in the interpretation of the results, an attempt has been made to provide further proof of their existence by a series of infra-red quenching experiments in the 7 to 11 μm region. No change in the photo-response could be detected; this was not altogether unexpected, because estimates indicated that with the infra-red intensity available the effect might only just be detectable.

However, in his P.E.M. studies on CdS, Auth [11] showed convincingly the presence of a level of centres lying between 0.1 and 0.2 eV above ε_v with densities of the order of 10^{16} cm^{-3}. Values of $\beta_{n_1} \simeq 10^{-10}$ cm^3 s^{-1} (and as low as 2×10^{-12} cm^3 in some crystals) were obtained. It is therefore very likely that this centre has class II properties, although the value of β_{p_1} is not known from Auth's results.

It is generally accepted that the luminescent centre should be a native crystal defect rather than an impurity, but the nature of this site still remains a controversial problem. Kulp and Kelley [24], on the basis of their electron bombardment experiments, assigned it to a S interstitial. Collins [9], on the other hand, concluded that the luminescent centre was a S vacancy, because his experiments indicated that edge emission was enhanced by the generation of S-vacancies. But according to Niekisch [10], the latter should lie about 0.12 eV below ε_c. Auth suggested, therefore, that the luminescent centre was only indirectly associated with a S vacancy, and was formed by the interaction of a V^{--} site with neighbouring S^{--} ions. Evidently all that can be said at present is that centre 1 is most probably connected with a S defect.

With regards to the results obtained after heat treatment of the crystals (Section 3.5) it is quite possible that the observed loss of edge emission is not directly connected with a reduction in the density of the luminescent centres.

Introducing additional hole trapping levels closer to the middle of the forbidden band could substantially reduce the hole population in centres 1, and, thereby, the intensity of the emission. Fairly conclusive evidence exists [21, 22, 25], to show that Cd vacancies form sensitising centres of the class II type, situated about 1.1 eV above the valence band. The observed infra-red quenching with 1 μm radiation (Section 3.4), should depend on the density of these centres. It can be seen from Table 1 that in each case the heat treatment led to an increase in the quenching effect. This is most probably connected with an increase in the density of Cd vacancies and lends support to the suggested mechanism. It is also likely that after the comparatively short heat treatment at 350 °C most of these centres are formed near the surface, that is, in the region where luminescence is excited by the u.v. radiation. This explains the fact that in our heat treated crystals luminescence could be restored by etching, which removed a surface layer of about 10 μm thickness.

8. Conclusions

1. On the basis of the experimental results in the luminescent temperature range it is concluded that in pure CdS crystals a centre of the class II type is situated 0.13 to 0.15 eV above the valence band.

2. It is shown by computations based on a three centre model (which includes the above centre), that the observed photoconductive properties between 77 °K and 130 °K can be explained satisfactorily both before and after heat treatment of the crystals.

3. The observed luminescent properties, such as the temperature and excitation dependence of the emission, can be predicted correctly, but only if the radiative recombination takes place between a free electron and a hole trapped in the above class II centre. This leads to a consistent interpretation of both the photoconductive and the luminescent properties within the framework of the

model. It is therefore suggested that the Schön-Klasens mechanism provides the correct explanation for the edge emission in pure CdS crystals.

4. The electron trapping centre, 0.15 to 0.16 eV below ε_c found by drift mobility measurements, is not directly connected with the luminescent process.

5. The gradient ε_L, associated with the temperature dependence of the luminescence, is critically dependent on centre densities. The experimental values of ε_L cannot in general be identified with the activation energy of the luminescent centre.

6. It is possible that the observed loss of emission after moderate heat treatment (350 °C) is not directly connected with the luminescent centre, but is caused by the introduction of additional Cd vacancies in the surface region.

Acknowledgements

The authors wish to thank Dr. P. D. Fochs of A. E. I., Mr. E. F. McGill of E. M. I., and Dr. G. T. Wright of Birmingham University for the crystals used in this work. We are indebted to Miss C. J. Stratford for help with the computations. The support of this investigation by the European Research Office of the U.S. Army is gratefully acknowledged.

References

[1] F. A. Kröger, Physica **7**, 1 (1940).
[2] J. J. Lambe, C. C. Klick, and D. L. Dexter, Phys. Rev. **103**, 1715 (1956).
[3] M. Schön, Z. Phys. **119**, 463 (1942).
[4] H. A. Klasens, Nature **158**, 306 (1940).
[5] J. J. Lambe and C. C. Klick, Phys. Rev. **98**, 909 (1955).
[6] J. J. Lambe, Phys. Rev. **98**, 985 (1955).
[7] I. Broser, R. Broser-Warminsky, G. Klipping, R. Ross, and H. J. Schulz, J. Phys. Chem. Solids **22**, 213 (1961).
[8] L. S. Pedrotti and D. C. Reynolds, Phys. Rev. **119**, 1897 (1960); **120**, 1664 (1960).
[9] R. J. Collins, J. appl. Phys. **30**, 1135 (1959).
[10] E. A. Niekisch, Z. phys. Chem. (Leipzig) **217**, 110 (1961).
[11] J. Auth, Z. phys. Chem. (Leipzig) **217**, 159 (1961).
[12] H. Friedrich, phys. stat. sol. **3**, K56 (1963).
[13] R. E. Halsted and B. Segall, Phys. Rev. Letters **10**, 392 (1963).
[14] M. R. Lorenz and H. H. Woodbury, Phys. Rev. Letters **10**, 215 (1963).
[15] B. S. Razbirin, Fiz. tverd. Tela **6**, 318 (1964).
[16] W. E. Spear and J. Mort, Proc. Phys. Soc. **81**, 130 (1963).
[17] R. Seiwert, Abh. Dtsch. Akad. Wiss. Berlin, Kl. Math. Phys. Tech., No. 7, 84 (1960).
[18] G. W. Bradberry and W. E. Spear, Brit. J. appl. Phys. **15**, 1127 (1964).
[19] A. Rose, Phys. Rev. **97**, 322 (1955).
[20] R. H. Bube, J. Phys. Chem. Solids **1**, 234 (1957); see also: Photoconductivity of Solids, John Wiley and Sons, Inc., New York 1960 (pp. 171, 338).
[21] See for example R. H. Bube, Solid State Phys. **11**, 223 (1960); J. appl. Phys. **32**, 1707 (1961).
[22] R. H. Bube and F. Cardon, J. appl. Phys. **35**, 2712 (1964).
[23] H. A. Klasens, J. Phys. Chem. Solids **9**, 185 (1958).
[24] B. A. Kulp and R. H. Kelley, J. appl. Phys. **31**, 1057 (1960).
[25] F. A. Kröger, H. J. Vink, and J. Boomgard, Z. phys. Chem. **203**, 1 (1954).

(Received December 9, 1964)

phys. stat. sol. 8, 663 (1965)

Institute of Physics of the Roumanian Academy, Bucharest

Absorption Spectra of Solid (Polycrystalline) and Liquid (Melt) Compounds of Bivalent Cobalt

By

A. TRUŢIA and M. MUSA

Measurements are made of the visible spectra of pure $CoCl_2$, $CoBr_2$ polycrystalline layers, and $CoCl_26H_2O$ and $CoBr_26H_2O$ monocrystals between 77 and 1050 °K. A good degree of correspondence is established between the spectral components of polycrystalline $CoCl_2$ and $CoBr_2$. The spectra of the two hydrates also show similarities which lead to the conclusion that the first coordination sphere around Co^{++} at 77 °K has an octahedral-type symmetry with two halide ions and four water molecules.

Modification of the $CoCl_2$ and $CoBr_2$ spectra when passing from the solid to liquid phase (melt) is also investigated. This indicates a transformation from octahedral to tetrahedral symmetry.

Im sichtbaren Bereich wurden die Spektren von polykristallinen $CoCl_2$- und $CoBr_2$-Schichten sowie von $CoCl_26H_2O$- und $CoBr_26H_2O$-Einkristallen zwischen 77 und 1050 °K gemessen. Die spektralen Komponenten von polykristallinem $CoCl_2$ und $CoBr_2$ entsprechen einander recht gut. Die Spektren der beiden Hydrate zeigt ebenfalls Ähnlichkeiten und lassen vermuten, daß die erste Koordinationssphäre von Co^{++} bei 77 °K eine oktaedrische Symmetrie mit zwei Halogen-Ionen und vier Wassermolekülen besitzt.

Die Modifizierung der $CoCl_2$- und $CoBr_2$-Spektren beim Übergang von der festen zur flüssigen Phase (Schmelze) wurde ebenfalls untersucht. Sie weist auf eine Transformation von der oktaedrischen zur tetraedrischen Symmetrie hin.

1. Introduction

Several authors [1, 2, 3, 4, 5, 6] have recently studied the spectral behaviour of bivalent cobalt on octahedral lattice-sites of different crystalline media. The most interesting of these paper is that of Ferguson, Wood, and Knox [5], who carefully analysed the low-temperature spectra of $KCoF_3$, $CoCl_2$, $CoBr_2$, and $CoWO_4$ single crystals from the theoretical and experimental points of view. They arrived at the conclusion that, generally, there is no good agreement between the corresponding calculated and observed energy levels. Another result of their work,

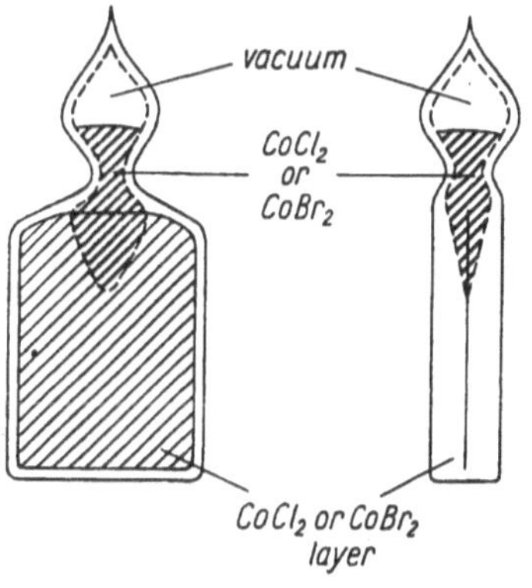

Fig. 1. Quartz cell containing the cobalt halide polycrystalline layer viewed from two sides

concerning $CoCl_2$ and $CoBr_2$, is that although there is a perfect resemblance between the two calculated spectra, the observed ones are rather different.

Within the scope of a more general program for studying the spectra of bivalent cobalt (also in solutions and vapour phase, to be published) and the hope of clarifying the mentioned discrepancy of the experimental data, we partially repeated the measurements of these spectra and investigated their temperature dependence as well as a small symmetry deformation effect. To this end spectra of pure $CoCl_2$ and $CoBr_2$ polycrystalline layers as well as $CoCl_2\,6\,H_2O$ and $CoBr_2\,6\,H_2O$ single crystals were measured in the range 77 to 1050 °K (including the melting point) for the pure cobalt halides, and the range 77 to 300 °K for the water-containing samples.

2. Experimental

The polycrystalline layers of pure $CoCl_2$ and $CoBr_2$ ("Carlo Erba" and "Serva" products, respectively) were obtained by melting them in quartz cells, having thicknesses from several microns to one tenth of a millimetre (Fig. 1). Before melting, the substances introduced in the cells were dried in vacuum during some hours at about 150 °C. Afterwards the cells were fused during continuous evacuation, keeping the samples at the drying temperature.

The $CoCl_2\,6\,H_2O$ and $CoBr_2\,6\,H_2O$ single crystals, about one millimetre thick, with one well deve loped face, were obtained by slow evaporation of the corresponding aqueous solutions in free atmosphere at room temperature.

For measurements in the range 77 to 300 °K the samples were closed in a big metallic holder (having holes for the light beam) which was introduced into liquid nitrogen (whose level rised up to about 5 mm under the holes). The temperature was increased by decreasing the liquid-nitrogen level (or its total evaporation) in the cryostat. For measurements in the range 300 to

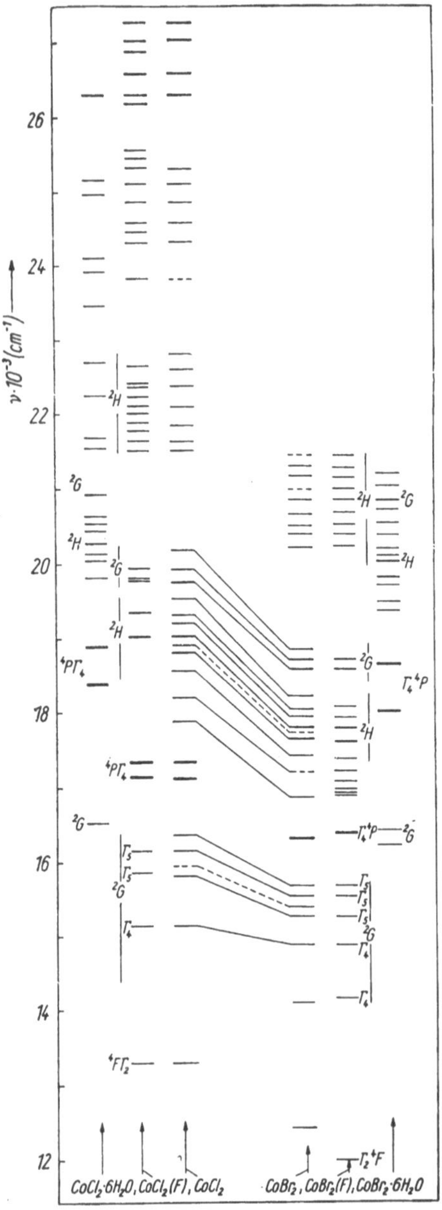

Fig. 2. Peak position in the low-temperature (77 °K) spectra of $CoCl_2$ and $CoBr_2$ compared with those obtained by Ferguson, Wood, and Knox (F) for single crystals at 20 °K. Spectra of $CoCl_2\,6\,H_2O$ and $CoBr_2\,6\,H_2O$ crystals, at 77 °K, are also presented here for an easier comparison (broken lines represent very weak peaks)

1050 °K we introduced the quartz cells containing the sample directly into a furnace.

The spectra were measured between 12000 and 28000 cm^{-1} using a Zeiss monochromator and M 13 gek. Zeiss or Soviet EFU 17A type photomultipliers connected with a "Sefram" GRVAC recording galvanometer. A tungsten lamp was used as a light source, the samples being always located in the focus of its beam. The sample image was projected on the entrance slit of the monochromator. We used a "single-beam" arrangement as we were interested in the measurements of the peak positions only. We could not determine the corresponding absorption coefficients because the samples did not have plane surfaces.

3. Results and Discussion

Fig. 2 shows the low-temperature (77 °K) spectra of our polycrystalline $CoCl_2$ and $CoBr_2$ layers compared with those obtained by Ferguson, Wood, and Knox for $CoCl_2$ and $CoBr_2$ single crystals at 20 °K. It can be seen that there is nearly perfect similarity between the spectra of pure polycrystalline $CoCl_2$ and $CoBr_2$. At this temperature the peaks of the absorption bands are at the same position as those reported by the above mentioned authors ((F) in the Figure) except for the transitions

$$^4F\,\Gamma_4 \rightarrow {}^4F\,\Gamma_2, \quad {}^4F\,\Gamma_4 \rightarrow {}^2G\,\Gamma_4, \quad \text{and} \quad {}^4F\,\Gamma_4 \rightarrow {}^4P\,\Gamma_4$$

in $CoBr_2$.

The transition to the first of the two $^2G\,\Gamma_4$ levels (Ferguson's assignment for $CoBr_2$) was not seen by Ferguson (nor by us) in $CoCl_2$; perhaps it is masked by the $^4F\,\Gamma_4 \rightarrow {}^4F\,\Gamma_2$ transition. Between the following five peaks of the two samples assigned to the ^2G level one may establish a good correspondence. The 17880 cm^{-1} and 20200 cm^{-1} range in $CoCl_2$ is richer in our case, but the components which appear in our case only are very weak, for that at 18820 cm^{-1} which is rather intense. Instead, our $CoBr_2$ spectrum is poorer in a small region around 17000 cm^{-1} (three components are not seen separately).

Between the maxima of $CoCl_2$ in the above mentioned region and those of $CoBr_2$ situated between 16600 and 18900 cm^{-1} there is again a nearly perfect correspondence. In addiation, except for the first two red side peaks, the splitting undergoes a linear variation, with different slopes for the seven ^2H components and the last three components which are attributed to the ^2G initial level (Fig. 3).[1]

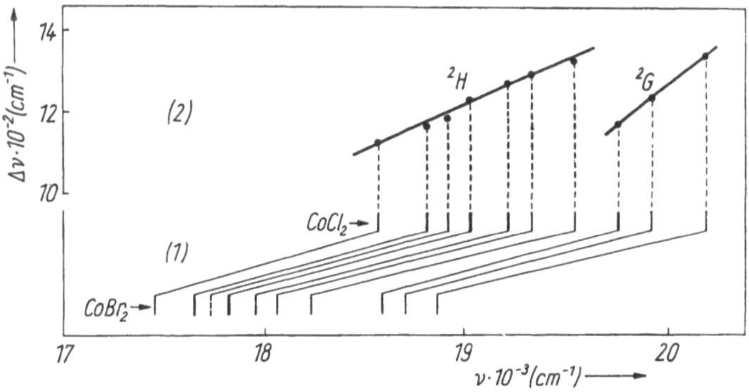

Fig. 3. Components of the ^2H and ^2G levels of polycrystalline $CoCl_2$ and $CoBr_2$ at 77 °K: (1) Peak positions; (2) the $\Delta\nu = f(\nu)$ representation

[1]) Such a linarity may be observed for the first ^2G level (below 16500 cm^{-1}) too, but the peaks are rather wide and, consequently, their localisation is difficult.

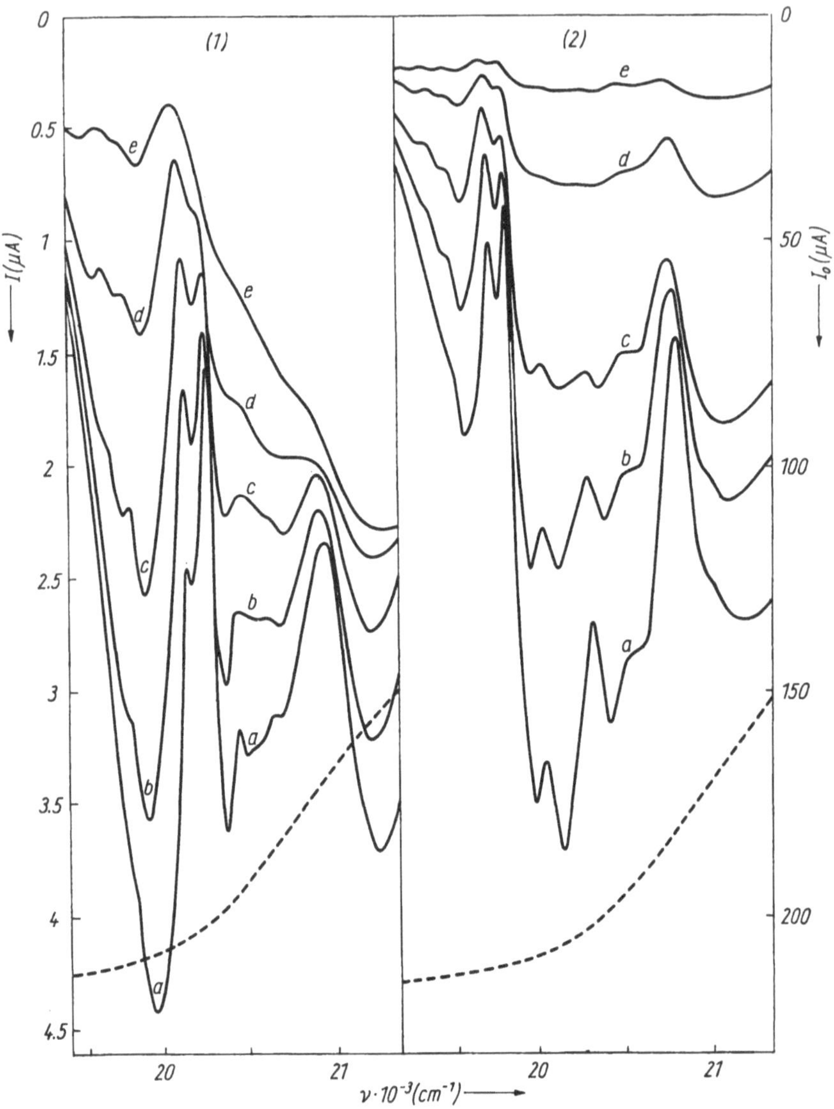

Fig. 4. Temperature dependence of the spectra of $CoCl_2$ 6 H_2O (1) and $CoBr_2$ 6 H_2O (2): (a) 77 °K, (e) room temperature (single beam spectrophotometer curves). Broken lines represent I_0

Taking into account the above mentioned resemblance of the spectra of our polycrystalline $CoCl_2$ and $CoBr_2$ layers at 77 °K on both sides of the ^4P Γ_4 level, one may assume that in this case the peaks at 17890 cm^{-1} and 18200 cm^{-1} of $CoCl_2$ and those at 16870 cm^{-1} and 17215 cm^{-1} of $CoBr_2$ correspond to the same level. Furthermore, the two ^4P Γ_4 levels of $CoCl_2$ would correspond to the single ^4P Γ_4 level at 16350 cm^{-1} of $CoBr_2$. Now, if the above mentioned affirmation holds true, there are two possibilities: either the 16350 cm^{-1} peak of $CoBr_2$ involves all the four double-group components Γ_6, Γ_7, Γ_8, and Γ_8, and the 16870 cm^{-1} and 17215 cm^{-1} peaks of $CoBr_2$ (and the corresponding peaks in $CoCl_2$) belong to another level, or these last two peaks are due to the Γ_6 sub-level in both substances, and then the 16350 cm^{-1} peak of $CoBr_2$ involves the Γ_7, Γ_8, and Γ_8 of the ^4P level, and the 17130 cm^{-1} and 17365 cm^{-1} peaks of $CoCl_2$ are due to the Γ_7, Γ_8 sub-levels and the Γ_8 sub-level, respectively. Further experimental data is needed to clarify assignments in this region of the spectrum.

The absorption bands at higher energies, around 22000 cm^{-1} in $CoCl_2$ and 21000 cm^{-1} in $CoBr_2$, exhibit different intensity-distributions and therefore a group correspondence could be established only. Here (and also for the higher energy levels of $CoCl_2$) the agreement between our and Ferguson's results may be seen in Fig. 2.

Furthermore, we studied the low-temperature (77 °K) absorption spectra of $CoCl_2\,6\,H_2O$ and $CoBr_2\,6\,H_2O$ single crystals. The positions of these peaks are shown also in Fig. 2, together with those of polycrystalline $CoCl_2$ and $CoBr_2$. As expected, these spectra are displaced towards higher energies compared with those of the water-free compounds, and also show a mutual displacement. Taking into account spectra of $Co^{++}6O^{--}$ [7] and $Co^{++}6X^-$ ($X = Cl$, Br), the positions and intensity distribution between the peaks and of those belonging to the water-containing crystals, we may conclude that in all these cases we have to deal with

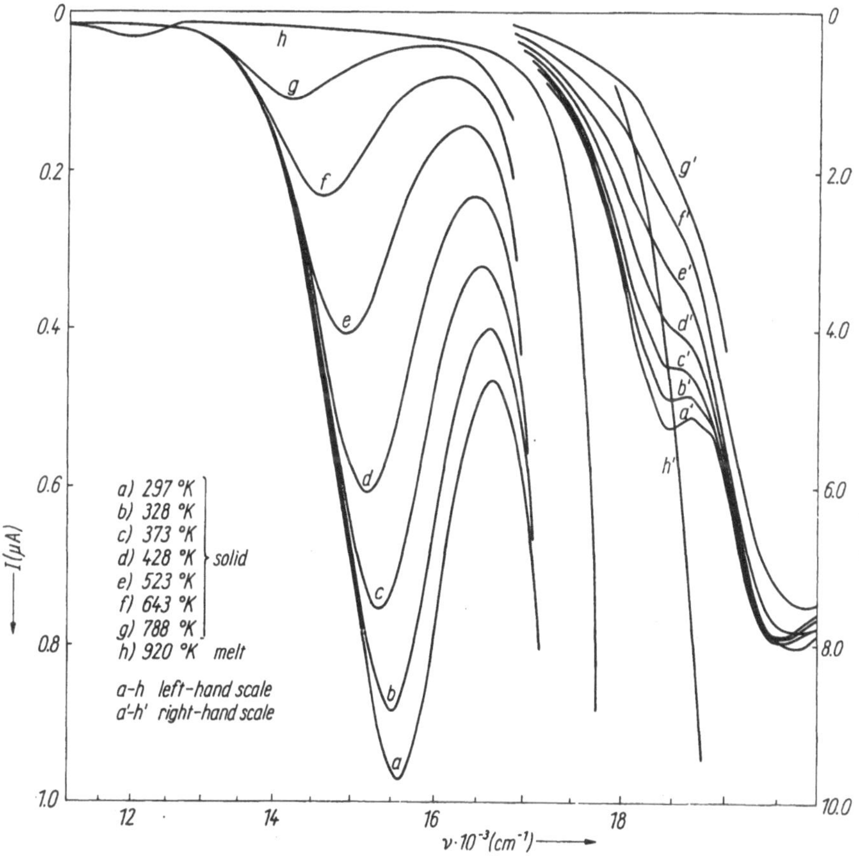

Fig. 5. The change of octahedral to tetrahedral symmetry in $CoCl_2$ during melting (single-beam spectrophotometer curves)

the same octahedral symmetry, but in the last ones the ligands are no more identical. Assuming, according to Fine [8], that the contribution to the crystal field of each ligand is approximately independent of the type of the other ligands in the coordination sphere, the mentioned position of the peaks lead to the conclusion that in CoX_2 6 H_2O crystals the first coordination sphere contains two halogen ions and four water molecules. This fact was also established for $CoCl_2$ 6 H_2O at room temperature by X-ray diffraction [9].

The room temperature spectra of the $CoCl_2$ 6 H_2O and $CoBr_2$ 6 H_2O crystals are very similar, and therefore we may conclude that they conserve their octahedral character and the chemical composition of the first coordination sphere in the entire temperature-range investigated.

The temperature dependence of the $CoCl_2$ 6 H_2O and $CoBr_2$ 6 H_2O spectra between 77 °K and room temperature is presented in Fig. 4. Comparing these curves we arrive at the conclusion that the splitting of the levels situated around 20500 cm^{-1} in $CoBr_2$ 6 H_2O (corresponding to the ^2H and ^2G groups in $CoBr_2$) is larger than the corresponding one in $CoCl_2$. At the same time their centers of gravity have normal relative positions.

The spectra of $CoCl_2$ and $CoBr_2$ at high temperatures (including the melting point) were measured especially in order to study the liquid state. We needed these spectra for the interpretation of the spectra of CoX_2 in the vapour phase [10]. Fig. 5 shows the temperature variation of the spectrum of a $CoCl_2$ layer. We give the $^4F\ \Gamma_4 \rightarrow {}^4P\ \Gamma_4$ transition region only because the other transitions vanish at high temperatures. One may see that the spectrum conserves its octahedral character as long as the sample remains solid (curves a to g). Passing to the melt, the spectrum suddenly changes: the old absorption band disappears and a new one (h), much more intense and displaced towards the red part of the spectrum appears. The position and the intensity of the new band, in comparison with that of the solid phase as well as with that of the $[CoCl_4]^=$ complex in solutions [8], indicates the presence of a tetrahedral complex: $[CoCl_4]^=$. The same is true for liquid $CoBr_2$.

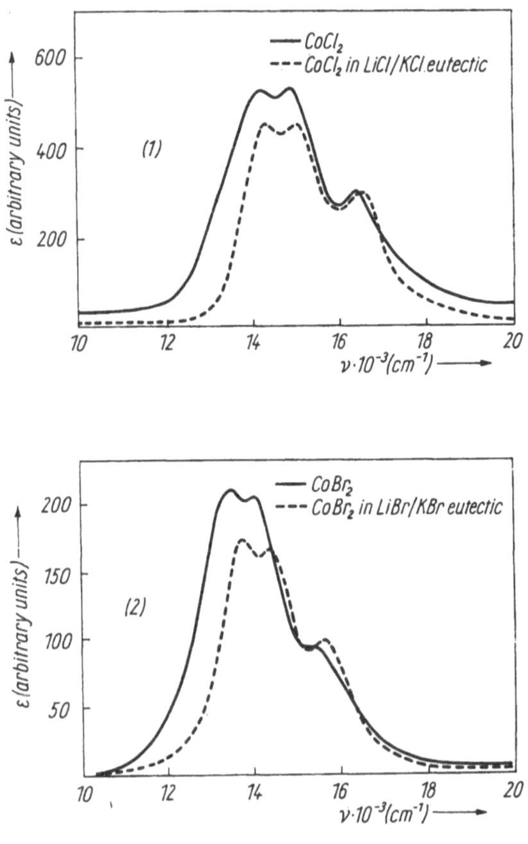

Fig. 6. Absorption curves of pure $CoCl_2$ (1) and $CoBr_2$ (2) melts (full lines) compared with those obtained by Sundheim and Kukk [11] (broken curve)

For liquid $CoCl_2$ in a LiCl/KCl eutectic and liquid $CoBr_2$ in a LiBr/KBr eutectic, Sundheim and Kukk [11] observed spectra which are identical with those obtained by us for the two pure CoX_2 melts. (The fact that the peaks are slightly displaced towards the red part of the spectrum in our case may be explained by a temperature effect) (Fig. 6).

4. Conclusions

Absorption spectra of polycrystalline $CoCl_2$ and $CoBr_2$ layers at liquid nitrogen temperature were measured and compared with those obtained by Ferguson, Wood, and Knox for single crystals at liquid hydrogen temperature. Generally the spectra are equal, but small details which appear in our case give the possibility to establish a good correspondence between the spectra of the two substances; except for the $^4F\ \Gamma_4 \rightarrow {}^4P\ \Gamma_4$ transition which requires further experimental work to be done.

The spectra of $CoCl_2\ 6\ H_2O$ and $CoBr_2\ 6\ H_2O$ indicate an octahedral symmetry of the first coordination sphere around a Co ion: two halogen ions and four water molecules.

A change from octahedral to tetrahedral symmetry during melting was observe for polycrystalline $CoCl_2$ and $CoBr_2$.

References

[1] W. Low, Phys. Rev. **109**, 256 (1958).
[2] I. V. Tananaev and B. F. Djurinski, Dokl. Akad. Nauk SSSR **139**, 120 (1961).
[3] J. C. Zahner and H. G. Drickamer, J. chem. Phys. **35**, 1483 (1961).
[4] T. R. Sliker, Phys. Rev. **130**, 1749 (1963).
[5] J. Ferguson, D. L. Wood, and K. Knox, J. Chem. Phys. **39**, 881 (1963).
[6] A. Truţia and A. Bohun, Czech. J. Phys. **B 13**, 45 (1963).
[7] W. Low, Phys. Rev. **109**, 247 (1958).
[8] D. A. Fine, J. Amer. Chem. Soc. **84**, 1139 (1962).
[9] J. Mizuno, J. Phys. Soc. Japan **15**, 1412 (1960).
[10] A. Truţia and M. Musa, to be published.
[11] B. R. Sundheim and M. Kukk, Disc. Faraday Soc. **32**, 49 (1961).

(Received December 16, 1964)

phys. stat. sol. 8, 671 (1965)

Institut für technische Physik „A. F. Joffe" der Akademie der Wissenschaften der UdSSR, Leningrad

Zur Beweglichkeit der Stromträger in Indiumantimonidkristallen

Von

V. V. Galavanov, D. N. Nasledov und A. S. Filipchenko

Zur Untersuchung des Streumechanismus für die Elektronen wurde die Beweglichkeit an n-leitenden InSb-Kristallen im Temperaturbereich von 77 bis 773 °K und bei Störstellenkonzentration von 10^{15} bis $7 \cdot 10^{18}$ cm^{-3} gemessen. Die erhaltenen Ergebnisse werden mit theoretischen Überlegungen, die die nichtparabolische Form des Leitungsbandes berücksichtigen, verglichen.

Die Ergebnisse zeigen, daß die Elektronen im n-leitenden InSb an ionisierten Störstellen, optischen Gitterschwingungen und Löchern gestreut werden. Der Einfluß einer Streuung an akustischen Gitterschwingungen wird für verschiedene Werte der Deformationspotentialkonstanten diskutiert.

Für die Elektronenbeweglichkeit bei gemischter Streuung (an ionisierten Störstellen, optischen Gitterschwingungen und Löchern) wird eine exakte Beziehung hergeleitet. Der Gültigkeitsbereich der verwendeten Näherungsformeln wird überprüft. Für das anomale Anwachsen des Hallkoeffizienten bei hohen Temperaturen in stark dotierten Proben werden mehrere Gründe angeführt.

С целью выяснения природы рассеяния электронов исследовалась холловская подвижность в монокристаллах n — InSb в интервале температур (77—773 °К) и концентраций примеси (10^{15}—$7 \cdot 10^{18}$ см$^{-3}$). Полученные результаты сравнивались с теорией, учитывающей непараболичность зоны проводимости.

Показано, что электроны в n — InSb рассеиваются на ионах примеси, оптических колебаниях решётки и на дырках. Обсуждается возможность влияния рассеяния на тепловых акустических колебаниях решётки при различных значениях постоянной потенциала деформации.

Выведена точная формула для подвижности электронов в случае смешанного механизма рассеяния (ионы примеси, оптические колебания решётки и дырки). Проверена справедливость использованной в расчётах приближённой формулы. Обсуждаются возможные причины аномального роста постоянной Холла при высоких температурах в сильно легированных кристаллах.

1. Einleitung

Indiumantimonid besitzt eine Reihe sehr interessanter Eigenschaften (sowohl vom Standpunkt seiner technischen Verwendbarkeit als auch vom rein wissenschaftlichen Gesichtspunkt gesehen). Wegen dieser Eigenschaften (geringe Breite der verbotenen Zone, kleine effektive Masse und außerordentlich hohe Beweglichkeit der Elektronen) hat InSb sehr große Perspektiven als Material für die Herstellung vieler Halbleitergeräte (Hall-Generator, Empfänger für UR-Strahlung, Tunneldioden, Transistoren, Laser usw.). Gleichzeitig erschweren jedoch diese Eigenschaften die Untersuchung der physikalischen Parameter von Indiumantimonid. Infolge der geringen Breite der verbotenen Zone ist die Wechselwirkung zwischen dem Leitungsband und dem Valenzband groß [1] und führt dazu, daß das Leitungsband nicht-parabolisch ist. Dadurch wird die theoretische Begründung der Meßergebnisse erschwert.

Insbesondere der Streumechanismus der Stromträger in Indiumantimonidkristallen ist Gegenstand vieler Untersuchungen. Zur Klärung dieser Frage wurden Untersuchungen der thermomagnetischen Effekte [2, 3], der effektiven Masse der Elektronen [4, 5], der elektrischen und galvanomagnetischen Effekte [6, 7] sowie vieler anderer Erscheinungen durchgeführt. Die meisten Experimente führten zu außerordentlich widersprüchlichen Ergebnissen bezüglich des Streumechanismus für Elektronen.

Ursprünglich nahm man in Analogie zum Germanium an, daß Streuung an akustischen Gitterschwingungen der Hauptmechanismus der Elektronenstreuung in InSb ist. Später wurde jedoch von vielen Autoren Streuung an optischen Gitterschwingungen als der dominierende Mechanismus vorgeschlagen. In einer kürzlich erschienenen Arbeit [8] wird jedoch wiederum Streuung an akustischen Gitterschwingungen angenommen.

Diese unterschiedlichen Auffassungen sind unserer Meinung nach darauf zurückzuführen, daß 1. viele Parameter von InSb (die effektive Ladung, das Deformationspotential u. a.) nicht genau bekannt sind, 2. in vielen Fällen nicht oder nicht genügend berücksichtigt wird, daß das Leitungsband nicht parabolisch ist und 3. in vielen Arbeiten die Untersuchungen in einem relativ engen Temperatur- und Konzentrationsbereich der Störstellen vorgenommen wurden.

Die theoretischen und experimentellen Arbeiten [9 bis 11], in denen versucht wird, die Abweichung des Leitungsbandes von der parabolischen Form zu berücksichtigen, ermöglichen recht genaue Schlußfolgerungen über den Mechanismus der Elektronenstreuung in Indiumantimonidkristallen.

In der vorliegenden Arbeit wird versucht, den Mechanismus der Elektronenstreuung in InSb in einem großen Temperatur- und Konzentrationsbereich der Störstellen durch Untersuchung der Elektronenbeweglichkeit als Funktion der Konzentration und der Temperatur sowie durch genaue Berechnung der Abweichung des Leitungsbandes von der parabolischen Form zu bestimmen.

2. Meßergebnisse

Es wurden die elektrische Leitfähigkeit σ und der Hall-Koeffizient R an n-leitenden InSb-Einkristallen (Störstellenkonzentration 10^{14} bis $7 \times 10^{18}\,cm^{-3}$) im

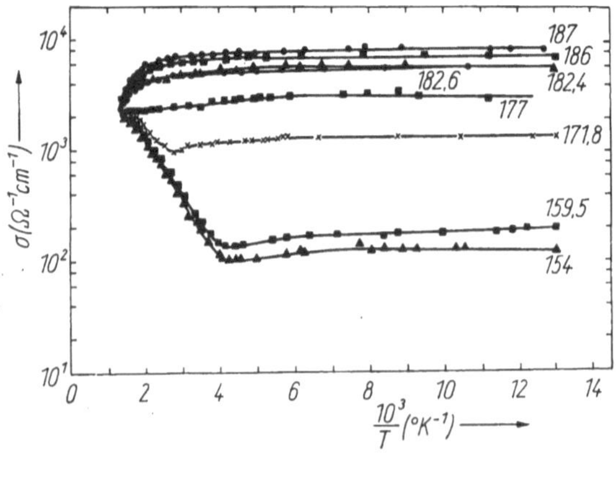

Fig. 1. Temperaturabhängigkeit der elektrischen Leitfähigkeit σ für einige untersuchte Proben

Fig. 2. Temperaturabhängigkeit des Hall-Koeffizienten R für einige untersuchte Proben

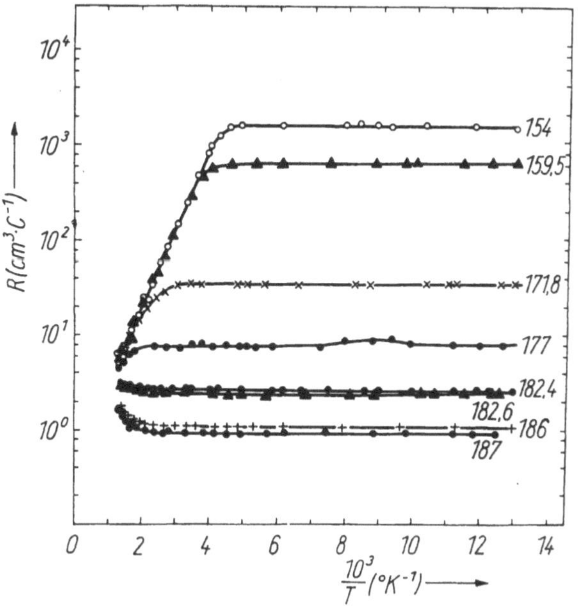

Temperaturbereich von 77 bis 773 °K untersucht. In Fig. 1 und 2 sind die Meßergebnisse graphisch dargestellt. Bei stark dotierten Proben ($n > 10^{18}$ cm^{-3}) ist bei hohen Temperaturen ($T > 500$ °K) eine anomale Zunahme des Hallkoeffizienten zu beobachten (Fig. 2). Bei den Proben mit Störstellenkonzentrationen von $2{,}4 \times 10^{18}$, $2{,}6 \times 10^{18}$, 6×10^{18} bzw. 7×10^{18} cm^{-3} nimmt der Hallkoeffizient um das 1,23-, 1,29-, 1,64-, bzw. 1,88fache zu.

Fig. 3 zeigt die Temperaturabhängigkeit der Hallbeweglichkeit $R\sigma$ für eine Reihe untersuchter Prüflinge.

3. Diskussion der Meßergebnisse

Die Elektronenbeweglichkeit wurde für alle untersuchten Störstellenkonzentrationen und Temperaturen für die folgenden Streumechanismen berechnet: Streuung an ionisierten Störstellen, an akustischen Gitterschwingungen, an optischen Gitterschwingungen und für die Elektron-Loch-Streuung. Die verwendeten

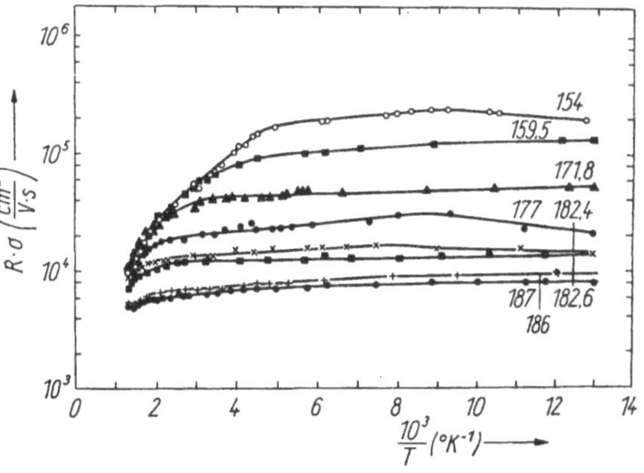

Fig. 3. Temperaturabhängigkeit der Hallbeweglichkeit $R\sigma$ für einige untersuchte Proben

Beziehungen berücksichtigen, daß das Leitungsband nicht parabolisch ist. Die Formeln wurden teilweise den Arbeiten von Kolodziejczak [11, 12] entnommen und teilweise von uns abgeleitet.

Bei Streuung der Elektronen an ionisierten Störstellen und für Elektron-Loch-Streuung wurde die Beweglichkeit nach folgender Formel bestimmt:

$$U_i = \frac{32\,\varkappa^2\,(k_0\,T)^3\,m_n}{3\,h^3\,e^3\,n^2\,B(\xi)}\,L^3_{-2}(\eta,\beta) \tag{1}$$

mit

$$B(\xi) = \ln(1+\xi) - \frac{\xi}{1+\xi} \quad \text{und} \quad \xi_{\max} = \frac{\varkappa\,h\,\sqrt{k_0\,T}}{e^2\sqrt{2\,m_n}}\,\frac{L_0^{5/2}}{L_0^{3/2}\,L_1^{1/2}}\,. \tag{1a}$$

Darin sind

$\varkappa$ die Dielektrizitätskonstante,
m_n die effektive Masse des Elektrons am unteren Rand des Leitungsbandes,
e die Ladung des Elektrons,
n die Elektronenkonzentration,
k_0 die Boltzmannkonstante,
h die Plancksche Konstante.

Bei Streuung der Elektronen an optischen Gitterschwingungen ($T > \Theta_D = 290\ °K$) gilt

$$U_o = \frac{4}{3}\,\frac{M\,V_a\,\gamma_0^2\,k_0\,T}{h\,e\,e^{*2}\,n}\,L^2_{-2}(\eta,\beta)\,. \tag{2}$$

Darin sind

M die reduzierte Masse der Atome,
V_a das Volumen der Elementarzelle des Kristalls,
γ_0 die Frequenz der longitudinalen optischen Gitterschwingungen,
e^* die effektive Ladung.

Bei Streuung der Elektronen an akustischen Gitterschwingungen gilt

$$U_{ak} = \frac{h}{3\,\pi^2}\,\frac{e\,\varrho\,U_l^2}{E_1^2}\,\frac{1}{m_n\,n}\,L^1_{-2}(\eta,\beta)\,, \tag{3}$$

wobei ϱ die Dichte, U_l die Schallgeschwindigkeit im Kristall und E_1 die Konstante des Deformationspotentials sind.

Bei starker Entartung des Elektronengases vereinfachen sich die Formeln zu

$$U_i = \frac{1}{48}\left(\frac{3}{\pi}\right)^2 \frac{h^2\,\varkappa^2}{e^3\,B\,m^{*2}} \tag{4}$$

mit

$$B = \ln(1+\xi) - \frac{\xi}{1+\xi}\,, \qquad \xi = \left(\frac{3}{\pi}\right)^{1/3} \frac{\varkappa\,h^2\,n^{1/3}}{4\,e^2\,m^*}\,, \tag{4a}$$

$$U_o = \frac{3^{1/3}}{16\,\pi^{4/3}}\,\frac{M\,V_a\,\gamma_0^2\,h^3\,n^{1/3}}{e\,e^{*2}\,k_0\,T\,m^{*2}}\,, \tag{5}$$

$$U_{ak} = \left(\frac{\pi}{3}\right)^{1/3} \frac{e\,\varrho\,U_l^2\,h^3\,n^{-1/3}}{E_1^2\,k_0\,T\,m^{*2}}\,, \tag{6}$$

und m^* der effektiven Masse eines Elektrons.

Die Integrale vom Typ $L_m^n(\eta, \beta)$ in den Formeln (1) bis (3) haben folgende Form [11]:

$$L_m^n(\eta, \beta) = \int_0^\infty \left(-\frac{\partial f_0}{\partial x}\right) (x + \beta x^2)^n (1 + 2\beta x)^m dx .$$

Diese Integrale berechneten wir nach der in [13] angegebenen Näherungsmethode. Außerdem standen uns Integraltabellen zur Verfügung, die auf unsere Bitte im Institut für Physik der Polnischen Akademie der Wissenschaften berechnet wurden. Die Differenz zwischen unseren Berechnungen und den Werten, die die polnischen Wissenschaftler erhielten, betrug nicht mehr als 1 bis 2%.

Die Gesamtbeweglichkeit für einen gemischten Streumechanismus wurde nach der Näherungsformel

$$\frac{1}{U} = \sum_K \frac{1}{U_K} \tag{7}$$

bestimmt.

Zuerst wurden die Berechnungen für stark dotierte Proben ($n = 2{,}5 \times 10^{18}$ und 7×10^{18} cm^{-3}) durchgeführt. Berechnet wurde die Elektronenbeweglichkeit bei Streuung an ionisierten Störstellen und an optischen Gitterschwingungen, ferner wurde auch die Elektron-Loch-Streuung berücksichtigt. Die Gesamtbeweglichkeit ergibt sich aus Gleichung (7).

In Fig. 4 ist die theoretisch berechnete Temperaturabhängigkeit der Beweglichkeit für diese Proben dargestellt. Wie aus der Figur hervorgeht, liegen die Meßpunkte recht gut auf der Kurve für die Gesamtbeweglichkeit U_{i+o}.
Wir sehen also, daß die Elektronen in stark dotierten Indiumantimonidproben hauptsächlich an ionisierten Störstellen und zu einem geringen Teil an optischen Gitterschwingungen gestreut werden.

Die Fig. 5 und 6 zeigen die theoretisch berechnete Konzentrationsabhängigkeit der Beweglichkeit für verschiedene Temperaturen (77, 300, 500 und 773 °K). Die Rechnungen wurden für Elektronenstreuung an ionisierten Störstellen und an optischen Gitterschwingungen durchgeführt. Man sieht, daß die Meßpunkte ebenfalls gut auf den Kurven für die Gesamtbeweglichkeit U_{i+o} liegen.

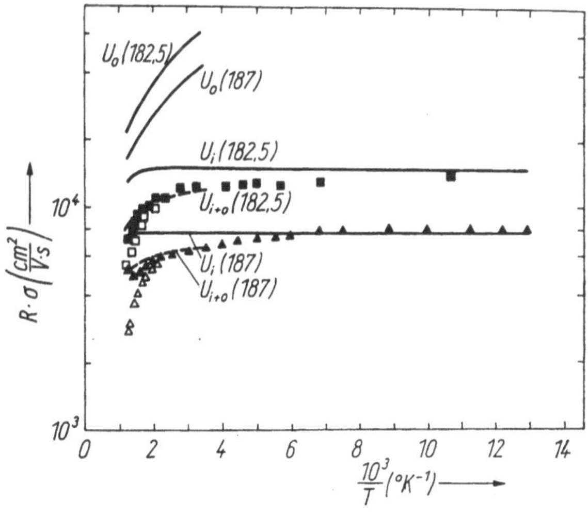

Fig. 4. Temperaturabhängigkeit der Elektronenbeweglichkeit für Proben mit den Störstellenkonzentrationen $2{,}5 \times 10^{18}$ cm^{-3} (182,5) und 7×10^{18} cm^{-3} (187). U_i theoretische Werte für die Elektronenbeweglichkeit bei Streuung der Elektronen an ionisierten Störstellen, U_o theoretische Werte für die Beweglichkeit bei Streuung der Elektronen an optischen Gitterschwingungen, ■, ▲ experimentell erhaltene Werte für $R\sigma$

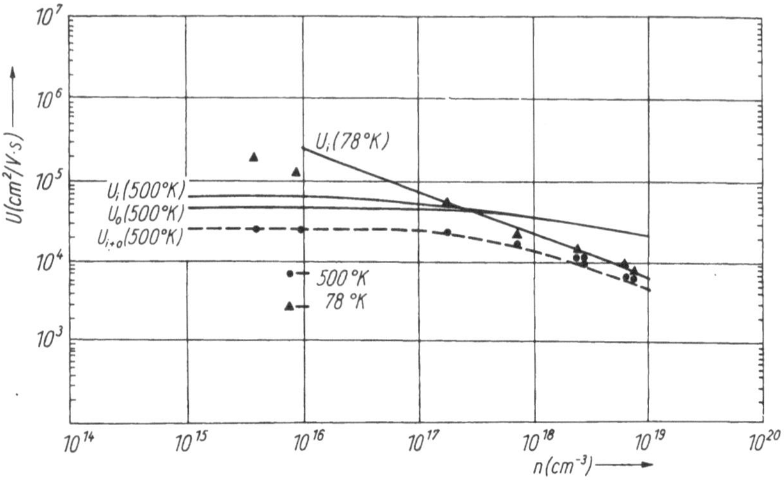

Fig. 5.

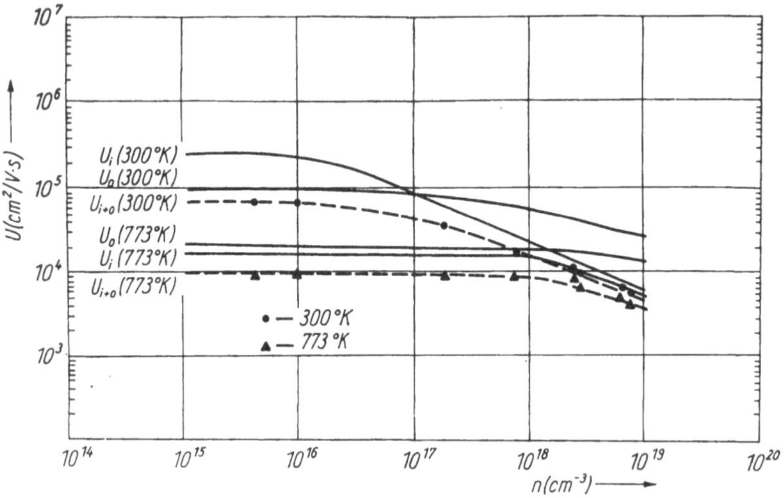

Fig. 6.

Fig. 5 und Fig. 6. Abhängigkeit der Elektronenbeweglichkeit von der Störstellenkonzentration bei verschiedenen Temperaturen (Bezeichnungen wie bei Fig. 4)

Man kann daher folgern, daß für alle untersuchten Temperaturen und Störstellenkonzentrationen ein gemischter Streumechanismus (an ionisierten Störstellen und an optischen Gitterschwingungen) vorliegt. Der Anteil der einzelnen Mechanismen läßt sich anhand der Fig. 5 und 6 ermitteln.

Haga und Kimura [8] indessen nehmen jedoch an, daß die Streuung an akustischen Gitterschwingungen überwiegt. Sie vermuten, daß der Einfluß der akustischen Gitterschwingungen auf die Elektronenstreuung bisher deshalb als gering angenommen wurde, weil die von Ehrenreich angegebene Konstante des Deformationspotentials, $E_1 = 7{,}2$ eV, ihrer Meinung nach nicht richtig ist. Die Autoren [8] bestimmten die Konstante des Deformationspotentials neu und fanden $E_1 = 30$ eV. Wir versuchten dieses Ergebnis anhand unserer Werte für die Elektronenbeweglichkeit zu überprüfen.

Die Elektronenbeweglichkeit bei Streuung an akustischen Gitterschwingungen wurde nach den Gleichungen (3) und (6) für die beiden Werte der Konstanten des Deformationspotentials 7,2 und 30 eV berechnet. Die Berechnung erfolgte für eigenleitende Kristalle und für Kristalle mit einer Störstellenkonzentration von 7×10^{18} cm^{-3} für die Temperaturen 300 und 773 °K.

Wir betrachten zunächst qualitativ die Abhängigkeiten der Beweglichkeiten U_i, U_o und U_{ak} von n, m^* und T. Aus den Gleichungen (4), (5) und (6) folgt:

$$U_i \sim \frac{1}{m^{*2}}, \quad U_o \sim \frac{n^{1/3}}{m^{*2}\,T}, \quad U_{ak} \sim \frac{1}{m^{*2}\,T\,n^{1/3}}.$$

Es ist bekannt, daß $m^* \sim n^{1/3}$ ist, d. h. alle Beweglichkeiten (U_i, U_0 und U_{ak}) werden mit zunehmender Konzentration kleiner, U_{ak} nimmt jedoch stärker als U_i ab und U_i wiederum stärker als U_o. Wenn die akustischen Gitterschwingungen eine wesentliche Rolle spielen würden, müßte sich das also in erster Linie in stark dotierten Proben und bei hohen Temperaturen zeigen. Die Rechnungen für $E_1 = 7{,}2$ eV bestätigten Ehrenreichs Folgerungen, daß der Anteil der Streuung an akustischen Gitterschwingungen unbedeutend ist.

Mit dem Wert $E_1 = 30$ eV ergibt sich, wie aus Tabelle 1 ersichtlich ist, daß nur für den Fall der Eigenleitfähigkeit, bei $T = 300$ °K, die Streuung an akustischen Gitterschwingungen gleich wahrscheinlich mit der Streuung an optischen Gitterschwingungen ist. In allen übrigen Fällen zeigt die Rechnung, daß die für den Fall der Streuung an akustischen Gitterschwingungen erhaltene Beweglichkeit unter dem experimentellen Wert liegt. In noch größerem Maße gilt das für die Gesamtbeweglichkeit U_{i+o+ak}. Somit führt der Wert $E_1 = 30$ eV für die Konstante des Deformationspotentials zu einem Widerspruch zu den Meßergebnissen.

Durch Streuung an akustischen Gitterschwingungen läßt sich auch dann ein so starkes Absinken der Beweglichkeit mit wachsender Temperatur in stark dotierten Proben nicht erklären, wenn man $R =$ const annimmt (Proben mit anomaler Zunahme des Hall-Koeffizienten, Fig. 4, helle Quadrate und Dreiecke). Daraus folgt, daß man entweder den Anteil der akustischen Gitterschwingungen bei der Elektronenstreuung in Indiumantimonid vernachlässigen kann oder daß der in der Arbeit [8] angegebene Wert der Konstanten des Deformationspotentials unrichtig (zu groß) ist.

Zur Berechnung der Beweglichkeit bei gemischtem Streumechanismus verwendeten wir bisher den Näherungsausdruck (7). Wegen der sehr unterschiedlichen

Tabelle 1

Probe	E_1 (eV)	T (°K)	U_i (cm²/V s)	U_o (cm²/V s)	U_{ak} (cm²/V s)	U_{i+o+ak} (cm²/V s)	$R\sigma$ (cm²/V s)
Eigenleitfähigkeit	7,2	300	$2{,}7 \cdot 10^5$	$1{,}0 \cdot 10^5$	$2{,}3 \cdot 10^6$	$7{,}1 \cdot 10^4$	$7{,}0 \cdot 10^4$
7×10^{18} cm^{-3}	7,2	300	$7{,}7 \cdot 10^3$	$4{,}3 \cdot 10^4$	$7{,}0 \cdot 10^4$	$6{,}0 \cdot 10^3$	$6{,}5 \cdot 10^3$
Eigenleitfähigkeit	30	300	$2{,}7 \cdot 10^5$	$1{,}0 \cdot 10^5$	$1{,}3 \cdot 10^5$	$4{,}7 \cdot 10^4$	$7{,}0 \cdot 10^4$
7×10^{18} cm^{-3}	30	300	$7{,}7 \cdot 10^3$	$4{,}3 \cdot 10^4$	$4{,}0 \cdot 10^3$	$2{,}5 \cdot 10^3$	$6{,}5 \cdot 10^3$
Eigenleitfähigkeit	30	773	$1{,}8 \cdot 10^4$	$2{,}4 \cdot 10^4$	$5{,}8 \cdot 10^3$	$3{,}7 \cdot 10^3$	$1{,}1 \cdot 10^4$
7×10^{18} cm^{-3}	30	773	$7{,}6 \cdot 10^3$	$1{,}66 \cdot 10^4$	$1{,}2 \cdot 10^3$	$9{,}8 \cdot 10^2$	$5{,}2 \cdot 10^3$

Abhängigkeit der Relaxationszeiten τ_i und τ_0 von der Energie ist die Gleichung (7), besonders für einen nichtentarteten Halbleiter nur in sehr starker Näherung gültig (τ_i und τ_0 sind die Relaxationszeiten für die Streuung der Elektronen an ionisierten Störstellen bzw. optischen Gitterschwingungen). Die Elektronenstreuung ist ein Wahrscheinlichkeitsprozeß, und zur Bestimmung des Wertes der Gesamtbeweglichkeit für den gemischten Streumechanismus muß man die Streuwahrscheinlichkeiten für die einzelnen Mechanismen addieren [14]

$$\frac{1}{\tau_{CM}} = \frac{1}{\tau_i} + \frac{1}{\tau_0}, \qquad \tau_{CM} = \frac{\tau_i}{1 + \frac{\tau_i}{\tau_0}}. \tag{8}$$

Ist die Ralaxationszeit bekannt, so kann man die Beweglichkeit nach folgender Beziehung berechnen [11]

$$U = \frac{c_1}{e\,n}\int\limits_0^\infty \left(-\frac{\partial f_0}{\partial \varepsilon}\right) \tau\, k^4 \left(\frac{1}{m^*}\right)^2 dk\,. \tag{9}$$

Setzt man hier den Wert τ_{CM} aus Gleichung (8) ein, so erhält man

$$U_{CM} = \frac{32\,\varkappa^2\,(k_0\,T)^3\,m_n}{3\,\hbar^3\,e^3\,n^2}\int\limits_0^\infty \left(-\frac{\partial f_0}{\partial x}\right)\frac{1}{B(\xi)}\,\frac{(x+\beta\,x^2)^3}{(1+2\,\beta\,x)^2}\,\frac{dx}{1+\frac{\tau_i^0}{\tau_0^0}\,\frac{2\,m_n\,E_g}{\hbar^2}\,\beta\,(x+\beta\,x^2)}\,. \tag{10}$$

Der Faktor $B(\xi) = \ln(1+\xi) - \frac{\xi}{1+\xi}$ hängt wenig von der Energie ab und wird gewöhnlich für $\xi = \xi_{max}$ aus dem Integral herausgenommen (siehe Gleichung 1a). Wir haben $B(\xi)$ im Integral gelassen, ξ aber in allgemeinerer Form verwendet

$$\xi = \frac{\varkappa\,h\,(x+\beta\,x^2)\sqrt{k_0\,T}}{e^2\sqrt{2\,m_n}\,L_1^{1/2}(\eta,\beta)}\,. \tag{11}$$

τ_i^0 und τ_0^0 sind durch folgende Ausdrücke gegeben

$$\tau_i^0 = \frac{\hbar\,\varkappa^2}{2\,\pi\,e^4\,n\,B(\xi)}, \qquad \tau_0^0 = \frac{M\,V_a\,\gamma_0^2\,h}{4\,(e\,e^*)^2\,k_0\,T}\,. \tag{12}$$

In Tabelle 2 sind die nach Gleichung (10) für einige Proben berechneten Ergebnisse angegeben. Das Integral L_{CM} in Gleichung (10) wurde numerisch nach dem in [13] angegebenen Näherungsausdruck bestimmt. Zum Vergleich werden in Tabelle 2 die nach der Näherungsformel (7) errechneten Werte der Beweglichkeit für dieselben Proben angeführt.

Wie aus der Tabelle hervorgeht, beträgt sowohl bei stark entarteten als auch bei nicht entarteten Proben die Differenz zwischen den Werten für die Elektronen-

Tabelle 2

n (cm^{-3})	T (°K)	η	L_{CM}	U_{CM} (cm²/V s)	U_{i+0} (cm²/V s)
$2{,}5 \cdot 10^{16}$	300	−0,57	0,177	68800	66000
$1{,}0 \cdot 10^{17}$	300	1,4	2,33	56600	54500
$1{,}0 \cdot 10^{18}$	300	6,15	85	20700	18500
$1{,}37 \cdot 10^{18}$	773	2,5	7,7	13300	12250
$1{,}0 \cdot 10^{19}$	773	7,2	171	5550	5260

beweglichkeit U_{CM} bzw. U_{i+o}, die nach den Beziehungen (10) bzw. (7) errechnet wurden nicht mehr als 7%. Das weist darauf hin, daß 1. die Näherungsformel (7) im gesamten Temperatur- und Konzentrationsbereich der Störstellen anwendbar ist und 2. für alle untersuchten Störstellenkonzentrationen und Temperaturen die Elektronen in n-leitendem InSb an ionisierten Störstellen und an optischen Gitterschwingungen gestreut werden.

Abschließend wird auf die in Fig. 2 gezeigte anomale Zunahme des Hall-Koeffizienten an stark dotierten Proben bei Temperaturen oberhalb 500 °K eingegangen.

Es ist schwer, die Zunahme von R auf einen Mechanismus zurückzuführen, der auf einer Verringerung der Elektronenkonzentration beruht.

Zur Erklärung der Zunahme von R kann man mehrere Hypothesen aufstellen:

1. *Einfluß von Inhomogenitäten des Kristalls*

Unserer Meinung nach kann man das als Ursache ausschließen, da alle Proben auf Homogenität geprüft wurden. Die Messungen erfolgten mehrmals im gesamten Temperaturbereich, es wurde jedoch keine Hysterese in der Temperaturabhängigkeit des Hall-Koeffizienten beobachtet. Außerdem wurden jeweils mehrere Proben mit annähernd gleicher Konzentration, die aus verschiedenen Einkristallen herausgeschnitten wurden, gemessen und ergaben übereinstimmende Ergebnisse.

2. *Aufhebung der Entartung*

Da das Fermi-Niveau bei diesen Proben (nach ziemlich genauen Schätzungen) nicht unter 4 (für die Probe mit der Störstellenkonzentration von $2{,}5\times10^{18}$ cm^{-3}) bzw. 5 (für die Probe mit der Störstellenkonzentration von 7×10^{18} cm^{-3}) bei 773 °K absinkt, ist auch das als Ursache auszuschließen.

3. *Zweite Zone*

In der Literatur gibt es keinerlei Hinweise über ihren Einfluß auf n-leitendes InSb.

4. Am wahrscheinlichsten scheint uns folgende Annahme: Bisher wurde vorausgesetzt, daß der Parameter A in der Formel $R = A/enc$ gleich Eins ist. Das ist nur für ein parabolisches Band gültig. Bei InSb mit einem stark nicht-parabolischem Leitungsband kann man, wie Kolodziejczak [11] gezeigt hat, A mit ausreichender Genauigkeit ebenfalls gleich Eins annehmen (für den Fall der Streuung an ionisierten Störstellen). Wie sich der Parameter A bei einem gemischten Streumechanismus und stark nicht-parabolischem Leitungsband verhält, wissen wir nicht. Diese Frage ist schwer zu beantworten und erfordert komplizierte Rechnungen, die zur Zeit noch nicht durchgeführt sind. Es scheint jedoch möglich, daß in diesem Falle $A > 1$ ist.

Literatur

[1] E. J. Kane, J. Phys. Chem. Solids **1**, 249 (1957).
[2] V. P. Zhuse und I. M. Zidilkovskii, Zh. tekh. Fiz. **28**, 2373 (1958).
[3] M. Rodot, J. Phys. Radium **19**, 140 (1958).
[4] O. M. Emelyanenko, F. P. Kesamanli und D. N. Nasledov, Fiz. tverd. Tela **3**, 1256 (1961).
[5] R. W. Keyes, Phys. Rev. **99**, 490 (1955).
[6] D. J. Horwarth, R. H. Jone und E. H. Putley, Proc. Phys. Soc. **B 70**, 124 (1957).

[7] V. V. GALAVANOV, D. N. NASLEDOV und A. S. FILIPCHENKO, Isv. Akad. Nauk SSSR, ser. fiz. **28**, 6, 959 (1964).
[8] E. HAGA und H. KIMURA, J. Phys. Soc. Japan **18**, 777 (1963).
[9] H. EHRENREICH, J. Phys. Chem. Solids **9**, 129 (1959).
[10] J. KOLODZIEJCZAK und L. SOSNOWSKI, Acta phys. Polon. **21**, 4, 399 (1962).
[11] J. KOLODZIEJCZAK, Acta phys. Polon. **20**, 4, 289 (1961).
[12] J. KOLODZIEJCZAK, Acta phys. Polon. **20**, 5–6, 379 (1961).
[13] V. I. KRILOV, Priblizhonnoe vichislenie integralov, GIFML, Moskau 1959.
[14] A. I. ANSELM, Vvedenie v teoriyu poluprovodnikov, Fizmatgiz, Moskau, 1962.

(Received October 30, 1964)

phys. stat. sol. 8, 681 (1965)

Laboratoire d'Electronique et de Physique du Solide de l'Université de Lyon

Mesure de la permittivité complexe du titanate de baryum entre 100 et 1000 MHz sous champ statique

Par

A. Deguin et G. Mesnard

On présente une méthode permettant en ondes décimètriques d'étudier l'influence d'un champ électrique continu superposé, sur la valeur de la permittivité diélectrique d'échantillons ferroélectriques. L'échantillon est placé à l'extrémité d'une ligne terminée par un court-circuit pour la haute fréquence. On discute l'erreur introduite par la source de tension continue sur l'impédance de ce court-circuit. On présente des résultats obtenus sur $BaTiO_3$ en l'absence et en présence de ce champ statique.

Es wird eine Methode beschrieben, bei der der Einfluß eines überlagerten Gleichfeldes auf die Dielektrizitätskonstante ferroelektrischer Proben mit Dezimeterwellen untersucht wird. Die Probe befindet sich dabei am äußersten Ende der Leitung, die durch einen Kurzschluß-widerstand für hohe Frequenzen abgeschlossen wird. Der durch die Gleichspannungsquelle bedingte Fehler auf die Kurzschlußimpedanz wird diskutiert. Es werden Ergebnisse für $BaTiO_3$ mit und ohne statischem Feld angegeben.

1. Introduction

L'application à un cristal ferroélectrique d'un champ électrique continu E_0 permet de le placer en un point donné de son cycle d'hystérésis $P_0 = f(E_0)$. En y superposant un champ variable à haute fréquence de faible amplitude, la polarisation spontanée P en fonction du champ instantané E décrit un microcycle autour du point (P_0, E_0) assimilable à une ellipse. Il en résulte une constante diélectrique complexe $\varepsilon^* = \varepsilon' - j\,\varepsilon'' = \varepsilon'\,(1 - j \operatorname{tg} \delta)$. ε' et ε'' sont des fonctions de E_0 et de la fréquence f.

La variation avec f de ε' et $\operatorname{tg} \delta$ permet de déterminer l'inertie des diverses contributions à la variation élémentaire de polarisation quand E varie (mouvements de parois de domaines, nucléation de nouveaux domaines, variation de la polarisation induite). En pratique l'échantillon, métallisé, forme un condensateur plan et l'on suit l'évolution de son impédance Z en fonction de la fréquence.

Aux fréquences peu élevées (quelques kHz), le champ continu est appliqué (Fig. 1) à travers une self L ou une résistance R telles que la dérivation correspondante ne perturbe pas la mesure de Z (on doit avoir $L\,\omega$ ou $R \gg |Z|$). Cette mesure est effectuée à l'aide d'un pont BF; la capacité C bloque la tension continue;

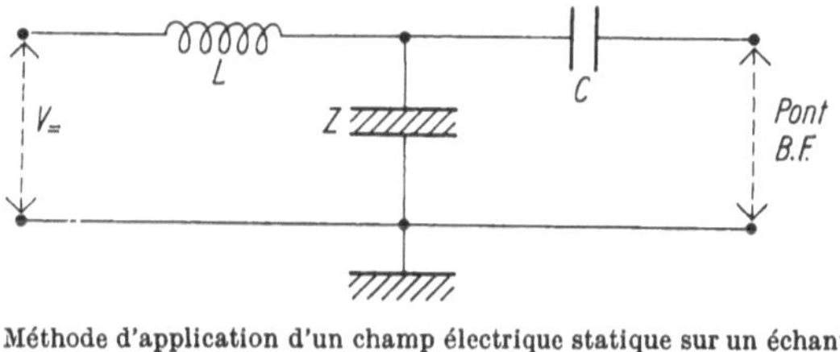

Fig. 1. Méthode d'application d'un champ électrique statique sur un échantillon lors de mesures en basse fréquence

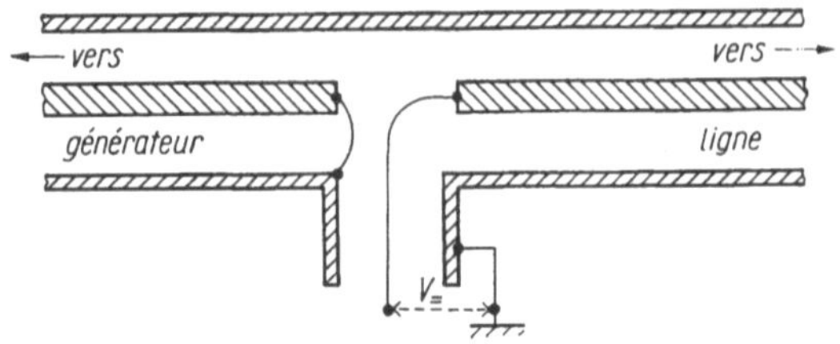

Fig. 2. Introduction, d'un champ statique sur une ligne de mesure en ondes décimètriques

sa valeur est suffisamment élevée pour que l'impédance correspondante soit négligeable $\left(\frac{1}{C\omega} \ll |Z|\right)$.

Ce circuit à « constantes groupées » s'adapte mal aux fréquences élevées. Nous avons envisagé le cas des mesures en ondes décimétriques avec des lignes coaxiales. Fousek [1] a proposé le montage suivant, pour un appareillage utilisant la technique de la ligne fendue. La ligne, fermée sur l'échantillon à étudier, est alimentée par un générateur UHF couplé par l'intermédiaire d'un atténuateur. Afin d'introduire le champ statique, la boucle de couplage de l'atténuateur côté ligne est prolongée vers l'extérieur (Fig. 2) par un passage à travers le conducteur externe. La capacité de ce passage est rendue suffisante pour assurer un court-circuit en UHF.

La tension appliquée peut être soit continue, pour une étude par points, soit périodique en basse fréquence, par exemple de forme carrée [2].

Ce dispositif, qui a l'avantage d'être apériodique, ne peut pas être employé avec une méthode de mesure par impédance ramenée sur un pont. Nous décrirons un montage, que nous avons mis au point, qui ne présente pas cet inconvénient et nous donnerons un aperçu des résultats obtenus avec des monocristaux de titanate de baryum.

2. Méthode d'application du champ statique

La technique générale de mesure consiste à placer l'échantillon à l'extrémité court-circuitée d'une ligne coaxiale, entre le conducteur interne et le conducteur externe. Un appareil classique (admittancemètre General Radio) permet de mesurer l'admittance ramenée à une distance électrique connue l_e sur la ligne. Il convient d'appliquer la tension continue entre les 2 faces du cristal en maintenant l'état de court-circuit haute-fréquence sur une face du cristal et en protégeant le pont de mesure.

On est conduit à effectuer successivement plusieurs mesures d'admittances (Fig. 3):

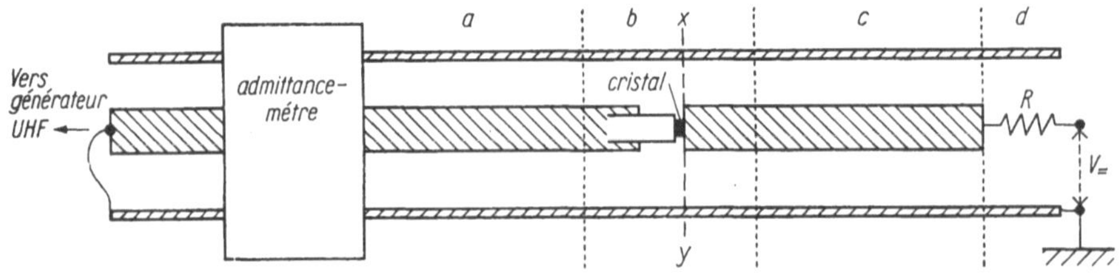

Fig. 3. Schéma de l'appareillage utilisé en ondes décimètriques

1. On connecte une ligne ajustable a terminée par un court-circuit amovible (type WN3 General Radio). La longueur de cette ligne est ajustée de façon que l'admittance ramenée soit $0 + j\,0$. Soit XY le plan du court-circuit.

2. On remplace le court-circuit par un tronçon de ligne b dont le conducteur interne comporte un piston mobile permettant d'intercaler le cristal dans le plan XY. On prolonge ce tronçon par une deuxième ligne ajustable c terminée par un circuit ouvert amovible (type WO3 General Radio). On règle la longueur de cette ligne de façon que l'admittance mesurée soit toujours $0 + j\,0$; au-delà du plan XY on a alors une longueur $\lambda/4$, à $k\,\lambda/2$ près.

3. Le circuit ouvert amovible est remplacé par une terminaison d comportant une résistance au graphite R de 20 kΩ en série avec le conducteur interne; cette terminaison comporte deux bornes pour l'application de la tension continue. Le pont de mesure se trouve légèrement déséquilibré (avec notre montage l'admittance ramenée devenait $0 + j\,0{,}6\ \mathrm{m}\Omega^{-1}$); on ramène à zéro en agissant sur la ligne ajustable c.

Dans ces conditions, dans le plan XY tout se passe du point de vue haute-fréquence comme si l'on avait un court-circuit effectif. En outre la tension continue est effectivement appliquée à l'échantillon, car le conducteur interne et le conducteur externe se trouvent réunis en un point de la ligne d'alimentation en énergie par la présence d'un atténuateur à piston et le pont proprement dit n'est soumis à aucune tension continue.

Dans cette technique, le conducteur interne d'un tronçon coaxial se trouve porte à la haute-tension. Il ne paraît pas que ceci perturbe sensiblement le régime UHF.

3. Discussion de la méthode

1. *Du point de vue continu*:

La résistance continue du cristal doit être très supérieure à R.

2. *Du point de vue haute-fréquence*:

Il convient de calculer dans quelle mesure le court circuit reste réalisé dans le plan XY quand on introduit la source. Soit $Z_e = R_e + j\,X_e$ l'impédance terminale de la ligne (R_e comprend la résistance au graphite). L'impédance ramenée dans le plan XY est

$$Z = R_0 \frac{Z_e \operatorname{ctg} \dfrac{2\pi l}{\lambda} + j\,R_0}{R_0 \operatorname{ctg} \dfrac{2\pi l}{\lambda} + j\,Z_e},$$

R_0 étant l'impédance caractéristique de la ligne (50 Ω) et l la longueur de la ligne ajustable c, voisine de $\lambda/4$.

Posons $l = \dfrac{\lambda}{4}(1 + \varepsilon)$. $\dfrac{R_0}{Z_e}$ est très petit, d'après la valeur de R_e choisie. On trouve alors sensiblement pour les parties réelle et imaginaire de Z:

$$R(Z) = \frac{R_0^2\,R_e}{R_e^2 + X_e^2}, \qquad I(Z) = R_0\left(\frac{-R_0\,X_e}{R_e^2 + X_e^2} + \frac{\pi}{2}\varepsilon\right).$$

On voit qu'il est possible d'annuler $I(Z)$ par une modification convenable de la longueur de la ligne ajustable. Le réglage devrait être changé chaque fois que la source est modifiée. Pratiquement on ne l'effectue pas. Evaluons l'erreur commise.

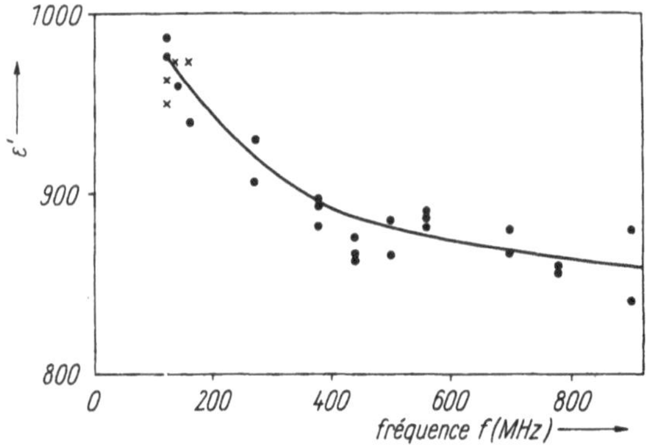

Fig. 4. Variation de ε' avec f

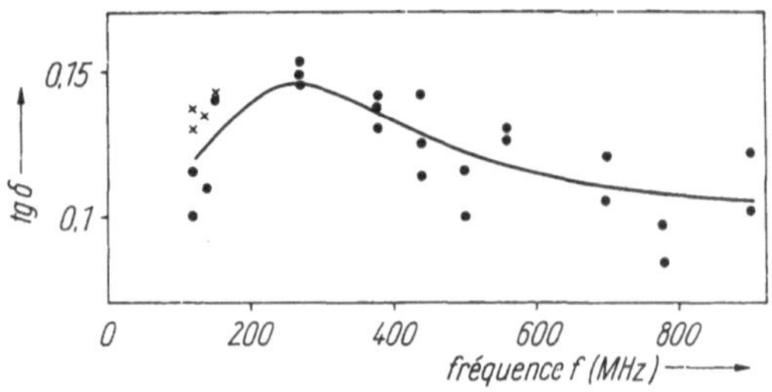

Fig. 5. Variation de tg δ avec f

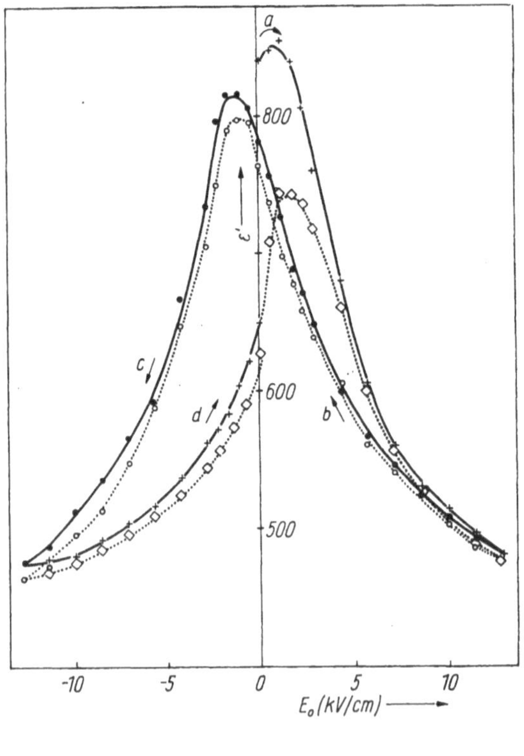

Fig. 6. Variation de ε' avec E_0
(f = 500 MHz).
Premier cycle: Traits pleins; sens a, b, c, d.
Deuxième cycle: courbe en pointillé

a) Terme réel: $R(Z) \leqq \frac{R_0^2}{R_e}$.

b) Terme imaginaire: La borne supérieure de $I(Z)$ est

$$R_0 \left(\frac{\pi}{2} |\varepsilon| + \frac{R_0}{2\,R_e} \right)$$

car $R_0/2\,R_e$ est la valeur maximale de $\frac{R_0 X_e}{R_e^2 + X_e^2}$ obtenue pour $R_e = X_e$.

On obtient en définitive dans le plan XY l'admittance $0 + j_0$ avec l'erreur

$$\frac{R_0^2}{R_e} \pm j\, R_0 \left(\frac{\pi}{2} |\varepsilon| + \frac{R_0}{2\,R_e} \right).$$

Avec $R_0 = 50\ \Omega$, $R_e = R = 20\ \mathrm{k}\Omega$ et $\varepsilon = 1/50$ (correspondant à la modification de la longueur de ligne de l'expérience 3), l'erreur est

$$(0{,}1 \pm j\, 1{,}6)\ \Omega\,.$$

4. Résultats obtenus avec le titanate de baryum

On a opéré avec des monocristaux de titanate de baryum formés de domaines a et c. Ils ont été préparés par croissance à partir d'un bain de titanate et de chlorure de baryum fondus. Ce sont des parallélépipèdes rectangles avec une épaisseur de l'ordre de 0,3 mm, les faces ayant une surface de l'ordre de 0,5 mm².

Les faces ayant été métallisées à l'argent à l'aide d'un vernis, le cristal est placé dans le plan XY entre les 2 éléments du conducteur central; il est maintenu grâce au ressort dont est muni le piston mobile. On déterminait l'admittance ramenée à la distance $l_e = \lambda/4$. On peut en tirer l'admittance de l'échantillon, puis les valeurs correspondantes de ε' et $\mathrm{tg}\,\delta = \varepsilon''/\varepsilon'$.

On doit tenir compte du fait que le cristal n'occupe pas la totalité de l'espace compris entre les 2 éléments du conducteur central. On peut admettre que l'admittance qui a été déterminée comprend, en parallèle, l'admittance de l'échantillon et l'admittance dûe à la capacité à air correspondant au reste de l'espace compris entre les conducteurs. Celle-ci est facile à évaluer; elle était d'ailleurs relativement peu importante à cause de la valeur élevée de ε' pour le titanate de baryum.

Nous présentons maintenant quelques résultats, afin de montrer que la méthode est assez fine pour donner beaucoup de renseignements expérimentaux, sans insister ici sur leur interprétation.

Les Fig. 4 et 5 donnent les variations de ε' et $\mathrm{tg}\,\delta$ obtenues en l'absence de champ statique ($E_0 = 0$), en fonction de la fréquence, entre 100 et 1000 MHz. Dans la zone 100 à 150 MHz des mesures de contrôle (représentées par des croix) ont été faites avec un pont VHF General Radio type 1601 A. Malgré la dispersion des points, on voit l'influence très nette de la fréquence. Elle est liée à l'inertie du renversement de la polarisation.

Les Fig. 6 et 7 donnent un exemple de résultats obtenus à 500 MHz pour ε' et $\mathrm{tg}\,\delta$ en décrivant un cycle statique (pour la clarté des figures les points experimentaux sont représentés différemment sur les diverses branches des courbes). Dans le cas de la Fig. 6 on a représenté deux cycles statiques successifs; on voit que la courbe s'est décalée d'un cycle à l'autre; des effets « de temps » de ce genre sont courants dans les corps ferroélectriques [3]. On remarque aussi l'asymétrie des courbes pour les 2 sens du champ, particulièrement marquée pour $\mathrm{tg}\,\delta$; il est très difficile d'obtenir des échantillons ne présentant pas d'asymé-

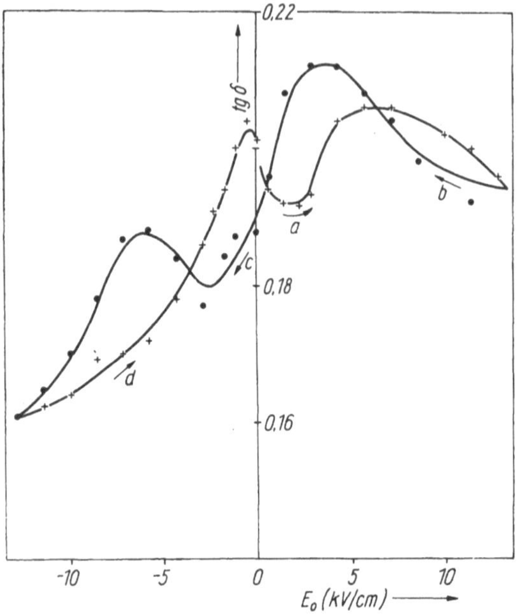

Fig. 7. Variation de tg δ avec E_0 (f = 500 MHz). Cycle décrit selon les flèches a, b, c …

trie importante. Un résultat concernant ε', analogue à celui de la Fig. 6, a été indiqué par Fousek [4].

Les courbes rappellent celles que l'on obtient avec des céramiques au titanate de baryum, car nous opérions avec des cristaux polydomaines; elles rappellent aussi celles que l'on obtient en basse fréquence. Il y a toutefois des différences du point de vue quantitatif. On peut interpréter l'ensemble de ces résultats.

Bibliographie

[1] J. Fousek, Czech. J. Phys. 8, 732 (1958).
[2] E. Stern et A. Lurio, Phys. Rev. **123**, 117 (1961).
[3] Y. Takagi et E. Sawaguchi, J. Phys. Soc. Japan **4**, 363 (1949).
[4] J. Fousek, Czech. J. Phys. **8**, 254 (1958).

(Received November 2, 1964)

phys. stat. sol. 8, 687 (1965)

Институт физики Академии наук Латвийской ССР, Рига

К термодинамике гайзенберговского ферромагнетика (II)

В. Н. Кащеев

Построена термодинамика изотропного гайзенберговского ферромагнетика в окрестности точки Кюри на основе результатов для коррелятора $\langle S^z(\boldsymbol{q})\, S^z(-\boldsymbol{q})\rangle$, полученного в работе Томиты и Танака [5]. Точка Кюри T_c, восприимчивость и спиновая теплоемкость выше T_c оказываются такими же, как при непосредственном применении расцепления Тябликова, в то время как термодинамика магнетика слева от T_c построить не удается.

The thermodynamic of an isotropic Heisenberg ferromagnet near the Curie point is constructed on the basis of the results for the correlation $\langle S^z(q)\, S^z(-q)\rangle$, obtained by Tomita and Tanaka [5] . The Curie point, susceptibility, and the spin heat-capacity are the same as obtained from direct application of the splitting introduced by Tyablikov, but for $T < T_c$ a thermomagnetic theory of a substance cannot be derived.

1. Введение

В предыдущих работах [1, 2] было показано, как влияет изменение характера расцепления уравнений для спиновых функций Грина — переход от расцепления Тябликова [3] к расцеплению Коллена [4] — на температурный ход термодинамических характеристик гайзенберговского ферромагнетика в окрестности точки Кюри. Недавно появилась работа Томиты и Танаки [5], где была предложена новая процедура расцепления этих уравнений. Помимо обычных двухвременных спиновых функций Грина [3] там были введены так называемые кумулянтные спиновые гринианы, причем взаимосвязь первых с последними аналогична соотношению между моментами и кумулянтами в статистике [6] — в этом и заключается эвристичность идеи расцепления Томиты-Танаки [5]. Если ограничиться в уравнениях для спиновых кумулянтных гринианов низшим приближением, то для остающегося гриниана получается такой же результат [5], как в случае расцепления Тябликова для обычной функции Грина. Томита и Танака в этой [5] и последующих работах [7, 8] получили выражения для спиновых гринианов для изучения формы линии ферромагнитного резонанса. Кроме этих основных результатов, в работе [5] было предложено выражение для коррелятора $\langle S^z(\boldsymbol{q})\, S^z(-\boldsymbol{q})\rangle$, справедливое при высоких температурах (окрестность точки Кюри и парамагнитная область), которое было вспомогательным для целей авторов работы [5]. Как мы покажем ниже, это выражение может стать основой для систематического построения термодинамики ферромагнетика в окрестности точки Кюри. Ввиду чрезвычайной сложности проблемы фазовых переходов второго рода нам представляется заслуживающим внимания рассмотрение — наряду с уже имеющимися [1 до 4, 9, 10] — и нового варианта теории особенностей термодинамических характеристик ферромагнетика в окрестности точки Кюри, основанной на использовании спиновых функций Грина.

2. Точка Кюри и спонтанная намагниченность

Гамильтониан изотропного гайзенберговского ферромагнетика в однородном внешнем магнитном поле $\boldsymbol{H}(0, 0, H)$ запишем в виде:

$$\mathcal{H} = \mathcal{H}_z + \mathcal{H}_e, \quad \mathcal{H}_z = -g\,\mu_0\,H \sum_n S_n^z, \quad \mathcal{H}_e = -\sum_{nm} J(\boldsymbol{R}_n - \boldsymbol{R}_m)\, S_n\, S_m. \quad (1)$$

Здесь $\mathcal{H}_z$ — зеемановская, а $\mathcal{H}_e$ — обменная энергия; g — факторов Ландэ, μ_0 — магнетон Бора, $S_n^\alpha = N^{-1/2} \sum_{\boldsymbol{q}} S^\alpha(\boldsymbol{q}) \times \exp(i\,\boldsymbol{q}\boldsymbol{R}_n)$, α — компонента спинового оператора n-го узла решетки (N — их общее число), $J(\boldsymbol{R}_n - \boldsymbol{R}_m) = N^{-1} \sum_{\boldsymbol{q}} J(\boldsymbol{q}) \exp[-i(\boldsymbol{R}_n - \boldsymbol{R}_m)]$ — обменный интергал для n-го и m-го узлов. Очевидно, что

$$\mathcal{H}_z = -g\,\mu_0\,H\sqrt{N}\,S^z(0), \qquad \mathcal{H}_e = -\sum_{\boldsymbol{q}} J(\boldsymbol{q})\,[S^-(-\boldsymbol{q})\,S^+(\boldsymbol{q}) + S^z(\boldsymbol{q})\,S^z(-\boldsymbol{q})],$$

$$S_n^\pm = S_n^x \pm i\,S_n^y. \quad (2)$$

Основополагающее для нашего дальнейшего рассмотрения выражение для коррелятора $\langle S^z(\boldsymbol{q})\,S^z(-\boldsymbol{q})\rangle = \mathrm{Sp}\,[S^z(\boldsymbol{q})\,S^z(-\boldsymbol{q})\exp(-\beta\,\mathcal{H})] \times [\mathrm{Sp}\exp(-\beta\,\mathcal{H})]^{-1}$ было получено в работе [5] следующим образом. К энергии $\mathcal{H}_z$ (2) добавлялось слагаемое $-\sum_{\boldsymbol{k}} \omega_0(\boldsymbol{k})\,S^z(\boldsymbol{k})$, так что, по определению,

$$\lim_{\omega_0(\boldsymbol{k})\to 0} \frac{\partial\langle S^z(\boldsymbol{k})\rangle}{\partial\omega_0(\boldsymbol{k})} \sim \int_0^\beta dt\,\{\langle S^z(\boldsymbol{k})|_0\,S^z(-\boldsymbol{k})|_{i\hbar t}\rangle - N\langle S^z\rangle^2\,\Delta(\boldsymbol{k})\}, \quad (3)$$

$$A|_\lambda = e^{i\lambda\mathcal{H}}\,A|_0\,e^{-i\lambda\mathcal{H}}. \quad (4)$$

Вследствие этого при $\beta = (k\,T)^{-1}$ правая часть (3) становится равной

$$\beta\,\gamma(\boldsymbol{k}, -\boldsymbol{k}) = \beta\,\{\langle S^z(\boldsymbol{k})\,S^z(-\boldsymbol{k})\rangle - N\,\Delta(\boldsymbol{k})\,\langle S^z\rangle^2\} \quad (5)$$

(операторы $S^z(\pm\boldsymbol{k})$ в (5) берутся в один и тот же момент времени λ). Величина же $\langle S^z(\boldsymbol{k})\rangle$ находилась из соотношения

$$\langle (S_n)^2\rangle = S\,(S+1), \quad (6)$$

в котором корреляторы $\langle S^\mp(\mp\boldsymbol{k})\,S^\pm(\pm\boldsymbol{k})\rangle$ вычислялись с помощью известной в теории двухвременных гринианов спектральной теоремы [11] на основе полученных в [5] (при гамильтониане $\mathcal{H} - \sum_{\boldsymbol{k}} \omega_0(\boldsymbol{k})\,S^z(\boldsymbol{k})$) кумулянтных функций Грина $(S^+(\boldsymbol{k});\ S^-(-\boldsymbol{k}))$. Поскольку последние в [5] брались в низшем приближении, то как уже говорилось в п. 1, в этом случае

$$\left.\begin{aligned}(S^+(\boldsymbol{k});\ \ S^-(-\boldsymbol{k})) = \langle\!\langle S^+(\boldsymbol{k})|_t,\ \ S^-(-\boldsymbol{k})|_0\rangle\!\rangle = -\,i\,\theta(t)\,\langle[S^+(\boldsymbol{k})|_t\,,\\ S^-(-\boldsymbol{k})|_0]\rangle = \int\limits_{-\infty}^{\infty} d\omega\, e^{-i\omega t}\langle\!\langle S^+(\boldsymbol{k})\rangle\!\rangle;\, S^-(-\boldsymbol{k})\rangle\!\rangle_\omega\,.\end{aligned}\right\} \tag{7}$$

Гриниан же $\langle\!\langle S^+(\boldsymbol{k});\, S^-(-\boldsymbol{k})\rangle\!\rangle_\omega$, по Тябликову [3], равен

$$\langle\!\langle S^+(\boldsymbol{k});\, S^-(-\boldsymbol{k})\rangle\!\rangle_\omega = \frac{1}{\pi}\,\frac{\langle S^z\rangle}{\omega - \omega_{\boldsymbol{k}}}\,, \tag{8}$$

$$\hbar\,\omega_{\boldsymbol{k}} = g\,\mu_0\,H + 2\,\langle S^z\rangle\,[J(0) - J(\boldsymbol{k})] \equiv E_{\boldsymbol{k}} = E_{-\boldsymbol{k}}\,. \tag{9}$$

В итоге для величины $\gamma(\boldsymbol{k}, -\boldsymbol{k})$ (5) в работе [5] получено выражение[1]):

$$\gamma(\boldsymbol{k}, -\boldsymbol{k}) = -\frac{1}{2\,\beta}\,\frac{\sum\limits_{\boldsymbol{q}} \dfrac{1}{J(\boldsymbol{q}) - J(\boldsymbol{q}-\boldsymbol{k})}\left[\operatorname{cth}\dfrac{\beta\,E_{\boldsymbol{k}-\boldsymbol{q}}}{2} - \operatorname{cth}\dfrac{\beta\,E_{\boldsymbol{q}}}{2}\right]}{\sum\limits_{\boldsymbol{q}}\left[\dfrac{J(\boldsymbol{k}) - J(\boldsymbol{k}-\boldsymbol{q})}{J(\boldsymbol{q}) - J(\boldsymbol{k}-\boldsymbol{q})}\operatorname{cth}\dfrac{\beta\,E_{\boldsymbol{k}-\boldsymbol{q}}}{2} + \dfrac{J(\boldsymbol{q}) - J(\boldsymbol{k})}{J(\boldsymbol{q}) - J(\boldsymbol{k}-\boldsymbol{q})}\operatorname{cth}\dfrac{\beta\,E_{\boldsymbol{q}}}{2}\right]}\,. \tag{10}$$

Мы будем использовать выражения (8) до (10) для нахождения точки Кюри $T_c = (k\,\beta_c)^{-1}$, а также намагниченности, восприимчивости и спиновой теплоемкости по обе стороны от нее в ее окрестности.

Найдем сначала точку Кюри. Поскольку, согласно [3], по использовании спектральной теоремы с гринианом (8)

$$\left.\begin{aligned}\langle S^+(\boldsymbol{k})\,S^-(-\boldsymbol{k})\rangle = 2\,\langle S^z\rangle\,(N_{\boldsymbol{k}} + 1) = \langle S^-(-\boldsymbol{k})\,S^+(\boldsymbol{k})\rangle + 2\,\langle S^z\rangle\,,\\ N_{\boldsymbol{k}} = (e^{\beta\hbar\omega_{\boldsymbol{k}}} - 1)^{-1} = \frac{1}{2}\left(\operatorname{cth}\frac{\beta\,E_{\boldsymbol{k}}}{2} - 1\right),\end{aligned}\right\} \tag{11}$$

то после подстановки (10), (11) в условие (6) получается уравнение

$$\langle S^z\rangle^2 + \frac{\langle S^z\rangle}{N}\sum_{\boldsymbol{k}}\operatorname{cth}\frac{\beta\,E_{\boldsymbol{k}}}{2} - S\,(S+1) -$$

$$-\frac{1}{2\,\beta\,N}\sum_{\boldsymbol{k}}\frac{\dfrac{\langle S^z\rangle}{N}\sum\limits_{\boldsymbol{q}}\dfrac{1}{J(\boldsymbol{q}) - J(\boldsymbol{q}-\boldsymbol{k})}\left[\operatorname{cth}\dfrac{\beta\,E_{\boldsymbol{q}-\boldsymbol{k}}}{2} - \operatorname{cth}\dfrac{\beta\,E_{\boldsymbol{q}}}{2}\right]}{\dfrac{\langle S^z\rangle}{N}\sum\limits_{\boldsymbol{q}}\left[\dfrac{J(\boldsymbol{k}) - J(\boldsymbol{k}-\boldsymbol{q})}{J(\boldsymbol{q}) - J(\boldsymbol{k}-\boldsymbol{q})}\operatorname{cth}\dfrac{\beta\,E_{\boldsymbol{q}-\boldsymbol{k}}}{2} + \dfrac{J(\boldsymbol{q}) - J(\boldsymbol{k})}{J(\boldsymbol{q}) - J(\boldsymbol{k}-\boldsymbol{q})}\operatorname{cth}\dfrac{\beta\,E_{\boldsymbol{q}}}{2}\right]} = 0\,. \tag{12}$$

При $H = 0$, $\beta = \beta_c$, $\langle S^z\rangle = 0$, и уравнение (12) дает

$$\beta_c = \frac{3}{2}\,\frac{F(-1)}{J(0)\,S\,(S+1)}\,,\quad F(-1) = \frac{1}{N}\sum_{\boldsymbol{k}}\frac{J(0)}{J(0) - J(\boldsymbol{k})}\,. \tag{13}$$

---

[1]) Надо сказать, что в [5] выражение [10] используется и при низких температурах, когда справедлива теория спиновых волн, что, конечно уже совсем не обосновано.

Точка Кюри оказалась такой же, как в случае расцепления Тябликова, примененного непосредственно, без необходимости знания коррелятора $\langle S^z(\boldsymbol{q})\, S^z(-\boldsymbol{q})\rangle$, для получения $\langle S^z\rangle$ как функционала энергетического спектра системы (9) в работе [4]

$$\langle S^z\rangle = \frac{(S-\Phi)\,(1+\Phi)^{2S+1} + (S+1+\Phi)\,\Phi^{2S+1}}{(1+\Phi)^{2S+1} - \Phi^{2S+1}}. \tag{14}$$

Здесь S — спин парамагнитных ионов на узлах решетки, $\Phi = N^{-1} \sum_{\boldsymbol{k}} N_{\boldsymbol{k}}$.

Если бы выражение (10) для $\gamma(\boldsymbol{k}, -\boldsymbol{k})$ было не приближенным, а строгим,[2]) то в совпадении (13) с выражением для точки Кюри, полученным при непосредственном применении расцепления Тябликова [3, 12], не было бы ничего удивительного. Теперь же это совпадение говорит о том, что, по крайней мере для самой температуры Кюри, приближение (4) и (5) является удовлетворительным.[3]) Это позволяет надеяться получить выражения для термодинамических величин в окрестности T_c, правдоподобные в той же мере, что и получаемые с помощью расцепления Тябликова непосредственно [3, 12].

Рассмотрим, однако, спонтанную намагниченность $M_z(H=0,\beta) = g\,\mu_0\, N\langle S^z\rangle|_{H=0}$ слева от точки Кюри. Теперь $\langle S^z\rangle \ll 1$, $a = \beta\, E_{\boldsymbol{k}} \ll 1$, $\operatorname{cth} a \to \frac{1}{a} + \frac{a}{3}$ и уравнение (12), в котором (в том числе в числителе и знаменателе дроби) удерживаются члены порядка не выше $0(\langle S^z\rangle^2)$, принимает вид

$$\frac{F(-1)}{J(0)} + \beta\langle S^z\rangle^2\left[1+\frac{1}{3}\beta\, J(0)\right] - \beta\, S\,(S+1) + \\ + \frac{1}{2N}\sum_{\boldsymbol{k}} \frac{3\,\varphi(\boldsymbol{k}) - \beta^2\,\langle S^z\rangle^2}{3\,\varphi(\boldsymbol{k})\,[J(0)-J(\boldsymbol{k})] + \beta^2\,\langle S^z\rangle^2\,[J(0)+J(\boldsymbol{k})]} = 0\,, \tag{15}$$

$$\varphi(\boldsymbol{k}) = \frac{1}{N}\sum_{\boldsymbol{q}} \frac{1}{J(0)-J(\boldsymbol{q}-\boldsymbol{k})}\,\frac{1}{J(0)-J(\boldsymbol{q})}. \tag{16}$$

В работе [9] показано, что при малых $|\boldsymbol{k}|$, когда $c\,|\boldsymbol{k}| \ll 1$, в приближении ближайших соседей для кубических кристаллов

$$\varphi(\boldsymbol{k}) = \frac{v}{8\,c^3}\,\frac{1}{J^2\,c\,|\boldsymbol{k}|}, \tag{17}$$

где J — обменный интеграл, v — объем элементарной ячейки, c — постоянная решетки.

Поскольку при $c\,|\boldsymbol{k}| \ll 1$ $\varphi(\boldsymbol{k}) \sim |\boldsymbol{k}|^{-1}$, то, если разлагать дробь в (15) по степеням $\langle S^z\rangle^2$, интеграл по $\boldsymbol{k}$, являющийся коэффициентом при $\langle S^z\rangle^2$ в этом разложении, будет расходиться при малых $|\boldsymbol{k}|$. Таким образом, следует решать трансцендентное относительно $\langle S^z\rangle^2$ уравнение (15) более точно, не разлагая дроби в (15) по степеням $\langle S^z\rangle^2$. В приложении

---

[2]) В пределах применимости расцепления Тябликова.

[3]) Напоминаем, что использованные для получения (5), (10) низшие кумулянтные спиновые гринианы [5] совпадают с обычными [3, 12] (см. (7)).

показано, что (15) удовлетворяется лишь в самой T_c, но не слева от неё. Таким образом, с помощью выражения Томиты-Танаки для $\langle S^z(\boldsymbol{k})\, S^{-z}(-\boldsymbol{k})\rangle$ нельзя рассматривать термодинамику ферромагнетика слева от T_c; оно смраведливо лишь в парамагнитой области.

3. Восприимчивость и спиновая теплоемкость

В соответствии с определением (5) в общем случае параллельная восприимчивость $\chi_{zz}(H, \beta) = \dfrac{\partial M_z(H, \beta)}{\partial H}$ имеет вид

$$\chi_{zz}(H, \beta) = (g\,\mu_0)^2\, \beta\, N\, \gamma(0, 0)\,, \tag{18}$$

а по (10)

$$\gamma_{00}(H, \beta) = \frac{2\,\langle S^z\rangle \sum\limits_{\boldsymbol{q}} N_{\boldsymbol{q}}\,(N_{\boldsymbol{q}} + 1)}{\sum\limits_{\boldsymbol{q}} \{1 + 2\,N_{\boldsymbol{q}} - 4\,\beta\,\langle S^z\rangle\,[J(0) - J(\boldsymbol{q})]\,N_{\boldsymbol{q}}\,(N_{\boldsymbol{q}} + 1)\}}\,. \tag{19}$$

Как и в случае непосредственного применения расцепления Тябликова [9], спонтанная восприимчивость $\chi^{\Phi}_{zz}(0, \beta) \equiv \chi_{zz}(0, \beta > \beta_c)$ оказывается расходящейся для изотропных магнетиков со спиновым гамильтонианом (2); эту аналогию со случаем расцепления по Тябликову [9] можно продолжить, если учесть, что выражение (19) слева от T_c становится конечным при $H \neq 0$ или при $H = 0$ и учете магнитной анизотропии в спиновом гамильтониане $\mathcal{H}$.

Рассмотрим теперь восприимчивость $\chi^{\text{п}}_{zz}(0, \beta) \equiv \chi_{zz}(0, \beta < \beta_c)$ справа от точки Кюри, в парамагнитной области. Полагая в (19) при малых H и $\beta < \beta_c$

$$\langle S^z\rangle = \frac{\chi^{\text{п}}_{zz}(0, \beta)}{g\,\mu_0\,N}\,H \equiv \frac{g\,\mu_0\,H}{2\,\zeta^{\text{п}}(\beta)} \tag{20}$$

и при перехода к $\lim\limits_{H\to 0}$, после подстановки (19) в (18) приходим к тождеству. Это говорит о том, что приближение (4), (5) для $\gamma(0, 0)$ является очень хорошим в парамагнитной области. Чтобы получить выражение для $\chi^{\text{п}}_{zz}(0, \beta)$, подставим (20) в (12), положим $H = 0$ и выразим $F(-1)$ через β_c по (13). Тогда уравнение (12) примет вид

$$\frac{1}{N}\sum_{\boldsymbol{k}}\left[\frac{1}{\zeta^{\text{п}}(\beta) + J(0) - J(\boldsymbol{k})} - \frac{\beta}{\beta_c}\,\frac{1}{J(0) - J(\boldsymbol{k})}\right] = 0\,. \tag{21}$$

Решение его относительно $\zeta^{\text{п}}(\beta)$ приведено в [2]; возвращаясь к восприимчивости по (20), находим

$$\chi^{\text{п}}_{zz}(0, \beta) = \frac{N\,\pi^2}{8\,J}\left(\frac{g\,\mu_0}{c\,k_m}\right)^2\left(\frac{\beta_c}{\beta} - 1\right)^{-2}, \qquad k_m = \max|\boldsymbol{k}|\,. \tag{22}$$

При этом использовалось приближение ближайших соседей для кубических кристаллов (П. 3). Выражение (22) совпадает с получающимся [2] при непосредственном использовании расцепления Тябликова;

парамагнитная восприимчивость (22) изотропного магнетика оказывается имеющей в точке Кюри полюс второго порядка.

Таким образом, метод Томиты и Танака [5] дает для параллельной восприимчивости в окрестности точки Кюри те же результаты, что и непосредственное использование расцепления Тябликова.

Вычислим теплоемкость спиновой системы по справа от точки Кюри. Пусть снова $H = 0$. Ввиду $J(\boldsymbol{R}_n = \boldsymbol{R}_m) = 0$ $\sum_{\boldsymbol{q}} J(\boldsymbol{q}) = 0$, так что средняя энергия спиновой системы по (2) и (5) равна

$$\langle \mathscr{H} \rangle = \langle \mathscr{H}_e \rangle = -J(0) \langle S^z \rangle^2 N - \sum_{\boldsymbol{q}} J(\boldsymbol{q}) \left[\langle S^-(-\boldsymbol{q}) S^+(\boldsymbol{q}) \rangle + \gamma(\boldsymbol{q}, -\boldsymbol{q}) \right]. \tag{23}$$

Рассмотрим правую окрестность точки Кюри T_c. Подставив (9) и (20) в формулы (11) и (19), находим, что при $\beta < \beta_c$, $H = 0$

$$\langle S^\alpha(\boldsymbol{q}) S^\alpha(-\boldsymbol{q}) \rangle = \gamma(\boldsymbol{q}, -\boldsymbol{q}) = \frac{1}{2\beta} \frac{1}{\zeta^{\text{п}}(\beta) + J(0) - J(\boldsymbol{q})} \tag{24}$$

независимо от α, т.е. в парамагнитной области коррелятор $\langle S^\alpha(\boldsymbol{q}) \times \times S^\alpha(-\boldsymbol{q}) \rangle$ изотропен, как и должно быть. Поскольку по (20) и (22) $\zeta^{\text{п}}(\beta) \sim (\beta_c/\beta - 1)^2$, то этой величиной $\Big($и ее производной по β, появляющейся в выражении для спиновой теплоемкости $C_v(\beta) = -k\beta^2 \times \times \frac{\partial \langle \mathscr{H} \rangle}{\partial \beta}\Big)$ в (23), куда подставляется формула (24), можно пренебречь. Таким образом, для энергии спиновой системы в парамагнитной области имеем следующее выражение

$$\frac{\langle \mathscr{H} \rangle^{\text{п}}}{N} = \frac{3}{2\beta} \left[1 - F(-1) \frac{\beta}{\beta_c} \right]. \tag{25}$$

В точке Кюри

$$\left. \frac{\langle \mathscr{H} \rangle^{\text{п}}}{N} \right|_{\beta_c} = -\frac{3}{2\beta_c} [F(-1) - 1] < 0\,. \tag{26}$$

Спиновая теплоемкость в парамагнитной области равна:

$$C_V^{\text{п}}(\beta) = \frac{3}{2} N k\,. \tag{27}$$

Результаты (25) до (27) в точности соответствуют получающимся [10] при непосредственном использовании расцепления Тябликова [12], [4]. Это подтверждает вывод, сделанный о приближении (4) и (5) Томиты-Танака [5] при обсуждении результата (22) (см. выше).

Резюмируя все вышеизложенное, можно отметить, что теория Томиты-Танаки [5] дает для точки Кюри и термодинамических характеристик ферромагнетика выше ее те же результаты, что и непосредственное расцепление Тябликова [3, 9, 12]; для намагниченности, энергии и теплоёмкости слева от T_c не получается физически приемлемого результата. Очевидно, что приближения [5], положенные в основу получения (10), оказались слишком грубыми. Мы имеем ввиду не только приближение (4) и (5), все улучшающееся с повышением температуры, но и уравнение (12), определяющее функционал $\langle S^z\rangle\{E_k\}$: в работе Томиты и Танаки [5] при получении этого выражения с заданным $\gamma(\boldsymbol{k}, -\boldsymbol{k})$ (5) пренебрегалось выражением $\lim\limits_{\omega_0(\boldsymbol{k})\to 0} \frac{\partial\langle S^z(\boldsymbol{k})\, S^z(-\boldsymbol{k})\rangle}{\partial\omega_0(\boldsymbol{k})}$, где усреднение производится по ансамблю с гамильтонианом $\mathcal{H} - \sum\limits_{\boldsymbol{k}} \omega_0(\boldsymbol{k})\, S^z(\boldsymbol{k})$.

Выделить роль этого приближения (4) и (5) при получении выражений для термодинамических характеристик ферромагнетика в окрестности точки Кюри затруднительно.

Приложение

Для решения уравнения (15) после подстановки туда $F(-1)$ из (13) перепишем его в виде

$$\langle S^z\rangle^2 \left[1 + \frac{1}{3}\beta\, J(0)\right] - S\,(S+1)\left(1 - \frac{\beta_c}{\beta}\right) - \beta\, J(0)\,\langle S^z\rangle^2\, F(\langle S^z\rangle) = 0\,, \quad (\Pi.1)$$

$$F(\langle S^z\rangle) = \frac{1}{N}\sum_{\boldsymbol{k}} \frac{1}{J(0) - J(\boldsymbol{k})}\, \frac{1}{3\,\varphi(\boldsymbol{k})\,[J(0) - J(\boldsymbol{k})] + \beta^2\,\langle S^z\rangle^2\,[J(0) + J(\boldsymbol{k})]}\,. \quad (\Pi.2)$$

Поскольку $\langle S^z\rangle^2 \ll 1$, то в $N^{-1}\sum\limits_{\boldsymbol{k}} \to (2\,\pi)^{-3}\, v \int d\boldsymbol{k}$ в (П.2) существенны малые $|\boldsymbol{k}|$. В случае кубических кристаллов при малых $|\boldsymbol{k}|$ можно положить

$$J(0) - J(\boldsymbol{k}) = J(c\,\boldsymbol{k})^2, \qquad J = z^{-1}\, J(0)\,, \quad (\Pi.3)$$

где z — число ближайших соседей. Тогда

$$F(\langle S^z\rangle) = -\frac{4}{3\,\pi^2}\ln\,(R\,\langle S^z\rangle^2)\,, \qquad R = \frac{16}{3\,z}\,\frac{c^3}{v}\,\frac{[\beta\, J(0)]^2}{c\, k_m}\,. \quad (\Pi.4)$$

Теперь уравнение (П.1) имеет вид

$$A\langle S^z\rangle^2 - B\langle S^z\rangle^2 \ln\,\langle S^z\rangle^2 = S\,(S+1)\,\varepsilon\,, \quad (\Pi.5)$$

$$1 \gg \varepsilon = 1 - \frac{\beta_c}{\beta} > 0\,, \quad A = 1 + \frac{1}{3}\beta\, J(0)\left[1 + \frac{4}{\pi^2}\ln R\right], \quad B = \frac{-4}{3\pi^2}\beta\, J(0)\,. \quad (\Pi.6)$$

Нетрудно видеть, что так как даже при не очень малых $\langle S^z \rangle^2$

$$\langle S^z \rangle^2 \left| \ln \langle S^z \rangle^2 \right| > \langle S^z \rangle^2 ,$$

то (П.5) не имеет решений для таких $\langle S^z \rangle^2 > 0$, ибо $\varepsilon > 0$, $B < 0$, но (П.5) удовлетворяется в самой T_c, где $\langle S^z \rangle = 0$.

Литература

[1] В. Н. Кащеев, phys. stat. sol. **5**, 373 (1964).
[2] В. Н. Кащеев, phys. stat. sol. **7**, 777 (1964).
[3] С. В. Тябликов, Укр. матем. ж. **11**, 287 (1959).
[4] H. B. Callen, Phys. Rev. **130**, 890 (1963).
[5] K. Tomita and M. Tanaka, Progr. theor. Phys. **29**, 528 (1963).
[6] R. Kubo, J. Phys. Soc. Japan **17**, 1100 (1957).
[7] M. Tanaka and K. Tomita, Progr. theor. Phys. **29**, 651 (1963).
[8] M. Tanaka, Progr. theor. Phys. **31**, 177 (1964).
[9] K. Kawasaki and H. Mori, Progr. theor. Phys. **28**, 690 (1962).
[10] R. A. Tahir-Kheli, Phys. Rev. **132**, 689 (1963).
[11] Н. Н. Боголюбов и С. В. Тябликов, Докл. Акад. наук СССР, **126**, 53 (1959).
Д. Н. Зубарев, Успехи физ. наук **71**, 71 (1960).
В. Л. Бонч-Бруевич и С. В. Тябликов, Метод функций Грина в статистической механике, Физматгиз, Москва 1961.
[12] R. A. Tahir-Kheli and D. ter Haar, Phys. Rev. **127**, 88 (1962).
[13] Л. Д. Ландау и Е. М. Лифшиц, Электродинамика сполшных сред. ГИТТЛ, Москва 1957.
[14] Л. Д. Ландау и Е. М. Лифшиц, Статистическая физика, ГИТТЛ, Москва 1951.

(Received November 9, 1964)

phys. stat. sol. 8, 695 (1965)

Institut für Experimentalphysik der Technischen Universität Dresden

Über die Beobachtung einer Korngrenzenwanderung auf der Wolframkathode im Feldelektronenmikroskop

Von

H. HESSE

Es wird über eine Korngrenzenwanderung auf einer Wolfram-Spitzenkathode bei ca. 2600 °K berichtet, die im Feldelektronenmikroskop beobachtet wurde. Die beiden Kristallite sind um [011] um ca. 71° gegeneinander verdreht. Diese Orientierung stimmt mit dem Koinzidenzgitter-Modell überein. Die Abweichung der Korngrenze von einer dichtgepackten Ebene des Koinzidenzgitters wird durch Stufenbildung gedeutet. Die Entstehung eines speziell orientierten zweiten Kristallbereichs im Wolframeinkristall ist eine besondere Form der Rekristallisation.

Observations are made of grain boundary migration at a tungsten tip cathode at about 2600 °K, by using a field electron microscope. The crystallites are rotated about [011] toward one another at approximately 71°. This orientation is in accordance with the coincidence-site lattice model. The deviation of the grain boundary from a close-packed plane of the conincidence-site lattice is interpreted in terms of step-formation. The formation of a specially orientated second grain in the tungsten single crystal is a particular type of recrystallization.

1. Einleitung

Schon bald nach Einführung des Feldelektronenmikroskops in die Metallographie konnte E. W. Müller [1] über die Beobachtung von Korngrenzen auf der Wolfram-Spitzenkathode berichten. Caspary und Krautz [2] erzielten eine Korngrenzenverschiebung im Bild einer Palladiumspitze, die vermutlich durch Abtragen des Oberflächenmaterials hervorgerufen wurde. Stark [3] veröffentlichte eine Bildserie, die einzelne Stadien der Korngrenzenwanderung auf einer Rhodiumspitze zeigt. Da das Feldelektronenmikroskop (FEM) im allgemeinen keine atomaren Strukturen auflösen kann, läßt sich aus diesen Aufnahmen keine Aussage über die Atomanordnung in der Korngrenze gewinnen. Dagegen reicht das Auflösungsvermögen des Feldionenmikroskops (FIM) dafür aus. Es sei hier nur auf die Arbeiten von E. W. Müller [4] und Brandon et al. [5] hingewiesen, denen die Abbildung atomarer Einzelheiten von Kleinwinkel-Korngrenzen in Platinspitzen bzw. von Klein- und Großwinkelkorngrenzen in Spitzen aus Wolfram und Wolfram-Rhenium-Legierungen gelang. Die Lage der Korngrenze in der Spitze kann dabei durch sukzessives Abtragen der Oberfläche mittels Feldverdampfung aufgeklärt werden. Dagegen ist in diesen Arbeiten [4, 5] keine Korngrenzenverschiebung bei hohen Temperaturen verfolgt worden. Obwohl also die FEM-Untersuchung von Korngrenzen der FIM-Untersuchung bezüglich der direkten Abbildung der Korngrenzenstruktur unterlegen ist, lassen sich aus ersterer doch Informationen über die kristallographische Orientierung der Körner und der Korngrenze entnehmen. In der vorliegenden Mitteilung wird über eine Korngrenzenwanderung auf einer Wolframspitze berichtet, die als Kathode eines

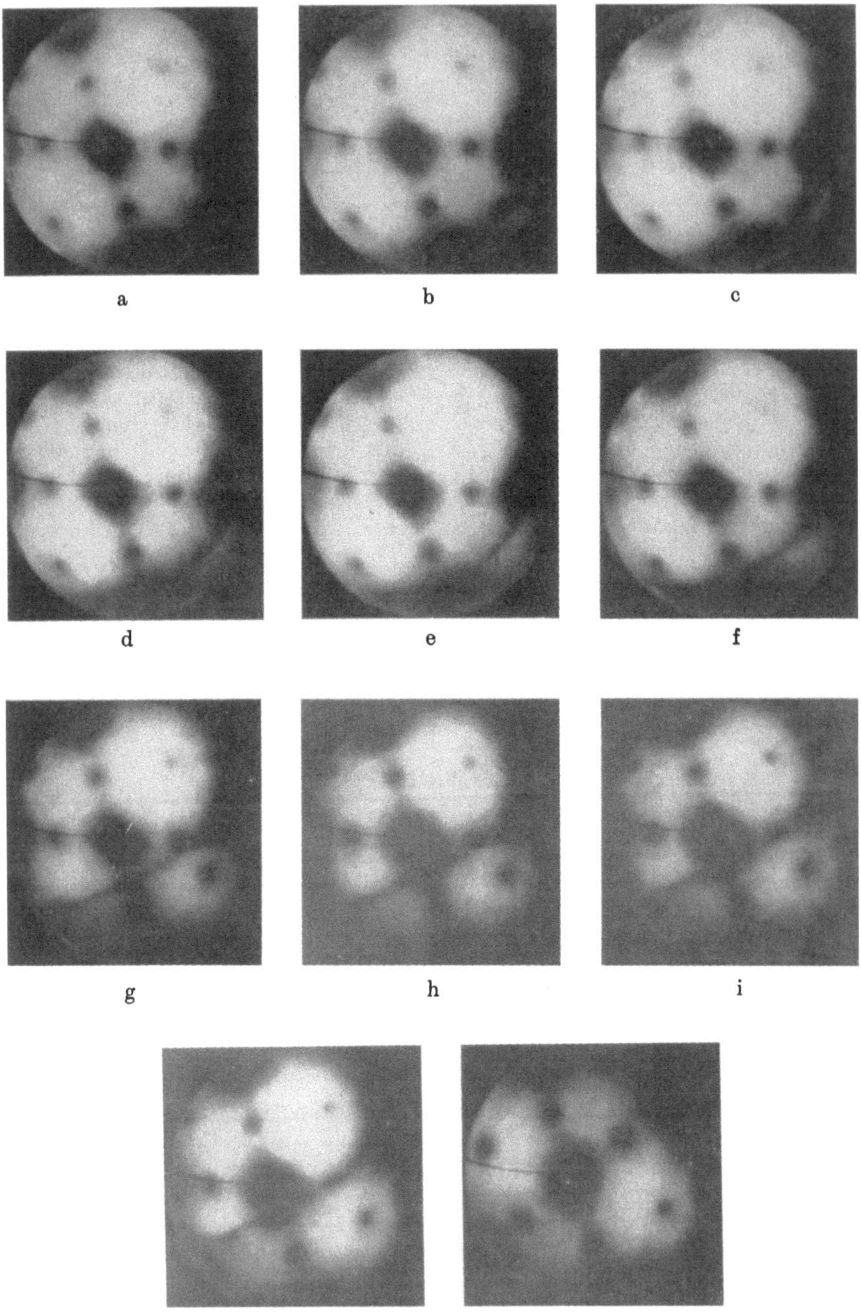

Fig. 1. Korngrenzenwanderung auf der Wolframkathode im Feldelektronenmikroskop bei ca. 2600 °K. a) Wolfram-Einkristall, b) bis j) fortschreitende Korngrenzenwanderung, k) Wolfram-Einkristall, Orientierung im Vergleich zu Fig. 1a verändert. Zwischen jeweils zwei Bildern, die bei Zimmertemperatur aufgenommen sind, wurde die Kathode für 60 s auf ca. 2600 °K erhitzt

FEM diente. Die festgestellte Orientierung der beiden Kristallite wird nach dem Koinzidenzgitter-Modell von Frank [6] und Dunn [7] erklärt und die Lage der Korngrenze nach einem Modell von Brandon et al. [5] gedeutet.

2. Beschreibung der Experimente

Fig. 1a bis k zeigt die Korngrenzenwanderung auf der Wolframspitze. In Fig. 1a ist das bekannte Bild einer einkristallinen Wolframkathode ohne Restgasbedeckung wiedergegeben, die sich während der Abbildungsdauer auf Zimmertemperatur befand. Bei den wenig emittierenden Gebieten handelt es sich um {001}-, {011}- und {112}-Flächen des kubisch-raumzentrierten Kristalls. Die Fig. 1a zuzuordnende orthogonale Projektion ist in Fig. 2a dargestellt. Zwischen den Aufnahmen zu Fig. 1a und Fig. 1b wurde die Kathode 60 s auf ca. 2600 °K erhitzt. Als Ergebnis ist im rechten unteren Bildteil der Fig. 1b eine Veränderung in Form einer hellen Linie angedeutet. Zwischen jeweils zwei aufeinanderfolgenden Bildern der Fig. 1 wurde nun immer die Spitzenkathode ohne elektrisches Feld für 60 s bei ca. 2600 °K geglüht und anschließend wieder auf Zimmertemperatur abgekühlt. Man sieht in Fig. 1b bis j, wie sich mit wachsender Zahl der Heizzyklen eine Korngrenze vom Bildrand rechts unten bis zur Bildmitte verschiebt. Nach Erhitzen der Kathode auf ca. 2600 °K für 60 s im Anschluß an ihre Abbildung in Fig. 1j ist die Korngrenze verschwunden (Fig. 1k): Die Spitze wird wieder von einem Einkristall gebildet, dessen Orientierung aber im Vergleich zum Einkristall Fig. 1a verändert ist. Fig. 2a und b stellt die beiden orthogonalen Projektionen gegenüber.

Die Spitzenkathode wurde aus gezogenem Wolframdraht NS von 0,1 mm Durchmesser hergestellt und vor dem Abschmelzen des FEM am Pumpstand 1 h bei ca. 2700 °K entgast. Der Versuch wurde nach dem Abschmelzen bei einem Restgasdruck $< 10^{-9}$ Torr durchgeführt. Die in allen Aufnahmen zu Fig. 1 sichtbare schwarze Linie im linken Bildteil ist durch den Farbstrich des zum Rezipientenbau verwendeten Glases bedingt. Die Abbildungsspannung am FEM und der Elektronenstrom waren während der ganzen Versuchsreihe konstant. Daraus geht hervor, daß die Korngrenzenverschiebung im Bild auf eine echte Korngrenzenwanderung und nicht auf Abtragung zurückzuführen ist, denn bei dieser müßte sich gleichzeitig der Spitzenradius und damit der Strom-Spannungs-Zusammenhang ändern.

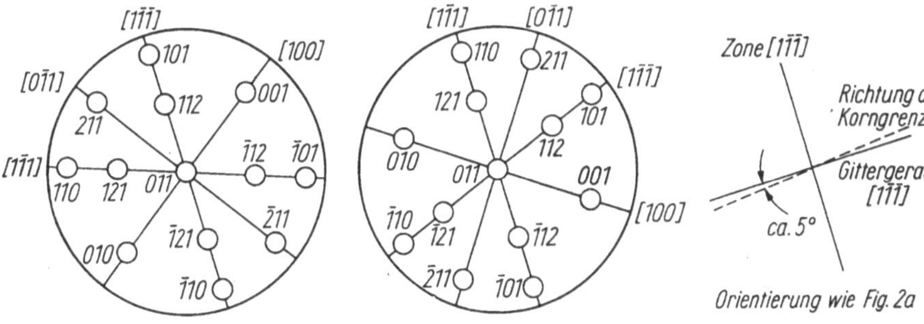

a b c

Fig. 2

a) Orthogonale Projektion zu Fig. 1a
b) Orthogonale Projektion zu Fig. 1k
c) Zur Richtungsbeziehung zwischen der Korngrenze und dem Koordinatensystem Fig. 2a

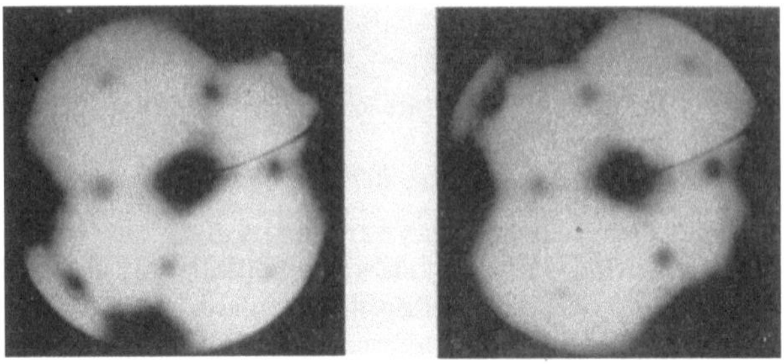

a b

Fig. 3. Orientierungsänderung der Wolfram-Spitzenkathode infolge Korngrenzenwanderung bei ca. 2700 °K. a) Wolfram-Einkristall, b) Wolfram-Einkristall mit neuer Orientierung, die nach Erhitzen der Kathode auf ca. 2700 °K für 60 s erhalten wurde

Aus Fig. 1 können folgende quantitative Angaben entnommen werden: Die Richtung der Korngrenze weicht um ca. 5° von der Gittergeraden $[1\bar{1}\bar{1}]$ ab (Fig. 2c). (Diese Orientierungsangabe bezieht sich auf das Koordinatensystem der Projektion Fig. 2a.) Der nach der Korngrenzenwanderung abgebildete Einkristall (Fig. 1k) ist gegen den ursprünglich vorhandenen (Fig. 1a) um die Achse [011] gedreht. Der Drehwinkel beträgt $(70 \pm 2)°$.

Fig. 3a zeigt die FEM-Abbildung einer anderen Wolfram-Spitzenkathode. Erhitzen der Spitze auf ca. 2700 °K für 60 s führte zum Bild Fig. 3b. Hier ist die Korngrenzenwanderung während der Heizzeit von 60 s vollständig abgelaufen, so daß als Ergebnis nur die Orientierungsänderung zwischen den Einkristallen festgestellt wird. Drehachse ist wieder [011], der Drehwinkel beträgt $(72 \pm 2)°$.

3. Diskussion

Das plötzliche Verschwinden der Korngrenze beim Übergang von Fig. 1j zu Fig. 1k läßt sich unabhängig von einem speziellen Korngrenzenmodell verstehen. Fig. 4 ist ein schematischer Schnitt durch den vorderen Teil der Wolframspitze. Die Korngrenze sei eben; ihre Projektion in die Zeichenebene wird durch die Gerade AA′ dargestellt. Ihr Schnitt mit der Kathodenoberfläche erscheint nur teilweise im FEM-Bild, weil der emittierende Spitzenbereich begrenzt ist. Dieser ist durch die stark ausgezogene Linie in Fig. 4 hervorgehoben. Der Korngrenzenverlauf auf der Spitzenoberfläche wird im Schnitt Fig. 4 durch die Punkte S und S′ wiedergegeben. Nimmt man nun bei einer festen Temperatur konstante Beweglichkeit der Korngrenze an, so verschiebt sich diese nach jedem Heizzyklus gleichbleibender Dauer um eine konstante Strecke, die in der Projektion als Δl erscheint. Die in Fig. 4 parallel zu AA′ eingezeichneten Geraden geben die Lage der Korngrenze an, die diese zwischen jeweils zwei Heizzyklen einnimmt. Die rechten Schnittpunkte S gehören zum abgebildeten Teil der Spitze, während die linken S′ nicht mit abgebildet werden. Nachdem die Korngrenze die Lage BB′ eingenommen hat (Verlauf etwa in der Bildmitte auf dem FEM-Leuchtschirm, vgl. Fig. 1j), ist sie nach dem nächsten Heizzyklus verschwunden (hypothetische Lage CC′; Fig. 1k): Das eine Korn ist durch das andere aufgezehrt worden. Damit ist erklärt, warum sich bei ausreichendem Winkel zwischen Korngrenzennormale und Spitzenachse ihre Bewegung nur in einer Bildhälfte widerspiegelt. Natürlich ist diese Deutung nicht an die Annahme konstanter Korngrenzenbeweglichkeit gebunden.

Fig. 4. Schematischer Schnitt durch die Spitzenkathode: im FEM abgebildeter Bereich stark ausgezogen. Die parallelen Geraden stellen die jeweilige Lage der Korngrenze zwischen aufeinanderfolgenden Heizzyklen dar. Bezeichnungen vgl. Text

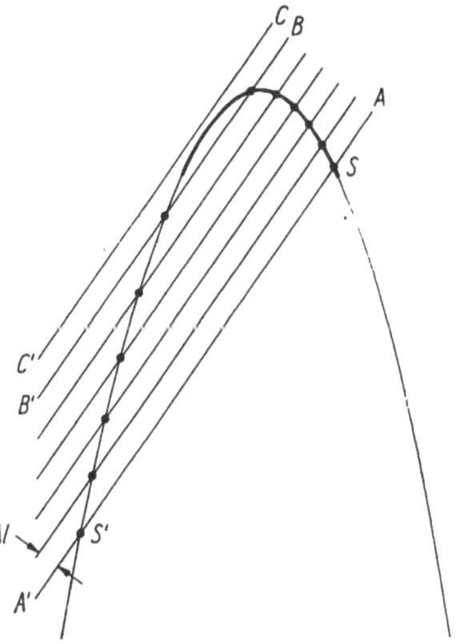

Bei veränderlicher Beweglichkeit werden lediglich die Verschiebungen Δl pro Heizzyklus verschieden.

Die gegenseitige Orientierung der Kristallite in Fig. 1 und Fig. 3 entspricht einer Verdrehung ihrer Gitter um rund 71° um [011]. Diese spezielle gegenseitige Orientierung läßt sich nach dem Koinzidenzgitter-Modell für kubische Kristalle von Frank [6] und Dunn [7] erklären, das auf Vorschläge von Kronberg und Wilson [8] zurückgeht. Danach koinzidiert bei geeigneter Orientierung zweier Gitter 1 und 2 ein Teil ihrer Gitterplätze miteinander. In Tabelle 1 sind für kubisch-raumzentrierte Gitter fünf mögliche diskrete Drehwinkel um [011] eingetragen, für die ein Koinzidenzgitter gebildet wird [5]. Es wurden nur solche Winkel ausgewählt, für die die relative Dichte der Koinzidenzplätze groß ist.

Tabelle 1

Relative Dichte der Koinzidenzplätze	Winkel zwischen Gitter 1 und 2 für Drehung um [011]
1: 3	70,5°
1: 9	38,9°
1:11	50,5°
1:17	86,6°
1:19	26,5°

An Wolframspitzen wurden von Brandon et al. [5] mittels Abbildung im FIM bereits Orientierungsunterschiede um [011] von ca. 50° und ca. 87° nachgewiesen (vgl. Tabelle 1). Zu der hier mitgeteilten 71°-Drehung konnte in der Literatur kein Beispiel gefunden werden.

Die Lage der Korngrenze in Fig. 1 fällt nicht genau mit einer dichtgepackten Ebene des zugeordneten Koinzidenzgitters zusammen: Im Schnitt mit der (011)-Ebene beträgt die Winkelabweichung ca. 5° (vgl. Fig. 2c). In der Bildserie Fig. 1 ist zu erkennen, daß die Korngrenze nicht achsenparallel verläuft. Das ist besonders deutlich in Fig. 1f sichtbar. Eine bessere Orientierungsangabe für die Korngrenzenfläche ist leider nicht möglich, da nur der Schnitt mit der in ihrer genauen Form unbekannten Spitzenoberfläche beobachtet wird. Diese Abweichung der realen Großwinkelkorngrenze von einer dichtgepackten Koinzidenzgitterebene kann nach Brandon, Ralph, Ranganathan und Wald [5] durch Stufenbildung erklärt werden, die den Übergang zwischen parallelen Koinzidenzgitterebenen ermöglicht.

In gezogenen Wolframdrähten haben die Einkristallbereiche bekanntlich Ausdehnungen in Achsenrichtung von der Größenordnung mm. Deshalb beobachtet man die Wolframspitze im FEM fast stets als Einkristall. Die Bildung eines Bereichs mit spezieller Orientierung zu diesem Einkristall kann als besondere Form der Rekristallisation angesehen werden, bei der ein bestimmter Teil der Atome seine Lage beibehält. Die beobachtete Korngrenzenwanderung in der Spitzenkathode zeigt den Wachstumsvorgang des rekristallisierten Bereichs auf Kosten des ursprünglich vorhandenen Einkristalls an. Auf die gleiche Weise wurden bereits von Stiegler et al. [9] Beobachtungen an deformierten und geglühten Niob-Einkristallen gedeutet.

Literatur

[1] E. W. MÜLLER, Z. Phys. **106**, 541 (1937).
[2] E. K. CASPARY und E. KRAUTZ, 4. Internat. Kongr. Elektronenmikroskopie 1958, Verhdlg. Bd. I, Berlin/Göttingen/Heidelberg 1960 (S. 780).
[3] D. STARK, Z. Phys. **172**, 503 (1963).
[4] E. W. MÜLLER, 4. Internat. Kongr. Elektronenmikroskopie 1958, Verhdlg. Bd. I, Berlin/Göttingen/Heidelberg 1960 (S. 820).
[5] D. G. BRANDON, B. RALPH, S. RANGANATHAN und M. S. WALD, Acta metall. **12**, 813 (1964).
[6] F. C. FRANK, Research Laboratory Lecture, Sept. 1958; zitiert in [5].
[7] C. G. DUNN, Annual A.I.M.E. Meeting, San Francisco, February 19, 1959; zitiert in [5].
[8] M. L. KRONBERG und F. H. WILSON, Trans. AIME **185**, 501 (1949).
[9] J. O. STIEGLER, C. K. H. DUBOSE, R. E. REED SR. und C. J. MCHARGUE, Acta metall. **11**, 851 (1963).

(Received December 21, 1964)

phys. stat. sol. 8, 701 (1965)

Institut für experimentelle Physik der Martin-Luther-Universität Halle-Wittenberg

Untersuchung der Verteilung von Versetzungen in plastisch deformierten Cu-Einkristallen mit Gitterquellen-Interferenzen (Kossel-Interferenzen)

Von

O. BRÜMMER, W. SCHÜLKE und H. BÖHNEL

Es wird mit Hilfe von Gitterquelleninterferenzen (Kossel-Interferenzen) gezeigt, daß bei plastischer Deformation von Cu-Einkristallen im Bereich I und II der Verfestigungskurve für vorliegende Orientierungen eine anisotrope Verbreiterung der Kossel-Linien auftritt. Diese richtungsabhängige Verbreiterung kann mit der Wirkung einer $[\bar{1}\bar{1}2]$- und [111]-Vorzugsdrehachse der Gitterrotation (senkrecht zur Gleitrichtung bzw. senkrecht zur Gleitebene des Hauptgleitsystems) erklärt werden. Es wird berechnet, daß diese beiden Drehachsen dann bevorzugt auftreten, wenn die Versetzungsdichte in den Nebengleitsystemen zwar kleiner, aber von gleicher Größenordnung ist wie die Versetzungsdichte in dem Hauptgleitsystem. Die Möglichkeit eines Aufstaus von primären Versetzungen gleichen Vorzeichens wird ebenfalls diskutiert.

By means of interferences from lattice sources (Kossel-patterns) it is shown that plastic deformation of Cu single crystals in the ranges I and II of the strain-hardening curve leads to an anisotropic broadening of the Kossel lines for preferred orientations. This can be explained by the occurrence of $[\bar{1}\bar{1}2]$ and [111] as preferred axes of lattice rotation (perpendicular to the slip direction and perpendicular to the slip-plane of the primary slipping system, respectively). Calculations show that both these rotational axes become preferred when the dislocation density in the secondary slipping system is smaller than, but of the same order of magnitude as the dislocation density in the primary slipping system. The possibility of a pile-up of primary dislocations of the same sign is also discussed.

1. Einleitung

Die Untersuchung der Versetzungsverteilung in plastisch deformierten kubisch-flächenzentrierten Kristallen ist bereits Gegenstand einer Anzahl von Veröffentlichungen (u. a. [1 bis 6]). Bisher sind Gitterquelleninterferenzen (Kossel-Interferenzen) noch nicht zu solchen Untersuchungen herangezogen worden. In der vorliegenden Arbeit sollen mit Hilfe von Kossel-Interferenzen aus der Bestimmung von Achsen bevorzugter Gitterrotation Aussagen über die Versetzungsverteilung gewonnen werden. Röntgenographisch lassen sich solche Vorzugsachsen im sogenannten vollständigen Reflexsystem [7] des Kristalls bzw. einem großen Ausschnitt desselben aus der richtungsabhängigen Verbreiterung der Kossel-Linien nachweisen.

2. Entstehung der anisotropen Verbreiterung von Kossel-Linien

Fig. 1a zeigt schematisch in reziproker Gitterdarstellung die Entstehung von Kossel-Kegeln. Die Schnittlinie einer Kugel mit dem Radius $\mathfrak{S}_0 = 1/\lambda$ um den reziproken Gitterpunkt $O = (000)$ mit einer Kugel um den Gitterpunkt $H = (h\,k\,l)$ von gleichem Radius ergibt die Umrandung des Kossel-Kegels, dessen Spitze in O liegt. Zur Veranschaulichung der richtungsabhängigen Linienverbreiterung ist

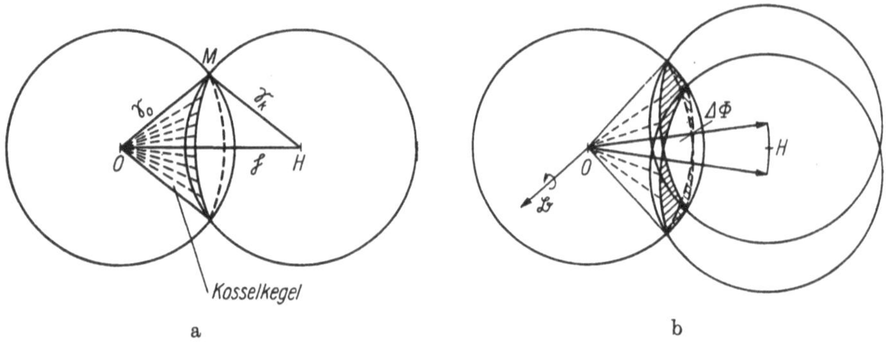

Fig. 1
a) Zur Entstehung der Kossel-Kegel in reziproker Gitterdarstellung.
b) Zur Entstehung der anisotropen Verbreiterung der Kossel-Kegel. Erläuterungen im Text

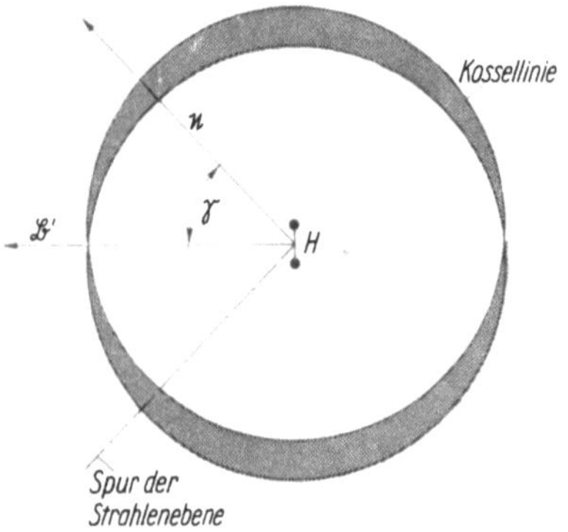

Fig. 2. Schnitt eines anisotrop verbreiterten Kossel-Kegels mit einer Ebene parallel zur Netzebene. Erklärung der Abkürzungen im Text

der reziproke Gitterpunkt H in Fig. 1b unter dem Einfluß der Drehachse $\mathscr{B}$ zu einem Kreisbogen deformiert. Das bewirkt an zwei Stellen der Kossel-Linie (Schnitt des Kossel-Kegels mit einer senkrecht auf seiner Achse stehenden Ebene) maximale Verbreiterung, während zwei Stellen der Linie unverbreitert bleiben. In Fig. 2 ist der Schnitt des anisotrop verbreiterten Kossel-Kegels mit einer Ebene parallel zur Netzebene $(h\,k\,l)$ dargestellt. $\mathscr{B}'$ ist die Projektion von $\mathscr{B}$ in die Zeichenebene. Ein bestimmtes Linienstück wird durch die Spur der Strahlenebene aus einfallendem und reflektiertem Strahl festgelegt. Dieses Linienstück wird durch den Winkel γ zwischen $\mathscr{B}'$ und $\mathfrak{n}$, der Normalen auf der Strahlenebene, festgelegt. Schließen $\mathscr{B}$ und $\mathscr{B}'$ den Winkel δ ein, so ist die durch die Gitterrotation $\Delta\Phi$ um $\mathscr{B}$ hervorgerufene Verbreiterung der Kossel-Linie gegeben durch

$$\Delta\beta = \cos\gamma \cos\delta \cdot \Delta\Phi\,. \tag{1}$$

Nach den Angaben von Fig. 2 und Gleichung (1) kann man bei Vorgabe einer Drehachse die verbreiterten und unverbreiterten Stellen der Kossel-Linien in ihre stereographische Projektion eintragen (siehe Fig. 6 und 7).

3. Experimentelles

Es ist möglich, mit Gitterquelleninterferenzen [7 bis 9] auf wenigen Aufnahmen auch an kompakten Proben wesentliche Auschnitte des vollständigen Reflexsystems zu erhalten. Im Röntgen-Mikroanalysator [10] wird mit einem nur auf ca. 0,3 mm ∅ gebündelten Elektronenstrahl die charakteristische Strahlung der

Fig. 3. Lichtmikroskopische Oberflächenaufnahme der Probe III bei $a = 0{,}133$. Gleitbänder und Spuren der Elektronensonde (100×)

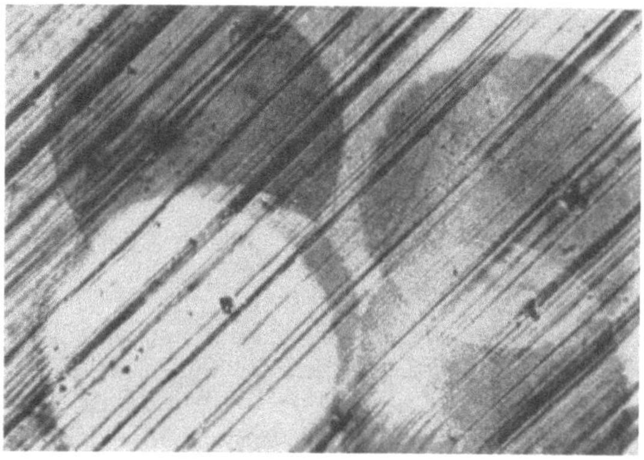

Cu-Einkristalle angeregt.[1]) Die unter dem Braggschen Winkel im Kristall reflektierte Strahlung liegt auf den Kossel-Kegeln. Der Schnitt dieser Kegel mit einem in ca. 30 cm Entfernung senkrecht zum Mittelstrahl des auffallenden Bündels aufgestellten Film erscheint als dunkle, helle oder hell–dunkle Interferenzlinie auf dem Untergrund der nicht interferierenden Strahlung [7, 8]. Durch entsprechende Bewegung der Proben wurden mit wenigen Aufnahmen von derselben Kristallstelle die zur Bestimmung der Drehachsen wesentlichen Teile des vollständigen Reflexsystems erhalten. Die Ausmessung der Linienbreite geschah fotometrisch, die Trennung der α_1- und α_2-Komponenten nach einer Methode von Rachinger [11]. Die untersuchten Zugproben ($40 \times 6 \times 2{,}6$ mm^3) wurden aus einem nach dem Bridgeman-Verfahren gewonnenen Cu-Einkristall (OFHC) ausgesägt, die mechanisch polierte Oberflächenschicht in einer Dicke von mindestens 0,6 mm chemisch abgetragen und dann elektrolytisch poliert. (Über die Dicke von mechanisch bearbeiteten Schichten, die Gitterquelleninterferenzen noch stören, siehe [8].) Zur Gewährleistung von Einfachgleitung lag die Orientierung der Zugachse im Inneren des Orientierungsdreiecks (Fig. 6 und 7). Die Proben wurden ohne Messung der Spannung plastisch gedehnt und die Abgleitung aus der Dehnung unter der Annahme von Einfachgleitung ausgerechnet. Die Probe III wurde in sechs Schritten bis zu einer Abgleitung von $a = 0{,}61$, die Probe IV in zwei Schritten bis zu einer Abgleitung von $a = 0{,}083$ untersucht. Nach jedem Verformungsschritt wurden Gitterquellen-Interferenzaufnahmen von der Umgebung der wichtigsten kristallographischen Pole angefertigt. Die Justiereinrichtung des Mikroanalysators gestattete eine reproduzierbare Einstellung derselben Kristallstelle. Gleichzeitig wurden die Gleitlinien lichtoptisch verfolgt. Fig. 3 zeigt das Gleitlinienbild der Probe III bei $a = 0{,}133$ mit den durch die Elektronensonde verursachten Oberflächenspuren. Zur Erfassung noch größerer Oberflächenbereiche wurden bei jedem Verformungsschritt zusätzliche Aufnahmen von benachbarten Oberflächengebieten angefertigt, deren Spuren in Fig. 3 erkennbar sind. Diese

[1]) Der Durchmesser des Elektronenstrahls wurde absichtlich in dieser Größe gewählt, um die lokalen Gitterverkrümmungen über ein ausreichend großes Gebiet zu mitteln, gleichzeitig aber auch den Einfluß der Brennfleckgröße auf die Linienbreite in Grenzen zu halten. Um Spannungsfelder von einzelnen Versetzungen oder Versetzungsgruppen zu untersuchen, kann der Durchmesser der Elektronensonde bis auf 1 μm zusammengezogen werden. Entsprechende Untersuchungen sind in Vorbereitung.

Aufnahmen unterscheiden sich nur im Ausmaß, nicht aber in der Richtungsabhängigkeit der anisotropen Verbreiterung. Die Gleitlinienbeobachtungen und ein Vergleich der Abgleitungswerte mit Verfestigungskurven von Diehl [4] an Cu-Kristallen nahezu gleicher Orientierung legt nahe, daß man sich bis $a = 0{,}231$ noch im Bereich I bzw. II der Verfestigungskurve befindet.

4. Ergebnisse

Bei der Probe III zeigten sich auf den Gitterquellen-Interferenzaufnahmen von $a = 0{,}026$ an erste Anzeichen einer anisotropen Linienverbreiterung, die bei $a = 0{,}069$ und $a = 0{,}133$ noch zunimmt. Erst bei $a = 0{,}231$ und ausgeprägter bei $a = 0{,}601$ ist wegen der starken Zunahme der Verbreiterung aller Linien die anisotrope Verbreiterung wieder weniger deutlich. Bei der Probe IV konnte sowohl bei $a = 0{,}024$ als auch bei $a = 0{,}083$ eine deutliche anisotrope Verbreiterung festgestellt werden. Beispiele sind in Fig. 4 und 5 wiedergegeben. Zur Ermittlung der die anisotrope Verbreiterung verursachenden Drehachsen wurden die nach Gleichung (1) errechneten Verbreiterungen für eine Reihe von Achsen in die stereografische Projektion des vollständigen Reflexsystems eingetragen und mit den

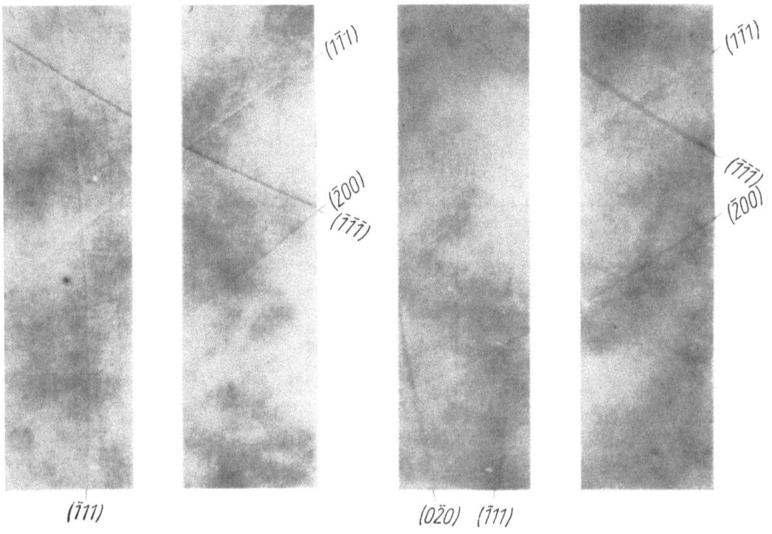

a b

Fig. 4

a) Ausschnitt einer Gitterquellen-Interferenzaufnahme der Probe III in der Nähe des $[\bar{1}\bar{1}1]$-Poles bei $a = 0$ Abstand Probe—Film ca. 30 cm. Alle Linien sind gleichmäßig scharf. Auf den Originalaufnahmen erkennt man schon an den (111)-Linien die K_α-Dublett-Aufspaltung. Wegen des relativ geringen Kontrastes der Kossel-Linien wurden die Aufnahmen mehrfach umkopiert. Dadurch gehen Einzelheiten verloren, und der Untergrund erscheint fleckig. Fig. 4 und 5 sind fotografische Negative

b) Ausschnitt einer Gitterquellen-Interferenzaufnahme derselben Probe wie in Fig. 4a in der gleichen Umgebung des $[\bar{1}\bar{1}1]$-Poles, jedoch bei $a = 0{,}133$. Die Linien erscheinen nicht mehr gleich scharf. Die Linien $(\bar{1}\bar{1}\bar{1})$, $(0\bar{2}0)$ und $(\bar{1}11)$ sind stark verbreitert, während $(\bar{2}00)$ und $(1\bar{1}1)$ relativ scharf bleiben. (Siehe zum Vergleich Fig. 6, Gegend des $[\bar{1}\bar{1}1]$-Poles.) Die $(0\bar{2}0)$-Linie erscheint auf dem Original aufgespalten. Der helle Streifen in der Mitte aller Aufnahmen ist der Schatten eines Steges im Austrittsfenster der Röhre. Die Verschiebung der Linien relativ zu dem Schatten rührt von der Umorientierung der Probe im Verlauf der plastischen Deformation her

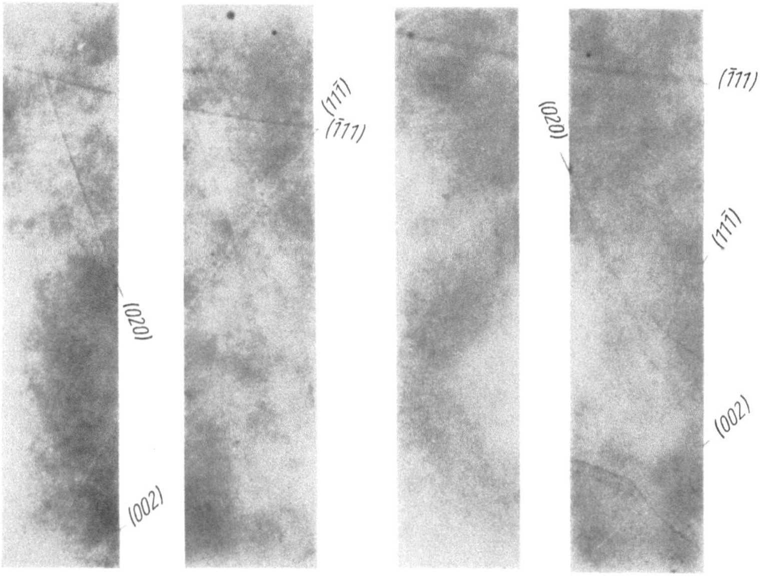

a b

Fig. 5

a) Ausschnitt der Gitterquellen-Interferenzaufnahme der Probe IV in der Nähe des [111]-Poles bei $\alpha = 0$. Die Linien sind gleichmäßig scharf

b) Ausschnitt einer Gitterquellen-Interferenzaufnahme derselben Probe wie in Fig. 5a in der gleichen Umgebung des [111]-Poles, jedoch bei $\alpha = 0{,}083$. Die $(\bar{1}11)$- und (020)-Linien sind stark verbreitert. (Dabei wird die Verbreiterung der $(\bar{1}11)$-Linie an dieser Stelle ausschließlich von der [111]-Vorzugsdrehachse, die der (020)-Linie nur von der $[\bar{1}\bar{1}2]$-Drehachse verursacht. Siehe zum Vergleich Fig. 7, Gegend des [111]-Poles). Dagegen bleiben die $(11\bar{1})$- und (002)-Linien relativ scharf

experimentell ermittelten Verbreiterungen verglichen. Übereinstimmung mit dem Experiment erhält man nur, wenn man für die Probe III eine $[2\bar{1}1]$-Achse (Richtung der Stufenversetzungen des primären Gleitsystems) und eine $[\bar{1}\bar{1}1]$-Achse (Normale auf der Hauptgleitebene) bzw. für die Probe IV die entsprechende $[\bar{1}\bar{1}2]$- und [111]-Achse annahm.[2]) Die zu diesen Achsen gehörenden Verbreiterungen sind zusammen mit der Gleitrichtung $\mathfrak{g}$, der Zugachse $\mathfrak{a}$ und der Normalen $\mathfrak{n}^*$ auf der Hauptgleitebene für die beiden Proben in den Fign. 6 und 7 in die stereografische Projektion des Reflexsystems eingezeichnet. Nur an den zur $[2\bar{1}1]$-Achse gehörenden Linienverbreiterungen tritt bisweilen eine Aufspaltung der Linien auf, was darauf schließen läßt, daß die Gitterrotation in definierten Kleinwinkelkorngrenzen konzentriert ist. Andere Vorzugsdrehachsen konnten nicht ermittelt werden.

[2]) Die unterschiedliche Indizierung der Hauptgleitebene in den Proben III und IV ist dadurch bedingt, daß die Orientierung beider Proben relativ zu der Orientierung des Kristalls angegeben wurde, aus dem sie ausgesägt waren. Im folgenden wird bei der Diskussion der anisotropen Linienverbreiterung immer nur die Indizierung des Gleitsystems der Probe IV genannt.

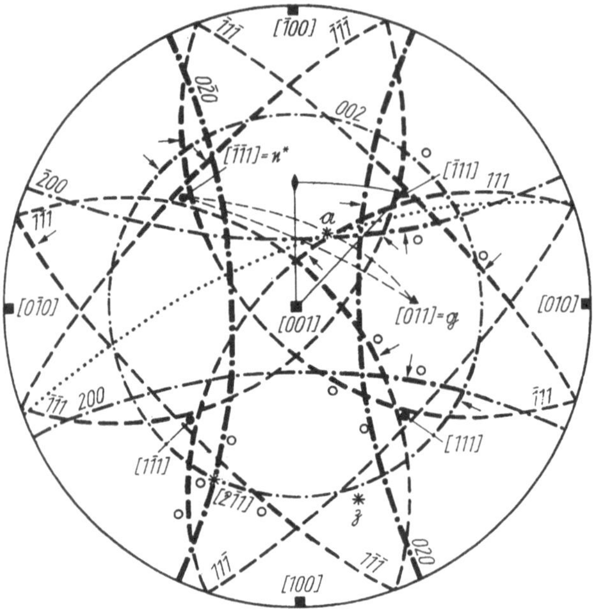

Fig. 6. Stereografische Projektion des Reflexsystems von Cu mit Cu K_α-Strahlung. Die durch die Verdrehung um die $[2\bar{1}1]$- und $[\bar{1}\bar{1}1]$-Achse (Probe III) hervorgerufenen anisotropen Verbreiterungen sind eingetragen (angedeutet durch stärkere Linien). Verbreiterungen durch die $[2\bar{1}1]$-Achse sind durch ↑, Verbreiterungen durch die $[\bar{1}\bar{1}1]$-Achse durch ○ angedeutet. Zusätzlich sind mit eingezeichnet: $\mathfrak{a}$ Zugrichtung, $\mathfrak{g}$ Gleitrichtung sowie $\mathfrak{n}^*$ Gleitebenennormale. . . . gibt die Spur der Probenoberfläche an

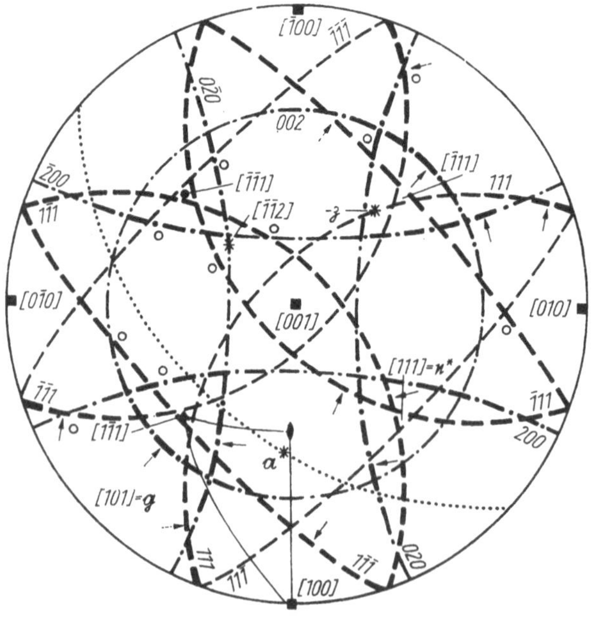

Fig. 7. Stereografische Projektion des Reflexsystems von Cu mit Cu K_α-Strahlung. Die durch die Verdrehung um die $[\bar{1}\bar{1}2]$- und $[111]$-Achse (Probe IV) hervorgerufenen Verbreiterungen sind eingetragen. Weitere Abkürzungen siehe Fig. 6

5. Bestimmung der Versetzungsverteilung aus den Vorzugsdrehachsen

Zur Bestimmung der Versetzungsverteilung aus den bevorzugten Drehachsen soll zunächst eine statistisch regellose Verteilung angenommen werden. Man denkt sich den untersuchten Kristallbereich in Unterbereiche eingeteilt. Die Größe dieser Unterbereiche soll so gewählt sein, daß jeweils nur *eine* Versetzung eines bestimmten Burgersvektors in diesem Bereich zu liegen kommt und eine Gitterkrümmung hervorruft. Da dann das Vorzeichen der Burgersvektoren der Versetzungen von Bereich zu Bereich in zufälliger Folge wechselt, kann man annehmen, daß die durch die Gitterkrümmung der Unterbereiche hervorgerufene Desorientierung nach einer Gauß-Verteilung um eine mittlere Orientierung erfolgt. Ist die Verdrehung in einem Unterbereich gleich Φ, so ist die röntgenografisch erfaßbare Gesamtverschwenkung nach Dunn et al. [12] gleich

$$\beta = 2{,}1\,\Phi\,. \tag{2}$$

Betrachtet man eine bestimmte Art von Versetzungen, so ist im allgemeinen Φ abhängig von der Richtung, in der man im Kristall fortschreitet. Der Zusammenhang zwischen der Gitterverschwenkung $\Delta\Phi_i$ um drei Achsen eines rechtwinkligen Koordinatensystems und den Vektoren Δx_i, die das Fortschreiten in Richtung der Koordinatenachsen angeben, ist über einen Krümmungstensor nach Nye [13] wie folgt gegeben:

$$\Delta\Phi_i = \varkappa_{ij}\,\Delta x_j\,. \tag{3}$$

Dabei bilden die $\Delta\Phi_i$ die Komponenten eines axialen Vektors. Der Krümmungstensor $\varkappa_{ij}$ ergibt sich aus dem „Tensor der Versetzungsdichte" α_{ij} wie folgt:

$$\varkappa_{ij} = \alpha_{ji} - \frac{1}{2}\,\delta_{ij}\,\alpha_{kk}\,, \qquad \begin{array}{l} \delta_{ij} = 1 \quad \text{für} \quad i = j\,, \\ \delta_{ij} = 0 \quad \text{für} \quad i \neq j\,, \end{array} \tag{4}$$

wobei die α_{ij} in folgender Weise zu erhalten sind:

$$\alpha_{ij} = n\,b_i\,r_j \qquad \text{bzw.} \qquad \alpha_{ij} = \sum_l n_l\,b_{il}\,r_{jl}\,. \tag{5}$$

Dabei gilt die erste Gleichung von (5), wenn nur eine Art von Versetzungen vorliegt, die andere, wenn Versetzungen mit l unterschiedlichen Burgersvektoren in Rechnung zu setzen sind. b_i ist die i-Komponente des Burgersvektors und r_j die j-Komponente des Einheitsvektors $\mathfrak{r}$ in Richtung der Versetzungslinie, n ist die Anzahl der Versetzungen, die die Einheitsfläche senkrecht zu $\mathfrak{r}$ durchsetzen. Da in den Unterbereichen jeweils nur *eine* Versetzung eines bestimmten Burgersvektors angenommen wird, brauchen in (5) jeweils nur Burgersvektoren *eines* Vorzeichens berücksichtigt zu werden. Nimmt man noch in der statistischen Betrachtungsweise die $\Delta x_{i,j,k}$ als gleich groß an, so sind die $\Delta\Phi_{i,j,k}$ wegen (2) ein Maß für die röntgenografisch nachweisbare Gitterverdrehung um die i, j, k-Achsen. Bei geeigneter Wahl dieser Koordinatenachsen kann man dann an Hand der Gleichungen (3 bis 5) untersuchen, bei welcher Versetzungsverteilung die genannten [111]- und [211]-Achsen bevorzugt auftreten. Man faßt zu diesem Zweck die 12 möglichen Gleitsysteme zu 4 Tripeln aus drei Gleitsystemen mit gemeinsamer Gleitebene zusammen, nimmt für jedes Gleitsystem gleiche Dichte von Schrauben -und Stufenversetzungen und zusätzlich für alle Nebengleitsysteme die gleiche Versetzungsdichte n_N an. Wählt man die in der Gleitebene senkrecht zu einer Gleitrichtung des Tripels liegende [211]-Richtung zur i-Achse, die genannte Gleitrichtung zur j-Achse und die Normale auf der Gleitebene zur k-Achse des Koordinatensystems in Gleichung (3), so erhält man für jedes Tripel, das das Hauptgleitsystem nicht enthält, folgenden Verdrehungszustand, wobei $\mathfrak{b}$ der Betrag des Burgersvektors 1/2 ⟨110⟩ im k.f.z. Gitter bedeutet:

$$\left.\begin{aligned} \alpha_{ii} &= 3/2\, n_N\, b\,, & \alpha_{ij} &= -\,3/2\, n_N\, b\,,\\ \alpha_{jj} &= 3/2\, n_N\, b\,, & \alpha_{ji} &= 3/2\, n_N\, b\,,\\ \alpha_{kk} &= 0\,, & \alpha_{ik} &= \alpha_{jk} = \alpha_{ki} = \alpha_{kj} = 0\,, \end{aligned}\right\} \quad (6)$$

$$\left.\begin{aligned} \Delta\Phi_i &= 3/2\, n_N\, b\, \Delta x_j\,,\\ \Delta\Phi_j &= -\,3/2\, n_N\, b\, \Delta x_i\,,\\ \Delta\Phi_k &= 3/2\, n_N\, b\, \Delta x_k\,. \end{aligned}\right\} \quad (7)$$

Bezeichnet man mit n_H die Versetzungsdichte im Hauptgleitsystem, so erhält man für das Tripel, das das Hauptgleitsystem enthält, folgenden Verdrehungszustand (j-Achse parallel zur Gleitrichtung des Hauptgleitsystems):

$$\left.\begin{aligned} \alpha_{ii} &= 3/2\, n_N\, b\,, & \alpha_{ij} &= -\,3/2\, n_N\, b\,,\\ \alpha_{jj} &= n_H\, b + 1/2\, n_N\, b\,, & \alpha_{ji} &= n_H\, b + 1/2\, n_N\, b\,,\\ \alpha_{kk} &= 0\,, & \alpha_{ik} &= \alpha_{jk} = \alpha_{ki} = \alpha_{kj} = 0\,, \end{aligned}\right\} \quad (8)$$

$$\left.\begin{aligned} \Delta\Phi_i &= 1/2\,(n_N - n_H)\, b\, \Delta x_i + (n_H + 1/2\, n_N)\, b\, \Delta x_j\,,\\ \Delta\Phi_j &= -\,3/2\, n_N\, b\, \Delta x_i + 1/2\,(n_H - n_N)\, b\, \Delta x_j\,,\\ \Delta\Phi_k &= -\,1/2\,(n_H + 2\, n_N)\, b\, \Delta x_k\,. \end{aligned}\right\} \quad (9)$$

Bei gleichen $\Delta x_{i,j,k}$ besitzen in Gleichung (7) die Komponenten des Vektors $\Delta\Phi$ gleiche Beträge, d. h., für jedes Tripel existieren drei senkrecht aufeinander stehende Drehachsen mit gleicher Verdrehung. Für Versetzungen in solchen Nebengleitsystemen, die mit dem Hauptgleitsystem die Gleitebene nicht gemeinsam haben, sollte sich also unter den angegebenen Voraussetzungen röntgenografisch keine anisotrope Linienverbreiterung nachweisen lassen. Bei dem Vektorpaar Δx_i und Δx_j kommt ebenso häufig gleiches Vorzeichen wie entgegengesetztes Vorzeichen vor. Ist daher $n_N \ll n_H$, d. h. kann man n_N gegen n_H vernachlässigen, so erhält man nach (9) für das Tripel mit dem Hauptgleitsystem nur die $[\bar{1}\bar{1}2]$-Achse in Richtung der Stufenversetzungen des Hauptgleitsystems als röntgenografisch wirksame Vorzugsdrehachse. Für den Fall, daß $n_N = n_H$ ist, ergibt sich wie bei den anderen Tripeln keine Vorzugsachse. Wenn dagegen $n_N < n_H$ ist, aber beide von derselben Größenordnung sind, sind sowohl die [111]-Achse senkrecht zur Hauptgleitebene als auch die $[\bar{1}\bar{1}2]$-Achse senkrecht zur Hauptgleitrichtung deutlich gegenüber der $[1\bar{1}0]$-Achse in Hauptgleitrichtung bevorzugt und können als Verzugsdrehachsen röntgenografisch wirksam werden. Die experimetell gefundenen Verbreiterungen durch $[\bar{1}\bar{1}1]$- und [112]-Achsen sind also unter den angegebenen Voraussetzungen mit einer Versetzungsverteilung verträglich, bei der die Versetzungsdichte in den Nebengleitsystemen zwar kleiner, aber doch von gleicher Größenordnung ist wie die Dichte im Hauptgleitsystem. Da die anisotrope Verbreiterung mit den zwei Vorzugsachsen über mehrere Verformungsschritte zu verfolgen ist, muß dieses Verhältnis der Versetzungsdichten auch im Verlauf der plastischen Verformung im Bereich I und II bestehen bleiben, d. h., die Versetzungsdichte in den Nebengleitsystemen muß etwa in demselben Maß zunehmen wie die Versetzungsdichte des Hauptgleitsystems. Schließlich kann man das Verschwinden der anisotropen Verbreiterung bei größeren Abgleitungen im Sinne der Gleichung (9) so deuten, daß die Versetzungsdichte in den Nebengleitsystemen die des Hauptgleitsystems erreicht hat. Das wäre in Übereinstimmung mit Messungen von Basinski et al. [3], die die Versetzungsdichte an plastisch deformierten Cu-Kristallen mit der Ätztechnik bestimmt haben, sowie elektronenmikroskopischen Beobachtungen von Essmann [2] und röntgenografischen Messungen von Wilkens et al. [6], die in den „Versetzungssträngen" parallel zu den primären Gleitebenen in beträchtlicher Dichte auch Versetzungen der Nebengleitsysteme fanden.

Die Betonung der $[\bar{1}\bar{1}2]$-Drehachse braucht nicht allein von einer größeren Versetzungsdichte des Hauptgleitsystems verursacht zu sein, sondern kann zum Teil auch mit einem Aufstau von Stufen- und Schraubenversetzungen gleichen Vorzeichens aus dem Hauptgleitsystem an den von der Seegerschen Verfestigungstheorie geforderten Gleithindernissen [14] (Lomer–Cottrell-Versetzungen) erklärt werden. Man rechnet dann mit den p aufgestauten Versetzungen wie mit einer Superversetzung des Burgersvektors $p\,\mathfrak{b}$. Nach Gay et al. [15] und verbessert durch Rechnungen von Dunn et al. [12] ist unter der Annahme statistischer Verteilung der Versetzungen der Zusammenhang zwischen der mittleren Gitterverdrehung β und der Versetzungsdichte gegeben durch

$$n = \frac{\beta}{4{,}3 \cdot b^2}. \tag{10}$$

Nimmt man jedoch an, daß bei gleicher Anzahl n_1 von Versetzungen pro cm^2 die Versetzungen in „pile ups" von je p Versetzungen zusammengefaßt sind, dann besteht die folgende Beziehung:

$$n_1 = \frac{\beta_1^2}{4{,}3 \cdot b^2} = p\, n_2 = \frac{p\, \beta_2^2}{4{,}3 \cdot p^2\, b^2}. \tag{11}$$

Dabei ist n_1 die Versetzungsdichte ohne Zusammenfassung in Superversetzungen, n_2 die Dichte der Superversetzungen, β_1 die zu der normalen, β_2 die zu der aufgestauten Versetzungsanordnung gehörende mittlere Verdrehung. β_2 ist also $\sqrt{p}$ mal so groß wie β_1. Dieser Versetzungsaufstau könnte aber wieder nur die Bevorzugung der $[\bar{1}\bar{1}2]$-Achse erklären, da er in seiner Wirkung einer größeren Versetzungsdichte im Hauptgleitsystem gleichkommt. Wegen des Auftretens der [111]-Vorzugsachse ist jedoch auch in diesem Fall mit einer zur Dichte im Hauptgleitsystem vergleichbaren Dichte in den Nebengleitsystemen zu rechnen, so daß auch bei dieser Deutung die vorangegangenen Schlußfolgerungen bestehenbleiben, nur daß die Versetzungsdichte in dem Hauptgleitsystem nicht ausdrücklich größer zu sein braucht als die in den Nebengleitsystemen.

Abschließend sei darauf hingewiesen, daß man wegen der Möglichkeit einer anisotropen Verbreiterung bei der Bestimmung von Versetzungsdichten an plastisch deformierten Kristallen aus Linienbreiten (Rocking-Kurven-Breiten; siehe z. B. Hordon et al. [16]) zu berücksichtigen hat, an welcher Stelle des Kossel-Kegels die Breite der Röntgenreflexe vermessen wird. Will man zwischen der Versetzungsdichte in Haupt- und Nebengleitsystemen unterscheiden, wird man ohne mehrere Messungen längs des Kossel-Kegels nicht auskommen, wie hier experimentell mit Hilfe der Gitterquelleninterferenzen gezeigt werden konnte. Ähnliche Untersuchungen lassen sich selbstverständlich auch mit normalen Weitwinkelinterferenzen in Transmission (Pseudo Kossel pattern) durchführen (siehe [17] und [18]).

Herrn Prof. Dr. W. Messerschmidt danken wir für das dieser Arbeit entgegengebrachte fördernde Interesse.

Literatur

[1] J. D. LIVINGSTON, Acta Metall. **10**, 229 (1962).
[2] U. ESSMANN, phys. stat. sol. **3**, 932 (1963).
[3] Z. S. BASINSKI und S. J. BASINSKI, Phil. Mag. **9**, 51 (1964).
[4] J. DIEHL, Z. Metallk. **47**, 331 und 412 (1956).
[5] S. MADER, Z. Phys. **149**, 73 (1957).
[6] M. WILKENS und K. ECKERT, Z. Naturf. **19a**, 459 (1964).
[7] W. KOSSEL, Erg. exakt. Naturwiss. **16**, 295 (1937).
[8] O. BRÜMMER und W. SCHÜLKE, Z. Naturf. **17a**, 203 (1962).
[9] W. SCHÜLKE und O. BRÜMMER, Z. Naturf. **17a**, 208 (1962).
[10] O. BRÜMMER, K. H. BRAUER und G. SUWALSKI, Z. angew. Phys. **16**, 27 (1963).
[11] W. A. RACHINGER, J. sci. Instrum. **25**, 254 (1948).
[12] C. O. DUNN und E. F. KOCH, Acta metall. **5**, 548 (1957).
[13] J. F. NYE, Acta metall. **1**, 153 (1953).
[14] A. SEEGER, Kristallplastizität, Handbuch der Physik, Bd. VII/2, Springer-Verlag, Berlin/Göttingen/Heidelberg 1958.
[15] P. GAY, P. B. HIRSCH und A. KELLY, Acta metall. **1**, 315 (1953).
[16] M. J. HORDON und B. L. AVERBACH, Acta metall. **9**, 237 (1961).
[17] O. BRÜMMER, Z. Naturf. **13a**, 571 (1958); **15a**, 876 (1960).
[18] O. BRÜMMER, G. DRÄGER und W. SCHÜLKE, Exper. Techn. Phys., in Vorbereitung.

(Received December 1, 1964)

phys. stat. sol. 8, 711 (1965)

Department of Physics, University of Missouri, Columbia

Higher Order Transitions in the EPR Spectrum of Fe^{3+} in MgO

By

J. L. Kolopus and L. V. Holroyd

The spectrum of the $\Delta m_s = 3$, 4, and 5 transitions of Fe^{3+} in single crystals of MgO has been measured. A computer was used to solve the secular equation of the spin Hamiltonian appropriate for Fe^{3+} in a cubic site in MgO. A table of the angular dependence of the energy levels as a function of magnetic field in the dimensionless quantities W/a vs. $g\,\beta\,H/2a$ is given for $g\,\beta\,H/2a$ between 1.30 and 2.50. The experimental data for the angular dependence of these lines is compared with the calculated values and the agreement found to be within experimental error.

Es wurde das Spektrum der $\Delta m_s = 3$, 4 und 5-Übergänge von F^{3+}-Ionen in MgO-Kristallen untersucht. Die Säkulargleichung des Spin-Hamilton-Operators für das Fe^{3+}-Ion auf einem kubischen Gitterplatz in MgO wurde mit einem Rechenautomaten gelöst. Die Winkelabhängigkeit der Energieniveaus als Funktion vom Magnetfeld wird in den dimensionslosen Größen W/a über $g\,\beta\,H/2a$ für die $g\,\beta\,H/2a$-Werte zwischen 1,30 und 2,50 in einer Tabelle angegeben. Die experimentellen Werte für die Winkelabhängigkeit dieser Linien werden mit den berechneten Werten verglichen. Die Übereinstimmung ist innerhalb der experimentellen Fehlergrenzen befriedigend.

1. Introduction

The paramagnetic resonance spectrum of Fe^{3+} in MgO has been investigated [1], but the principal concern has been with the $\Delta m_s = \pm 1$ transitions. Because of the cubic crystal field term in the Hamiltonian, transitions with $\Delta m_s = 3$, 4, or 5 are allowed, giving rise to a set of 6 lines at low magnetic fields corresponding to the transitions $+5/2$ to $-5/2$, $\pm 5/2$ to $\mp 3/2$, $\pm 5/2$ to $\mp 1/2$, and $+3/2$ to $-3/2$. Some of these lines were identified by Griffiths and Orton [2] using the solutions given by Kronig and Bouwkamp [3] for the energy levels of this system with the magnetic field along the [100] and [111] crystal directions. The solutions for H along any other direction, could only be approximated by a series expansion. The expansion for large magnetic field presumes $\beta H \gg a$ and this condition is satisfied for lines corresponding to $\Delta m_s = \pm 1$ which lie in the 3000 gauss region at X-band frequencies. However, the set of lines investigated in this paper extends from 500 to 1200 G at X-band frequencies and the convergence of the series solution is slow.

An accurate comparison between experiment and theory of the angular dependence of these lines in the [100] plane is possible only if the solutions of the secular equation are obtained directly. Using a computer, we have obtained numerical solutions for the energy levels as a function of angle in the [100] plane, which for the remainder of this paper will be referred to as "exact" solutions. The energy levels or transitions between levels will be denoted by their m_s quantum numbers.

2. Theory

The effective spin Hamiltonian appropriate for Fe^{3+} in a cubic site in MgO can be written:

$$\mathcal{H} = g\,\beta\,H\,S + (1/6)\,a\,[S_x^4 + S_y^4 + S_z^4 - (1/5)\,S\,(S+1)\,(3\,S^2 + 3\,S - 1)]$$

or in terms of the operators:

$$S_z, \qquad S_+ = S_x + i\,S_y, \qquad \text{and} \qquad S_- = S_x - i\,S_y,$$

$$\mathcal{H} = g\,\beta\,H_z\,S_z + (g\,\beta/2)\,[H_x\,(S_+ + S_-) - i\,H_y\,(S_+ - S_-)] + \\ + (a/120)\,[35\,S_z^4 - 30\,S\,(S+1)\,S_z^2 + 25\,S_z^2 - 6\,S\,(S+1) + 3\,S^2(S+1)^2] + \\ + (a/48)\,[S_+^4 + S_-^4]\,.$$

Using the abbreviations $q = g\,\beta\,H/2$; $\alpha = (l + i\,m)$; $\alpha^* = (l - i\,m)$; where l, m, n are directions cosines, the operation of $\mathcal{H}$ on the manifold $|5/2, m_s\rangle$ leads to the secular determinant (see following page)

and the corresponding secular equation

$$W^6 - (6\,a^2 + 35\,q^2)\,W^4 + 4\,a^3\,W^3 + \\ + (9\,a^4 + 800\,a^2\,q^2 + 259\,q^4)\,W^2 - \\ - [12\,a^5 + 20\,a^3\,q^2 + a\,q^4\,(384 - 1920\,\varphi)]\,W + \\ + 4\,a^6 - 25\,a^4\,q^2 - 225\,q^6 + a^2\,q^4\,(246 - 960\,\varphi) = 0\,.$$

For convenience we define the dimensionless quantities $E = W/a$ and $X = q/a = g\,\beta\,H/2\,a$ and obtain the equation

$$E^6 - C_4\,E^4 + 4\,E^3 + C_2\,E^2 - C_1\,E + C_0 = 0\,,$$

where

$$C_4 = 6 + 35\,X^2\,, \qquad C_1 = 12 + 20\,X^2 + (384 - 1920\,\varphi)\,X^4\,,$$

$$C_2 = 9 + 80\,X^2 + 259\,X^4\,, \qquad C_0 = 4 - 25\,X^2 - 225\,X^6 + (246 - 960\,\varphi)\,X^4\,,$$

and

$$\varphi = l^2\,m^2 + m^2\,n^2 + n^2\,l^2\,.$$

The Appendix gives the computer solutions of the secular equation in terms of E vs. X as a function of the angle θ, between the magnetic field direction and the [100] crystalline axis. For comparison, the high field approximations to the energy levels were computed including terms in $a^2/(g\,\beta\,H)^2$. These results listed

Table 1

Difference in gauss between the magnetic field positions calculated using the series approximation solution and those obtained from the exact solution for all lines at several angles

θ	5/2 to −5/2	5/2 to −3/2	3/2 to −5/2	1/2 to −5/2	3/2 to −3/2	5/2 to −1/2
0	+0.1	−1.2	+5.7	0.0	+3.2	0.0
15	+2.2	+1.2	+1.0	−0.3	+0.8	−0.2
25	+0.3	+0.1	−2.5	+0.9	−1.0	+0.5
35	−1.0	−0.5	−0.9	+0.3	−0.5	−2.0
45	−1.1	−0.2	−8.5	−0.1	+0.1	−1.2

in Table 1 show the approximation to be least reliable for transitions involving the $\pm 3/2$ energy levels at low fields.

3. Experiment

The MgO crystals used for this investigation were grown from a large melt prepared in an arc furnace. The melt gave a number of large single crystals, pale yellow in color and quite clear. A typical sample size cleaved from these crystals was $2 \times 5 \times 6$ mm. Spectroscopic analysis showed the impurities presene were: iron — 300 ppm., manganese — 30 ppm., and chromium — 10 ppm. All other paramagnetic impurities were absent within the limits of detection (approx 1 ppm.).

There were an unusually large number of lines present in these samples, some of which were identified as due to Fe^{3+}, Mn^{2+}, and Cr^{3+} in cubic sites, and Cr^{3+} in a [100] axial site [1, 4, 5]. In the neighborhood of the $\pm 5/2$ to $\mp 1/2$ and $+3/2$ to $-3/2$ transitions, there were several unidentified strong lines which made measurements of the weaker iron lines rather difficult and for some angles the lines of interest could not be measured at all. There were, in fact, a number of lines whose origins have not as yet been established.

The spectrometer used to obtain the experimental data was an X-band reflection type utilizing a rectangular cavity operating in the TE 011 mode at a frequency near 9.2 GHz. At this frequency the magnetic field positions of the lines investigated, fell in the range of 500 to 1200 G. All the data were taken at 77 °K.

Since the computer solution of the secular equation gives the energy levels in terms of $E = W/a$ as a function of $X = g \beta H / 2a$, a choice of the crystal field splitting parameter, a, is necessary to compare the experimental data with the calculated values. The analytic solutions for the magnetic field along a [100] crystalline axis given by Kronig and Bouwkamp [3] show that the splitting between the $+3/2$ to $+1/2$ and $-3/2$ to $-1/2$ lines is very nearly equal to $5\,a/g\beta$ and the splitting between the $+5/2$ to $-3/2$ and $-5/2$ to $+3/2$ lines is exactly equal to $a/g\beta$. Measurements of both of these splittings gave a value of $a = 224.0 \pm 0.5$ in electron gauss units or $a = 209.0 \pm 0.5 \times 10^{-4}\ \mathrm{cm}^{-1}$ for the samples used. This value was then used to obtain the magnetic field positions of the lines from the computer data to compare with the experimentally measured positions.

$$\begin{vmatrix} a/2 + 5qn - W & (5)^{1/2} q\alpha^* & 0 & 0 & (5/4)^{1/2} a & 0 \\ (5)^{1/2} q\alpha & -(3/2)a + 3qn - W & (8)^{1/2} q\alpha^* & 0 & 0 & (5/4)^{1/2} a \\ 0 & (8)^{1/2} q\alpha & a + qn - W & 3q\alpha^* & 0 & 0 \\ 0 & 0 & 3q\alpha & a - qn - W & (8)^{1/2} q\alpha^* & 0 \\ (5/4)^{1/2} a & 0 & 0 & (8)^{1/2} q\alpha & -(3/2)a - 3qn - W & (5)^{1/2} q\alpha^* \\ 0 & (5/4)^{1/2} a & 0 & 0 & (5)^{1/2} q\alpha & a/2 - 5qn - W \end{vmatrix} = 0$$

4. Results

The agreement between the calculated and experimental values of H is shown in Table 2. For the reasons given above, only partial data is presented for the $\pm 5/2$ to $\mp 1/2$ and $+3/2$ to $-3/2$ transitions. Fig. 1 is a plot of H vs. θ showing

Table 2

A comparison between the magnetic field positions calculated from the exact solution for the energy levels and those measured experimentally. Spectrometer frequency $\nu = 9.188$ GHz

θ	5/2 to −5/2 $H_{calc.}$	5/2 to −5/2 $H_{exp.}$	5/2 to −3/2 $H_{calc.}$	5/2 to −3/2 $H_{exp.}$	3/2 to −5/2 $H_{calc.}$	3/2 to −5/2 $H_{exp.}$
0	647.4	—	701.4	700.6	924.0	925.8
5	645.4	645.6	704.4	703.4	916.5	916.6
10	641.0	641.7	716.2	715.9	897.2	897.8
15	636.5	636.7	732.7	732.8	872.0	872.1
20	633.1	633.5	752.0	752.7	845.6	845.7
25	630.8	631.2	772.7	772.9	821.0	820.7
30	630.0	630.1	793.0	793.0	800.4	799.2
35	629.9	630.1	810.6	810.3	784.3	784.3
40	630.1	630.0	822.9	822.6	774.3	774.7
45	630.3	630.3	827.3	827.3	771.0	770.9

θ	1/2 to −5/2 $H_{calc.}$	1/2 to −5/2 $H_{exp.}$	5/2 to −1/2 $H_{calc.}$	5/2 to −1/2 $H_{exp.}$	3/2 to −3/2 $H_{calc.}$	3/2 to −3/2 $H_{exp.}$
0	1052.6	1052.4	1128.6	—	1082.6	—
5	1055.6	—	1127.6	—	1083.8	1084.6
10	1062.6	1065.1	1124.7	—	1079.0	1079.4
15	1070.8	1074.1	1119.9	1122.7	1074.4	—
20	1078.2	1080.5	1113.2	1111.8	1071.6	—
25	1084.2	1085.8	1105.1	1104.4	1070.8	—
30	1088.6	1089.6	1096.3	1092.2	1072.2	—
35	1091.5	1090.9	1090.6	1090.9	1074.5	—
40	1093.0	1093.1	1082.1	1084.0	1076.8	—
45	1093.5	1094.3	1079.9	—	1077.7	—

the computed curves and the experimental points for the six lines. The spectrum is symmetric about $\theta = 45°$ and the higher angles are omitted for simplicity. The angle on the scale of the magnet which corresponded to the crystalline [100] direction was determined using the angular dependence of the $\Delta m_s = \pm 1$ iron lines. In addition there were a number of sharp, fast moving lines from the [100] axial spectrum of chromium and other centers with [100] axial symmetry whose crossing at $\theta = 45°$ also gave a sensitive means of determining the origin of the angular scale.

The experimental error for the three lower field lines from all causes is estimated to be about ± 0.7 G. In most of the cases where the error is considerably larger, the line being measured was close to, or under, another stronger line. This is especially true of the three high field lines where, because of their weaker intensities and the numerous other lines in the same region, the experimental error is as much as ± 2 G.

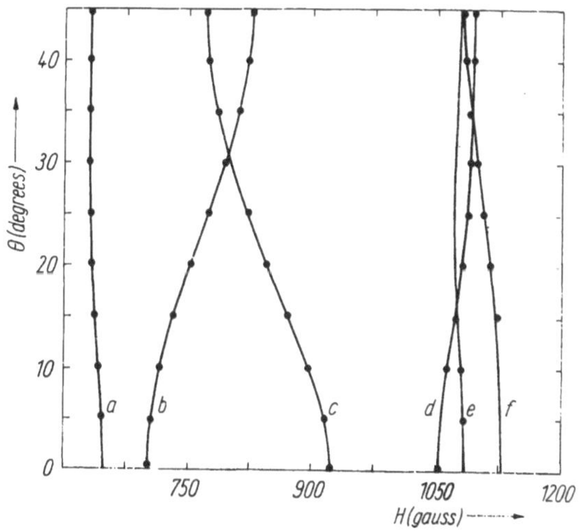

Fig. 1. The angular dependence of the higher order Fe^{3+} transitions in MgO. The solid lines are drawn through the values calculated from the computer solution of the secular equation while the data points are those measured experimentally. The numerical data are listed in Table 2. The curves labeled a, b, c, d, e, and f near $\theta = 0$ correspond to the 5/2 to −5/2, 5/2 to −3/2, 3/2 to −5/2, 1/2 to −5/2, 3/2 to −3/2 and 5/2 to −1/2 transitions respectively

A qualitative investigation of the intensities of the $\Delta m_s = 4$ and 5 lines revealed that, with the magnetic field H_0 directed along a [100] crystal direction ($\theta = 0$) and perpendicular to the microwave field H_{rf}, all three lines were quite weak. As H_0 was rotated towards $\theta = 90°$ their intensities increased and were a maximum for the $\pm 5/2$ to $\mp 3/2$ transitions when H_{rf} was parallel to H_0 at $\theta = 90°$. The intensity of the 5/2 to −5/2 transition increased until H_0 was almost parallel to H_{rf} where it dropped rapidly to zero as θ reached 90°.

The relative integrated intensities of the −3/2 to −1/2, +5/2 to −3/2, −5/2 to +3/2 and +5/2 to −5/2 transitions for H_0 directed along a [100] crystalline axis were 1:0.140:0.140:0.065, respectively. This is in good agreement with Kittel [6] for the intensities of these lines at their respective magnetic field positions.

The widths of these lines near the [100] direction were observed to be about 2 G for the $\Delta m_s = 5$ line and 6 G for the $\Delta m_s = 4$ line, compared with a width for the 1/2 to −1/2 central iron line of 5 G.

5. Conclusions

The spin Hamiltonian used to calculate the resonance positions contained terms appropriate only to cubic symmetry. The fit between the experimental data and the position calculated from the computer solution for the energy levels is within experimental error. Table 2 shows that the "high field" approximate solution would, in general, not fit the experimental data as well, with the greater error occurring in the fit to the $\Delta m_s = 4$ and 5 transitions at lower magnetic fields. This suggests that either the exact solution of the Hamiltonian, or a series approximation to it, which is carried to higher orders in a/H, is necessary to describe the experimental data when the investigation of these lines is carried out at X-band frequencies.

Distortions from cubic symmetry can be described by additional terms in the effective spin Hamiltonian. The data appears to set an upper limit on the coefficient D of the term $D\,[S_z^2 - (1/3)\,S\,(S+1)]$ of $\pm 0.5 \times 10^{-4}$ cm^{-1}.

Table 3

θ	X	5/2	3/2	1/2	−1/2	−3/2	−5/2
0	1.30	7.10000	2.54632	2.30000	−0.30000	−5.49999	− 6.14600
	1.50	8.08875	3.12348	2.50000	−0.50000	−6.08872	− 7.12384
	1.70	9.07974	3.70680	2.70000	−0.70000	−6.67972	− 8.10693
	2.10	11.06627	4.88399	3.10000	−1.10000	−7.86626	−10.08403
	2.50	13.05668	6.06918	3.50000	−1.50000	−9.05667	−12.06918
5	1.30	7.08991	2.75600	2.10198	−0.32016	−5.41496	− 6.21276
	1.50	8.07788	3.27227	2.36335	−0.52148	−6.01747	− 7.17477
	1.70	9.06826	3.82653	2.59283	−0.72261	−6.61512	− 8.14998
	2.10	11.05381	4.97775	3.01955	−1.12446	−7.80764	−10.11905
	2.50	13.04348	6.15149	3.43161	−1.52592	−9.00051	−12.10015
10	1.30	7.05989	3.05456	1.83936	−0.38157	−5.22021	− 6.35203
	1.50	8.04587	3.55739	2.11535	−0.58644	−5.83658	− 7.29571
	1.70	9.03451	4.09208	2.36546	−0.79066	−6.44315	− 8.25830
	2.10	11.01728	5.21339	2.82396	−1.19759	−7.64506	−10.21201
	2.50	13.00483	6.36895	3.25573	−1.60302	−8.84209	−12.18438
15	1.30	7.01189	3.35298	1.60350	−0.48578	−4.98007	− 6.50250
	1.50	7.99470	3.86579	1.87036	−0.69542	−5.59621	− 7.43928
	1.70	8.98069	4.40150	2.12045	−0.90387	−6.20403	− 8.39477
	2.10	10.95926	5.51603	2.58771	−1.31781	−7.40807	−10.33715
	2.50	12.94364	6.66342	3.02938	−1.72870	−8.60599	−12.30176
20	1.30	6.94898	3.63686	1.40864	−0.63143	−4.72169	− 6.64136
	1.50	7.92792	4.16607	1.65888	−0.84603	−5.32822	− 7.57865
	1.70	8.91069	4.71153	1.89949	−1.05889	−5.93023	− 8.53260
	2.10	10.88428	5.83485	2.35897	−1.48008	−7.12761	−10.47042
	2.50	12.86499	6.98470	2.79926	−1.89654	−8.32163	−12.43078
25	1.30	6.87618	3.89582	1.25911	−0.81003	−4.46090	− 6.76036
	1.50	7.85105	4.44214	1.49088	−1.02924	−5.05366	− 7.70118
	1.70	8.83054	4.99983	1.71853	−1.24599	−5.64619	− 8.65672
	2.10	10.79918	6.13842	2.16218	−1.67320	−6.83168	−10.59490
	2.50	12.77634	7.29667	2.59398	−2.09397	−8.01869	−12.55434
30	1.30	6.80085	4.11898	1.15247	−1.00349	−4.21371	− 6.85514
	1.50	7.77213	4.68033	1.36727	−1.22684	−4.79232	− 7.80056
	1.70	8.74879	5.24950	1.58202	−1.44655	−5.37489	− 8.75887
	2.10	10.71330	6.40386	2.00793	−1.87748	−6.54777	−10.69984
	2.50	12.68764	7.57203	2.42852	−2.30030	−7.72754	−12.66035
35	1.30	6.73314	4.29382	1.08269	−1.18417	−4.00146	− 6.92402
	1.50	7.70180	4.86654	1.28427	−1.41081	−4.56841	− 7.87339
	1.70	8.67647	5.44460	1.48849	−1.63221	−5.14298	− 8.83437
	2.10	10.63824	6.61169	1.89906	−2.06413	−6.30626	−10.77860
	2.50	12.61082	7.78830	2.30923	−2.48652	−7.48095	−12.74088
40	1.30	6.68512	4.40672	1.04361	−1.31672	−3.85294	− 6.96578
	1.50	7.65234	4.98633	1.23695	−1.54526	−4.41261	− 7.91773
	1.70	8.62596	5.56984	1.43442	−1.76711	−4.98253	− 8.88058
	2.10	10.58638	6.74486	1.83481	−2.19802	−6.14078	−10.82724
	2.50	12.55818	7.92687	2.23778	−2.61860	−7.31326	−12.79097
45	1.30	6.66788	4.44595	1.03107	−1.36618	−3.79868	− 6.97992
	1.50	7.63440	5.02784	1.22162	−1.59526	−4.35597	− 7.93262
	1.70	8.60773	5.61316	1.41678	−1.81705	−4.92447	− 8.89613
	2.10	10.56778	6.79082	1.81362	−2.24716	−6.08138	−10.84368
	2.50	12.53939	7.97466	2.21405	−2.66671	−7.25342	−12.80796

Appendix

Computer solution for the energy levels of the secular equation of a ${}^6S_{5/2}$ state ion in a cubic crystalline field. Table 3 (on the preceding page) gives the values of the dimensionless quantity W/a for each level, designated by its m_s quantum number, as a function of $X = g\,\beta\,H/2\,a$, where W is a root of the secular equation and H is the magnetic field. The cubic crystalline field splitting parameter a is assigned a value appropriate to the ion and compound under investigation. $\theta = 0°$ corresponds to H along a [100] crystalline axis.

Acknowledgements

The crystals used in this investigation were grown in the laboratories of the Department of Physics, University of Bristol in England during a visit by one of the authors (L.V.H.). The assistance of Professor M. H. L. Pryce and Dr. P. M. Llewellyn is acknowledged. Mr. K. Mann of this department helped in the taking and the analysis of the data. This work was supported in part by the University of Missouri Research Council.

References

[1] W. Low, Proc. Phys. Soc. (London) **B 69,** 1169 (1956).
[2] J. H. E. Griffiths and J. W. Orton, Proc. Phys. Soc. (London) **73,** 948 (1959).
[3] R. de L. Kronig and C. J. Bouwkamp, Physica **VI,** 290 (1939).
[4] W. Low, Phys. Rev. **105,** 793 (1956).
[5] J. E. Wertz and P. Auzins, Phys. Rev. **106,** 484 (1957).
[6] C. Kittel and J. M. Luttinger, Phys. Rev. **73,** 162 (1948).

(Received December 2, 1964)

phys. stat. sol. 8, 719 (1965)

II. Physikalisches Institut der Humboldt-Universität, Berlin

Thermische Akzeptoren in einkristallinem Tellur

Von

K. H. HERRMANN, R. LINK und W.-D. RENTSCH

Die Eigenschaften abgeschreckter Tellurproben werden untersucht. Durch schnelle Abkühlung von 150 bis 360 °C auf −196 °C werden thermische Akzeptoren in Konzentrationen zwischen $4 \cdot 10^{14}$ und $2 \cdot 10^{16}$ cm^{-3} eingefroren. Parallel dazu erfolgt eine Erhöhung der Gitterbeweglichkeit. Die experimentellen Aussagen deuten auf eine Analogie im Verhalten abgeschreckter Proben zu Proben unter hydrostatischem Druck.

The properties of quenched Tellurium specimens are investigated. By rapid cooling of the crystals from temperatures between 150 and 360 °C to −196 °C thermal acceptors are frozen in with a concentrations from 4×10^{14} to 2×10^{16} cm^{-3}. This treatment increases the lattice mobility. The experimental results point to an analogy between the quenched speciment and those under hydrostatic pressure.

1. Einleitung

Bisherige Untersuchungen haben ergeben, daß eine große Zahl von Elementen in Tellur als Beimischgung keine Ladungsträger erzeugt [1 bis 6]. Lediglich für Arsen, Antimon, Wismut und in geringerem Maße für Sauerstoff konnte elektrisch eine Wirkung als Akzeptor nachgewiesen werden [1, 2, 6, 7 u. a.]. Da alle Beimengungen aus dem Tellur durch wiederholte Anwendung der Reinigungsverfahren Destillation, Sublimation und Zonenschmelzen mit hohem Wirkungsgrad entfernt werden können [8], ist anzunehmen, daß die elektrischen Eigenschaften von sorgfältig gereinigtem Tellur nicht nur durch die restlichen Fremdatome, sondern in starkem Maße durch strukturelle Fehlordnungserscheinungen (Eigenfehlordnung) bestimmt sind.

Die vorliegende Arbeit ist der Untersuchung der thermischen Defekte des Tellurgitters, die durch Abschrecken eingefroren und durch Tempern wieder ausgeheilt werden können, gewidmet.

2. Durchführung der Experimente

Die für die Untersuchungen verwendeten Kristalle wurden nach dem Czochralski-Verfahren [9] aus zonengereinigtem Ausgangsmaterial [8] hergestellt. Die aus dem Hallkoeffizienten im Störleitungsgebiet ermittelten Akzeptorenkonzentrationen der Ausgangskristalle lagen zwischen $1 \cdot 10^{14}$ und $2 \cdot 10^{14}$ cm^{-3}. Spektralanalytisch zeigten die Kristalle keine Verunreinigungen.

Nach sorgfältiger Oberflächenbehandlung durch Abätzen mit heißer konzentrierter Schwefelsäure oder durch chemisches Polieren [10] erfolgte der Einbau der Meßproben (Abmessungen etwa $2 \times 2 \times 10$ mm) in einen Kristallhalter, in dem sie nur mit Graphit- und Wolframteilen in Berührung kamen. Der thermische Kontakt zu einem Heizöfchen wurde durch die Auflagefläche und den als Schutzgas verwendeten Stickstoff hergestellt. Der Kristallhalter war so konstruiert, daß in ihm Messungen im Störleitungsgebiet (bis zur Temperatur des flüssigen Stickstoffs) und die thermische Behandlung bei Temperaturen bis 360 °C vorgenommen werden konnten.

Durch Einbringen des Kristallhalters mit dem Kristall in flüssigen Stickstoff konnten Abschreckzeiten von einigen Sekunden verwirklicht werden. Vor dem Abschrecken erwies sich etwa einstündiges Tempern der Kristalle bei der Abschrecktemperatur als notwendig.

Der Halleffekt und die elektrische Leitfähigkeit wurden nach einer Gleichspannungs-Kompensationsmethode gemessen. Magnetfeld, Kristallstrom und kristallographische Hauptachse waren so orientiert, daß sich die Größen ϱ_{33} und R_{11} ergaben.

3. Meßergebnisse

3.1 Der Hallkoeffizient

Der Hallkoeffizient R_H abgeschreckter Tellurproben (Fig. 1) zeigt das übliche Verhalten: Im Störleitungsgebiet besitzt er eine geringe Temperaturabhängigkeit; aus seiner Größe wurde die Konzentration ionisierter Akzeptoren p_A nach der Formel $p_A = \frac{3\pi}{8\,e\,R_H}$ bestimmt. Beim Übergang zur Eigenleitung wechselt der Hallkoeeffizient in den stets p-leitend gefundenen Proben sein Vorzeichen bei der Temperatur, bei der

$$n\,b^2 = p \tag{1}$$

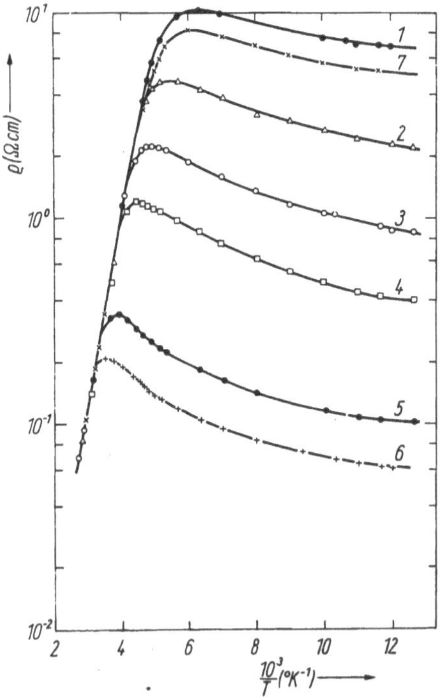

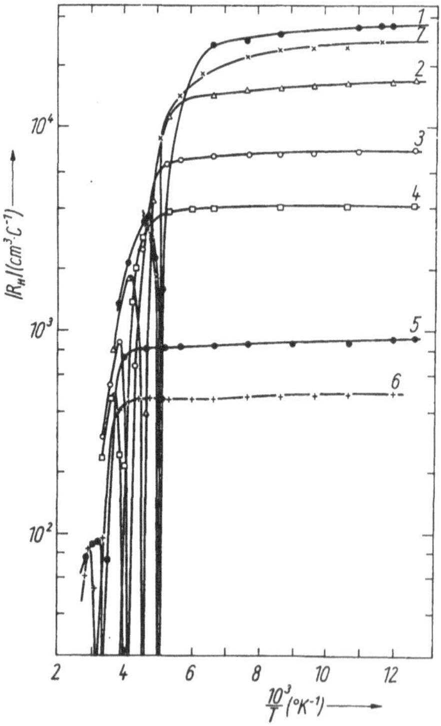

Fig. 1. a) spezifischer Widerstand ϱ b) Hallkoeffizient R_H als Funktion der reziproken Temperatur für eine von verschiedenen Temperaturen abgeschreckte Tellurprobe 1 – getemperte Ausgangsprobe, 2 – T_A = 190 °C, 3 – T_A = 220 °C, 4 – T_A = 250 °C, 5 – T_A = 310 °C, 6 – T_A = 340 °C, 7 – Kristall nach Beendigung der Messungen von 340 °C langsam abgekühlt

ist. Bei bekanntem Verhältnis zwischen Elektronen- und Defektelektronenbeweglichkeit $\mu_n/\mu_p = b$ kann aus (1) die Inversionsdichte n_i bestimmt werden. Bei weiterer Temperaturerhöhung durchläuft R_H ein Maximum, dessen Höhe die Bestimmung der Größe b zuläßt. Danach ändert der Hallkoeffizient ein zweites Mal sein Vorzeichen. Diese zweite Vorzeichenumkehr ist für Tellur eigentümlich und nur wenig vom Fremdatomzusatz abhängig.

Aus Fig. 1 ist ersichtlich, daß beim Abschrecken im Tellur zusätzliche Akzeptoren reversibel erzeugt werden und daß die durch Verunreinigung der Kristalle selbst nach mehrstündiger Wärmebehandlung eingebrachten Akzeptoren eine untergeordnete Rolle spielen. Ferner ist bemerkenswert, daß der Hallkoeffizient von Proben mit hoher Konzentration thermischer Akzeptoren im Gegensatz zu Proben mit geringer Akzeptorenkonzentration im Störleitungsgebiet temperaturunabhängig ist.

3.2 Die Konzentration thermischer Akzeptoren in Abhängigkeit von der Abschrecktemperatur

Wenn die aus dem Hallkoeffizienten im Störleitungsgebiet (bei 85 °K) ermittelten Akzeptorenkonzentrationen p_A über der reziproken Abschrecktemperatur T_A aufgetragen werden, erhält man den in Fig. 2 dargestellten Zusammenhang. Wenn man die Punktfolge durch eine Gerade annähert, erhält man für die Konzentration der thermischen Akzeptoren den Ausdruck $p_A^{\text{th}} = P \cdot \exp\left(-\frac{\delta E}{k\,T}\right)$. Aus dem Anstieg dieser Geraden wird die Aktivierungsenergie der thermischen Akzeptoren zu 0,62 eV bestimmt.

Es ist bemerkenswert, daß die von uns gefundene Akzeptorenkonzentration in einem weiten Bereich wesentlich größer als die von Kronmüller [6] angegebene ist und statt zweier nur eine Aktivierungsenergie aufgetreten ist.

In Fig. 2 sind auch Werte p_A für einen nach dem Czochralskiverfahren hergestellten Einkristall mit einem Selengehalt von 0,7 Atomprozent eingetragen worden. Diese Selenmenge hat offensichtlich keinen meßbaren Einfluß auf die Aktivierungsenergie der thermischen Fehlordnung.

Entgegen früheren Ergebnissen [6] setzen sich die thermischen Akzeptoren auch bei Zimmertemperatur mit dem Gitter ins Gleichgewicht. So führte die Lagerung eines von 320 °C abgeschreckten Kristalls mit $1{,}2 \cdot 10^{16}\,\text{cm}^{-3}$ thermischen Akzeptoren bei Zimmertemperatur innerhalb von 16,5 Stunden zu einer Abnahme der

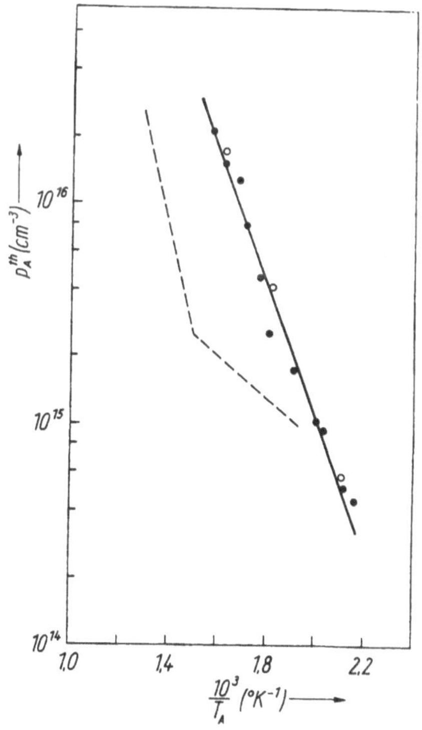

Fig. 2. Konzentration der thermischen Akzeptoren p_A^{th} als Funktion der reziproken Abschrecktemperatur T_A^{-1}
● Meßwerte an reinen Tellurproben
○ Meßwerte an einem selendotierten Kristall (0,7 At.-% Se)
- - - Werte nach Kronmüller u. a. [6]

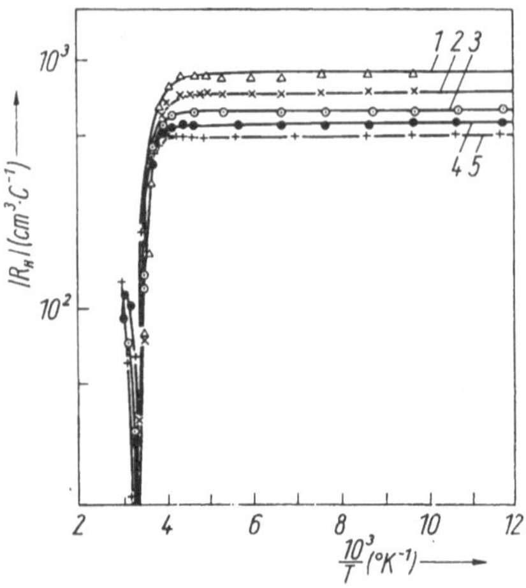

Fig. 3. Temperaturabhängigkeit des Hallkoeffizienten R_H für eine Probe, die vor dem Abschrecken von 320 °C verschieden lange bei dieser Temperatur getempert wurde. 1 — 5 min, 2 — 10 min, 3 — 30 min, 4 — 60 min, 5 — 100 min

Akzeptorenkonzentration auf $1{,}5 \cdot 10^{15}$ cm^{-3}. Unerwartet ist hiernach, daß bei erhöhter Temperatur (z. B. bei 320 °C) die thermischen Akzeptoren sich nicht augenblicklich mit dem Tellurgitter ins Gleichgewicht setzen, sondern erst im Verlaufe mehrerer Stunden (Fig. 3).

3.3 *Die Abhängigkeit der Beweglichkeit von der Abschrecktemperatur*

Aus dem spezifischen elektrischen Widerstand und dem Hallkoeffizienten im Störleitungsgebiet wurde die Hallbeweglichkeit der Defektelektronen bestimmt. Bekanntlich setzt sich diese im fraglichen Temperaturintervall aus einem temperaturunabhängigen, von der Streuung an Versetzungen [11] herrührenden Anteil und der Gitterstreuung zusammen [12]. Die reziproke Beweglichkeit der Meßproben wurde dementsprechend in zwei Summanden zerlegt:

$$\mu^{-1} = a_g \, T^{3/2} + \mu_v^{-1} \, . \qquad (2)$$

Das führt zu folgenden Schlüssen:

a) In nicht vorbehandelten Ausgangskristallen ist die temperaturunabhängige Komponente μ_v deutlich kleiner als in abgeschreckten Kristallen.

b) Der Wert der Gitterbeweglichkeit μ_g bei einer bestimmten Temperatur ist für die Proben mit verschiedenem Gehalt an thermischen Akzeptoren unterschiedlich. Zur Verdeutlichung dieses Tatbestandes sind in Fig. 4 die Werte von a_g für eine Anzahl von Kristallen über der Konzentration thermischer Akzeptoren aufgetragen worden. Trotz der Streuung der Meßwerte kann man aus der Figur als eindeutiges Ergebnis entnehmen, daß mit zunehmender Konzentration thermischer Akzeptoren die Beweglichkeit zunimmt und diese Änderung im untersuchten Konzentrationsintervall etwa den Faktor 2 ausmacht.

Die Gitterbeweglichkeit des selendotierten Kristalls unterscheidet sich deutlich von derjenigen der reinen Kristalle.

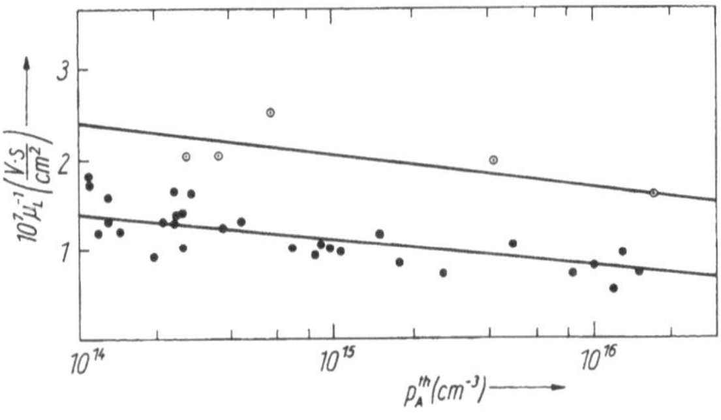

Fig. 4. Abhängigkeit der Gitterbeweglichkeit von der Konzentration der thermischen Akzeptoren p_A^{th} (dargestellt ist die Größe a_g nach Gleichung (2))
● Meßwerte an reinen Tellurproben
○ Meßwerte an einem selendotierten Kristall (0,7 At.-% Se)

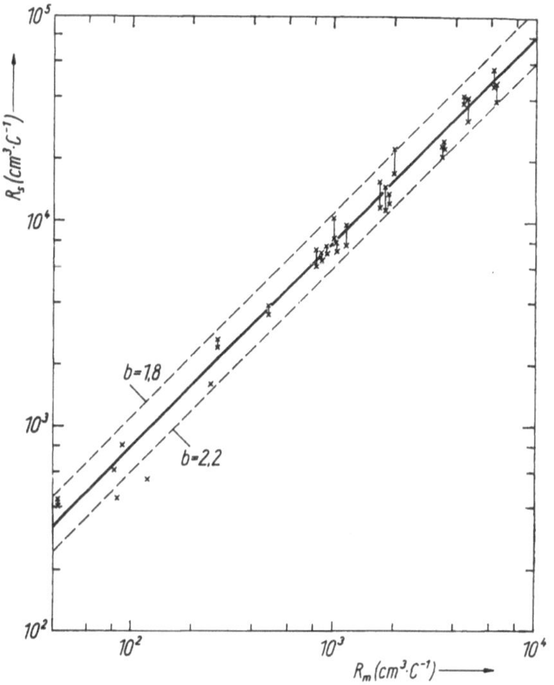

Fig. 5. Sättigungswert des Hallkoeffizienten R_s als Funktion des Maximalwertes R_m im Übergangsgebiet zur Eigenleitung. Die bei der Extrapolation des Sättigungswertes des Hallkoeffizienten R_s auf die Temperatur des Maximums entstehende Unsicherheit ist angedeutet

Aus dem Verhältnis zwischen Maximalwert des Hallkoeffizienten R_m im Übergangsgebiet von der Störleitung zur Eigenleitung und Sättigungswert im Störleitungsgebiet R_s läßt sich das Beweglichkeitsverhältnis b bestimmen. Unter der durch die Ergebnisse gerechtfertigten Annahme $\frac{\partial b}{\partial T} \ll \frac{\partial n_i}{\partial T}$ gilt

$$R_m = -\frac{(b-1)^2}{4\,b}\,R_s$$

oder

$$b = 1 - 2\frac{R_m}{R_s}\left(1 + \sqrt{1 + \frac{R_s}{R_m}}\right). \qquad (3)$$

Dabei ist vorausgesetzt, daß die Konzentration der aus Akzeptoren stammenden Löcher in diesem Bereich temperaturunabhängig ist (Störstellenerschöpfung). Andernfalls muß ein auf die Temperatur des Hallmaximums R_m extrapolierter Wert für R_s eingesetzt werden. Wie aus Fig. 1 ersichtlich ist, war eine solche Korrektur namentlich bei Proben mit geringer Konzentration thermischer Akzeptoren notwendig. Aus Fig. 5 ergibt sich für das Verhältnis R_m/R_s (mit korrigiertem R_s) der Mittelwert $-\frac{1}{8}$. Nach Formel (3) wird dann unabhängig von der Temperatur oder unabhängig von der Konzentration thermischer Akzeptoren $b = 2$.

Es ist zu erwähnen, daß die Proben mit geringem Wert von R_m bzw. R_s jene sind, in denen die Beweglichkeit der Defektelektronen um den Faktor 2 gegenüber langsam abgekühlten Kristallen vergrößert ist. Die Veränderung der Defektelektronenbeweglichkeit ist also von einer entsprechenden Veränderung der Elektronenbeweglichkeit begleitet, da das Beweglichkeitsverhältnis unverändert bleibt.

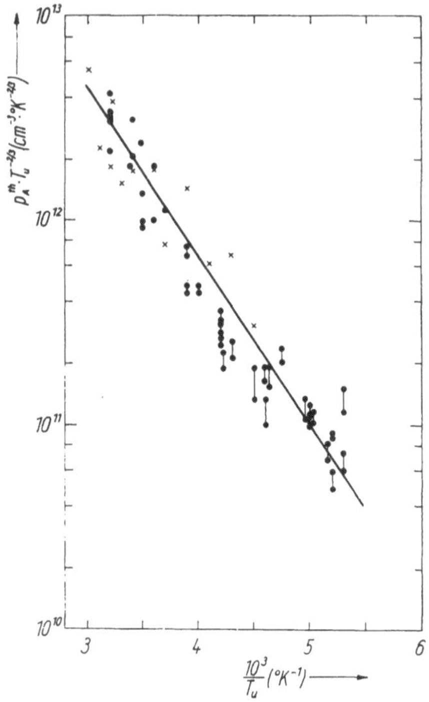

Fig. 6. $p_A^{\mathrm{th}} \cdot T_u^{-2/3} = f\left(T_u^{-1}\right)$ (T_u Temperatur des unteren Hallumkehrpunktes). Die durch die Extrapolation von R_H auf T_u entstehende Unsicherheit ist angedeutet. Die Gerade entspricht den Gleichungen (4) und (5). Die mit × bezeichneten Punkte wurden der Arbeit [29] entnommen

3.4 Die Breite der verbotenen Zone und die Temperaturabhängigkeit der Inversionsdichte

Durch die Analyse des Leitfähigkeitsverhaltens möglichst vollkommener Kristalle wurde die Inversionsdichte zu

$$n_i(T) = 7{,}1 \cdot 10^{14} \cdot T^{3/2} \exp\left(-\frac{0{,}32}{2\,k\,T}\right) \tag{4}$$

in guter Übereinstimmung mit den Meßwerten anderer Autoren [13] bestimmt. In Formel (4) wird als Breite der verbotenen Zone bei 0 °K 0,32 eV angesetzt, da wir an Kristallen hoher Vollkommenheit stets diesen Wert aus dem Anstieg der Eigenleitungsgeraden ermittelt haben.

Eine experimentelle Bestimmung der Inversionsdichte n_i ist auch nach Gleichung (1) möglich, wenn für n, p und b entsprechende Werte eingesetzt werden. Es folgt für die Umkehrtemperatur die Beziehung

$$n_i(T_U) = p_A^{\mathrm{th}} \frac{b}{b^2 - 1} = 2/3\, p_A^{\mathrm{th}} \,. \tag{5}$$

Fig. 6 zeigt neben den Meßpunkten für die Größe $T^{-3/2}\, p_A^{\mathrm{th}}$ in Abhängigkeit von der reziproken Temperatur die nach den Gleichungen (4) und (5) erwartete Gerade. Es ergibt sich eine gute Übereinstimmung.

Vergleicht man die experimentell bestimmten Werte mit den für isotrope parabolische Bänder unter Berücksichtigung der Temperaturabhängigkeit der Breite der verbotenen Zone [23] berechneten, so ergibt sich $\sqrt{m_n\, m_p} = 0{,}25\, m_0$.

Unter Vernachlässigung der Anisotropie und ohne Berücksichtigung der Verschiedenheit der Deformationspotentialkonstanten von Valenz- und Leitfähigkeitsband gilt für Streuung an akustischen Phononen (die gefundene $T^{-3/2}$-Abhängigkeit der Gitterbeweglichkeit deutet auf diesen Streumechanismus hin) $m_n/m_p = b^{-5/2}$. Daraus lassen sich als Orientierungswerte $m_n = 0{,}11\, m_0$ und $m_p = 0{,}6\, m_0$ berechnen.

4. Diskussion

4.1 „Grenzkonzentration“ 10^{14} cm^{-3}

Es ist bisher nicht gelungen, Tellureinkristalle mit einer Akzeptorenkonzentration deutlich unter 10^{14} cm^{-3} herzustellen. Die restliche Akzeptorenkonzentration und auch das bisherige Fehlen von n-leitendem Tellur ist von verschiedenen Autoren mit thermischen Akzeptoren in Zusammenhang gebracht worden [5, 6]. (Das von Mooser und Pearson [14] vorgeschlagene spezielle Bändermodell soll hier nicht diskutiert werden, da es beim Vergleich mit den Experimenten sowohl beim Selen als auch beim Tellur gewisse Unstimmigkeiten gibt.)

Extrapoliert man die Gerade in Fig. 2 durch die Meßpunkte bis zu Zimmertemperatur, so ergibt sich eine Gleichgewichtskonzentration der thermischen Akzeptoren von ca. 10^{11} cm^{-3}. Es ist ferner nach dem unter 3.1 Gesagten auszuschließen, daß die Akzeptoren schon bei Zimmertemperatur in einer Konzentration, die einer höheren Temperatur entspricht, eingefroren sind, weil die Defekte bei Zimmertemperatur noch ausheilen.

v. Lint und Roth haben Defekte des Tellurgitters untersucht, die durch Beschuß der Kristalle mit 20-MeV-Elektronen entstehen [15]. Auch diese strahlungsinduzierten Akzeptoren sind bei Zimmertemperatur beweglich, so daß man den gleichen Defekt vermuten kann.

Die geringe Temperaturabhängigkeit des Hallkoeffizienten abgeschreckter Proben im Gegensatz zu getemperten Proben spricht ebenfalls gegen eine Erklärung der Grenzkonzentration 10^{14} cm^{-3} durch die beim schnellen Abkühlen in den Kristall eingefrorenen Defekte.

Nach diesen Überlegungen gibt es bei 10^{14} cm^{-3} keine „Grenzkonzentration" auf Grund der beim Abschrecken eingefrorenen thermischen Akzeptoren. Vielmehr müssen alle, auch die nicht akzeptorbildenden Fremdatome in die Diskussion einbezogen werden. Die häufig im Tellur enthaltenen Verunreinigungen Mg und Cu z. B. haben so kleine Atomvolumina ($^2/_3$ bzw. $^1/_3$ desjenigen des Tellurs), daß ein Einbau auf Zwischengitterplätzen anzunehmen ist. Wie weiter unten diskutiert wird, könnten diese Zwischengitteratome Kettenrisse induzieren und damit mittelbar akzeptorbildend wirken. Reichliche Hinweise auf eine solche katalytische Wirksamkeit kleiner Mengen primär neutraler Fremdatome im Tellur liefern die Messungen von Lange u. Mitarb. [3, 4, 16].

Möglicherweise spielt aber auch eine andere, bisher noch nicht beobachtete Art gittereigener Defekte beim Zustandekommen der Grenzkonzentration eine Rolle. Weitere Versuche zu diesem Problemkreis sind in Vorbereitung.

4.2 Der obere Vorzeichenwechsel des Hallkoeffizienten

Auch die zweite Vorzeichenumkehr des Hallkoeffizienten in Tellur ist mit der Wirkung thermischer Akzeptoren in Zusammenhang gebracht worden [17 bis 19].

Aus Fig. 2 ergibt sich, daß die bei unseren Versuchen ermittelte Konzentration thermischer Akzeptoren p_A^{th} bei 500 °K (der Temperatur der oberen Hallumkehr) nur 10^{15} cm^{-3} beträgt, während die Inversionsdichte nach (4) um mehr als zwei Größenordnungen höher ist. Die Konzentration der thermischen Akzeptoren ist bei der Umkehrtemperatur also zu gering, um das Umschlagen des Leitungstyps zu bewirken.

Die zunächst erwartete starke Auswirkung der in Fig. 4 dargestellten Änderung der Löcherbeweglichkeit abgeschreckter Proben auf die Größe und das Vorzeichen des Hallkoeffizienten tritt nicht ein, wie aus der Temperaturabhängigkeit von b (Fig. 5) folgt.

4.3 Die Konzentrationsabhängigkeit der Gitterbeweglichkeit

Fig. 4 zeigt, daß der Faktor a_g von der Konzentration der eingefrorenen Akzeptoren abhängt. Eine Ladungsträgerkonzentrationsabhängigkeit der Gitterbeweglichkeit ist möglich,

a) wenn die effektive Masse der Ladungsträger etwa wie beim InSb [20] konzentrationsabhängig ist. Dagegen spricht die viel größere Termdichte beim Tellur, so daß sich durch 10^{16} cm^{-3} Akzeptoren der Auffüllungsgrad des Valenzbandes nicht wesentlich ändert. Ultrarotmessungen [21] zeigen ebenfalls keine derartige Tendenz für das Tellur.

b) wenn etwa das Valenzband aus zwei Teilbändern mit unterschiedlicher Beweglichkeit besteht und sich der relative Auffüllungsgrad mit zunehmender Löcherkonzentration ändert. Gegen eine solche Deutung durch Bandstruktureffekte spricht die Tatsache, daß das Beweglichkeitsverhältnis b beim Einfrieren bis zu 10^{16} cm^{-3} thermischer Akzeptoren unverändert bleibt.

c) wenn durch die Defekte, die die thermischen Akzeptoren erzeugen, die Beweglichkeit der Löcher erhöht wird. Das soll im folgenden diskutiert werden.

4.4 Die Art der Störstellen

Die Meßergebnisse an abgeschreckten Tellurproben können gut erklärt werden, wenn man dem Abschreckprozeß folgende Wirkungen zuordnet:

a) Beim Abschrecken entstehen lokalisierte Niveaus dicht oberhalb des Valenzbandes, deren Konzentration nach Fig. 2 durch den Ausdruck

$$p_A^{\mathrm{th}} = 1{,}8 \cdot 10^{21}\, e^{-\frac{0{,}62}{kT}} \tag{6}$$

gegeben ist.

b) Beim Abschrecken entsteht ein Spannungszustand im Kristall, der gewisse Analogien zu einer mechanischen Zugspannung in Richtung der kristallographischen Hauptachse und ebenso zu einem hydrostatischen Druck aufweist.[1])

Im einzelnen lassen sich die folgenden Versuchsergebnisse zur Bestätigung dieser Eigenschaften anführen:

a) Von Grison [24] ist festgestellt worden, daß in Tellurkristallen nach dem Abschrecken das kristallographische Achsenverhältnis c/a größer ist als dem Gleichgewicht entspricht. Dieselbe Wirkung kann durch eine mechanische Zugspannung in Richtung der c-Achse hervorgerufen werden.

b) Durch mechanischen Zug oder hydrostatischen Druck kann die Defektelektronenbeweglichkeit vergrößert werden. So hat z. B. Long unter der Wirkung von 2000 atm eine Zunahme auf das 1,23fache des Normalwertes beobachtet [25]. Außerdem schloß er, daß die Verteilung der Elektronen auf die Teilbänder in seinem Modell (bzw. die Verteilung der Defektelektronen auf die Teilbänder) sich nicht unter dem Einfluß des Druckes verändert. Demnach ist auch keine Änderung des Beweglichkeitsverhältnisses b zwischen Elektronen und Defektelektronen zu erwarten.

Zu ganz analogen Ergebnissen führen die von uns beschriebenen Abschreckversuche: Die Beweglichkeit nimmt zu (Fig. 4), das Beweglichkeitsverhältnis bleibt unverändert (Fig. 5). Wie aus Fig. 4 hervorgeht, entspricht der Zunahme der Beweglichkeit auf das 1,23fache eine Zunahme der eingefrorenen Akzeptorenkonzentration um den Faktor 6. Hiernach ist die Zunahme von p_A^{th} um den Faktor 6 einer Erhöhung des hydrostatischen Drucks um 2000 atm äquivalent.

---

[1]) Als Besonderheit für das Tellurgitter ist die Kompressibilität in Richtung der kristallographischen Hauptachse negativ, während sie senkrecht dazu positiv ist [22]. Daraus ergibt sich die auch experimentell bestätigte Ähnlichkeit in der Wirkung von allseitigem Druck und einachsigem Zug in Richtung der c-Achse [22].

Andererseits ist bekannt, daß die Einwirkung von hydrostatischem Druck [25] oder einer Zugspannung in Richtung der kristallographischen Hauptachse [26] beim Tellur zu einer Verringerung des aus dem Anstieg der Eigenleitungsgeraden ermittelten Abstandes zwischen dem oberen Rand des Valenzbandes und dem unteren Rand des Leitungsbandes führt. Die Verschiebung beträgt nach Messungen von Long [25] etwa 10% bei 2000 atm.

Eine direkte Bestimmung von ΔE_0 aus dem Anstieg der Eigenleitungsgeraden ist in unserem Fall nicht möglich, da Meßproben mit hoher Konzentration thermischer Akzeptoren erst bei Temperaturen eigenleitend werden, bei denen während der Aufnahme der Eigenleitungsgeraden bereits eine verfälschende Änderung der Größe p_A^{th} erfolgen muß. Die experimentell auf diese Weise ermittelten Werte streuten ganz beträchtlich und liefern offensichtlich kein reales Bild der Größe ΔE_0. Zur Kontrolle wurden Absorptionsmessungen durchgeführt, es konnte jedoch nicht mit Sicherheit eine Veränderung der Lage der Absorptionskante beim Abschrecken beobachtet werden. Die Messungen ergaben, daß die eventuelle Verschiebung der Kante um mehr als eine Größenordnung unter dem Effekt liegt, der sich auf Grund der parallel dazu durchgeführten Messungen des spezifischen Widerstandes nach der oben angeführten Abschätzung ergibt.

Die von Lange u. Mitarb. [3, 4, 16, 27] gefundenen Effekte an selendotiertem Tellur (Extrema der Dichte, der Mikrohärte, der Beweglichkeit, des Hallkoeffizienten und der Breite der verbotenen Zone bei einer bestimmten Selenkonzentration) zeigen gewisse Analogien zu den an abgeschrecktem Tellur beobachteten. Wir vermuten daher, daß bei den kritischen Selenkonzentrationen die Bedingungen für die Bildung thermischer Akzeptoren während der Kristallherstellung besonders günstig waren und es sich bei den gefundenen Anomalien um die Wirkung thermischer Akzeptoren handelt. Für eine derartige Deutung spricht die Tatsache, daß an einer Serie nach dem Czochralskiverfahren gezüchteter Kristalle mit unterschiedlichem Selengehalt innerhalb des kritischen Bereichs in unseren Laboratorien keine Anomalien (der spektralen Verteilung der Photoleitfähigkeit, der Leitfähigkeit und des Halleffekts) gefunden worden sind. Erst bei höheren Konzentrationen zeigten sich — zunächst in der Ladungsträgerbeweglichkeit — Abweichungen vom Verhalten des reinen Tellurs (siehe Fig. 2).

Es sind verschiedene Modelle der Atomanordnung im Tellur denkbar, die die beiden oben angeführten, beim Abschrecken in den Kristall eingebrachten Veränderungen plausibel machen. Wegen dieser Vielfalt der Möglichkeiten, unter denen durch die Experimente noch keine eindeutige Auswahl getroffen werden kann, sei hier auf eine detaillierte Beschreibung eines Modells verzichtet. Es lassen sich lediglich folgende Aussagen machen:

Freie Bindungen wirken im Gitter anderer Elementhalbleiter als Akzeptoren, und ähnliches wird auch für Tellur erwartet [28]. Kettenrisse, die zur Bildung solcher freier Bindungen führen, spielen wegen der besonderen Kristallstruktur des Tellurs wahrscheinlich eine wichtige Rolle als Fehlordnung. Während bei tiefen Temperaturen in getemperten Proben die Leerstellen zusammengeballt sein werden und deshalb nur in geringem Maße zu freien Bindungen und damit zu Akzeptoren führen, ist bei höherer Temperatur und auch nach dem Abschrecken jede Tellurkette in eine Reihe von kurzen Abschnitten zerlegt, die durch eine Anzahl Vakanzen getrennt sind.

Dies erklärt die starke Zunahme der Konzentration thermischer Akzeptoren mit der Temperatur. Die Erhöhung der Gitterbeweglichkeit muß man möglicherweise ganz ähnlich erklären wie Lange und Regel [4] in selendotierten Kristallen, wobei aber Zwischengitter-Telluratome oder neutrale Fremdatome an die Stelle der Selenatome im Lange-Regelschen Modell treten müßten.

Literatur

[1] G. FISCHER und F. T. HEDGCOCK, J. Phys. Chem. Solids **17**, 246 (1961).
[2] C. H. CARTWRIGHT und M. HABERFELD-SCHWARZ, Proc. Roy. Soc. **148 A**, 648 (1935).
[3] V. N. LANGE und A. R. REGEL', Fiz. tverd. Tela **1**, 562 (1959).
[4] V. N. LANGE und A. R. REGEL', Fiz. tverd. Tela **2**, 2439 (1960).
[5] R. LINK, phys. stat. sol. **3**, 1098 (1963).
[6] H. KRONMÜLLER, J. JAUMANN und K. SEILER, Z. Naturf. **11a**, 243 (1956).
[7] T. FUKUROI, S. TANUMA und S. TOBISAWA, Sci. Rep. RITU **A 4**, 283 (1952).
[8] R. v. KUJAWA, phys. stat. sol. **3**, 1089 (1963).
[9] R. LINK, Z. phys. Chem. (Leipzig) **221**, 274 (1962).
[10] J. S. BLAKEMORE, J. W. SCHULTZ und K. C. NOMURA, J. appl. Phys. **31**, 2226 (1960).
[11] R. LINK, Z. phys. Chem. (Leipzig) **224**, 269 (1963).
[12] R. V. PARFEN'EV, I. I. FARBŠTEJN und S. S. ŠALYT, Fiz. tverd. Tela **2**, 2923 (1960).
[13] R. W. MCKAY und W. E. GRAVELLE, Canad. J. Phys. **39**, 534 (1961).
[14] E. MOOSER und W. B. PEARSON, Canad. J. Phys. **34**, 1369 (1956); J. Phys. Chem. Solids **7**, 65 (1958).
[15] V. v. LINT und H. ROTH, J. appl. Phys. **30**, 1235 (1959).
[16] V. I. VERAKSA, V. N. LANGE und T. I. LANGE, Zh. fiz. chim. **37**, 2308 (1963). V. I. VERAKSA, V. N. LANGE, T. I. LANGE und A. R. REGEL', Sbornik Issledovanija po poluprovodnikam, Kišinev 1964.
[17] H. FRITZSCHE, Science **115**, 571 (1952).
[18] A. E. MIDDLETON, Ph. D. thesis, Purdue Univ. 1944.[2])
[19] S. TANUMA, Sci. Rep. RITU **A 6**, 159 (1954).
[20] F. R. KESSLER und E. SUTTER, Z. Naturf. **16a**, 1173 (1961).
[21] R. S. CALDWELL und H. Y. FAN, Phys. Rev. **114**, 664 (1959).
[22] J. S. BLAKEMORE, D. LONG, K. C. NOMURA und A. NUSSBAUM, Progr. Semiconductors **6**, 38 (1961).
[23] J. S. BLAKEMORE und K. C. NOMURA, Phys. Rev. **127**, 1024 (1962).
[24] E. GRISON, J. chem. Phys. **19**, 1109 (1951).
[25] D. LONG, Phys. Rev. **101**, 1256 (1956).
[26] G. SYRBE, Ann. Phys. (Germany) **4**, 132 (1959).
[27] V. N. LANGE und A. R. REGEL', Fiz. tverd. Tela **1**, 559 (1959).
[28] G. BUSCH und O. VOGT, Helv. phys. Acta **30**, 224 (1957).
[29] E. A. MOVČAN und E. G. MISELJUK, Ukr. fiz. Zh. **8**, 1174 (1963).

(Received December 4, 1964)

[2]) Diese Arbeit konnte nicht im Original eingesehen werden.

phys. stat. sol. 8, 729 (1965)

II. Physikalisches Institut der Universität München

Experimentelle Untersuchungen zum plastischen Verhalten von vielkristallinem Nickel

Von

CH. SCHWINK und D. KNOPPIK

Es wird die Verfestigung von vielkristallinem Nickel bis zu rund 10% Dehnung mit mechanischen und magnetischen Messungen verfolgt, mechanisch durch Bestimmung von Dehnung und Dehnungszunahme und durch Temperaturwechselversuche, magnetisch durch Messung von Koerzitivkraft und Anfangspermeabilität. Für letztere wird eine Meßanordnung näher erläutert. Eine früher für Zimmertemperatur angegebene Einteilung der Verfestigungskurve [3] läßt sich bis zur Temperatur der flüssigen Luft nachweisen, der Temperaturgang der charakteristischen Übergangsspannungen wird angegeben und besprochen. Temperaturwechselvorgänge zeigen die Bedeutung von Schneidvorgängen vor allem im Anfangsstadium der Verformung, wo auch die Abweichungen vom Cottrellschen Gesetz am größten sind.

The strain hardening of polycrystalline nickel is investigated up to about 10% straining a) with mechanical measurements — strain and strain increment; variation of flow stress with temperature —, b) with magnetic measurements – coercive force; initial permeability, for which the experimental device are fully described. A subdivision of the strain hardening curve, recently found at room temperature [3], exists down to 77 °K. The variation with temperature of the characteristic flow-stresses at the transition points from one part of the strain-hardening curve to the next is measured and discussed. Measurements of the variation of flow stress with temperature show the importance of cutting processes, especially during the first stages of formation, where the deviations from Cottrell's law are greatest.

1. Die Bereicheinteilung der vielkristallinen Verfestigungskurve

Ausgangspunkt der vorliegenden Arbeit sind Untersuchungen [1, 2], in denen der Wert magnetischer Messungen für das Studium der Vorgänge bei der Verformung von vielkristallinem Nickel aufgezeigt wurde. Die Kopplung solcher Messungen mit mechanischen und insbesondere elektronenmikroskopischen Oberflächenuntersuchungen hat dabei dazu geführt, daß eine Einteilung der Verfestigungskurve in verschiedene Bereiche gegeben werden konnte [3]. Diese ist nicht nur differenzierter als von anderen Autoren versuchte derartige Einteilungen [4 bis 7], sondern bei ihr werden vor allem den einzelnen Bereichen ganz bestimmte Vorgänge in den Kristalliten zugeordnet. Die Einteilung von Zankl [3], die sich auf Zimmertemperatur bezieht und kürzlich einer theoretischen Untersuchung der Vielkristallplastizität zugrunde gelegt wurde [8], läßt sich in Form einer Tabelle (Tabelle 1) darstellen.

Die getroffene Einteilung erstreckt sich bis zu mittleren Verformungsgraden. Die Frage, ob oberhalb der Spannung σ_3 von Tabelle 1 weitere charakteristische Vorgänge im Kristallgefüge stattfinden, die eine Fortsetzung der Unterteilung der Verfestigungskurve in Bereiche nach sich ziehen, bleibe offen. Wir beschränken uns auch in dieser Arbeit auf einen Bereich kleiner bis mittlerer Dehnungen, der vom theoretischen Standpunkt für den Vielkristall besonders interessant erscheint [9].

Tabelle 1

Bereiche	Einleitungsbereich E	Bereich 1	Bereich 2	Bereich 3
Dehnung (Zahlen für Ni)	$0 \cdots \varepsilon_0 \cdots \varepsilon_1$ $0 \ldots 2 \cdot 10^{-5} \ldots 10^{-3}$	$\varepsilon_1 \cdots \varepsilon_2$ $10^{-3} \ldots 10^{-2}$	$\varepsilon_2 \cdots \varepsilon_3$ $1 \cdot 10^{-2} \ldots 4 \cdot 10^{-2}$	$\varepsilon_3 \cdots$ $4 \cdot 10^{-2} \ldots$
Spannung	$0 \cdots \sigma_0 \cdots \sigma_1$	$\sigma_1 \cdots \sigma_2$	$\sigma_2 \cdots \sigma_3$	$\sigma_3 \cdots$
Oberflächenerscheinungen	Gleitlinien in Körnern, die sich um große Körner gruppieren	Gleitlinien in allen Körnern	Gruppierung der Gleitlinien (strukturierte Feingleitung), Bildung von Gleitbändern und Gleitbandgruppen	Verstärkte Ausbildung von Gleitbändern und Gleitbandgruppen, Einsetzen von Quergleitung
Aktive Gleitsysteme im Einzelkorn	Mehrfachgleitung	Mehrfachgleitung	Einfachgleitung im Korninnern	Einfachgleitung im Korninnern
Zahl der Gleitebenenspuren pro Korn	3 (im Mittel)	3 (im Mittel), gegen ε_2 hin abnehmend	1,5 (im Mittel)	1,5 (im Mittel)
Gleitlinienlänge	gleich dem Korndurchmesser	gleich dem Korndurchmesser	Wird kleiner als der Korndurchmesser und nimmt mit zunehmender Verformung weiter ab	Nimmt weiter ab
Stufenhöhe der Gleitlinien	75 ± 40 Å	75 ± 40 Å	75 ± 40 Å	75 ± 40 Å

2. Problemstellung

Wir stellen die Frage, ob und wie die gegebene Bereicheinteilung sich ändert, wenn man die Untersuchungen über einen größeren Temperaturbereich erstreckt. Hauptziel der im folgenden mitgeteilten Messungen sollte die Temperaturabhängigkeit der charakteristischen Spannungen σ_2 und σ_3 sein (vgl. Tabelle 1); diejenige von σ_0 und σ_1 wird lediglich gestreift.

Als Meßgrößen verwenden wir dazu neben Dehnung, Dehnungszunahme, und Koerzitivkraft – die Remanenz ziehen wir in dieser Arbeit, im Gegensatz zu früheren [1, 3] nicht heran – auch noch die Anfangspermeabilität. Eine dafür geeignete Wechselstrommethode wird beschrieben.

Besonders aufschlußreich für Verformungsvorgänge haben sich bereits bei Einkristallen Temperaturwechselversuche herausgestellt, die über den Anteil der stark temperaturabhängigen Schneidvorgänge an der Gesamtverfestigung eine Aussage liefern. Wir haben deshalb derartige Messungen auch an unseren Proben durchgeführt, über deren Ergebnisse wir ebenfalls berichten.

3. Die Messung der Anfangspermeabilität

Während die Meßmethoden der mechanischen Größen und der Koerzitivkraft bekannt sind und früher bereits geschildert wurden, werde hier kurz auf eine Anordnung zur Messung der Anfangspermeabilität eingegangen, die wir für unsere Fragestellung aufbauten.

Die Messung erfolgte nach der ballistischen Methode, bei der eine abmagnetisierte ferromagnetische Probe, die von einer Induktionsspule mit N Windungen umgeben ist, in eine Feldspule gebracht wird. Schickt man durch letztere einen sinusförmigen Wechselstrom, so wird in der Induktionsspule eine Spannung mit dem effektiven Mittelwert U (V) induziert, die der Suszeptibilität der Probe proportional ist, wenn der vom äußeren Feld herrührende Teil der Induktion mit einer weiteren Induktionsspule kompensiert wird. Man leitet für die Suszeptibilität $\varkappa$ und die mit ihr durch die Beziehung $\mu - 1 = 4\pi\varkappa$ verbundene Permeabilität μ für hinreichend kleine Wechselstromamplituden die Beziehung ab:

$$\mu - 1 = \frac{U \cdot 10^8}{2\pi^2 R^2 N C_F \nu i_{0F}}. \tag{1}$$

Neben den schon genannten Größen ist R (cm) der Radius der Probe, C_F (cm^{-1}) die Eichkonstante der Feldspule, ν (s^{-1}) die Frequenz des Wechselstroms und i_{0F} (A) dessen effektiver Mittelwert.

Führt man die geschilderte Messung aus, nachdem die Probe vorher sorgfältig abmagnetisiert wurde, so ist bei hinreichend kleiner Aussteuerung die durch (1) gegebene Permeabilität die Anfangspermeabilität.

Bei Messungen wie der vorliegenden ist der Skineffekt besonders zu beachten, der bewirkt, daß sowohl Amplitude als auch Phase des Wechselfeldes über den Querschnitt des Probendrahts nicht konstant sind. In Formeln, die sich bei Joos [10] finden, lautet der Zusammenhang:

$$H(r,t) = H(r)\, e^{i(\omega t + \varphi(r))}, \qquad H(r) = H(0)\sqrt{\mathrm{ber}^2(a\,r) + \mathrm{bei}^2(a\,r)},$$

$$\mathrm{tg}\,\varphi = \frac{\mathrm{bei}\,(a\,r)}{\mathrm{ber}\,(a\,r)}, \qquad a = \sqrt{\frac{4\pi\mu\omega}{c^2\varrho}},$$

ϱ spezifischer Widerstand,

$\mathrm{ber}\,(x) = \mathrm{Re}\, I_0(\sqrt{-i}\,x))$, $\mathrm{bei}\,(x) = \mathrm{Im}\,(I_0(\sqrt{-i}\,x))$,

I_0 Besselfunktion nullter Ordnung.

Für verschiedene Parameter a ist in Fig. 1 das Verhältnis $H(0)/H(R)$ (vgl. Fig. 2) und die Phase $\varphi(r)$ als Funktion des Probenradius R bzw. r aufgezeichnet.

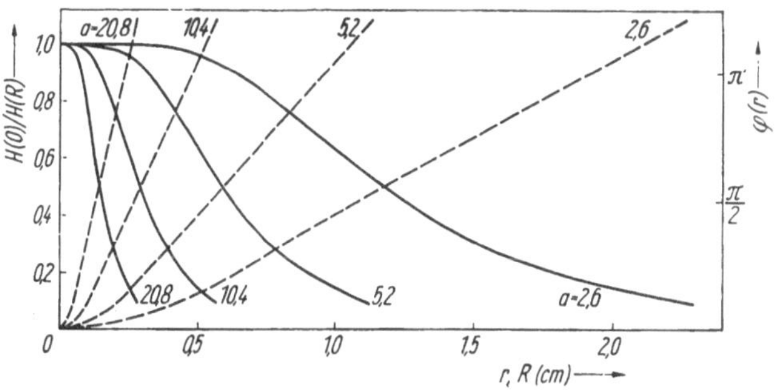

Fig. 1. Radialer Amplituden- und Phasengang des Feldes infolge des Skineffekts.

$a = \sqrt{\frac{4\pi\mu\omega}{c^2\varrho}}$

—— $= \frac{H(0)}{H(R)}$

- - - $= \varphi(r)$

Man hat bei vorgegebenem Probenradius R die Frequenz ν so niedrig zu wählen, daß das Verhältnis $H(0)/H(R)$ möglichst nahe an eins liegt und die Phase noch möglichst klein bleibt. In unseren Nickeldrähten haben wir im ausgeglühten Zustand Anfangspermeabilitäten von rund 350 zu erwarten, der Parameter a überschreitet demnach den Wert 10,4 nicht, wenn wir eine Frequenz von etwa 25 Hz wählen. Im Experiment haben wir immer mit der konstanten Frequenz $\nu = 24$ Hz an Drähten von 1,0 mm Radius gemessen, was nach den Kurven mit dem Parameter $a = 10{,}4$ von Fig. 1 bedeutet, daß $H(0)/H(R)$ nicht unter 0,983 sinkt und φ den Wert 0,26 nicht übersteigt.

Da die Permeabilität mit der Verformung rasch abnimmt, ergibt sich durch Mittelung über den Probenquerschnitt, die nicht näher ausgeführt sei, daß der vom Skineffekt hervorgerufene Meßfehler für unsere unverformten Drähte kleiner als 10%, für solche nach einer Reckung mit 4 kp/mm² aber bereits kleiner als 1% ist. Das war für unsere Aufgabe ausreichend.

Für die Durchführung der μ-Messung war es noch wichtig, störende Brummspannungen zu unterdrücken. Dazu wurde vor den Eingang des Röhrenvoltmeters, das die Induktionsspannung U anzeigte, ein Filter für 50 Hz geschaltet (Fig. 3).

Das Ergebnis der Strom-Spannungsmessungen zur Ermittlung von μ nach Formel (1) zeigt Fig. 4. Damit der Zusammenhang zwischen U und i_{OF} linear bleibt, darf der Strom und damit das Wechselfeld nur bis zu einer kritischen Feldstärke H_k ausgesteuert werden. Für unverformte Proben mit einer Koerzitivkraft H_c von rund 0,8 Oe war H_k etwa 0,03 Oe, für stark verformte Proben ($\sigma = 25$ kp/mm²) mit $H_c = 21$ Oe wuchs H_k auf 0,35 Oe.

Man sieht, daß in jedem Fall, auch für $\sigma = 0$, sich ein hinreichend langer linearer Bereich ergibt, dessen Extrapolation sehr gut durch Null geht, was verlangt werden muß. Die Abweichung vom geradlinigen Verlauf für sehr kleine i_{OF}-Werte, die vor allem bei den Kurven schwach verformter Drähte mit kleinem σ auftritt, rührt von der nicht zu unterdrückenden Brummspannung des Röhrenvoltmeters (0,05 mV) und der durch das Filter immer noch hindurchgelassenen Störspannung (0,03 mV) her.

Die Diskussion sämtlicher in die Messung eingehender Größen, zu denen insbesondere noch der Probenradius gehört, führte schließlich auf die Meßgenauigkeiten, die in Tabelle 2 zusammengestellt sind; diejenigen für die Koerzitivkraft

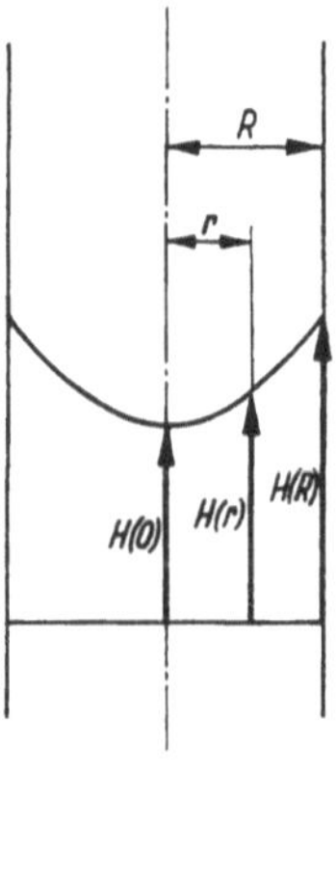

Fig. 2. Änderung der Amplitude $H(r)$ über den Probenquerschnitt, schematisch

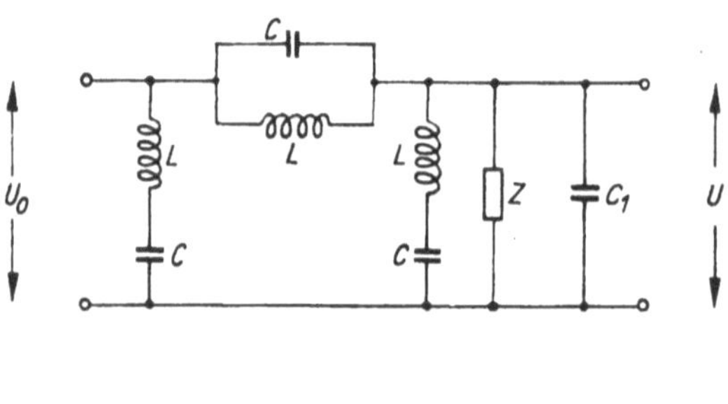

Fig. 3. 50 Hz-Filter. U_0 Eingangsspannung, U Ausgangsspannung. $L = 10$ H, $C = 1\,\mu$F, $C_1 = 1000$ pF, $Z = \sqrt{L/C} = 3{,}1$ kΩ

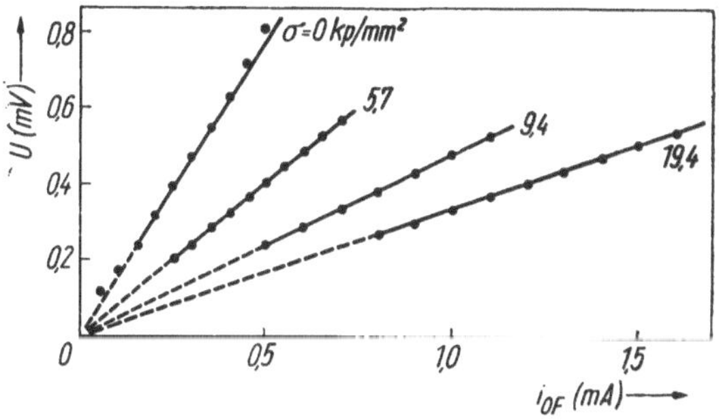

Fig. 4. Bestimmung von Anfangspermeabilitäten, die nach (1) den Steigungen der $U - i_{0F}$-Geraden proportional sind, für verschiedene Werte vorher angelegter Zugspannungen σ (kp/mm²)

sind ebenfalls mit eingetragen. Dabei geben die Spalten „absolut" die Unsicherheiten bei der Absolutmessung der betreffenden Größe an, die Spalten „relativ" entsprechend die Unsicherheiten bei der Relativmessung, wenn also μ oder H_c an ein und derselben Probe als Funktion zunehmender Verformung und damit Fließspannung gemessen wird.

Tabelle 2

	$\frac{\Delta H_c}{H_c}$ (absolut)	$\frac{\Delta H_c}{H_c}$ (relativ)	$\frac{\Delta\mu}{\mu}$ (absolut)	$\frac{\Delta\mu}{\mu}$ (relativ)
unverformt	2%	1%	15%	10%
$\sigma \geqq 4$ kp/mm²	2%	1%	7%	3%

4. Meßergebnisse

4.1 Probenmaterial, Vorbehandlung und Übersicht über die Meßkurven

Die Untersuchungen wurden mit Nickeldraht (Typ Ni–S) von 2 mm Durchmesser der Firma Vacuumschmelze durchgeführt. Die wesentlichen Verunreinigungen waren:

	C	Fe	Cu	Si	Mg
%	0,01	0,002	0,002	0,001	Spur

Das in hartem Zustand angelieferte Material wurde in Längen von rund 300 mm zunächst bei 500 °C im Wasserstoffstrom vorgeglüht, dann für die magnetischen Messungen gerade gerollt und anschließend einer dreistündigen Hauptglühung bei 820 °C unterworfen. Der mittlere Korndurchmesser des rekristallisierten Materials betrug dann $D = 0{,}06$ mm.

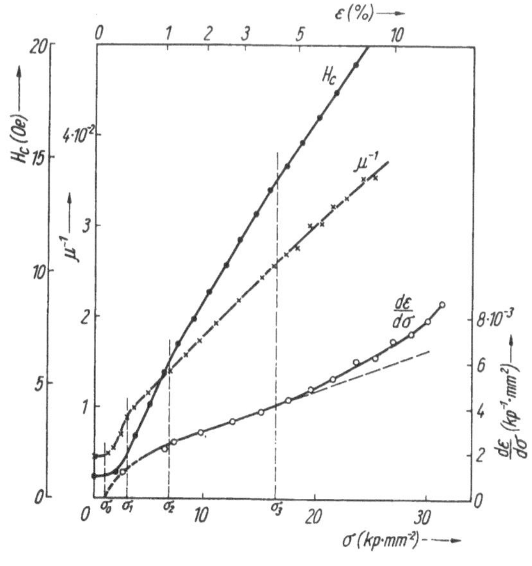

Fig. 5. Bereicheinteilung der Verfestigungskurve nach Messungen der Koerzitivkraft H_c, der reziproken Anfangspermeabilität μ^{-1} und der Dehnungszunahme $d\varepsilon/d\sigma$

Wir geben in Fig. 5 eine Zusammenstellung von Meßkurven der Koerzitivkraft H_c, der reziproken Anfangspermeabilität μ^{-1} und der Dehnungszunahme $d\varepsilon/d\sigma$ als Funktion der vor der Messung angelegten äußeren Spannung σ. Die charakteristischen Punkte σ_ν sind eingezeichnet, die Entsprechung zwischen den einzelnen Kurven ist gut erkennbar. Man sieht, daß sich die verschiedenen Meßgrößen ergänzen: Bei der μ^{-1}-Kurve treten die Punkte σ_0, σ_1 und σ_3 klar auf, besonders deutlich σ_1, ein Knick bei σ_2 ist dagegen meist nur sehr schwach ausgeprägt und manchmal gar nicht festzustellen. Er tritt viel deutlicher in der H_c-Kurve hervor, die auch die Spannung σ_3 sehr gut bestimmen läßt, während σ_1 lediglich als diejenige Stelle erscheint, von der ab die Koerzitivkraft linear mit zunehmender Spannung wächst [3]. Die Dehnungskurve steigt, wie schon bei Zankl [3], monoton ohne erkennbaren geradlinigen Anfangsteil an, dagegen besitzt die Kurve der Dehnungszunahme zwischen σ_2 und σ_3 einen linearen Verlauf, der — ebenfalls wie bei Zankl — oberhalb σ_3 verlassen wird. Das gibt uns, vor allem für Verformungen bei tiefen Temperaturen, die Möglichkeit, σ_3 unabhängig von den magnetischen Messungen festzulegen.

4.2 Der Temperaturgang der Übergangsstellen σ_0 bis σ_3

Den Temperaturgang der Spannung σ_0 bei Fließbeginn haben wir nicht eingehender untersucht. Unsere Messungen liefern σ_0 (360 °K) $= 1{,}0 \pm 0{,}3$ kp/mm², σ_0 (77 °K) $= 1{,}3 \pm 0{,}2$ kp/mm². Bei diesen Spannungen beträgt die Gesamtdehnung der Proben, die sich hier sehr inhomogen über die Probenlänge verteilt, etwa $5 \cdot 10^{-5}$. Die kleinen σ_0-Werte lassen sich quantitativ verstehen, wenn man die Beobachtung heranzieht, daß die ersten Gleitspuren in den größten, am günstigsten orientierten Körnern auftreten und die Versetzungen an den Grenzen dieser Körner gestaut werden [3, 8]. Ein schwaches Ansteigen mit abnehmender Temperatur ist erkennbar, doch ermöglichen die Messungen keine Aussage über Feinheiten des Temperaturverlaufs.

Der Temperaturgang der Spannung σ_1, der sich durch die μ-Messung besonders gut verfolgen läßt, ist bereits stärker ausgeprägt und läßt sich durch eine Gerade darstellen (Fig. 6). Die Werte liegen zwischen σ_1 (360 °K) $= 2{,}65 \pm 0{,}15$ kp/mm² und σ_1 (77 °K) $= 3{,}5 \pm 0{,}2$ kp/mm² und werden für alle Temperaturen bei einer Dehnung von rund $1 \cdot 10^{-3}$ erreicht. Sie stimmen gut mit Spannungswerten überein, die Macherauch und Vöhringer an vergleichbarem Nickelmaterial (Probe Ni 1 in [7]) mit $D = 0{,}05$ mm als Streckgrenzen σ_{St} gefunden haben (σ_{St} (360 °K) $=$ $= 2{,}6$ kp/mm²; σ_{St} (77 °K) $= 3{,}5$ kp/mm²). Die Streubreite unserer Meßpunkte, die von unvermeidlichen Korngrößenschwankungen herrührt, und deren geringe Zahl gestatten es uns nicht, die von den genannten Verfassern entdeckte Stufe im Temperaturgang von σ_{St} bestätigen zu können.

Noch stärker als σ_1 wächst die Spannung σ_2 mit abnehmender Temperatur. Sowohl in linearer wie in logarithmischer Auftragung liegen unsere Meßpunkte an-

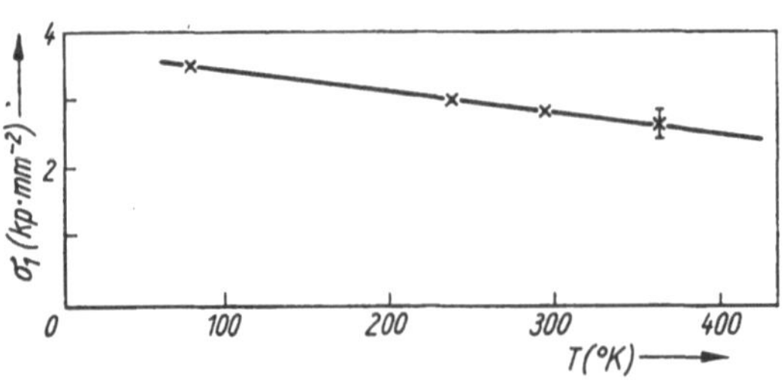

Fig. 6. Temperaturgang der Spannung σ_1

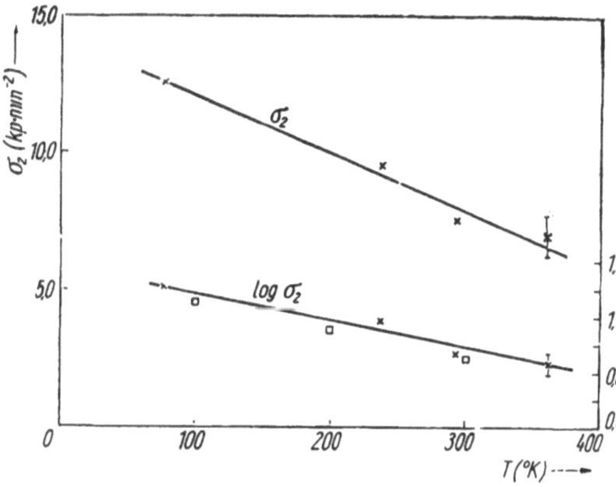

Fig. 7. Temperaturgang der Spannung σ_2 in linearer und logarithmischer Auftragung; ☐ Meßpunkte der Spannung σ_{III} von Macherauch [7]

nähernd auf einer Geraden (Fig. 7a und b). In letzterer Darstellung sind noch die von Macherauch und Vöhringer [7] mitgeteilten Werte der von ihnen σ_{III} genannten Größe eingezeichnet. Die Übereinstimmung sowohl im Absolutbetrag wie in der Steigung legt die Vermutung nahe, daß es sich in beiden Fällen um dieselbe Größe handeln könnte. Es besteht dann jedoch ein Unterschied in der Interpretation dieses Punkts, bei dem nach Macherauch die Quergleitung der Schraubenversetzungen einsetzen soll, während das nach den elektronenmikroskopischen Befunden von Zankl nicht hier, sondern erst bei der höheren Spannung σ_3 der Fall ist (vgl. Tabelle 1).

Zur weiteren Klärung dieses Sachverhalts ist auch die Korngrößenabhängigkeit von σ_2 zu beachten. In Übereinstimmung mit Macherauch und Vöhringer [7] finden wir nämlich, daß σ_2 mit zunehmendem Korndurchmesser abnimmt. Bei Zimmertemperatur ist σ_2 für $D = 0{,}06$ mm nach unseren Messungen im Mittel 7,5 kp/mm², für $D = 0{,}2$ mm erniedrigt sich dieser Wert auf rund 5,0 kp/mm². Nach kürzlich veröffentlichten theoretischen Vorstellungen [8] ist das dadurch zu verstehen, daß die Verkürzung der Gleitlinienlänge unter den Korndurchmesser – was den Punkt σ_2 physikalisch bestimmen soll – bei größerem mittlerem Korn eher, d. h. bei kleineren Spannungen, erfolgen muß als bei kleinerem.

Besonders interessant ist schließlich der Temperaturverlauf von σ_3, wo nach Tabelle 1 die thermisch aktivierte Quergleitung einsetzen soll. Fig. 8 zeigt eine Schar von Steigungskurven, die aus dynamisch gemessenen Spannungs-

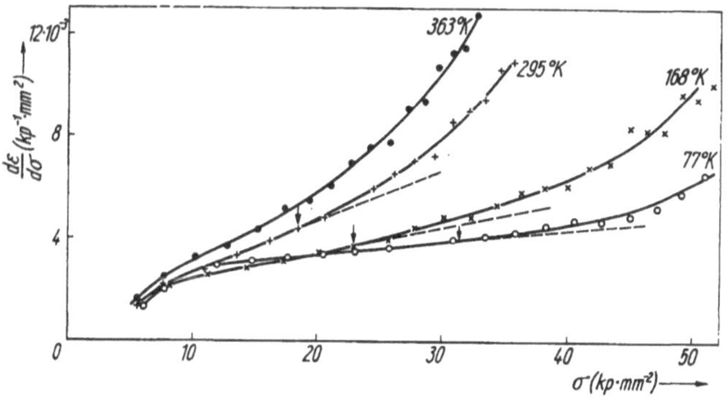

Fig. 8. Kurven der Dehnungszunahme für verschiedene Verformungstemperaturen; die Pfeile geben jeweils die Stelle σ_3 an

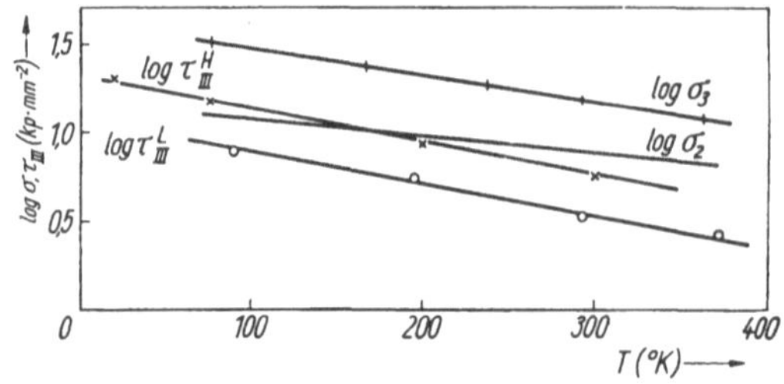

Fig. 9. Temperaturgang der Spannung σ_3 in logarithmischer Auftragung; zum Vergleich σ_2 nach Fig. 7, τ^H_{III} nach Haasen [11] und τ^L_{III} nach Leitz [12]

Dehnungskurven abgeleitet wurden, und Fig. 9 den aus diesen sowie magnetischen Messungen gewonnenen Temperaturgang von σ_3 in logarithmischer Darstellung. Zum Vergleich ist hier noch der Gang der an Nickeleinkristallen von Haasen [11] und Mader u. Mitarb. [12] gemessenen Größe τ_{III} eingezeichnet. Die Steigung der Geraden, die noch größer ist als bei σ_2, ist für die Einkristalle und den Vielkristall nahezu dieselbe. Zusammen mit der Tatsache, daß sich σ_3 im Gegensatz zu σ_2 bisher als unabhängig von der Korngröße erwies, stellt das eine indirekte Stütze für die unmittelbare elektronenmikroskopische Beobachtung von Zankl [3] dar, nach der erst oberhalb von σ_3 merklich Quergleitspuren auftreten.

Endgültig würde diese Frage mit mechanischen Messungen aber erst durch Versuche zur Geschwindigkeitsabhängigkeit der Größen σ_3 und τ_{III} entschieden. Aus den Beziehungen [11, 13]

$$\ln \tau_{\text{III}}(T)/G(T) = \ln \tau_{\text{III}}(O)/G(O) - B\,T\,, \qquad B = \frac{k}{A(\gamma)} \ln \frac{\dot{a}_0}{\dot{a}}$$

($\dot{a}$ Abgleitgeschwindigkeit, γ Stapelfehlerenergie),
sieht man nämlich, daß aus gleicher Steigung B nur dann auf gleiches A geschlossen werden kann, wenn auch $\dot{a}_0$ jeweils dasselbe ist; das aber würde durch eine Untersuchung der $\dot{a}$-Abhängigkeit entschieden.[1])

Als Ergebnis dieses Abschnitts können wir festhalten, daß die von uns gegebene Einteilung der vielkristallinen Verfestigungskurve von Nickel in einzelne Bereiche nicht nur bei Zimmertemperatur auftritt, sondern sich über den ganzen bisher untersuchten Temperaturbereich nachweisen läßt. Die vier Punkte σ_0, σ_1, σ_2 und σ_3 treten stets deutlich getrennt voneinander auf, σ_1 und σ_2 dürften dabei mit charakteristischen Punkten, die von anderen Autoren angegeben werden [7], zusammenfallen. Die Übergangsstellen von einem Bereich zum nächsten steigen dabei sämtlich, σ_2 und σ_3 sogar sehr ausgeprägt, mit abnehmender Temperatur an.

4.3 Temperaturwechselversuche

Wir wollen nun noch über Temperaturwechselversuche berichten, die deshalb von besonderem Interesse sind, weil sie Aufschluß über den Anteil thermisch aktivierter Vorgänge an der jeweiligen Gesamtspannung geben und weil sie ebenso wie die magnetischen Messungen über das gesamte Probenvolumen mitteln.

Nach Cottrell und Mitarbeitern [14] sollte das Verhältnis der Fließspannungen bei verschiedenen Temperaturen längs der gesamten Verfestigungskurve konstant sein. Haasen [11], Rebstock [15], Diehl und Berner [16] zeigten, daß das für Nickel- und Kupfereinkristalle keineswegs zutrifft, sondern daß man das in Fig. 10 schematisch gezeichnete Ergebnis erhält. Wäre τ/τ' konstant, so würde das eine Konstanz des Verhältnisses τ_S (Schneidvorgänge) zu τ_G (Spannungsfelder) bedeu-

[1]) Den Hinweis auf diesen Punkt verdanken wir Herrn Prof. Haasen.

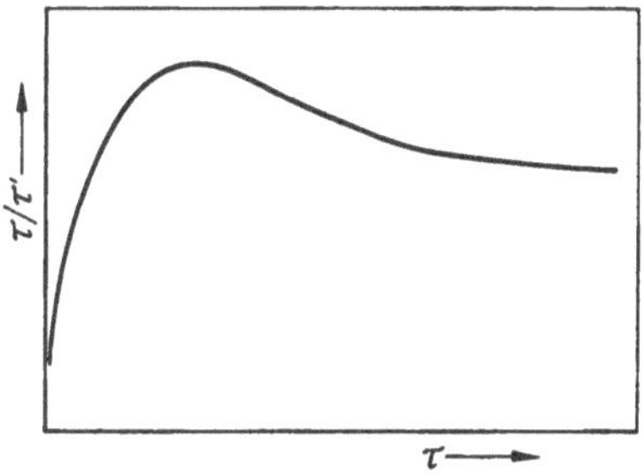

Fig. 10. Schubspannungsverhältnis nach Temperaturwechselversuchen an Einkristallen [16], schematisch; τ bezieht sich auf die höhere Temperatur, τ' auf die niedrigere

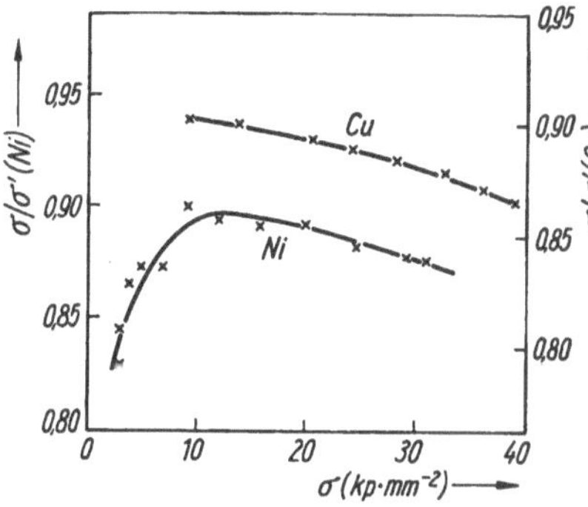

Fig. 11. Fließspannungsverhältnis bei Vielkristallen aus Nickel (eigene Messungen) und Kupfer (nach Bullen und Hutchison [18]); σ Fließspannung bei Zimmertemperatur, σ' Fließspannung bei 77 °K

ten [11, 16]. Je kleiner τ/τ' ist, desto größer ist der Anteil der Schneidvorgänge an der Gesamtfließspannung und umgekehrt.

An Vielkristallen wurden Temperaturwechselversuche bisher im wesentlichen von Derner und Kappler an Aluminium [17], von Bullen und Hutchison [18] und von Bullen und Rogers [19] an Kupfer, sowie von Macherauch, Mayr und Vöhringer [20] an Nickel, Kupfer und Silber durchgeführt. Während Derner und Kappler sowie Macherauch u. Mitarb. die Cottrellsche Behauptung praktisch bestätigen und lediglich für kleine Dehnungen eine mehr oder weniger stark ausgeprägte Abweichung finden, stellen die anderen Autoren an Kupfer auch für größere Dehnungen, die sie ausschließlich untersuchen, Abweichungen fest.

Unsere Messungen an Nickel zwischen den Temperaturen $T = 295$ °K und $T' = 77$ °K zeigt Fig. 11. Das Fließspannungsverhältnis steigt zunächst deutlich und stark an und nimmt nach Erreichen einer Spannung, die einem Mittelwert zwischen σ_2 (77 °K) und σ_2 (295 °K) entspricht, wieder leicht ab, und zwar mit derselben Steigung, wie sie Bullen und Hutchison [18] bei Kupfer finden (Fig. 11). Höchstens im Bereich zwischen σ_2 und σ_3, entsprechend 1 bis 5% Dehnung, kann man das Fließspannungsverhältnis als konstant bezeichnen. Die Beziehung von Cottrell gilt danach auch für Nickel nicht.

Aus dem charakteristischen Anstieg des Fließspannungsverhältnisses bei kleinen Verformungen schließen wir, daß bei Spannungen unterhalb σ_2, also insbesondere bei σ_1, noch ein erheblicher Schneidanteil zur Gesamtfließspannung beiträgt, dessen Verhältnis zum Anteil der Spannungsfelder aber bis σ_2 rasch abnimmt. Zwischen σ_2 und σ_3 ändert sich dann das Spannungsverhältnis nur sehr wenig, Schneid- und Spannungsanteil wachsen hier nahezu proportional zueinander. Von diesen Befunden wurde in der bereits erwähnten theoretischen Arbeit [8] Gebrauch gemacht.

Abschließend weisen wir noch darauf hin, daß der Maximalwert des Verhältnisses σ/σ' bei Vielkristallen mit rund 0,90 nicht ganz den bei Einkristallen mit $\tau/\tau' = 0{,}94$ gefundenen erreicht [16]. Das entspricht wegen des größeren Anteils von Mehrfachgleitung bei Vielkristallen ganz der Erwartung.

Literatur

[1] Ch. Schwink und G. Zankl, Z. Phys. **158**, 181 (1960).
[2] Ch. Schwink und G. Zankl, phys. stat. sol. **2**, K 241 (1962).
[3] G. Zankl, Z. Naturf. **18a**, 795 (1963).

[4] CH. CRUSSARD, Rev. Mét. **50**, 697 (1953).
[5] P. FELTHAM und J. D. MEAKIN, Phil. Mag. **2**, 105 und 1237 (1957).
[6] J. FRIEDEL, Proc. Symposium on Internal Stresses and Fatigue in Metals, Elsevier, Amsterdam 1959 (S. 220 ff.).
[7] E. MACHERAUCH und O. VÖHRINGER, phys. stat. sol. **6**, 491 (1964).
[8] CH. SCHWINK, phys. stat. sol. **8**, 457 (1965).
[9] E. KRÖNER, Acta metall. **9**, 155 (1961).
[10] G. JOOS, Lehrbuch der theoretischen Physik, 9. Aufl., 1956 (S. 290ff.).
[11] P. HAASEN, Phil. Mag. **3**, 384 (1958).
[12] S. MADER, A. SEEGER und CH. LEITZ, J. appl. Phys. **34**, 3368 (1963).
[13] R. BERNER, Z. Naturf. **15a**, 689 (1960).
[14] M. H. ADAMS und A. H. COTTRELL, Phil. Mag. **46**, 1187 (1955).
[15] H. REBSTOCK, Z. Metallk. **48**, 206 (1957).
[16] J. DIEHL und R. BERNER, Z. Metallk. **51**, 522 (1960).
[17] P. DERNER und E. KAPPLER, Z. Phys. **163**, 62 (1961).
[18] F. P. BULLEN und M. M. HUTCHISON, Phil. Mag. **8**, 461 (1963).
[19] F. P. BULLEN und C. B. ROGERS, Phil. Mag. **9**, 401 (1964).
[20] E. MACHERAUCH, P. MAYR und O. VÖHRINGER, phys. stat. sol. **5**, K 73 (1964).

(Received December 4, 1964)

phys. stat. sol. 8, 739 (1965)

Institute of Physics, Polish Academy of Sciences, Warsaw

Thermomagnetic Phenomena in Cubic Semiconductors Possessing Nonparabolic Energy Bands

By

W. ZAWADZKI

Thermomagnetic phenomena are considered in a semiconductor crystal having cubic symmetry and an arbitrary relation between energy and wave-number. The theory is formulated in terms of three basic transport tensors. Expressions are obtained for thermomagnetic effects in the limiting cases of weak and strong magnetic fields. The influence of anisotropy and non-parabolicity of the energy bands is considered. The results apply both to the many-ellipsoid cubic energy bands of Si and Ge and to the non-parabolic band structure of III-V compounds.

Es werden die thermomagnetische Eigenschaften in einem halbleitenden Kristall mit kubischer Symmetrie und einer beliebigen Abhängigkeit der Energie vom Wellenvektor untersucht. Die Theorie wird mit drei grundlegenden Transporttensoren formuliert. Es werden Ausdrücke für die thermomagnetischen Effekte in den Grenzfällen schwacher und starker Magnetfelder hergeleitet. Der Einfluß der Anisotropie und nicht parabolischer Energiebänder wird diskutiert. Die Ergebnisse sind sowohl auf die Energiebandstruktur von Si und Ge als auch auf die nicht parabolische Bandstruktur von III-V-Verbindungen anwendbar.

The transport phenomena in a cubic semiconductor possessing a many-ellipsoidal energy band, with an arbitrary dependence of energy on the magnitude of the wave vector in each ellipsoid, were considered theoretically in a previous publication [1]. This approach enabled us to treat the many-ellipsoidal conduction bands of Si and Ge and the nonparabolic spherical bands of III—V compounds from one point of view. This model may be necessary in order to explain the newest results concerning the transport properties of highly doped n-silicon [2]. Here we shall consider a few general properties of transport phenomena in cubic many-ellipsoidal energy bands. In particular, we shall deal with the thermomagnetic effects which were not treated in the previous paper. All the references concerning these problems and the definitions of notation can be found in [1].

1. Phenomenological Approach

Current density I_ν and heat flux Q_ν in the presence of electric field and temperature gradient in a magnetic field are given by

$$I_\nu = \sigma_{\nu\mu} E_\mu - \theta_{\nu\mu} \frac{\partial T}{\partial x_\mu}, \qquad Q_\nu = \chi_{\nu\mu} E_\mu - \xi_{\nu\mu} \frac{\partial T}{\partial x_\mu} - \varkappa^l_{\nu\mu} \frac{\partial T}{\partial x_\mu}. \tag{1.1}$$

In the equations (1.1) the currents are expressed through the field intensities.

The Onsager relation $\chi_{\nu\mu}(H) = T\,\theta_{\mu\nu}(-H)$ is fulfilled. Thus we have three basic transport-tensors $\sigma_{\nu\mu}$, $\theta_{\nu\mu}$ and $\xi_{\nu\mu}$. The tensor $\varkappa^{l}_{\nu\mu}$ denotes the thermal conductivity due to lattice vibrations.

We can transform the equations (1.1) in order to express the field intensities through the currents.

$$E_\nu = (\varrho_{\nu\mu} + \alpha_{\nu\alpha}\,\varkappa^{-1}_{\alpha\beta}\,\pi_{\beta\mu})\,I_\mu - \alpha_{\nu\alpha}\,\varkappa^{-1}_{\alpha\mu}\,Q_\mu\,, \tag{1.2}$$

$$\frac{\partial T}{\partial x_\nu} = \varkappa^{-1}_{\nu\alpha}\,\pi_{\alpha\mu}\,I_\mu - \varkappa^{-1}_{\nu\mu}\,Q_\mu \tag{1.3}$$

with

$$\left.\begin{array}{lll} & \varrho_{\nu\mu} = \sigma^{-1}_{\nu\mu} & \text{(resistivity tensor),} \\ & \alpha_{\nu\mu} = \sigma^{-1}_{\nu\beta}\,\theta_{\beta\mu} & \text{(thermoelectric tensor),} \\ & \pi_{\nu\mu} = \chi_{\nu\beta}\,\sigma^{-1}_{\beta\mu} & \text{(Peltier tensor),} \\ \text{and} & \varkappa_{\nu\mu} = \xi_{\nu\mu} - \chi_{\nu\beta}\,\alpha_{\beta\mu} + \varkappa^{l}_{\nu\mu} & \text{(thermal conductivity tensor).} \end{array}\right\} \tag{1.4}$$

The relation $\pi_{\nu\mu}(H) = T\,\alpha_{\mu\nu}(-H)$ is fulfilled.

Thus all the phenomenological transport-tensors in (1.2) and (1.3) can be obtained by multiplication and addition of the basic tensors $\sigma_{\nu\mu}$, $\theta_{\nu\mu}$, and $\xi_{\nu\mu}$ and their inverse tensors.

Assuming isothermal conditions $\frac{\partial T}{\partial x_\nu} \equiv 0$, we obtain from (1.2) $E_\nu = \varrho_{\nu\mu}\,I_\mu$. Thus the phenomenological tensor $\varrho_{\nu\mu}$ determines the galvanomagnetic phenomena (electrical resistance, transverse and planar Hall effects, and magnetoresistance) under isothermal conditions.

As can be seen from (1.2), the tensor $(\varrho_{\nu\mu} + \alpha_{\nu\alpha}\,\varkappa^{-1}_{\alpha\beta}\,\pi_{\beta\mu})$ determines the galvanomagnetic phenomena under adiabatic conditions $Q_\nu \equiv 0$.

The phenomenological tensor $\alpha_{\nu\mu}$ determines thermoelectric phenomena (thermoelectric power; longitudinal, planar and transverse Nernst-Ettingshausen effects).

The phenomenological tensor $\varkappa^{-1}_{\nu\mu}$ determines the thermal resistivity, and the longitudinal, planar and transverse Maggi-Righi-Leduc effects.

The phenomenological tensor $\varkappa^{-1}_{\nu\alpha}\,\pi_{\alpha\mu}$ determines the Nernst effect, and the longitudinal, planar and transverse Ettingshausen-Nernst effects under adiabatic conditions.

2. Basic Transport-Tensors

We consider transport phenomena in a cubic semiconductor having a many-ellipsoidal nonparabolic energy-band. One ellipsoid of this kind may be described by the relation

$$\gamma(\varepsilon) = a_{\alpha\beta}\,k_\alpha\,k_\beta\,. \tag{2.1}$$

For a parabolic band $\gamma(\varepsilon) \equiv \varepsilon$, and for a spherical band $a_{\alpha\beta} = a\,\delta_{\alpha\beta}$. The basic transport-tensors may be found from the solution of the Boltzmann equation, assuming the existence of a single relaxation time $\tau(\varepsilon)$. Then the expressions for a single ellipsoid are summed over all ellipsoids which constitute a cubic structure.

It was shown in [1] that in the weak-magnetic-field approximation the three basic transport-tensors have the following form (in the coordinate system connected with the principal cubic axes):

$$\left.\begin{aligned} \sigma_{\nu\mu} &= \sigma_1\,\delta_{\nu\mu} + \sigma_2\,\varepsilon_{\nu\mu\zeta}\,H_\zeta + \sigma_3\,H^2\,\delta_{\nu\mu} + \sigma_4\,H_\nu\,H_\mu + \sigma_5\,H^2_{(\nu)}\,\delta_{(\nu)\mu} \\ \theta_{\nu\mu} &= \theta_1\,\delta_{\nu\mu} + \theta_2\,\varepsilon_{\nu\mu\zeta}\,H_\zeta + \theta_3\,H^2\,\delta_{\nu\mu} + \theta_4\,H_\nu\,H_\mu + \theta_5\,H^2_{(\nu)}\,\delta_{(\nu)\mu} \\ \xi_{\nu\mu} &= \xi_1\,\delta_{\nu\mu} + \xi_2\,\varepsilon_{\nu\mu\zeta}\,H_\zeta + \xi_3\,H^2\,\delta_{\nu\mu} + \xi_4\,H_\nu\,H_\mu + \xi_5\,H^2_{(\nu)}\,\delta_{(\nu)\mu} \end{aligned}\right\}, \qquad (2.2)$$

where

$$\sigma_1 = q\,N\,K_1\,,\ \sigma_2 = q\,N\,K_2\,,\ \sigma_3 = q\,N\,K_3\,l_1\,,\ \sigma_4 = q\,N\,K_3\,l_2\,,\ \sigma_5 = q\,N\,K_3\,l_3 \qquad (2.3)$$

with

$$K_1 = \frac{2}{3}\frac{\langle\mu\rangle}{\langle 1\rangle}\,\mathrm{Sp}\,A\,, \qquad K_2 = \frac{4}{3\,c}\frac{\langle\mu^2\rangle}{\langle 1\rangle}\,A_{\alpha\alpha}\,, \qquad K_3 = \frac{8}{c^2}\frac{\langle\mu^3\rangle}{\langle 1\rangle}\,|A|\,, \qquad (2.4)$$

and

$$l_1 = \beta_{(\lambda\lambda\zeta\zeta)}\,,\quad l_2 = 2\,\beta_{(\lambda\zeta\lambda\zeta)}\,,\quad l_3 = \beta_{(\lambda\lambda\lambda\lambda)} - l_1 - l_2\,, \qquad (2.5)$$

where

$$\beta_{\nu\mu m n} = \frac{1}{2}\,(\delta_{\nu m}\,\delta_{\mu n} + \delta_{\nu n}\,\delta_{\mu m}) - \frac{1}{l}\sum_{i=1}^{l} a^{(i)}_{\nu\mu}\,a^{-1(i)}_{m n}\,. \qquad (2.6)$$

The summation in (2.6) is performed over the ellipsoids of a cubic energy band.

The θ_i and ξ_i in (2.2) are given by expressions analogous to (2.3) in which K_1, K_2, and K_3 are replaced by

$$P_1 = \frac{2}{3}\frac{\langle(z-\eta)\,\mu\rangle}{\langle 1\rangle}\,\mathrm{Sp}\,A\,,\quad P_2 = \frac{4}{3\,c}\frac{\langle(z-\eta)\,\mu^2\rangle}{\langle 1\rangle}\,A_{\alpha\alpha}\,,\quad P_3 = \frac{8}{c^2}\frac{\langle(z-\eta)\,\mu^3\rangle}{\langle 1\rangle}\,|A| \qquad (2.7)$$

for θ_i, and by

$$R_1 = \frac{2}{3}\frac{\langle(z-\eta)^2\,\mu\rangle}{\langle 1\rangle}\,\mathrm{Sp}\,A\,,\quad R_2 = \frac{4}{3\,c}\frac{\langle(z-\eta)^2\,\mu^2\rangle}{\langle 1\rangle}\,A_{\alpha\alpha}\,,\quad R_3 = \frac{8}{c^2}\frac{\langle(z-\eta)^2\,\mu^3\rangle}{\langle 1\rangle}\,|A| \qquad (2.8)$$

for ξ_i ($z = \varepsilon/k_0\,T$ reduced energy, $\eta = \zeta/k_0\,T$ reduced Fermi level, μ mobility, N total carrier density, q elementary electric charge ($\pm\,e$), c light velocity).

We use the definition

$$\langle X\rangle = \int_0^\infty \left(-\frac{\partial f_0}{\partial\varepsilon}\right)\hat{\chi}\,\gamma^{3/2}(\varepsilon)\,d\varepsilon\,. \qquad (2.9)$$

3. Thermomagnetic Phenomena

It is shown in the Appendix that the tensors of the type (2.2) are a field, i.e. every phenomenological tensor obtained by multiplication and addition of the basic transport-tensors and their inverse tensors (in the weak-magnetic-field approximation) has the form

$$\gamma_{\nu\mu} = \gamma_1\,\delta_{\nu\mu} + \gamma_2\,\varepsilon_{\nu\mu\zeta}\,H_\zeta + \gamma_3\,H^2\,\delta_{\nu\mu} + \gamma_4\,H_\nu\,H_\mu + \gamma_5\,H^2_{(\nu)}\,\delta_{(\nu)\mu}\,. \qquad (3.1)$$

The electrical resistivity tensor $\varrho_{\nu\mu}$ and the Nernst-Ettingshausen tensor $\alpha_{\nu\mu}$ (1.4) have been discussed in the previous paper [1].

We only note here that the resistivity tensor is expressed through the anisotropy constants l_1, l_2, and l_3, scattering factors a_r, b_r, and q_r (the last one being involved in expressions for high magnetic fields),

$$a_r = \frac{3\,A_{\alpha\alpha}}{(\mathrm{Sp}\,A)^2}\,\frac{\langle\mu^2\rangle\,\langle 1\rangle}{\langle\mu\rangle^2}\,,\qquad b_r = \frac{27\,|A|}{(\mathrm{Sp}\,A)^3}\,\frac{\langle\mu^3\rangle\,\langle 1\rangle^2}{\langle\mu\rangle^3}\,,\qquad q_r = \frac{\langle\mu\rangle\,\langle 1/\mu\rangle}{\langle 1\rangle^2}\,, \tag{3.2}$$

and the so-called ohmic mobility $\bar{u} = \frac{2}{3}\frac{\langle\mu\rangle}{\langle 1\rangle}\,\mathrm{Sp}\,A$.

The Nernst-Ettingshausen tensor is described by the above parameters and, in addition, the five quantities

$$\alpha_1 = \frac{\langle z\,\mu\rangle}{\langle\mu\rangle} - \eta\,,\qquad \alpha_2 = \frac{\langle z\,\mu^2\rangle}{\langle\mu^2\rangle} - \frac{\langle z\,\mu\rangle}{\langle\mu\rangle}\,,\qquad \alpha_3 = \frac{\langle z\,\mu^3\rangle}{\langle\mu^3\rangle} - \frac{\langle z\,\mu\rangle}{\langle\mu\rangle}\,, \tag{3.3}$$

$$\alpha_4 = \frac{\langle z\rangle}{\langle 1\rangle} - \eta\,,\qquad \alpha_5 = \frac{\langle z\rangle}{\langle 1\rangle} - \frac{\langle z/\mu\rangle}{\langle 1/\mu\rangle}\,. \tag{3.4}$$

α_4 and α_5 occur in expressions for high magnetic fields.

It is easy to see that all the quantities α_1 to α_5 represent various differences of the mean reduced energy.

Now we shall compute and discuss the thermal resistivity (Maggi-Righi-Leduc) tensor $\varkappa_{\nu\mu}^{-1}$ and the Ettingshausen-Nernst tensor $\varkappa_{\nu\alpha}^{-1}\,\pi_{\alpha\mu}$ (1.3).

Using the relations (1.4) and the rules for the multiplication, addition, and formation of inverse tensors (A.2 to A.4), we obtain after a little laborious but straightforward calculation

$$(\varkappa_{\nu\mu}^{-1})^a = -\frac{\varkappa^e}{(\varkappa^e + \varkappa^f)^2}\,a_r\,\frac{\psi_2}{\psi_1}\,\varepsilon_{\nu\mu\zeta}\,\frac{H_\zeta\,\bar{u}}{c}\,, \tag{3.5}$$

$$\begin{aligned}(\varkappa_{\nu\mu}^{-1})^s = \frac{1}{\varkappa^e + \varkappa^f}\Big[\delta_{\nu\mu} &- \frac{\varkappa^e}{\varkappa^e + \varkappa^f}\left(l_1\,b_r\,\frac{\psi_3}{\psi_1} + a_r^2\,\frac{\alpha_2^2}{\psi_1} + \frac{\varkappa^e}{\varkappa^e + \varkappa^f}\,a_r^2\,\frac{\psi_2^2}{\psi_1^2}\right)\frac{\bar{u}^2}{c^2}\,H^2\,\delta_{\nu\mu} - \\ &- \frac{\varkappa^e}{\varkappa^e + \varkappa^f}\left(l_2\,b_r\,\frac{\psi_3}{\psi_1} - a_r^2\,\frac{\alpha_2^2}{\psi_1} - \frac{\varkappa^e}{\varkappa^e + \varkappa^f}\,a_r^2\,\frac{\psi_2^2}{\psi_1^2}\right)\frac{\bar{u}^2}{c^2}\,H_\nu\,H_\mu - \\ &- \frac{\varkappa^e}{\varkappa^e + \varkappa^f}\,l_3\,b_r\,\frac{\psi_3}{\psi_1}\,\frac{\bar{u}^2}{c^2}\,H_{(\nu)}^2\,\delta_{(\nu)\mu}\Big]\,.\end{aligned} \tag{3.6}$$

Here $\varkappa^e$ is the electronic part of thermal conductivity,

$$\varkappa^e = \frac{k_0^2}{q}\,T\,N\,\psi_1\,\bar{u} \tag{3.7}$$

and

$$\psi_1 = \frac{\langle z^2\,\mu\rangle}{\langle\mu\rangle} - \frac{\langle z\,\mu\rangle^2}{\langle\mu\rangle^2}\,, \tag{3.8}$$

$$\psi_2 = \frac{\langle z^2\,\mu^2\rangle}{\langle\mu^2\rangle} + \frac{\langle z\,\mu\rangle^2}{\langle\mu\rangle^2} - 2\,\frac{\langle z\,\mu^2\rangle}{\langle\mu^2\rangle}\,\frac{\langle z\,\mu\rangle}{\langle\mu\rangle}\,,\qquad \psi_3 = \frac{\langle z^2\,\mu^3\rangle}{\langle\mu^3\rangle} + \frac{\langle z\,\mu\rangle^2}{\langle\mu\rangle^2} - 2\,\frac{\langle z\,\mu^3\rangle}{\langle\mu^3\rangle}\,\frac{\langle z\,\mu\rangle}{\langle\mu\rangle}\,. \tag{3.9}$$

We assume $\varkappa_{\nu\mu}^f = \varkappa^f\,\delta_{\nu\mu}$, which is valid for cubic crystals.

Thus we deal here with the differences of squares of mean energy. We have ψ_1, ψ_2, and ψ_3 in the expressions for weak magnetic fields (similarly to α_1, α_2, and α_3 in the Nernst-Ettingshausen tensor) and, in addition, ψ_4 and ψ_5 in the expressions for high magnetic fields. Both α_1 to α_5 and ψ_1 to ψ_5 are identically equal to zero for an absolutely degenerate electron gas.

The first term in the symmetric part of the thermal resistivity-tensor (3.6) gives the thermal resistivity in the absence of a magnetic field.

The antisymmetric part (3.5) determines the Righi-Leduc effect in a weak magnetic field. Traditionally the Righi-Leduc constant is defined as a proportionality coefficient between a transverse temperature-gradient and a longitudinal gradient multiplied by a magnetic field; e.g. $\partial T/\partial x_2 = P_{R-L} H_3 \partial T/\partial x_1$. In our formulation $\partial T/\partial x_2 = P'_{R-L} H_3 Q_1$ the Righi-Leduc effect is fully analogous to the Hall effect. Instead of the electric current $\vec{I}$ we have the heat flux $\vec{Q}$, and the electric field intensity $\vec{E}$ is substituted by the thermal field intensity grad T.

Thus the Righi-Leduc constant is

$$P'_{R-L} = -\frac{\varkappa^e}{(\varkappa^e + \varkappa^f)^2} a_r \frac{\psi_2}{\psi_1} \frac{\bar{u}}{c} \tag{3.10}$$

which differs from the usual Righi-Leduc constant P_{R-L} by the term $(\varkappa^e + \varkappa^f)^2$ appearing instead of $\varkappa^e + \varkappa^f$ in the denominator.

The thermal resistivity in a transverse magnetic field (sometimes called the Maggi-Righi-Leduc effect) can be obtained from (3.6);

$$\varkappa^{-1}_{(\alpha\alpha)}(H_\beta) = \frac{1}{\varkappa^e + \varkappa^f}\left[1 - \frac{\varkappa^e}{\varkappa^e + \varkappa^f}\left(l_1 b_r \frac{\psi_3}{\psi_1} + a_r^2 \frac{\alpha_2^2}{\psi_1} + \frac{\varkappa^e}{\varkappa^e + \varkappa^f} a_r^2 \frac{\psi_2^2}{\psi_1^2}\right) \frac{\bar{u}^2 H_\beta^2}{c^2}\right]. \tag{3.11}$$

The thermal resistivity in a longitudinal magnetic field is

$$\varkappa^{-1}_{(\alpha\alpha)}(H_\alpha) = \frac{1}{\varkappa^e + \varkappa^f}\left[1 - \frac{\varkappa^e}{\varkappa^e + \varkappa^f}(l_1 + l_2 + l_3)\, b_r \frac{\psi_3}{\psi_1} \frac{\bar{u}^2 H_\alpha^2}{c^2}\right]. \tag{3.12}$$

We shall consider also (in analogy to the planar Hall effect and the planar Nernst-Ettingshausen effect) the planar Righi-Leduc effect. The planar Righi-Leduc effect involves the creation of a temperature gradient in a plane determined by the directions of a heat flux and a magnetic field which are not perpendicular nor parallel to each other.

$$\varkappa^{-1}_{\alpha\beta}(H_\alpha, H_\beta) = -\frac{\varkappa^e}{(\varkappa^e + \varkappa^f)^2} \frac{1}{2}\left(l_2 b_r \frac{\psi_3}{\psi_1} - a_r^2 \frac{\alpha_2^2}{\psi_1} - \frac{\varkappa^e}{\varkappa^e + \varkappa^f} a_r^2 \frac{\psi_2^2}{\psi_1^2}\right) \frac{\bar{u}^2 H^2}{c^2} \sin 2\Psi \tag{3.13}$$

(Ψ is the angle between $\vec{Q}$ and $\vec{H}$).

All the above effects have been computed under the assumption that the heat-flux component perpendicular to the main heat-flux is equal to zero. We can also consider these effects assuming that the temperature gradient perpendicular to the main heat flux vanishes (with the exception of the Righi-Leduc effect). This is sometimes called the isothermal condition. The assumption is equivalent to putting the Righi-Leduc constant equal to zero.

Now the symmetric part of the thermal resistivity tensor takes the form

$$(\varkappa_{\nu\mu}^{-1})^s = \frac{1}{\varkappa^e + \varkappa^l}\left[\delta_{\nu\mu} - \frac{\varkappa^e}{\varkappa^e + \varkappa^l}\left(l_1\, b_r\, \frac{\psi_3}{\psi_1} + a_r^2\, \frac{\alpha_2^2}{\psi_1}\right)\frac{\bar{u}^2}{c^2} H^2\, \delta_{\nu\mu} + \right.$$

$$\left. - \frac{\varkappa^e}{\varkappa^e + \varkappa^l}\left(l_2\, b_r\, \frac{\psi_3}{\psi_1} - a_r^2\, \frac{\alpha_2^2}{\psi_1}\right)\frac{\bar{u}^2}{c^2} H_\nu\, H_\mu - \frac{\varkappa^e}{\varkappa^e + \varkappa^l}\, l_3\, b_r\, \frac{\psi_3}{\psi_1}\, \frac{\bar{u}^2}{c^2}\, H_{(\nu)}^2\, \delta_{(\nu)\mu}\right]. \quad (3.14)$$

On the basis of (3.14) the corresponding effects under isothermal conditions can be easily obtained. There is no difference between the adiabatic and isothermal longitudinal Maggi-Righi-Leduc effects.

Now we shall consider effects determined by the Ettingshausen-Nernst tensor $\varkappa_{\nu\beta}^{-1}\pi_{\beta\mu}$. In order to find the form of this tensor it is sufficient to multiply $\varkappa_{\nu\beta}^{-1}$ (3.6) with the Peltier tensor (1.4) (see [1]) which can easily be done. The final formulæ are, however, rather complicated. For this reason we shall consider only the symmetric part in the absence of a magnetic field and the antisymmetric part of the $\varkappa_{\nu\beta}^{-1}\pi_{\beta\mu}$ tensor.

As can be easily seen from (1.3) the effects determined by the Ettingshausen-Nernst tensor are strictly adiabatic (all $Q_\nu = 0$).

The symmetric part of the tensor $\varkappa_{\nu\beta}^{-1}\pi_{\beta\mu}$ for zero magnetic field describes the Nernst effect, i.e., the creation of a thermal gradient parallel to the electric current, when the longitudinal heat-flux is kept equal to zero. The effect is described by the formula

$$\varkappa_{\nu\beta}^{-1}\,\pi_{\beta\mu}(0) = \frac{k_0\, T}{q}\, \frac{\alpha_1}{\varkappa^e + \varkappa^l}\, \delta_{\nu\mu}\,. \quad (3.15)$$

The effect strongly depends upon thermoelectric properties and thermal conductivity of the material and is hardly observable in metals, but it may be of importance in semiconductors having a small thermal conductivity.

The antisymmetric part of $\varkappa_{\nu\beta}^{-1}\pi_{\beta\mu}$ determines the adiabatic Ettingshausen-Nernst effect

$$(\varkappa_{\nu\beta}^{-1}\,\pi_{\beta\mu})_{ad}^a = \frac{k_0\, T}{q}\, \frac{a_r}{\varkappa^e + \varkappa^l}\left(\alpha_2 - \frac{\varkappa^e}{\varkappa^e + \varkappa^l}\, \alpha_1\, \frac{\psi_2}{\psi_1}\right)\varepsilon_{\nu\mu\zeta}\, \frac{H_\zeta\, \bar{u}}{c}\,. \quad (3.16)$$

If the temperature gradient parallel to the direction of the electric current is equal to zero, we obtain

$$(\varkappa_{\nu\beta}^{-1}\,\pi_{\beta\mu})^a = \frac{k_0\, T}{q}\, \frac{a_r}{\varkappa^e + \varkappa^l}\, \alpha_2\, \varepsilon_{\nu\mu\zeta}\, \frac{H_\zeta\, \bar{u}}{c}\,. \quad (3.17)$$

The effect is closely related to the Nernst-Ettingshausen effect (see [1]).

The symmetric part of the tensor $\varkappa_{\nu\beta}^{-1}\pi_{\beta\mu}$ which is proportional to the square of magnetic field intensity describes the planar Ettingshausen-Nernst effect and the magnetic field dependence of the Nernst effect.

The transport phenomena in high magnetic fields can be considered only for special orientations of the magnetic field and the currents. In the previous paper [1] classical formulæ were derived for the galvanomagnetic and thermoelectric phenomena in the presence of large magnetic fields. Here we shall give the final results concerning the thermomagnetic phenomena.

The magnetic field is assumed to be parallel to the $\vec{x}_3$ axis. The electronic part of the thermal conductivity in a transverse magnetic field (with $\partial T/\partial x_2 \equiv 0$) is

$$\frac{\varkappa_{11}^e(\infty)}{\varkappa_{11}^e(0)} = \lambda_1\, q_r \frac{\psi_5}{\psi_1} \left(\frac{H\,\bar{u}}{c}\right)^{-2}, \tag{3.18}$$

where

$$\psi_5 = \frac{\langle z\rangle^2}{\langle 1\rangle^2} + \frac{\langle z^2/\mu\rangle}{\langle 1/\mu\rangle} - 2\,\frac{\langle z\rangle}{\langle 1\rangle}\,\frac{\langle z/\mu\rangle}{\langle 1/\mu\rangle}, \tag{3.19}$$

λ_2 is an anisotropy parameter depending on number and orientation of the ellipsoids in the cubic energy band.

In the case $Q_2 \equiv 0$ this effect is described by

$$\frac{\varkappa_{11}^e(\infty)}{\varkappa_{11}^e(0)} = \left(\lambda_1\, q_r \frac{\psi_5}{\psi_1} + \frac{\varkappa^e}{\varkappa^f}\,\frac{\psi_4^2}{\psi_1^2}\right)\left(\frac{H\,\bar{u}}{c}\right)^{-2}, \tag{3.20}$$

where

$$\psi_4 = \frac{\langle z^2\rangle}{\langle 1\rangle} - \frac{\langle z\rangle^2}{\langle 1\rangle^2}. \tag{3.21}$$

The thermal conductivity in a strong longitudinal magnetic field is

$$\frac{\varkappa_{33}^e(\infty)}{\varkappa_{33}^e(0)} = \frac{1}{\lambda_2}, \tag{3.22}$$

where λ_2 is another anisotropy parameter (there is no difference between the isothermal and adiabatic cases).

The Righi-Leduc effect and the Ettingshausen-Nernst effect in a strong magnetic field are

$$\frac{\partial T}{\partial x_2} = \frac{\varkappa^e}{(\varkappa^f)^2}\,\frac{\psi_4}{\psi_1}\left(\frac{H\,\bar{u}}{c}\right)^{-1} Q_1 \tag{3.23}$$

and

$$\frac{\partial T}{\partial x_2} = -\frac{k_0\,T}{q\,\varkappa^f}\,\alpha_5\left(\frac{H\,\bar{u}}{c}\right)^{-1} I_1, \tag{3.24}$$

respectively, where α_5 is given by (2.13).

4. Influence of Band Structure on the Thermomagnetic Phenomena in Cubic Materials

First we shall consider the influence of anisotropy upon the thermomagnetic phenomena.

In *weak magnetic fields* the anisotropy of an energy band affects the phenomena through the scattering factors a_r and b_r and the tensor $\beta_{\nu\mu mn}$ (2.6), i.e., the constants l_1, l_2, and l_3. All the formulæ are given in a coordinate system connected with the principal cubic axes. In order to obtain formulæ in an arbitrary coordinate-system we have to transform the constants l_1, l_2, and l_3.

Considering a spherical energy-band we have $a_{\alpha\beta} = a\,\delta_{\alpha\beta}$ and the coefficients $3\,A_{\alpha\alpha}/(\mathrm{Sp}\,A)^2$ and $27\,|A|/(\mathrm{Sp}\,A)^3$ in a_r and b_r (3.2), respectively, become equal to 1. Moreover we have in this case

$$l_1 = -1\,, \quad l_2 = 1\,, \quad l_3 = 0 \tag{4.1}$$

so that $l_1 + l_2 + l_3 \equiv 0$. Now it is easy to write down all the effects for a spherical energy band. The results are identical with those obtained earlier [3]. In particular all effects are isotropic, and all the effects involving a longitudinal magnetic field vanish identically. It should be emphasized, however, that this vanishing cannot serve as a proof for the isotropy of a band, since all the longitudinal effects vanish also in the case of a single ellipsoidal band and a single relaxation time $\tau(\varepsilon)$ [4].

In the case of an anisotropic cubic energy-band the constants l_1, l_2, and l_3 have to be calculated separately for every many-ellipsoidal structure. For the sake of

Table 1

Constant	Orientation of ellipsoids		
	⟨100⟩	⟨111⟩	⟨110⟩
l_1	$-\dfrac{K^2+K+1}{3\,K}$	$-\dfrac{(2\,K+1)\,(K+2)}{9\,K}$	$-\dfrac{(K+1)^2}{4\,K}$
l_2	1	$\dfrac{(2\,K+1)\,(K+2)}{9\,K}$	$\dfrac{K^2+4\,K+1}{6\,K}$
l_3	$\dfrac{(K-1)^2}{3\,K}$	$-\dfrac{2\,(K-1)^2}{9\,K}$	$-\dfrac{(K-1)^2}{12\,K}$

completeness we quote them here for three cases: a) 6 spheroids aligned along the ⟨100⟩ directions, b) 8 spheroids aligned along the ⟨111⟩ directions, and c) 12 spheroids aligned along the ⟨110⟩ directions (For the case b they can be found in Tsidilkovskii's book [5] in a different formulation. In many publications (see e.g. [6]) they are quoted for the galvanomagnetic effects), where $K = a_2/a_1$, a_1 and a_2 denoting an axis of rotation and an axis perpendicular to this axis, respectively.

For $K = 1$ we obtain in each case the values given by (4.1).

In *strong magnetic fields* it is convenient to consider separately the effects in transverse and longitudinal field.

In a high transverse magnetic field several effects are isotropic: the Hall effect, Righi-Leduc, Nernst-Ettingshausen, Ettingshausen-Nernst effect, and the thermoelectric power. Other effects involve the anisotropy factor λ_1. Table 2 gives values of λ_1 for a few special cases.

For $K = 1$ all the transverse effects are isotropic.

Some effects (e.g. thermal e.m.f.) are also isotropic in a longitudinal magnetic field.

The longitudinal electric and thermal magnetoresistivity are independent of the type of scattering and the $\gamma(\varepsilon)$ relation, but depend on the anisotropy parameter λ_2.

In principle it is possible to determine K directly by measuring the thermal conductivity of free carriers at zero and large magnetic field strengths. In practice, however, it is rather difficult to determine the saturation value $\varkappa^e\,(\infty)$.

Table 2
Values of λ_1

Direction of current (gradient)	Direction of magnetic field	Orientation of ellipsoids ⟨100⟩	⟨111⟩	⟨110⟩
⟨100⟩	arbitrary $\perp$ $\vec{I}, \vec{Q}$ or grad T	$\frac{(2K+1)(K+2)}{9K^2}$	$\frac{(2K+1)^2}{3K(K+2)}$	$\frac{(2K+1)(K^2+8K+3)}{18K(K+1)}$
⟨111⟩	arbitrary $\perp$ $\vec{I}, \vec{Q}$ or grad T	$\frac{(2K+1)^2}{3K(K+2)}$	$\frac{(2K+1)(4K+5)}{3K(K+8)}$	$\frac{K^2+4K+1}{6K}$
⟨110⟩	⟨001⟩	$\frac{(2K+1)(5K+1)}{9K(K+1)}$	$\frac{(2K+1)(K+2)}{9K}$	$\frac{(2K+1)(K+11)(K+1)}{18K(K+3)}$

In the quoted papers (see references in [1] and [8]) the influence of the nonparabolicity of the energy band on transport phenomena is discussed in detail. For this reason we will stress here only the main points concerning deviations from the parabolic shape of a band.

Table 3
Values of λ_2

Direction of current (gradient)	Orientation of ellipsoids ⟨100⟩	⟨111⟩	⟨110⟩
⟨100⟩	1	$\frac{(2K+1)(K+2)}{9K}$	$\frac{(2K+1)(K+2)}{9K}$
⟨111⟩	$\frac{(2K+1)(K+2)}{9K}$	$\frac{(2K+1)(K+8)}{3(7K+2)}$	$\frac{(9K+1)^2}{3K(K+2)}$
⟨110⟩	$\frac{(2K+1)(K+1)}{K(K+5)}$	$\frac{(2K+1)^2}{3K(K+2)}$	$\frac{2(2K+1)(K+3)}{K^2+20K+3}$

In the general case, the effective mass in a nonparabolic band depends upon energy. The relaxation time cannot be expressed in this case by a simple power function of energy. Thus the resulting mobilities are largely different in the parabolic and nonparabolic cases. This must also be taken into account in evaluating the scattering factors a_r, b_r, and q_r. For first order nonparabolic bands $\gamma(\varepsilon) = \varepsilon\,(1 + \lambda\,\varepsilon)$ all the $\langle\ \rangle$ integrals may be expressed by generalized Fermi-Dirac functions [7]. The influence of nonparabolicity is of greatest importance in the presence of strong degeneracy. It turns out that the differences of mean energy α_1 to α_5 (3.3 and 3.4) are most sensitive to deviations from parabolicity, whereas these deviations have less influence on the differences of the squares of mean energy ψ_1 to ψ_5 (3.8, 3.9, 3.19, and 3.21). Thus, in the strongly degenerate case, we can divide all the effects investigated into two groups: 1) all effects expressed by the tensors $\varrho_{\nu\mu}$ and $\varkappa_{\nu\mu}^{-1}$ which can be changed quantitatively by non-parabolicity, 2) effects expressed by $\alpha_{\nu\mu}$, $\pi_{\nu\mu}$ and $\varkappa_{\nu a}^{-1}\,\pi_{a\mu}$ for which the influence of non-parabolicity is essential and can lead to changes in sign. Thermomagnetic effects in the presence of strong degeneracy have been considered in [3].

Appendix

We shall prove (neglecting terms higher than H^2) that tensors having the form

$$\alpha_{\nu\mu} = \alpha_1 \delta_{\nu\mu} + \alpha_2 \varepsilon_{\nu\mu\zeta} H_\zeta + \alpha_3 H^2 \delta_{\nu\mu} + \alpha_4 H_\nu H_\mu + \alpha_5 H^2_{(\nu)} \delta_{(\nu)\mu} \tag{A.1}$$

with real coefficients α_1, α_2, α_3, α_4, and α_5 are a field.

We shall denote a tensor of the form (A.1) as $\alpha_{\nu\mu} = (\alpha_1, \alpha_2, \alpha_3, \alpha_4, \alpha_5)$. It is easy to see that the tensors (A.1) constitute an Abelian additive group.

We have

$$\alpha_{\nu\mu} + \beta_{\nu\mu} = (\alpha_1 + \beta_1, \alpha_2 + \beta_2, \alpha_3 + \beta_3, \alpha_4 + \beta_4, \alpha_5 + \beta_5) = \beta_{\nu\mu} + \alpha_{\nu\mu} . \tag{A.2}$$

Thus G is commutative and closed under addition.

1. $(\alpha_{\nu\mu} + \beta_{\nu\mu}) + \gamma_{\nu\mu} = \alpha_{\nu\mu} + (\beta_{\nu\mu} + \gamma_{\nu\mu})$; the addition is associative.
2. $e_{\nu\mu} = (0, 0, 0, 0, 0)$; there is an identity element $e_{\nu\mu}$ such that

$$\alpha_{\nu\mu} + e_{\nu\mu} = \alpha_{\nu\mu} = e_{\nu\mu} + \alpha_{\nu\mu} .$$

3. $(-\alpha_{\nu\mu}) = (-\alpha_1, -\alpha_2, -\alpha_3, -\alpha_4, -\alpha_5)$; for each $\alpha_{\nu\mu}$ there is an inverse element $(-\alpha_{\nu\mu})$ having the property that

$$\alpha_{\nu\mu} + (-\alpha_{\nu\mu}) = e_{\nu\mu} = (-\alpha_{\nu\mu}) + \alpha_{\nu\mu} .$$

Now we shall prove that tensors having the form (A.1) constitute an Abelian multiplicative group, after deleting elements with $\alpha_1 = 0$ (in the approximation neglecting terms higher than H^2).

We have

$$\alpha_{\nu\gamma} \beta_{\gamma\mu} = (\alpha_1 \beta_1, \alpha_1 \beta_2 + \alpha_2 \beta_1, \alpha_1 \beta_3 + \alpha_3 \beta_1 - \alpha_2 \beta_2, \alpha_1 \beta_4 + \alpha_4 \beta_1 + \alpha_2 \beta_2, \alpha_1 \beta_5 + \\ + \alpha_5 \beta_1) = \beta_{\nu\gamma} \alpha_{\gamma\mu} . \tag{A.3}$$

Thus G is commutative and closed under multiplication.

1. $(\alpha_{\nu\zeta} \beta_{\zeta\eta}) \gamma_{\eta\mu} = \alpha_{\nu\zeta}(\beta_{\zeta\eta} \gamma_{\eta\mu})$; the multiplication is associative.
2. $e'_{\nu\mu} = (1, 0, 0, 0, 0)$; there is an identity element $e'_{\nu\mu}$ such that

$$\alpha_{\nu\zeta} e'_{\zeta\mu} = \alpha_{\nu\mu} = e'_{\nu\zeta} \alpha_{\zeta\mu} .$$

3. $$\alpha^{-1}_{\nu\mu} = \left[-\frac{1}{\alpha_1}, -\frac{\alpha_2}{\alpha_1^2}, -\frac{1}{\alpha_1^2}\left(\alpha_3 + \frac{\alpha_2^2}{\alpha_1}\right), -\frac{1}{\alpha_1^2}\left(\alpha_4 - \frac{\alpha_2^2}{\alpha_1}\right), -\frac{\alpha_5}{\alpha_1^2} \right]; \tag{A.4}$$

for every $\alpha_{\nu\mu}$ there is an inverse element $\alpha^{-1}_{\nu\mu}$ having the property that

$$\alpha_{\nu\zeta} \alpha^{-1}_{\zeta\mu} = e'_{\nu\mu} = \alpha^{-1}_{\nu\zeta} \alpha_{\zeta\mu} .$$

Properties 1. and 3. can easily be proved on the basis of (A.3). In calculating the inverse element $\alpha^{-1}_{\nu\mu}$ we used

$\alpha^{-1}_{\nu\mu} = \frac{1}{|\alpha|} \frac{1}{2} \varepsilon_{\nu\delta\sigma} \varepsilon_{\mu\gamma\varrho} \alpha_{\gamma\delta} \alpha_{\varrho\sigma}$, where $|\alpha| = \frac{1}{6} \varepsilon_{\gamma\delta\sigma} \varepsilon_{\mu\nu\varrho} \alpha_{\nu\delta} \alpha_{\varrho\sigma} \alpha_{\mu\gamma}$ and neglected all terms higher then H^2.

Since we have also

$\alpha_{\nu\zeta} (\beta_{\zeta\mu} + \gamma_{\zeta\mu}) = \alpha_{\nu\zeta} \beta_{\zeta\mu} + \alpha_{\nu\zeta} \gamma_{\zeta\mu}$; the multiplication is distributive over addition, we proved finally that tensors having the form (A.1) are a field. It should be noted that we had to delete not only the identity element of the additive group, $e_{\nu\mu}$, but all elements having $\alpha_1 = 0$. They do not have, however, any physical meaning.

Thus all tensors obtained by means of multiplication and addition of basic tensors and their inverse tensors belong to the field and have the form (A.1).

It can also be seen that the first two components of every tensor obtained by means of multiplication and addition of the basic tensors and their inverse tensors are always expressed by the two first components of these tensors. Since the two first components of the basic tensors $\sigma_{\nu\mu}$, $\theta_{\nu\mu}$ and $\xi_{\nu\mu}$ are invariants of rotations of a coordinate system, the same may be said about every phenomenological tensor.

Acknowledgements

The author is indebted to Prof. L. Sosnowski and Dr. J. Kołodziejczak for critical remarks on the subject.

References

[1] W. Zawadzki and J. Kołodziejczak, phys. stat. sol. **6**, 419 (1964).
[2] I. M. Tsidilkovskii, private communication.
[3] W. Zawadzki, phys. stat. sol. **2**, 385 (1962).
[4] J. Kołodziejczak and S. Żukotyński, phys. stat. sol. **5**, 145 (1964).
[5] I. M. Tsidilkovskii, Thermomagnetic Phenomena in Semiconductors, Fizmatgiz, Moscow 1960.
[6] E. H. Putley, The Hall Effect and Related Phenomena, Butterworths, London 1960.
[7] Properties and values of generalized Fermi-Dirac integrals are prepared and will be published.
[8] L. Sosnowski, Proc. Int. Conf. Phys. Semicond., Paris 1964.

(Received December 7, 1964)

phys. stat. sol. 8, 751 (1965)

Харьковский государственный университет им. А. М. Горького

Магнитострикция никель–кобальтовых ферритов

Б. Я. Пинес и Н. М. Гумен

Исследовано изменение магнитострикции насыщения λ_s у Ni-Co-ферритов разного состава после закалки от различных температур. $|\lambda_s|$ аномально возрастает у всех ферритов (кроме чисто никелевого) после закалки от температур лежащих несколько ниже точки Кюри, причем одновременно получается уменьшение постоянной решетки a.

Наибольшие изменения λ_s и a после закалки наблюдаются у ферритов наиболее богатых кобальтом, а у чисто никелевого феррита они обращаются в нуль. Отличия обусловлены разным количеством кислорода растворенного в решетке ферритов разного состава.

Рентгенографическим путем на поликристаллических образцах определена первая константа магнитострикции $\lambda_{[100]}$ во всем интервале концентраци CoO. Константа $\lambda_{[111]}$ у ферритов различного состава рассчитана с помощью формулы $\lambda_s = \dfrac{2\,\lambda_{[100]} + 3\,\lambda_{[111]}}{5}$.

The change of the magnetostrictive constant λ_s is investigated for Ni–Co ferrites of various composition after thermal quenching. $|\lambda_s|$ increases anomalously for all ferrites (except for pure Ni-ferrites) after quenching from temperatures which are lower than the Curie temperature, which also decreases the lattice constant. The largest change of λ_s and a are observed after quenching for ferrites which have a high Co-content. For pure Ni-ferrites no change occurs. The differences are due to the O_2-contents of the lattices of ferrites of various compositions. The first magnetostrictive constant $\lambda_{]100]}$ of polycrystalline samples is determined by an X-ray study for all CoO concentration. The constant $\lambda_{[111]}$ of ferrites of various compositions is calculated from the formula $\lambda_s = \dfrac{2\,\lambda_{[100]} + 3\,\lambda_{[111]}}{5}$.

1. Опыт показывает, что величина магнитострикции насыщения λ_s у кобальтовых ферритов меняется в зависимости от способа термической обработки. Закалка от разных температур фиксирует состояния феррита, различающиеся как величиной λ_s, так и значением постоянной решетки a.

Обычно применяемое после спекания феррита „охлаждение с печью" оказывается в ряде случаев недостаточно медленным, чтобы привести ферриты в равновесное состояние, отвечающее температурам близким к комнатным.

По этой причине значения λ_s, а также констант магнитострикции $\lambda_{[100]}$ и $\lambda_{[111]}$ могут иметь различные случайные значения соответствующие разным состояниям образца.

Для построения концентрационной зависимости λ_s и констант $\lambda_{[100]}$ и $\lambda_{[111]}$ у смешанных ферритов необходимо определить их значения у образцов различного состава в одинаковом состоянии.

Изучение зависимости λ_s от характера термической обработки у ферритов различного состава может способствовать нахождению тех условий обработки, при которых получаются одинаковые состояния образцов.

Большая величина $|\lambda_s|$ у число кобальтового и смешанных кобальт содержащих ферритов позволяет определить ,,кристалло-графические" константы магнитострикции рентгенографическим способом. Этим способом были определены константы $\lambda_{[100]}$ и $\lambda_{[111]}$ у Fe–Co-ферритов [1].

Несложный способ приготовления образцов и простая методика измерений позволяет применить этот способ для определения констант магнитострикции у кобальтсодержащих ферритов других систем.

В настоящей работе определялась магнитострикция насыщения λ_s у Ni–Co-ферритов различного состава в зависимости от температуры закалки. Рентгенографическим способом на поликристаллических образцах были определены значения константы $\lambda_{[100]}$ во всем интервале концентраций. Константа $\lambda_{[111]}$ находилась расчетным путем (по величине λ_s и $\lambda_{[100]}$).

2. Исходными материалами для приготовления образцов служили порошки окислов Co_2O_3 и Fe_2O_3 и никеля углекислого основного $2\,NiCO_3 \cdot 3\,Ni(OH)_2 \cdot 4\,H_2O$. Были приготовлены образцы Ni–Co-ферритов состава $Co_xNi_{1-x}F_2O_4$ при значениях $x = 0$: 0,2; 0,4; 0,6; 0,8; 1,0. Технология приготовления образцов в основном такая же как описано в работе [2].

Предварительный отжиг смесей порошков для осуществления реакции в твердой фазе производился при температуре 1100 °C в течение 2 часов с последующей закалкой от этой температуры.

Для лучшего взаимодействия образцов с внешней средой прессование ферритового порошка производилось под небольшим давлением ($\approx$ 60 кг/см²). Спекание осуществлялось путем быстрого нагрева образцов, в которых уже получился феррит до температуры 1200 °C и выдержки при этой температуре в течение 3 часов, после чего велось медленное охлаждение $\left(\frac{dT}{dt} \approx 50°/\text{час}\right)$ до температуры, от которой производилась закалка. Последней закалке предшествовала выдержка разной длительности в зависимости от температуры закалки (от 5 часов при 1000 °C до 250 часов при 100 °C).

Режим термомагнитной обработки (ТМО) и методика рентгенографических исследований были такие же как описано в работе [2].

Продольная магнитострикция определялась при комнатной температуре тензометрическим способом.

3. Было обнаружено, что у всех кобальт содержащих ферритов зависимость магнитострикции λ от величины приложенного поля H, а также значения λ_s получаются различными после разной обработки образцов.

У образцов закаленных от высоких температур максимальное значение $|\lambda|$ достигается при более низкой напряженности магнитного поля, чем у образцов закаленных от сравнительно невысоких температур. Отличие обоих значений H уменьшается с возрастанием содержания никеля в феррите.

Рис. 1. Зависимость магнитострикции насыщения λ_s от температуры закалки $T_з$ для ферритов состава $Co_xNi_{1-x}Fe_2O_4$: 1) $x=0$; 2) $x=0{,}2$; 3) $x=0{,}4$; 4) $x=0{,}6$; 5) $x=0{,}8$ 6) $x=1{,}0$

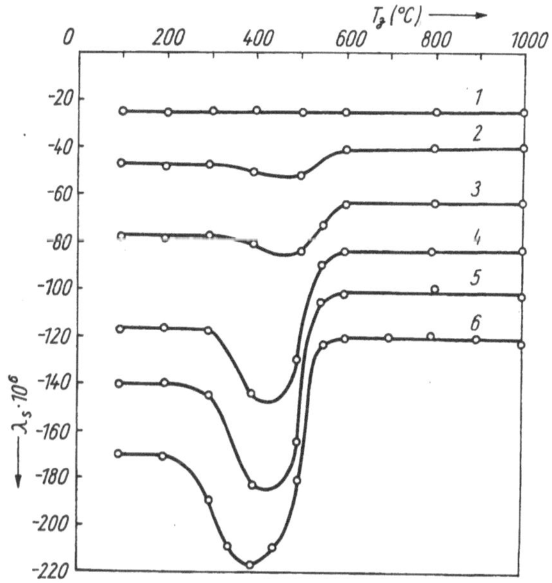

Образцы закаленные от высоких температур обнаруживают значительную об'емную положительную по знаку магнитострикцию в больших полях. Поэтому при увеличении поля $|\lambda|$ сначала растет, достигает максимального значения и при дальнейшем возрастании H уменьшается приблизительно по линейному закону.

На Рис. 1 приведены кривые зависимости значений λ_s от температуры закалки $T_з$ для образцов разного состава. Все ферриты, за исключением чисто никелевого, имеют немонотонную зависимость $|\lambda_s|$ от $T_з$ с максимумом при температурах ниже точки Кюри. При этом положение максимума по оси температур смещается к более высоким температурам при увеличении содержания никеля в феррите. Это смещение максимума находится в соответствии с ростом температуры Кюри при увеличении содержания никеля в Ni–Co-феррите.

С уменьшением содержания кобальта наблюдается снижение максимума на кривой $\lambda_s(T_з)$ и уменьшение отличия в значениях λ_s у образца закаленного от высоких температур ($T_з > \theta$) и образца медленно охлажденного до сравнительно низких температур ($T_з \leqq 200$ °C).

Изменению величины λ_s у Ni–Co-ферритов сопутствует уменьшение постоянной решетки a в указанном температурном интервале.

Обозначим через Δa разность значений постоянной решетки феррита в двух состояниях — после закалки от высоких температур ($T_з > \theta$) и после медленного охлаждения до комнатной температуры.

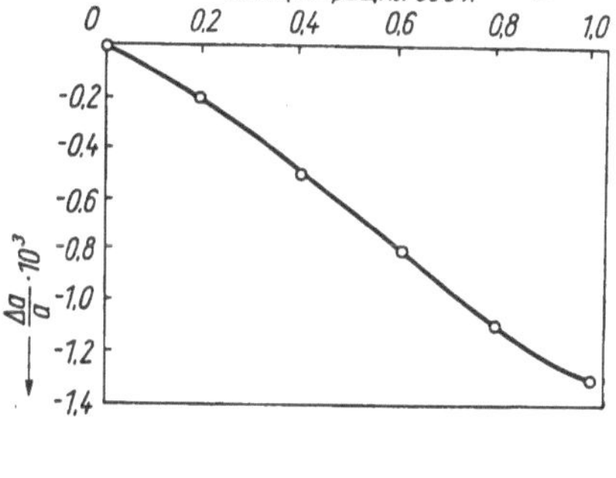

Рис. 2. Относительное изменение постоянной решетки у Ni–Co-ферритов различного состава

На Рис. 2 представлена зависимость относительного изменения $\frac{\Delta a}{a}$ от содержания CoO в феррите. Значение $\frac{\Delta a}{a}$ возрастает с увеличением содержания CoO от нуля для чисто никелевого до $1{,}3 \cdot 10^{-3}$ для чисто кобальтового феррита.

Отжиг в различных средах (азот, воздух) показывает, что изменение a наблюдаемое при медленном охлаждении вблизи точки Кюри у чисто кобальтового феррита, обусловленно растворением избыточного кислорода в решетке феррита. Повидимому, величина $\frac{\Delta a}{a}$ соответствует количеству растворенного кислорода и в случае ферритов различного состава. Следовательно, можно предполагать, что понижение максимума на кривой $\lambda_s(T_з)$ и уменьшение отличия в значениях λ_s для высокотемпературного и относительно низкотемпературного состояний обусловлено уменьшением количества кислорода, растворяющегося в феррите. Так как чисто никеленый феррит не присоединяет избыточного кислорода ($a = \text{const}$ для всех $T_з$), то для него и значение λ_s не зависит от способа термообработки.

Используя зависимость λ_s от $T_з$ для ферритов различного состава (Рис. 1), можно построить концентрационную зависимость магнитострикции насыщения для различных состояний ферритов, которая изображена на рис. 3. Здесь кривая 1 представляет значения λ_s для состояния ферритов, получающегося после закалки от высоких температур ($T_з > \theta$); кривая 2-значения λ_s у образцов медленно охлажденных до комнатной температуры (или закаленных от $T_з \leqq 200$ °C) и кривая 3-значения $|\lambda_s|$ в максимуме (т. е. соответствующие минимуму на кривой $\lambda_s(T_з)$ Рис. 1).

Отметим, что значения λ_s Ni–Co-ферритов разного состава представленные кривой 2 Рис. 3 согласуются с данными приведенными в работе [3], которые получены на образцах прошедших ТМО, причем направление поля при измерении было перпендикулярно направлению поля при отжиге. Упомянутые данные фактически фиксируют величину $\frac{\Delta l}{l} = \lambda_s(1 + \mu)$, где μ равно отношению поперечной магнитострикции насыщения к продольной.

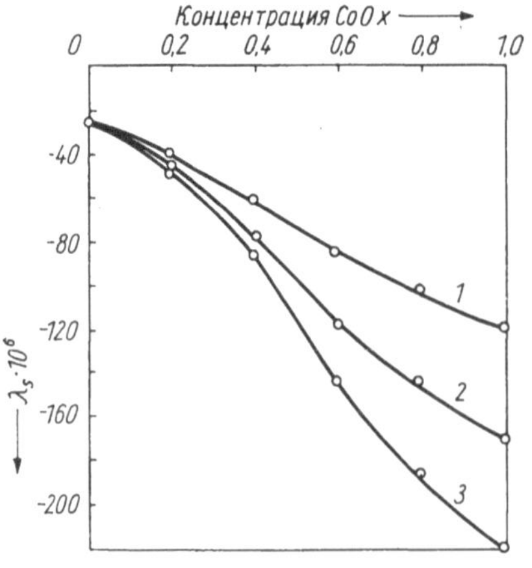

Рис. 3. Концентрационная зависимость магнитострикции насыщения λ_s у Ni–Co-ферритов в различных состояниях:
1. после закалки от высоких температур ($T_з > \theta$)
2. после медленного охлаждения до комнатной температуры
3. максимальные значения $|\lambda_s|$

Рис. 4. Зависимость продольной магнитострикции от напряженности магнитного поля у Ni-Co-ферритов различного состава: 1) $x=1{,}0$; 2) $x=0{,}8$; 3) $x=0{,}6$; 4) $x=0{,}4$; 5) $x=0{,}2$

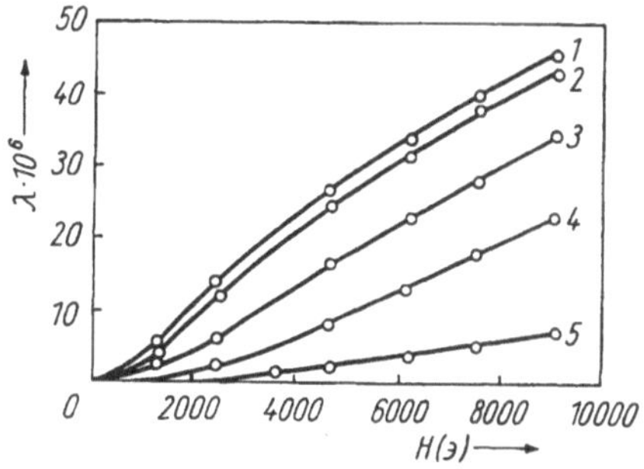

Сравнивая Рис. 2 и кривые 2 и 3 Рис. 3, можно заметить, что наибольшее изменение как максимальных значений $|\lambda_s|$ так и λ_s у медленно охлажденных образцов наблюдается для ферритов того состава, при котором происходит наибольшее изменение постоянной решетки.

4. Так как кобальт содержащие ферриты хорошо поддаются ТМО только в „окисленном" состоянии [2], то для исследования эффекта ТМО использовались образцы в состоянии, которое получается после медленного охлаждения до комнатной температуры в окислительной среде (на воздухе). Использовались образцы первоначально служившие для определения зависимости λ_s от $T_з$, которые в исходном состоянии (до ТМО) имели значения λ_s представленные на кривой 2 Рис. 3.

После ТМО (режим ТМО описан в [2]) все образцы (за исключением чисто никелевого феррита) имели значение λ_s равное или близкое нулю. На основании этого можно предположить, что в образцах установилась предельная одноосная магнитная текстура. Однако, было замечено, что изменение свойств ферритов в процессе ТМО продолжается и после того, как λ_s уменьшилось до нуля. Увеличение времени ТМО (например, с 3 до 6 часов) приводит к тому, что образцы после указанной обработки обнаруживают положительную по знаку продольную магнитострикцию, которая возрастает с увеличением поля и не обнаруживает склонности к насыщению вплоть до полей достигающих 16000 эрстед.[1])

На Рис. 4 представлена зависимость указанной продольной магнитострикции от напряженности поля для образцов различного состава после ТМО при 300 °C в течение 6 часов. Зависимость λ от H несколько отличается от линейной, особенно при малых и наибольших полях. Средний наклон кривых увеличивается с возрастанием содержания кобальта в феррите. Положительная магнитострикция возникает и после ТМО при более высоких температурах, однако в этом случае она накладывается на имеющуюся об'емную магнитострикцию в исходном (до ТМО) состоянии образца.

Подобные эффекты насколько известно в литературе ещё описаны не были и об'яснение их в настоящее время затруднительно.

5. Для определения констант магнитострикции рентгенографическим способом, использовались образцы медленно охлажденные до

---

[1]) При этом значении поля продольная магнитострикция у чисто кобальтового феррита получается разной $\lambda = +70 \cdot 10^{-6}$.

комнатной температуры и прошедшие ТМО. После ТМО все образцы (кроме $NiFe_2O_4$) имели значения λ_s равные нулю. В этом случае остаточная деформация кристаллической решетки в направлении оси магнитной текстуры равна полной магнитострикционной деформации.

Фиксируемая рентгенографически полная магнитострикционная деформация у кубического поликристалла по нормали к плоскости с индексами $h_1\,h_2\,h_3$ описывается формулой [2]

$$\left(\frac{\Delta d}{d}\right)_{h_1 h_2 h_3} = -\frac{a_1}{3} + a_2 \sin^2\theta + (a_1 - a_2) \times \\ \times \left[\left(b_1^2 + \frac{c_1^2}{2}\right)(\beta_1^2 - \beta_3^2) + \left(b_2^2 + \frac{c_2^2}{2}\right)(\beta_2^2 - \beta_3^2) + \beta_3^2\right]. \quad (1)$$

Здесь d — межплоскостное расстояние, $\left(\frac{\Delta d}{d}\right)_{h_1 h_2 h_3}$ — деформация в направлении нормали к плоскости $(h_1\,h_2\,h_3)$, $\beta_i = \frac{h_i}{\sqrt{\Sigma h_i^2}}$, $b_i = \beta_i \sin\theta$, $c_i = \sqrt{1-\beta_i^2}\cos\theta$, θ — брегговский угол, a_1 и a_2 — константы анизотропии магнитострикции.

Легко видеть, что определение константы $\lambda_{[100]} = \frac{2}{3}a_1$ по формуле (1) упрощается, если использовать отражения от плоскостей типа $(h\,0\,0)$. В этом случае $\beta_1 = 1$, $\beta_2 = \beta_3 = 0$, а сумма

$$S = \left(b_1^2 + \frac{c_1^2}{2}\right)(\beta_1^2 - \beta_3^2) + \left(b_2^2 + \frac{c_2^2}{2}\right)(\beta_2^2 - \beta_3^2) + \beta_3^2 = \sin^2\theta\,. \quad (2)$$

Поэтому выражение (1) преобразуется к виду

$$\lambda_{[100]} = \frac{2}{3}\,\frac{\Delta d/d}{\sin^2\theta - \frac{1}{3}}\,. \quad (3)$$

Видно, что $\lambda_{[100]}$ определяется по данным о смещении одной линии рентгенограммы. Для определения константы $\lambda_{[111]} = \frac{2}{3}a_2$ необходимо в общем случае знать смещение двух линий, составить и решить два уравнения.

Принципиально возможно подобрать такие условия получения рентгенограмм, при которых $\lambda_{[111]}$ также может быть определено по смещению одной линии. Для этого необходимо, чтобы сумма S оказалась равной 1/3. Такой подбор сложен, так как S зависит от многих параметров. Кроме того, данный прием может оказаться малоэффективным, если значение $\lambda_{[111]}$ мало. В этом случае смещение линии также будет незначительно, а так как измерение смещения линии производится с определенной погрешностью, то относительная ошибка в определении $\lambda_{[111]}$ может быть велика.

Если же величина λ_s определена с достаточной точностью и известна константа $\lambda_{[100]}$ можно рассчитать $\lambda_{[111]}$ воспользовавшись соотношением

$$\lambda_s = \frac{2\,\lambda_{[100]} + 3\,\lambda_{[111]}}{5}\,. \quad (4)$$

Рис. 5. Концентрационная зависимость констант магнитострикции $\lambda_{[100]}$ и $\lambda_{[111]}$ у Ni–Co-ферритов: 1. $\lambda_{[111]}$, 2. $\lambda_{[100]}$. (Светлые кружки — значения констант определенные непосредственно из рентгенографических данных, зачерненные — расчитанные по формуле (4))

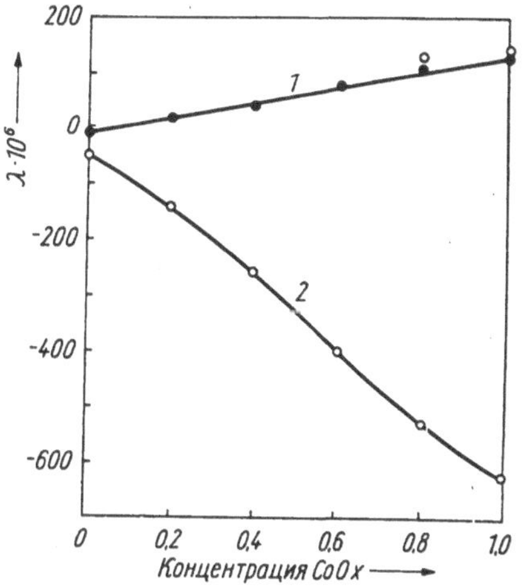

На Рис. 5 (кривая 2) представлена концентрационная зависимость константы $\lambda_{[100]}$ определенной из рентгенографических данных (отражения от плоскостей (800), $\theta \approx 67°$) по формуле (3). Значение $\lambda_{[100]}$ для чисто никелевого феррита получено экстраполяцией кривой 2 до пересечения её с осью ординат.

Из Рис. 5 видно, что $\lambda_{[100]}$ возрастает с увеличением содержания CoO в феррите от значения $\lambda'_{[100]} = -50 \cdot 10^{-6}$ для $NiFe_2O_4$ до $\lambda''_{[100]} = -630 \cdot 10^{-6}$ для $CoFe_2O_4$.

Константа $\lambda_{[111]}$ у ферритов различного состава определена расчетным путем с использованием соотношения (4). Найденные значения константы $\lambda_{[111]}$ (кривая 1, Рис. 5) практически ленейно изменяются с содержанием CoO в феррите от $\lambda'_{[111]} = -10 \cdot 10^{-6}$ для $NiFe_2O_4$ до $\lambda''_{[111]} = 130 \cdot 10^{-6}$ для $CoFe_2O_4$.

Сравнивая Рис. 3 (кривая 2) и Рис. 5 (кривая 2) можно заметить, что характер зависимости λ_s от содержания CoO в феррите, а также величина и знак λ_s в основном определяется константой $\lambda_{[100]}$.

Значения $\lambda_{[100]}$ и $\lambda_{[111]}$ представленные на Рис. 5 соответствуют состоянию Ni–Co-ферритов, получающемуся после медленного охлаждения их от высоких температур ($T > \theta$) до комнатной ($T_з \leqq 200$ °C). Иному состоянию ферритов должны соответствовать другие значения констант.

У Ni–Co-ферритов возможно определение констант магнитострикции рентгенографическим способом для состояний получающихся при закалке от температур выше 200 °C, так как после ТМО при этой и более высоких температурах λ_s равно нулю.

Наибольший интерес представляет состояние ферритов, при котором $|\lambda_s|$ имеет максимальное значение. Поэтому были выполнены определения $\lambda_{[100]}$ и $\lambda_{[111]}$ у образцов прошедших длительный отжиг и ТМО при тех температурах, закалка от которых фиксирует состояние ферритов с $|\lambda_s|_{max}$.

Константы $\lambda_{[100]}$ определенные рентгенографическим способом получились увеличенными для указанного состояния ферритов по сравнению с их значениями представленными на Рис. 5, в k раз, где k

равно отношению $|\lambda_s|_{max}$ к λ_s в состоянии с $T_з \leqq 200$ °C для феррита соответствующего состава.

У чисто кобальтового феррита $\lambda_{[100]} = -630 \cdot 10^{-6}$ ($T_з \leqq 200$ °C), а величина $(\lambda_{[100]})_{max} = -810 \cdot 10^{-6}$, поэтому $k \approx 1{,}3$. Отношение значений λ_s для этих состояний феррита также равно 1,3. Для феррита состава $x = 0{,}8$ получилось $\lambda_{[100]} = -530 \cdot 10^{-6}$ и $(\lambda_{[100]})_{max} = -670 \cdot 10^{-6}$, следовательно $k = 1{,}26$. Такое же значение k получается и из отношения λ_s и $(\lambda_s)_{max}$.

Значения констант $(\lambda_{[111]})_{max}$ можно и в этом случае определить по формуле (4) с большей точностью, чем рентгенографически.

Результаты

1. Исследовано изменение магнитострикции насыщения у Ni–Co-ферритов различного состава после закалки от разных температур. Закалка от высоких температур ($T_з > \theta$) фиксирует содержание кислорода в феррите близкое к стехиометрическому. В этом состоянии ферриты обнаруживают минимальную (измеренную при комнатной температуре) величину $|\lambda_s|$. При закалке от температур, лежащих ниже точки Кюри, $|\lambda_s|$ аномально возрастает и при $T_з$ приблизительно на 100° ниже точки Кюри достигает максимального значения, после чего снова уменьшается и при $T_з \leqq 200$ °C остается постоянной.

Аномальному изменению λ_s сопутствует уменьшение постоянной решетки (получаемое также при медленном охлаждении образца в указанном температурном интервале), которое обусловлено растворением избыточного кислорода в решетке феррита. Наибольшее изменение λ_s наблюдается у ферритов того состава, при котором происходит наиболее значительное изменение постоянной решетки (у ферритов с наибольшим содержанием кобальта). Эти ферриты растворяют большее, по сравнению с другими, количество кислорода.

2. Обнаружено, что при длительной ТМО, после того, как $|\lambda_s|$ уменьшилось до нуля, у Ni–Co-ферритов продолжают развиваться процессы, приводящие к возникновению значительной об'емной магнитострикции, даже если в исходном (до ТМО) состоянии она отсутствовала.

3. По изменению мехплоскостных расстояний кристаллической решетки у образцов прошедших ТМО определена концентрационная зависимость константы магнитострикции $\lambda_{[100]}$ для одного из состояний ферритов. Полученные этим путем величины $\lambda_{[100]}$ и измеренные тензометрическим способом значения λ_s использованы для расчета констант $\lambda_{[111]}$ у ферритов различного состава.

Литература

[1] Б. Я. Пинес и Н. М. Гумен, Физ. тверд. тела **5**, 3485 (1963).
[2] Б. Я. Пинес и Н. М. Гумен, Кристаллография **6**, 901 (1961).
[3] R. Weil, C. R. Acad. Sci. France **234**, 1351 (1952).

(Received December 14, 1964)

phys. stat. sol. 8, 759 (1965)

Институт физики металлов АН СССР, Свердловск

К квантовой теории термогальваномагнитных явлений в металлах и полупроводниках

III. Сравнение теории с экспериментом

Г. И. Гусева и П. С. Зырянов

Произведено сравнение теории [1, 2] с экспериментами Стила и Бабискина [3], Пури и Джеблла [4] по изучению термогальваномагнитных явлений в квантующем магнитном поле.

The theory [1, 2] is compared with experiments of Steele and Babiskin [3] and Puri and Geballe [4] on thermogalvanomagnetic phenomena in the high magnetic field quantum limit.

1. Введение

В предыдущих работах [1, 2] (впоследствии цитируемых как I и II) была предложена квантовая теория термогальваномагнитных явлений в металлах и полупроводниках. Настоящее сообщение посвящается теоретическому анализу имеющихся экспериментальных данных по термомагнитным явлениям. С сожалением приходится отмечать, что в настоящее время мы располагаем весьма незначительным экспериментальным материалом по термомагнитным явлениям в квантующем магнитном поле, тогда как по гальваномагнитным явлениям имеется большое количество тщательно выполненных экспериментов.

Термомагнитные явления в квантующем магнитном поле (теплопроводность, дифференциальная термоэлектродвижущая сила — ТЭДС) впервые экспериментально изучались в работе Стила и Бабискина [3] на очень чистых монокристаллах висмута.

В этой работе ими были установлены весьма общие закономерности, связывающие осцилляции коэффициентов ТЭДС, тепло- и электросопротивления при изменении магнитного поля, а также исследована температурная зависимость ТЭДС. Найденные в этих экспериментах закономерности можно объяснить исходя из весьма общих свойств электронного газа, сохраняющихся практически в любой реалистической модели металла. Именно поэтому опыты Стила и Бабискина [3] представляют наибольший интерес для теории. В § 2 будет показано, что опыты [3] можно объяснить исходя из двух предположений: а) сильное вырождение электронного газа и б) уширение энергетических уровней электронов в квантующем магнитном поле, обусловленное упругим рассеянием электронов на примесях и дефектах, является доминирующим.

Пури и Джеблл [4] экспериментально изучали ТЭДС в квантующем магнитном поле на образцах n-Ge. В этих опытах весьма существенным и оказались процессы неупругого рассеяния электронов фононами и эффекты взаимного увлечения электронов и фононов. Поэтому, как нам представляется, опыты Пури и Джеблла [4] могут послужить проверкой правильности теоретических представлений, положенных в основу расчёта II объемных плотностей тока проводимости и потока тепла, обусловленных силами статистической природы-градиентами температуры и химического потенциала.

В § 3 основные положения теории II распространяются на многодолинный анизотропный энергетический спектр электронов и учитывается эффект взаимного увлечения электронов и фононов. При этом в теории появляется чисто фононная характеристика — время релаксации длинноволновых фононов на коротковолновых тепловых.

В § 4 будет показано, что опыты Пури и Джеббла [4] хорошо объясняются теорией II, если положить в основу расчёта механизм релаксации длинноволновых фононов на коротковолновых, предложенный Саймонсом [5]. Суть дела при этом, заключается в том, что при рассеянии длинноволновых фононов на коротковолновых нельзя пренебрегать декрементом затухания последних, поскольку он оказывается величиной такого же порядка, что и частота колебаний длинноволнового фонона. Учёт этого обстоятельства допускает взаимодействие трех фононов, принадлежащих к одной ветви продольных колебаний. Эти процессы, как оказывается, вносят основной вклад в релаксацию продольных длинноволновых фононов на коротковолновых и приводят к температурной и частотной зависимости времени релаксации продольных фононов такой же, как и для длинноволновых поперечных фононов.

Таким образом, теоретический анализ экспериментов Пури и Джеблла может осветить интересную проблему о релаксации длинноволновых фононов.

2. Термомагнитные явления в вырожденном Ферми газе в приближении упругого рассеяния

В основу расчёта любого термогальваномагнитного коэффициента, как указано в книге Ландау и Лифшица [6], должны быть положены объемные плотности потока тепла Q_T и тока проводимости j_{np}

$$(j_{np})_i = \sigma_{ik} E_k - \beta_{ik} \nabla_k T\,, \qquad (Q_T)_i = \chi_{ik} E_k - \varkappa_{ik} \nabla_k T\,, \tag{2.1}$$

где

$$\chi_{ik} = T \beta_{ik}\,, \qquad E_k = -\nabla_k \left(\varphi + \frac{\zeta}{e}\right),$$

φ — потенциал электрического поля, а
ζ — химический потенциал.

В I были найдены общие формулы для кинетических коэффициентов, входящих в (2.1). Недиагональные компоненты этих тензоров, не зависящие от рассеяния, выражаются через термодинамические функции электронного газа, поэтому они сохраняют свой вид и в случае более общей модели с произвольной зависимостью энергии электронов от квазиимпульса. Это утверждение относится и к структуре формул для диагональных компонентов тензоров кинетических коэффициентов, обусловленных рассеянием.

Ниже будем рассматривать все тензоры кинетических коэффициентов в первом неисчезающем по вырождению приближении. Согласно I (см. формулы (14), (17) и (18)) находим в этом приближении[1])

$$\sigma_{xy} = c\,e\,N/H\,, \tag{2.2}$$

$$\beta_{xy} = T^{-1}\chi_{xy} = \frac{c}{H}\,S\,, \tag{2.3}$$

$$\varkappa_{xy} = \frac{\pi^2}{3}\left(\frac{k}{e}\right)^2 T\sigma_{xy}, \tag{2.4}$$

где $S = \frac{\pi^2}{3}\,\frac{2\,T\sqrt{2\,m}}{(2\,\pi\,\alpha)^2\,\hbar}\sum\limits_{u}[\zeta - E_n]^{-1/2}$, N — плотность электронов, E_n — энергия поперечного движения электрона в магнитном поле, равная для квадратичного закона дисперсии $\hbar\,\Omega\left(n+\frac{1}{2}\right)$. Поскольку $N = -\left(\frac{\partial\Phi}{\partial\zeta}\right)_{T,\,H}$, а $S = -\left(\frac{\partial\Phi}{\partial T}\right)_{\zeta,\,H}$ (Φ — термодинамический потенциал единицы объема), то

$$\beta_{xy} = \left(\frac{\partial}{\partial T}\right)_{\zeta_0}\int\limits_{-\infty}^{\zeta_0} d\zeta\,\frac{\sigma_{xy}}{e}\,. \tag{2.5}$$

Диагональные компоненты тензоров кинетических коэффициентов [см. I, формулы (22) и (23)] в пренебрежении спином можно представить в следующем виде[2])

$$\begin{pmatrix}\sigma_{xx}\\ \beta_{xx}\\ \varkappa_{xx}\end{pmatrix} = \frac{1}{(2\,\pi)^2\,T}\left(\frac{\alpha}{2\,\pi\,\hbar}\right)^2\sum_{n,\,n'}\int dp_z\,d\boldsymbol{q}\;|C_q\,q_y|^2\,F_{n'n}\left(\frac{\alpha^2\,q_\perp^2}{2}\right)\times$$

$$\times\, f_0\,(1-f_0)\,\delta\,(E\,(n', p_z+\hbar\,q_z) - E(n, p_z))\begin{pmatrix} e^2\\ e\,\dfrac{[E(n,p_z)-\zeta_0]}{T}\\ \dfrac{[E(n,p_z)-\zeta_0]^2}{T}\end{pmatrix}. \tag{2.6}$$

Из (2.5) и (2.6) в первом неисчезающем по вырождению приближении вытекает закон Видемана-Франца

$$\varkappa_{ik} = \frac{\pi^2}{3}\left(\frac{k}{e}\right)^2 T^0\,\sigma_{ik}\,. \tag{2.7}$$

Здесь T^0 в градусах, k — постоянная Больцмана. Если $|C_q|^2$ не зависит от температуры, то, как это видно из (2.6), β_{xx} также можно представить с помощью фомулы (2.5)

$$\beta_{ik} = \left(\frac{\partial}{\partial T}\right)_{\zeta_0}\int\limits_{-\infty}^{\zeta_0} d\zeta\,\frac{\sigma_{ik}}{e}\,. \tag{2.8}$$

---

[1]) Формула (2.3) впервые была найдена Образцовым [7] (см. также работу Барьяхтара и Пелетминского [8], где найдены как недиссипативные, так и диссипативные потоки тепла и ток проводимости).

[2]) Все обозначения совпадают с использованными в I.

В опытах Стила и Бабискина [3] ТЭДС, теплопроводность и электросопрот ивление измерялись в адиабатических условиях, поэтому и коэффициент теплопроводности электронов поперёк магнитного поля следовало бы вычислять в этих условиях. Однако теплопроводность электронного газа поперёк магнитного поля составляет в условиях опыта [4] величину порядка одного процента от теплопроводности решетки. Поэтому адиабатические поправки к изотермическим термомагнитным коэффициентам будут малы и ими в дальнейшем мы будем пренебрегать.

Изотермический коэффициент теплопроводности поперёк магнитного поля $\varkappa_\perp$ при токе равном нулю выражается фомулой:

$$\varkappa_\perp = \varkappa_{xx} + H\,\mathfrak{N}\,\chi_{xy} - \chi_{xx}\,Q(H)\,, \tag{2.9}$$

где

$$H\,\mathfrak{N} = \frac{\sigma_{xx}\,\beta_{xy} - \beta_{xx}\,\sigma_{xy}}{\sigma_{xy}^2}\,, \tag{2.10}$$

$$Q(H) = \beta_{xy}/\sigma_{xy}\,, \tag{2.11}$$

$\mathfrak{N}$ — изотермический коэффициент Нернста, $Q(H)$ — изотермическая дифференциальная ТЭДС. Из (2.8) видно, что в первом неисчезающем по вырождению приближении $\beta_{ik} \sim T$, а $\chi_{ik} \sim T^2$, поэтому, согласно (2.9) — (2.11), лишь первое слагаемое в $\varkappa_\perp$ пропорционально первой степени T, а второе и третье пропорциональны T^3. Пренебрегая в $\varkappa_\perp$ членами, содержащими T^3, находим, что

$$\varkappa_\perp = \varkappa_{xx} = \frac{\pi^2}{3}\left(\frac{k}{e}\right)^2 T^0\,\sigma_{xx}\,. \tag{2.12}$$

При неслишком малых значениях квантового числа n Адамс и Голстейн [9] показали, что зависимость осцилляций σ_{xx} от механизма рассеяния не играет большой роли и осциллирующую часть электропроводности можно преставить в виде

$$\sigma_{\text{осц}} \sim \sigma_0\left(\frac{\hbar\,\Omega}{\zeta}\right)\delta^{-1/2}\,, \tag{2.13}$$

где σ_0 — электропроводность при $\hbar = 0$, а

$$\delta^{-1/2} = 2\sqrt{\pi}\sum_{M>0}(-1)^M\,(2\pi M)^{-1/2}\cos\left(\frac{2\pi M\zeta}{\hbar\,\Omega} - \frac{\pi}{4}\right). \tag{2.14}$$

Согласно (2.3) и (2.11), осциллирующая часть дифференциальной ТЭДС $Q(H)$ также при неслишком малых квантовых числах n может быть представлена формулой

$$Q_{\text{осц}} = Q(0)\left(\frac{\hbar\,\Omega}{\zeta}\right)^{1/2}\delta^{-1/2}\,, \tag{2.15}$$

где

$$Q(0) = \frac{1}{e}\,\frac{\pi^2}{2}\,\frac{k\,T^0}{\zeta} \quad \text{— ТЭДС при } \hbar = 0\,.$$

Таким образом из (2.12), (2.13) и (2.15) следует вывод, что фаза и период осцилляций электропроводности а также и злектросопротивления ϱ_{xx} (поскольку при $\Omega\,\tau \gg 1$, $\varrho_{xx} = \varrho_\perp \approx \sigma_{xx}/\sigma_{xy}^2$), дифференциальной ТЭДС и теплопроводности одинаковы. Это следствие теории согласуется с опытами Стила и Бабискина [3].

Максимальные значения амплитуд осцилляций ТЭДС, тепло- и электропроводности при низких температурах зависят от уширения энергетических уровней электронов, обусловленного столкновениями. Если доминирующую роль в уширении энергетических уровней играет механизм, не зависящий от температуры, то зависимость амплитуд осцилляций от магнитного поля не будет чувствительна к изменениям температуры. Действительно, согласно [9], максимальное значение $\delta^{-1/2}$ равно по порядку величины $\sqrt{\Omega\tau_c}$ (τ_c — параметр обрезания, связанный с размытием уровней вследствие столкновений), и если τ_c не зависит от T, то справедливо вышеприведенное утверждение о нечувствительности зависимости амплитуд осцилляций от магнитного поля к изменениям температуры.

Это согласуется с экспериментами Стила и Бабискина [3]. В этих экспериментах изучалась также температурная зависимость ТЭДС и зависимость от напряженности магнитного поля огибающей минимумов осцилляций ТЭДС. Линейная зависимость ТЭДС от T, наблюдаемая на опыте, сразу же вытекает из (2.3) и (2.11). Для вычисления же огибающей минимумов осцилляций ТЭДС необходимо сначала изучить зависимость от магнитного поля H уширения энергетических уровней, обусловленного рассеянием электронов. Отсутствие решения такой задачи ограничило нас лишь качественным сравнением теории I с опытами [3]. В заключение этого раздела отметим, что обнаруженные в опытах Стила и Бабискина [3] закономерности вытекают из весьма общих свойств электронного газа: а) сильное вырождение, б) независимость уширения энергетических уровней электронов от температуры. Именно поэтому можно надеяться, что подтверждаемые экспериментом следствия теории I будут сохраняться и в более реалистической модели висмута, на котором были выполнены эксперименты [3].

3. Ток проводимости в многодолинном анизотропном кристалле и ТЭДС

В опытах Пури и Джеблла [4] измерялось отношение

$$\frac{Q(H)-Q(0)}{Q(0)}=\frac{\Delta Q(H)}{Q(0)}, \tag{3.1}$$

$Q(0)$ — дифференциальная ТЭДС при $H=0$. Замечательным оказалось то, что теоретическая формула для $\frac{\Delta Q(H)}{Q(0)}$ содержит величины хорошо известные из других независимых экспериментов за исключением времени релаксации фононов. Поэтому изучение $Q(H)$ является новым эффективным методом исследования времени релаксации различных ветвей фононного спектра, наряду с опытами по поглощению ультразвука и измерению теплопроводности.

При вычислении тока проводимости, необходимого для определения $\frac{\Delta Q(H)}{Q(0)}$ учтём, что в многодолинном анизотропном полупроводнике типа n-Ge закон дисперсии для ν — го минимума в произвольной системе координат имеет вид:

$$E^{\nu}(k) = \frac{\hbar^2}{2\,m} \sum_{ik} L^{\nu}_{ik}\,(k_i - k^{\nu}_{0\,i})\,(k_j - k^{\nu}_{0\,j})\,, \tag{3.2}$$

где L^{ν}_{ik}/m — тензор обратных эффективных масс, k — квазиимпульс электрона, k^{ν}_0 — центр ν – го минимума, а m — масса свободного электрона.

Собственные значения соответствующего (3.2) гамильтониана при наличии магнитного поля вычислены в [10] и имеют вид

$$E^{\nu}(n, p_z) = \hbar\, \Omega^{\nu}_c \left(n + \frac{1}{2}\right) + \frac{\hbar^2\, k_z^2}{2\, m^{\nu}_z}\,, \tag{3.3}$$

Ω_c—циклотронная частота электрона с массой $m_c = m/\sqrt{M^{\nu}_{33}}$; $m^{\nu}_z = m\; M_{33}/D$ — эффективная масса вдоль z, D — детерминант матрицы L^{ν}_{ik}, M^{ν}_{ik} минор элемента L^{ν}_{ik}. Следуя методике, изложенной в II, найдем для диссипативного тока формулу

$$J^{\mathrm{st}}_x = e\,\alpha^2 \sum_{\nu, q, \lambda} \left[2\left(\mathrm{ch}\,\frac{\hbar\,\omega_{q\lambda}}{T} - 1\right)(\omega_{ef}\langle 1\rangle + \omega_{ff})\right]^{-1} \left\{q^{*\,\nu}_y \left[\hbar\, v^2_{\lambda}\, \omega^{\nu}_{ef}\langle 1\rangle\, \boldsymbol{q}\, \nabla\, \frac{1}{T} + \right.\right.$$
$$\left.\left. + \,\omega_{ff}\, \omega^{\nu}_{ef}\langle D^{\nu}_z\, \eta'\rangle\right] + \sum_{\mu} \omega^{\nu}_{ef}\langle D^{\nu}_z\, \eta'\rangle\; \omega^{\mu}_{ef}\langle 1\rangle\; (q^{*\,\nu}_y - q^{*\,\mu}_y)\right\}. \tag{3.4}$$

Здесь ν, μ — номера долины, $\omega_{q\lambda}$, v_{λ} — частота и фазовая скорость λ-моды, ω_{ff} — частота релаксации фононов на фононах и границах образца,

$$q^{*\,\nu}_y = q_y + q_z\, \frac{M^{\nu}_{23}}{M^{\nu}_{33}}\,, \qquad q^{*\,\nu}_x = q_x + q_z\, \frac{M^{\nu}_{13}}{M^{\nu}_{33}}\,, \qquad \omega_{ef}\langle 1\rangle = \sum_{\nu} \omega^{\nu}_{ef}\langle 1\rangle\,, \tag{3.5}$$

$$D_z\, \eta' = \alpha^2 \left\{\left(q^*_x\, \frac{\partial}{\partial y} + q^*_y\, \frac{\partial}{\partial x}\right)\left(\frac{E(n', p_z + \hbar\, q_z) - \zeta}{T}\right)\right\}, \tag{3.6}$$

$\omega^{\nu}_{ef}\langle\cdots\rangle$ — определено в II формулой (25), причём все величины относятся к ν-долине, а аргумент $|X|^2$ в $|C_{\boldsymbol{q}\,\lambda}\, I_{n'n}(X)|^2 = |C_{\boldsymbol{q}\,\lambda}|^2\, F_{n'n}(|X|^2)$ следует взять равным $|X|^2 = \alpha^2 \left\{\sum_{ik} L^{\nu}_{ik}\, q_i\, q_k - D\, q_z^2/M^{\nu}_{33}\right\} \Big/ 2\sqrt{M^{\nu}_{33}}\,$.

Явный вид $|C_{\boldsymbol{q}\,\lambda}|^2$ для различных мод вычислялся по методу Херринга и Фогта [11]. Ниже рассматривается случай, когда магнитное поле направлено вдоль оси четвертого порядка и сохраняется кубическая симметрия. Такая ориентация наиболее детально исследована в экспериментах [4]. Система координат совпадает с кристаллографическими осями. Значения $|M(\boldsymbol{q}\,\lambda)|^2 = \dfrac{2\,T}{\hbar\,\omega_{\boldsymbol{q}\,\lambda}}\, |C_{\boldsymbol{q}\,\lambda}|^2$, усредненные по азимутальному углу и зависящие от угла Θ между $\boldsymbol{q}$ и $\boldsymbol{k}^{\nu}_0$ взяты из таблицы V работы [11] равными

$$\langle |M(\boldsymbol{q}, l)|^2\rangle = \frac{T\, C_2^2}{V\, C_l}\, G_l\,(\cos^2\Theta)\,, \tag{3.7}$$

$$\langle |M(\boldsymbol{q}, t_1)|^2 + |M(\boldsymbol{q}, t_2)|^2\rangle = \frac{T\, C_2^2}{V\, C_t}\, G_t\,(\cos^2\Theta)\,, \tag{3.8}$$

где C_l и C_t — средние значения упругих постоянных для продольных и поперечных волн, определяемые через компоненты тензора упругих постоянных формулами

$$C_l = C_{12} + 2\,C_{44} + \frac{3}{5}\,C^*, \qquad C_t = C_{44}\,\frac{(C_{44} + C^*/3)}{(C_{44} + C^*/8)}, \tag{3.9}$$

$C^* = C_{11} - C_{12} - 2\,C_{44}$ — характеризуют анизотропию упругости, а C_1 и C_2 — константы потенциала деформации,

$$\left.\begin{aligned} G_l &= \left(\frac{C_1}{C_2} + \cos^2\Theta\right)^2 \left[1 + \frac{2}{3}\,\frac{C^*}{C_l}\,(0{,}15 - 1{,}50\cos^2\Theta + 1{,}75\cos^4\Theta)\right], \\ G_t &= \cos^2\Theta\,(1 - \cos^2\Theta)\left(1 + \frac{3}{8}\,\frac{C^*\cos^2\Theta}{C_{44} + C^*/8}\right). \end{aligned}\right\} \tag{3.10}$$

В системе кристаллографических осей угол Θ между $\boldsymbol{q}(\vartheta, \varphi)$ и $\boldsymbol{k}_0^\nu(\vartheta_\nu, \varphi_\nu)$ равен

$$\cos\Theta = \cos\vartheta\cos\vartheta_\nu - \sin\vartheta\sin\vartheta_\nu\cos(\varphi - \varphi_\nu)\,. \tag{3.11}$$

Результаты, найденные в серии работ Самойловича и др. [12, 13] для упругого анизотропного рассеяния электронов на акустических фононах хорошо согласуются с [11], хотя в [11] и в [12, 13] используются различные методы апроксимации спектра фононов. В частности, в [13] показано, что анизотропный кристалл германия с достаточной точностью можно рассматривать как упругоизотропный, если надлежащим образом усреднить упругие постоянные. Введенные в [13] константы C'_{11} и C'_{44} хорошо согласуются со средними упругими постоянными C_l и C_t, определенными в (3.9). Поэтому в дальнейшем спектр фононов будем приближенно определять по формуле

$$\omega_{q\lambda} = q\sqrt{\frac{C_\lambda}{\varrho}}, \tag{3.12}$$

ϱ — плотность вещества.

Изотермическая дифференциальная ТЭДС складывается из двух частей: электронной и фононной. Электронная часть ТЭДС была вычислена для n-Ge в [10] и в условиях опытов Пури и Джеблла [4] оказывается малой по сравнению с фононной ТЭДС, обусловленной взаимным увлечением электронов и фононов.

В сильных магнитных полях ($\Omega\tau \gg 1$) ТЭДС увлечения $Q^{(p)}(H)$ определяется отношением коэффициента β_{xy}^{cm} в диссипативном токе (3.4) к $\sigma_{xy} = \dfrac{c\,e\,N}{H}$. Учитывая, что при $H \| [001]$ $E^\nu(n, p_z)$ одинаковы для всех долин, находим, что отличный от нуля вклад в β_{xy}^{st} дает лишь первое слагаемое в фигурной скобке (3.4)

$$\beta_{xy}^{st} = e\,\frac{\alpha^2 K \hbar}{T^2} \sum_{q,\lambda} \frac{v_\lambda^2\,\omega e_f^{(1)}\,\langle 1\rangle \left(q_y^2 + \dfrac{M_{23}^{(1)}}{M_{33}^{(1)}}\,q_y\,q_z\right)}{2\left(\operatorname{ch}\dfrac{\hbar\,\omega_{q\lambda}}{T} - 1\right)\left(K\,\omega_{ef}^{(1)}\langle 1\rangle + \omega_{ff}\right)}, \tag{3.13}$$

где K — число долин. В дальнейшем будем опускать индекс долины, предполагая что все параметры относятся к первой долине.

В зависимости от соотношения ω_{ff} и ω_{ef} различают

1. случай слабого увлечения $\omega_{ef} \ll \omega_{ff}$,
2. случай полного увлечения (или „насыщения" по Херрингу), когда $\omega_{ef} \gg \omega_{ff}$.

В квантующем магнитном поле основной вклад во взаимодействие с электронами вносят фононы с $q \sim q_H = \frac{1}{\alpha} = \sqrt{\frac{|e|\,H}{c\,\hbar}}$, при этом можно рассматривать следующие предельные случаи:

а) $q_H^2 \ll \tilde{q}_\lambda^2$, $\left(\tilde{q}_\lambda = \frac{T}{\hbar\, v_\lambda}\right)$, не очень сильные поля,

б) $q_H^2 \gtrsim \tilde{q}_\lambda^2$ предельно сильные поля.

В n-Ge при $T^0 \sim 20 \div 100\,^\circ\mathrm{K}$ и практически доступных значениях магнитного поля реализуется случай 1, а). Об этом случае ($\omega_{ef} \ll \omega_{ff}$ и $q_H^2 \ll \tilde{q}_\lambda^2$) и будет итти речь. Переходя в формуле (3.13) от суммирования к интегрированию и используя явный вид $\omega_{ef}^{(1)}\langle 1\rangle$, согласно II (25), находим для λ моды

$$\beta_{xy}^\lambda = \frac{e\,V\,K\,\hbar\,v_\lambda^2}{T^2\,(2\,\pi)^4} \int d\boldsymbol{q}\,dp_z \omega_{ff}^{-1}\,F_{00}\,|M(\boldsymbol{q}\,\lambda)|^2\,f_0 \left(q_y^2 + \frac{M_{23}}{M_{33}}\,q_y\,q_z\right) \times$$
$$\times\, \delta\,(E\,(p_z + \hbar\,q_z) - E(p_z) - \hbar\,\omega_q)\,. \tag{3.14}$$

В самом общем виде зависимость ω_{ff} от q и T, согласно [14], можно представить в виде

$$\omega_{ff}(q,\lambda) = \tau^{-1}(q,\lambda) = A_\lambda\,T^{3-S_\lambda-\gamma}\,q^{2+S_\lambda}\,,$$

где

$$A_\lambda = \frac{Z_\lambda\,P^{\gamma/5}}{\varrho\,\hbar^{2-S_\lambda}\,v_\lambda^{3-S_\lambda}}\,,$$

Z_λ — некоторое число, а

$$P = \varrho\,\hbar^{2-S_\lambda}\,v_\lambda^{4-S_\lambda}/L\,,$$

L — размер кристалла. Значения параметров S_λ и γ зависят от механизма релаксации фононов. Интегрируя в (3.14) по p_z, находим

$$\beta_{xy}^\lambda = \frac{e\,v_\lambda^2\,m_z\,K\,\exp\{\zeta'(H)/T\}}{16\,\pi^4\,\hbar^2\,A_\lambda\,C_\lambda\,T^{2-S_\lambda-\gamma}\,\alpha^{2-S_\lambda}} \left(\frac{C_2}{T}\right)^2 Y_\lambda\,, \tag{3.15}$$

где $\zeta'(H) = \zeta(H) - \hbar\,\Omega_c/2$, а Y_λ — безразмерный интеграл.[3]) Для параметров этого интеграла, соответствующих $H \| [001]$, находим

$$Y_\lambda = \frac{\pi\,\Gamma\left(\frac{2-S_\lambda}{2}\right)}{2\sqrt{a^{2-S_\lambda}}}\,F_\lambda\,, \tag{3.16}$$

[3]) Недостаток места не позволяет нам остановиться на вычислении Y_λ.

где $\Gamma(t)$ — гамма функция, $a = L_{11}/2\, M_{33}^{1/2}$, $F_t \approx 0{,}1\, R_t$; $F_z \approx 0{,}06\, R_z$

$$R_\lambda = \ln \left(\frac{\alpha\, T}{\hbar\, v_\lambda}\right)^2 \frac{9}{1 + 17{,}6\; T/\hbar\, \Omega_c} - 2\, C - \Psi\left(\frac{2 - S_\lambda}{2}\right),$$

C — постоянная Эйлера, $\Psi(x)$ — пси-функция Эйлера. Для H, удовлетворяющих неравенству $17{,}6\; T/\hbar\, \Omega_c \gg 1$, R_λ не зависит от H и равен

$$R_\lambda \approx \ln \frac{T}{2\, m_c\, v_\lambda^2} - 2\, C - \Psi\left(\frac{2 - S_\lambda}{2}\right).$$

Подставив (3.16) в (3.15) и поделив результат на σ_{xy}, находим для ТЭДС увлечения[4])

$$Q_\lambda^{(p)}(H) = \frac{\Gamma\left(\frac{2 - S_\lambda}{2}\right)}{e\, 2^{7/2}\, \pi^{3/2}\, 2^{S_\lambda/2}} \left(\frac{M_{33}}{D}\right)^{1/2} \frac{1}{\sqrt{L_{11}^{2-S_\lambda}}} \left(\frac{\hbar\, \Omega_c}{T}\right)^{\frac{2-S_\lambda}{2}} \left(\frac{m\, v_\lambda^2}{T}\right)^{\frac{3-S_\lambda}{2}} \times$$

$$\times \left(\frac{C_2}{T}\right)^2 \frac{T^\gamma\, F_\lambda}{P^{\gamma/5}\, C_\lambda}. \tag{3.17}$$

Оценка $Q_\lambda^{(p)}(H)$ и сравнение с $Q^e(H)$ из [10] показывает, что в условиях опытов Пури и Джеблла [4] $Q_\lambda^{(p)}(H) \gg Q^e(H)$.

Необходимую для вычисления $\frac{\Delta Q(H)}{Q(0)}$ величину $Q^{(p)}(0) = \sum_\lambda Q_\lambda^{(p)}(0)$ получим из общей формулы (A 8) работы Херринга [14] для $Q_\lambda^{(p)}(0)$ в многодолинном анизотропном кристалле:

$$Q_\lambda^{(p)}(0) = -\frac{9\, \Gamma\left(\frac{3 - S_\lambda}{2}\right) \chi_{\parallel}^\lambda\, D^{\frac{2+S_\lambda}{6}}}{\pi\, e\, (2 - S_\lambda) \sqrt{2^{5+3 S_\lambda}\, D\, p^{*\, 2+S_\lambda}} \left(2\, \alpha_1 \frac{\chi_{\parallel}^\lambda}{\chi_{\perp}^\lambda} + \alpha_3\right)} \left(\frac{C_2}{T}\right)^2 \left(\frac{m\, v_\lambda^2}{T}\right)^{\frac{3-S_\lambda}{2}} \frac{T^\gamma}{P^{\gamma/5}\, Z_\lambda}, \tag{3.18}$$

где $\chi_{\parallel, \perp}^\lambda$ — некоторые числа, определяемые по формулам (49) и (50) работы [11] для компонентов тензоров времени релаксации электронов, если последние представить в виде[5]):

$$\frac{1}{\tau_{\parallel, \perp}^\lambda} = \frac{3}{\pi} \left(\frac{m\, T}{2}\right)^{3/2} \frac{C_2^2\, \chi_{\parallel, \perp}^\lambda}{\hbar^4\, D^{1/2}\, C_\lambda} \left(\frac{E}{T}\right)^{1/2},$$

$$\alpha_1 = \frac{m}{m_\perp}, \qquad \alpha_3 = \frac{m}{m_\parallel}, \qquad \delta = \frac{\alpha_3}{\alpha_1},$$

p^* — некоторая константа, учитывающая анизотропию G_λ. Если G_λ изотропно, $p^* = 1$. Для акустического рассеяния, по оценкам

---

[4]) Переходя в (3.17) к изотропному рассеянию на продольных фононах ($S_\lambda = 0$) к квадратичному изотропному спектру электронов, получим результат работы Гуревича и Недлина [15].

[5]) Тензор времени релаксации электронов и тензор m записаны в системе главных осей долины.

Херринга [14], p^* порядка единицы. Величину p^* можно установить из сопоставления результатов теории с экспериментом. Отношение $Q_\lambda^{(p)}(H)$ из (3.17) к $Q_\lambda^{(p)}(0)$ из (3.18) равно

$$\frac{Q_\lambda^{(p)}(H)}{Q_\lambda^{(p)}(0)} = B\left(\frac{\hbar\,\Omega_c}{T}\right)^{\frac{2-S_\lambda}{2}}, \qquad (3.19)$$

где

$$B = \frac{\Gamma\left(\frac{2-S_\lambda}{2}\right)(2-S_\lambda)\,2^{S_\lambda}\,p^{*\frac{2+S_\lambda}{2}}\left(2\,\frac{\chi_{\parallel}^{\lambda}}{\chi_{\perp}^{\lambda}}+\delta\right)(1+2\,\delta)^{1/2}}{\sqrt{\pi}\;\Gamma\left(\frac{3-S_\lambda}{2}\right)\sqrt{3^{3+S_\lambda}\,(2+\delta)^{2+S_\lambda}\,\delta^{2+S_\lambda}}}\,\frac{F_\lambda}{\chi_{\parallel}^{\lambda}}, \qquad (3.20)$$

F_λ — содержит слабую логарифмическую зависимость от H и T.

4. Фонон-фононная релаксация ТЭДС увлечения в электронном германии

1. Из формулы (3.19) видно, что в квантовом пределе термоэдс, обусловленная увлечением электронов фононами λ — ветви, зависит от магнитного поля и температуры через параметры S_λ и γ, непосредственно связанные со временем релаксации фононов данной ветви. Экспериментально изучаемая величина

$$\frac{\Delta Q(H)}{Q(0)} = \sum_\lambda \Delta Q_\lambda(H)/\sum_\lambda Q_\lambda(0) \qquad (4.1)$$

содержит вклад от различных ветвей и более сложна для анализа. Сопоставим полученные теоретические результаты с экспериментом, во-первых, качественно сравнив характер зависимости $\frac{Q(H)}{Q(0)}$ от магнитного поля и температуры, и, во-вторых, количественно, поскольку в расчете учитывались детали, связанные с реальной структурой спектра электронов и колебаний решетки.

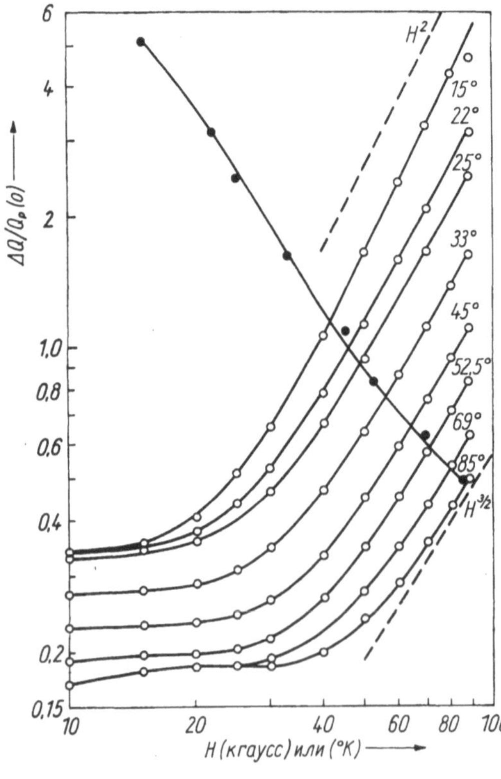

Рис.1. Зависимость $\frac{\Delta Q(H)}{Q(0)}$ от магнитного поля при различных температурах $H \parallel [100]$ и $\nabla T \parallel [010]$. Кривая, обозначенная ●-точками, отражает температурную зависимость $\frac{\Delta Q(H)}{Q(0)}$ при постоянном H (88 кгаусс)

На рис. 1 воспроизводятся экспериментальные результаты, полученные в [4], для зависимости $\frac{\Delta Q(H)}{Q(0)}$ от магнитного поля при различных температурах для $H \parallel [100]$ и $\nabla T \parallel [010]$. Измерения выполнены на чистых образцах электронного германия с концентрацией примеси $n_d \sim 10^{13}$ см$^{-3}$. На этом же графике приведена температурная зависимость термоэдс увлечения, полученная в поле $H = 88$ кгаусс. Из анализа экспериментальных данных вытекают следующие эмпирические формулы:
в интервале температур 20—80 °К

$$\frac{\Delta Q(H)}{Q(0)} = D_1 \left(\frac{H}{T}\right)^{3/2} \tag{4.2}$$

и для $T < 20$ °К

$$\frac{\Delta Q(H)}{Q(0)} = D_2 \left(\frac{H}{T}\right)^{2} \text{ }^{6)}. \tag{4.3}$$

D_1 и D_2 слабо меняются с температурой и полем.

В работе [14] приводятся формулы для времени релаксации длинноволновых фононов на тепловых в приближении трехфононных процессов, происходящих с выполнением законов сохранения энергии и импульса. При этом в кубическом кристалле при температуре значительно ниже дебаевской оказывается, что

$$\tau_e^{-1} = A_e q^2 T^3, \qquad \tau_t^{-1} = A_t q T^4. \tag{4.4}$$

Если температуры столь низки, что релаксация импульса фононов происходит за счет их рассеяния на границах образца, то $\tau_{\text{гр}}^{-1}$ не зависит от T и q и

$$\tau_{\text{гр}}^{-1} \sim \frac{v_\lambda}{L}. \tag{4.5}$$

Вклад той или иной λ моды в термоэдс, пропорциональный τ_λ, определяется соотношением соответствующих времен релаксации.

В условиях опыта [4] реализовались не слишком сильные магнитные поля $H < \frac{c\hbar}{|e|}\left(\frac{T}{\hbar v_\lambda}\right)^2$, когда во взаимодействии с электронами существенны фононы с $q < \left(\frac{|e| H}{c \hbar}\right)^{1/2}$. Согласно (4.4)

$$\frac{\tau_e(H)}{\tau_t(H)} = \left(\frac{v_e}{v_t}\right)^3 \left(\frac{T}{m v_t^2} \frac{T}{\hbar \Omega}\right) \gg 1 . \tag{4.6}$$

Из (4.6) следует, что существенный вклад в $\frac{Q(H)}{Q(0)}$ вносят лишь продольные фононы, и это приводит, согласно (3.19), к

$$\frac{Q(H)}{Q(0)} \sim \frac{H}{T}$$

---

[6]) Следует заметить, что при $T < 20$ °К график измерения температурной зависимости не является вполне надежным (слишком мало экспериментальных точек).

тогда как на опыте в интервале температур, где преобладает фонон-фононная релаксация, имеет место (4.2).

Последняя формула получилась бы только тогда, когда имело бы место неравенство, обратное (4.6), что противоречит условиям эксперимента [4]. Таким образом, τ_e и τ_t, вычисленные в приближении трехфононных процессов при условии выполнения закона сохранения импульса и энергии, невозможно согласовать с экспериментом [4], выполненным при наличии квантующего магнитного поля. Аналогичная ситуация возникала и ранее в работе Херринга и др. [16] при анализе экспериментов по термомагнитным явлениям в классической области полей.

Существуют и другие эксперименты [17] по изучению температурной зависимости τ_e и τ_t, которые также противоречат формулам (4.4). Если принять во внимание, что декремент затухания коротковолновых фононов может быть сравним с частотой колебаний длинноволновых, то закон сохранения энергии при взаимодействии длинноволновых фононов с коротковолновыми уже не будет выполнятьсея, и это, как показал Саймонс [5], приводит вместо формул (4.4) к следующим:

$$\tau_e^{-1} \sim q\,T^4\,, \qquad \tau_t^{-1} \sim q\,T^4\,. \tag{4.7}$$

Формулы (4.7) совместно с (3.19) приводят в температурном интервале, где преобладает фонон-фононная релаксация, к наблюдаемой на опыте зависимости (4.2). Это подтверждает представления о механизме релаксации длинноволновых фононов на коротковолновых, положенные в основу формулы (4.7).

Следует подчеркнуть, что результаты экспериментов [17] по изучению температурной зависимости поглощения продольного и поперечного эвука в диэлектриках, также хорошо согласуются с формулами (4.7).

При $T < 20$ °K формулы (4.5) и (3.19) приводят к

$$\frac{Q(H)}{Q(0)} \sim \left(\frac{H}{T}\right)^2,$$

что хорошо согласуется с экспериментальным результатом (4.3).

2. Результаты количественного сравнения теории с экспериментом для $\frac{Q(H)}{Q(0)}$ представлены в таблице 1. В области температур, где преобладает фонон-фононная релаксация (20—80 °K), мы воспользовались для τ_λ формулами Саймонса [5], полученными для изотропной среды, подставив в них средние упругие постоянные для германия, согласно (3.9). При этом оказалось, что $\frac{\tau}{\tau_t} = 0{,}3$. Для $T < 20$ °K использовали формулу (4.5). Поскольку теоретическая формула (3.19) для $\frac{Q_\lambda(H)}{Q_\lambda(0)}$ получена при $\frac{\hbar\,\Omega_c}{T} \gg 1$ $(n = 0)$, то наилучшего согласия теории с опытом следует ожидать именно в этой области H и T.

В интервале температур 20—80 °K теоретические формулы содержат неопределенную константу порядка единицы p^*. Выбирая p^* таким образом, чтобы точно совпадали теоретические и эксперимен-

Таблица 1
Сравнение теоретических и экспериментальных результатов для ТЭДС увлечения в n-Ge

T °K / H кгаусс	15		25		45		65		85	
	$\frac{Q(H)^{теор}}{Q(0)} \Big/ \frac{Q(H)^{эксп}}{Q(0)}$	$\frac{\hbar\Omega_c}{T}$	$\frac{Q(H)^{теор}}{Q(0)} \Big/ \frac{Q(H)^{эксп}}{Q(0)}$	$\frac{\hbar\Omega_c}{T}$	$\frac{Q(H)^{теор}}{Q(0)} \Big/ \frac{Q(H)^{эксп}}{Q(0)}$	$\frac{\hbar\Omega_c}{T}$	$\frac{Q(H)^{теор}}{Q(0)} \Big/ \frac{Q(H)^{эксп}}{Q(0)}$	$\frac{\hbar\Omega_c}{T}$	$\frac{Q(H)^{теор}}{Q(0)} \Big/ \frac{Q(H)^{эксп}}{Q(0)}$	$\frac{\hbar\Omega_c}{T}$
100	0,93	6,70	1,00	4,00	0,82	0,22	0,68	1,54	0,55	1,18
90	0,96	6,00	1,00	3,60	0,78	2,00	0,65	1,39	0,50	1,06
80	0,89	5,33	0,95	3,20	0,73	1,78	0,57	1,23	—	—
70	0,84	4,67	0,90	2,80	0,66	1,50	0,50	1,08	—	—
60	0,80	4,00	0,83	2,40	0,58	1,33	—	—	—	—
50	0,70	3,33	0,75	2,00	—	—	—	—	—	—

тальные значения $\frac{Q(H)}{Q(0)}$ при максимальном значении $\frac{\hbar\,\Omega_c}{T}$ $\Bigl($в данном эксперименте $\left(\frac{\hbar\,\Omega_c}{T}\right)_{\max} = 4\Bigr)$, получим $p^* = 0{,}8$. Это значение вполне согласуется с оценками Херринга [14] для небольшой анизотропии рассеяния.

Из таблицы видно, что асимптотическая формула для $\frac{\Delta Q(H)}{Q(0)}$ при $\frac{\hbar\,\Omega_c}{T} \gg 1$ находится в хорошем качественном и количественном согласии с опытом [4]. Этот вывод наряду с экспериментами [17] позволяет надеяться, что представления Саймонса, положенные в основу формул (4.7), и методика вычисления потоков, предолженная в I, II, правильны.

Литература

[1] П. С. Зырянов, phys. stat. sol. **6**, 401 (1964).
[2] П. С. Зырянов, phys. stat. sol. **7**, 223 (1964); Физ. металлов и металловедение **16**, 13 (1963).
[3] M. C. Steele and J. Babiskin, Phys. Rev. **98**, 359 (1955).
[4] S. M. Puri and T. H. Geballe, Phys. Rev. Letters **9**, 378 (1962).
[5] S. Simons, Proc. Phys. Soc. **83**, 749 (1964).
[6] Л. Д. Ландау и Е. М. Лифшиц, Электродинамика сплошных сред, Гостехиздат, Москва 1957.
[7] Ю. Н. Образцов, Физ. тверд. тела **6**, 414 (1964).
[8] В. Г. Барьяхтар и С. В. Пелетминский, Ж. экспер. теор. физ. (в печати).
[9] E. N. Adams and T. D. Holstein, J. Phys. Chem. Solids **10**, 254 (1959).
[10] Г. И. Гусева, Физ. металлов и металловедение **18**, 321 (1964).
[11] C. Herring and E. Vogt, Phys. Rev. **101**, 944 (1956).
[12] А. Г. Самойлович, И. Я. Коренблит, И. В. Даховский и В. Д. Искра, Физ. тверд. тела **3**, 2939 и 3285 (1961).
[13] А. Г. Самойлович и В. Д. Искра, Физ. тверд. тела **2**, 2827 (1960).
[14] C. Herring, Phys. Rev. **95**, 954 (1954); **96**, 1163 (1954).
[15] Л. Э. Гуревич и Г. М. Недлин, Физ. тверд. тела **3**, 2779 (1961).
[16] C. Herring, T. H. Geballe and J. E. Kunzler, Bell Syst. Techn. J. **38**, 657 (1959).
[17] H. E. Bömmel and K. Dransfeld, Phys. Rev. Letters **2**, 248 (1959) and Phys. Rev. **117**, 1245 (1960).
I. S. Ciccarello and K. Dransfeld, Phys. Rev. **134**, A 1517 (1964).

(Received December 16, 1964)

phys. stat. sol. 8, 773 (1965)

Institute for Atomic Physics, Bucharest

Electrical Properties of Non-Stoichiometric Nickel Oxide

By

M. NACHMAN[1]), L. N. COJOCARU, and L. V. RÎBCO

The temperature dependence of the electrical conductivity and of the Seebeck coefficient of NiO samples with various Ni^{3+} concentrations is determined. These samples are obtained from a nickel salt, using various decomposition temperatures.

The temperature dependence of the hole concentration, p_α, is calculated from the Seebeck coefficient. Since p_α is constant up to about 500 °K, the exponential temperature variation of the conductivity must be attributed to the temperature dependence of the mobility. It seems, therefore, that at least up to the Néel temperature non-stoichiometric NiO exhibits a similar conduction mechanism to Li-doped NiO, with holes hopping between Ni^{2+}-ions. The discrepancies between the values of the hole concentration determined from the Seebeck coefficient and those given by the chemical analysis are discussed in terms of a simplified model, taking into account the non-uniform distribution of the excess-oxygen atoms throughout the volume of the NiO grains. The results confirm that for NiO samples of high excess oxygen content the electrical conduction is mainly due to grain boundaries.

Es wurde die Temperaturabhängigkeit der elektrischen Leitfähigkeit und des Seebeck-Koeffizienten an NiO-Proben mit verschiedenen Ni^{3+}-Konzentrationen untersucht. Diese Proben wurden aus Nickelsalzen bei unterschiedlichen Zersetzungstemperaturen hergestellt. Die Temperaturabhängigkeit der Löcherkonzentration p_α wurde aus dem Seebeck-Koeffizienten bestimmt. Da p_α bis 500 °K konstant ist, muß die exponentielle Abhängigkeit der Leitfähigkeit von der Temperatur von der Temperaturabhängigkeit der Beweglichkeit herrühren. Es hat deshalb den Anschein, als ob mindestens bis zur Néel-Temperatur nichtstöchiometrisches NiO einen ähnlichen Leitfähigkeitsmechanismus besitzt wie Li-dotiertes NiO, bei dem Löcher von einem Li-Ion zu einem anderen springen. Die Diskrepanz zwischen den Werten für die Löcherkonzentrationen, die aus dem Seebeck-Koeffizienten und aus der chemischen Analyse erhalten wurden, werden in einem vereinfachten Modell diskutiert, wobei die ungleichmäßige Verteilung der überschüssigen Sauerstoffatome im Volumen der NiO-Körner berücksichtigt wurde. Das Ergebnis bestätigt, daß in NiO-Proben mit einem hohen Sauerstoffüberschuß die Korngrenzenleitfähigkeit überwiegt.

1. Introduction

The electrical properties of nickel oxide have been studied extensively by many investigators. Morin [1] and van Houten [2] have discussed the electronic structure of this oxide and have shown that NiO has no 3d band but full, localized $3d^8$ levels (Ni^{2+} levels) and empty $3d^9$ levels (Ni^+ levels).

According to van Houten [2] the $3d^9$ levels are situated 5.4 eV above the $3d^8$ levels which lie about 5.3 eV above the filled O^{2-} band.

The energy difference between the Ni^{2+} and Ni^+ levels is too large for pure stoichiometric NiO to behave like an intrinsic semiconductor at ordinary temperatures, and as electron transfer between the full localized $3d^8$ levels is not possible, pure NiO is an insulator at room temperature with a conductivity of $\sigma < 10^{-13}\ \Omega^{-1}\ cm^{-1}$.

[1]) Present address: Institute for Technical Documentation, Bucharest.

However, it becomes a semiconductor when acceptor levels, situated close above the $3d^8$ levels, are present. In this case electrons from the $3d^8$ levels can be excited into the acceptor levels, thus transforming some Ni^{2+} ions into Ni^{3+} ions. The holes on the $3d^8$ levels can migrate through the lattice, by hopping between Ni^{2+} ions. This migration gives rise to an electrical conduction, which may be regarded as a thermally activated diffusion of holes on Ni^{2+} levels [3].

According to Heikes and Johnston [3], the activation energy of the process is mainly connected with the self-trapping by the polarization, induced by the moving hole.

Acceptor levels are usually introduced by doping the nickel oxide with Li [4], i.e. by replacing some Ni^{2+} ions in the lattice by Li^+ ions. Each singly charged lithium ion is then compensated by a Ni^{3+} ion. The Ni^{3+} ions may be regarded as holes on the Ni^{2+} levels, and therefore, doping with Li gives p-type NiO.

It is, however, known that one can obtain Ni^{3+} ions in the lattice by creating Ni vacancies. Such vacancies occur in NiO with an excess of oxygen [5, 6]. To preserve the neutrality, two Ni^{3+} ions must be formed for each cation-vacancy, i.e. for each excess oxygen atom in the crystal. We will thus have $2\,n$ Ni^{3+} ions, i.e. $2\,n$ holes for n excess oxygen atoms in the lattice. Therefore NiO with an excess of oxygen behaves like a p-type semiconductor.

The oxygen content in a NiO sample depends on the salt from which it was prepared as well as on the heating temperature at which the salt vas heated to give NiO [6 to 9]. By varying the heating temperature of a given nickel salt, NiO samples with various Ni^{3+} concentrations may be obtained, without introducing foreign ions into the lattice.

The present investigation was undertaken in order to gain more knowledge about the electrical properties of non-stoichiometric NiO, prepared from nickel carbonate, using various decomposition temperatures.

2. Experimental

2.1 Sample preparation

Nickel oxide was prepared by thermal decomposition of chemically pure nickel carbonate. The salt was heated for 2 hours in air, at fixed temperatures ranging from 700 to 1300 °K. According to the heating temperature T_H five types of non-stoichiometric NiO were obtained (Table 1, columns 1 and 2). By pressing the NiO powder at a pressure of 1.5 t/cm², cylindrical samples resulted. Each sample was then heated for 20 hours in air at the same temperature at which it was prepared by the decomposition of $NiCO_3$. The mean densities ϱ of the samples after pressing and heating ranged from 5.2 to 5.9 g/cm³ (Table 1, columns 8 and 9).

The mean concentration of Ni^{3+} ions in nickel oxide was determined by a iodometric method. The values obtained with complete dissolution of the sample were close to those obtained with partial dissolution, except for the samples D and E, for which they were somewhat larger. The experimental results are summarized in Table 1, columns 3 and 4, where x represents the number of excess oxygen atoms per NiO molecule[2]), and p the mean hole concentration (i. e. the Ni^{3+} concentration) at about 300 °K (the temperature at which the chemical analysis was made). Iodometric determinations performed at two different temperatures (300 and 340 °K) gave essentially the same values for x.

[2]) This means that the molecular formula of non-stoichiometric nickel oxide may be written NiO_{1+x}.

Table 1

1	2	3	4	5	6	7	8	9
Sample type	T_H (°K)	x (Number of excess oxygen atoms per NiO molecule)	p (cm^{-3})	A (cm^2/g)	n (cm^{-2})	δ (cm)	ϱ (g/cm^3)	Ratio between sample density and NiO single crystal density (%)
A	873	9.7×10^{-4}	1.05×10^{20}	8.1×10^{4}	1.304×10^{15}	1.42×10^{-5}	5.21	76.5
B	973	3.7×10^{-4}	4.03×10^{19}	5.5×10^{4}	0.732×10^{15}	2.09×10^{-5}	5.20	76.5
C	1073	2.5×10^{-4}	2.72×10^{19}	2.4×10^{4}	1.130×10^{15}	4.05×10^{-5}	5.54	81.0
D	1173	0.7×10^{-4}	7.63×10^{18}	1.9×10^{4}	0.401×10^{15}	5.53×10^{-5}	5.72	84.0
E	1273	0.2×10^{-4}	2.18×10^{18}	0.9×10^{4}	2.420×10^{15}	11.26×10^{-5}	5.91	87.0

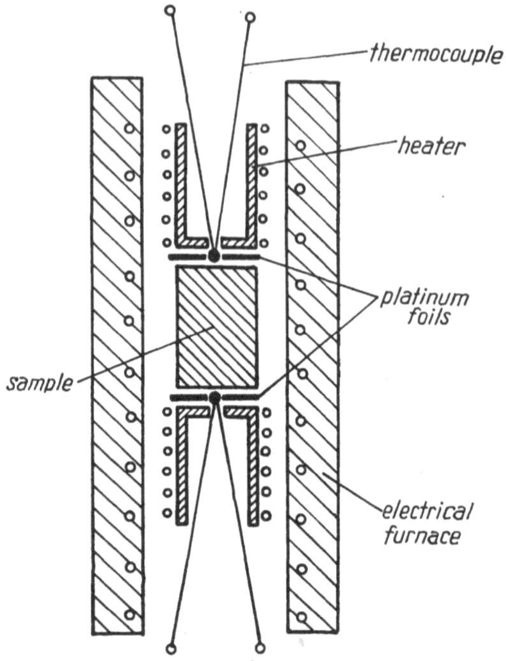

Fig. 1. Arrangement for thermoelectric power measurements

Since our samples were prepared in a similar way as in [7], we used these data for the surface area A of the non-stoichiometric NiO samples (Table 1, column 5). These data are in good agreement with those of [8]. From p and A the number n of Ni^{3+} ions per unit of surface area can be calculated according to $n = \frac{p}{A\varrho}$ (Table 1, column 6).

We have also calculated the mean grain-sizes of the samples, under the assumption that the grains may be assimilated to cubes (δ side of the cube, Table 1, column 7). The calculated δ values agree well with those resulting from [6].

2.2 Electrical measurements

The temperature dependences of the d.c. electrical conductivity (in the temperature range from 300 to 850 °K) and of the thermoelectric power (in the temperature range from 400 to 750 °K) were measured for the five sample types. The measurements were performed in air at atmospheric pressure.

A four-probe method was used for the conductivity measurements. The cylindrical sample was mounted in a holder, between two platinum foil electrodes in tight contact with the two opposite ends of the sample; the holder also supported the two platinum point-contacts for the potential-drop measurements along the sample. The entire device was similar to that described in [10].

For the measurements of the thermoelectric power the sample was placed between two platinum electrodes. Each electrode had a small opening in its center, through which Pt–PtRh thermocouples were pressed against the two opposite ends of the sample (Fig. 1). The sample was then introduced in a device similar to that described in [11]. The thermoelectric voltage of the sample was measured between the platinum wires of the two thermocouples (Pt–PtRh) with a high-resistance potentiometer. The temperatures of the two ends of the sample were measured by a second potentiometer. The temperature difference between the two ends of the sample was maintained between 18 and 24 °C.

3. Results and Discussion

The temperature dependence of the conductivity is shown in Fig. 2. Here $\log \sigma$ is plotted versus $1/T$ for the five samples with various Ni^{3+} concentrations. In Fig. 3 the Seebeck coefficient α is given as a function of absolute temperature for the same samples.

Fig. 2. log σ (σ conductivity) as a function of $1/T$

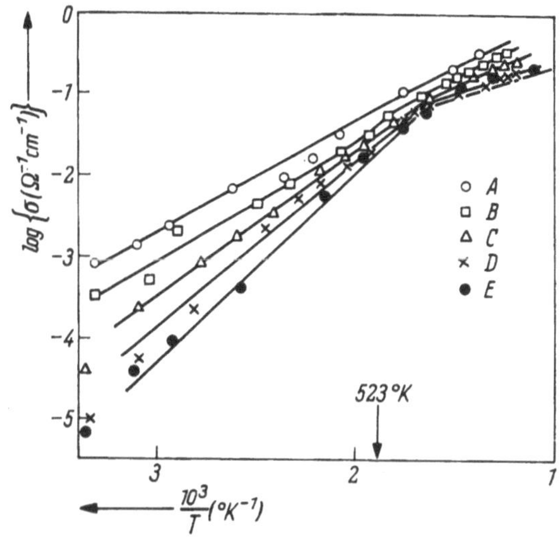

Under the assumption that the conduction process is due to a thermally activated hopping process of charge carriers from site to site, we have according to [12]

$$\sigma \sim T \exp\left(-\frac{\varepsilon}{k\,T}\right) \quad \text{for} \quad T < T_N\,,$$

where ε is the activation energy for electrical conduction, T the absolute temperature and T_N the Néel temperature of the given sample. From Fig. 4 it can be seen that up to the Néel point (523 °K)[3]) the slope of $\log \sigma/T$ versus $1/T$ is actually a straight line for all samples. The corresponding activation energies ε are given in Table 2.

Table 2

Sample type	ε (eV)
A	0.22
B	0.28
C	0.35
D	0.41
E	0.45

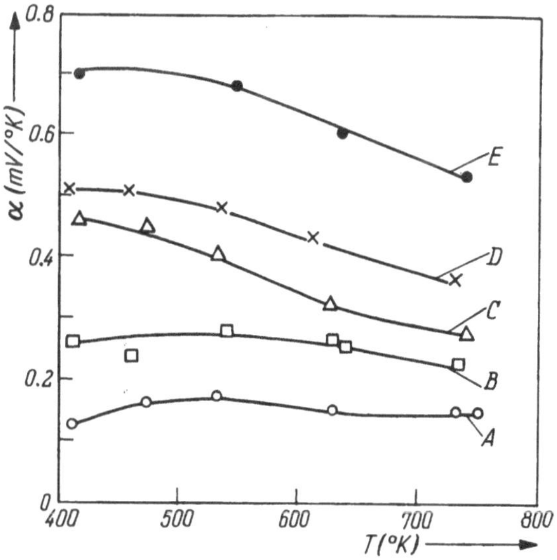

Fig. 3. Seebeck coefficient α as a function of T

[3]) According to Richardson and Milligan [13], the Néel temperature doesn't depend on the excess oxygen content of the NiO samples.

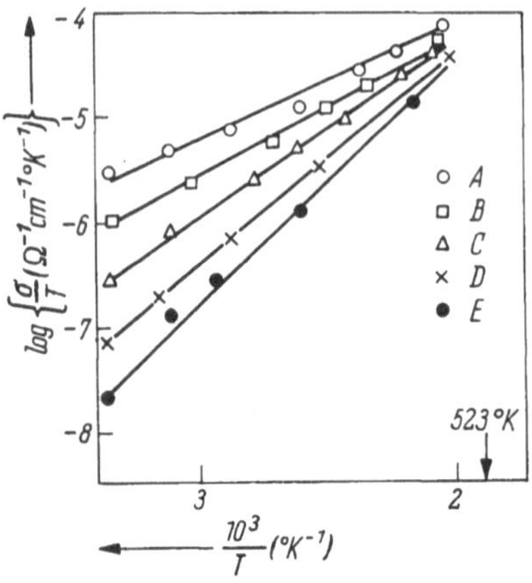

Fig. 4. log σ/T as a function of $1/T$

The Seebeck coefficient for NiO is given by the expression [12]

$$e\,\alpha = A_0 + A_1 + \frac{E_F}{T},$$

where A_0 is the kinetic energy, A_1 the contribution due to the activated process and E_F the distance of the Fermi level from the Ni^{2+} levels. As can be seen from [12] the contribution of the first two terms to α is normally small [4]) $\left(|A_0| < \frac{k}{2}\right.$ and A_1 is of the order of $\left.\frac{1}{20}\frac{\varepsilon}{T}\right)$. We can, therefore, write in a first approximation

$$e\,\alpha = \frac{E_F}{T}.$$

In Fig. 5 E_F is plotted as a function of absolute temperature for the five samples. From E_F the hole concentration p_α on the Ni^{2+} levels can be calculated,

$$p_\alpha = \frac{N}{\exp\left(\frac{E_F}{kT}\right) + 1};$$

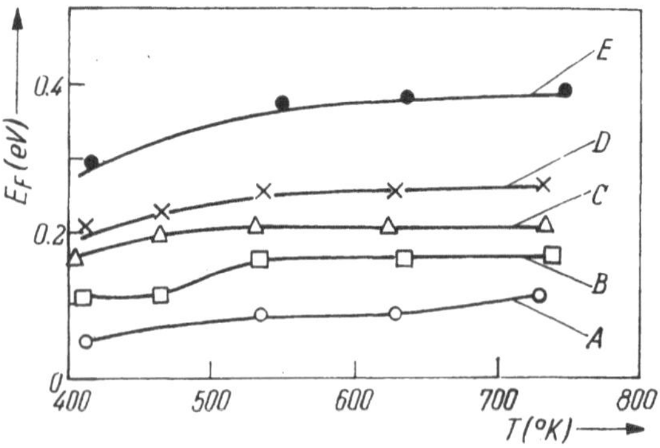

Fig. 5. Fermi level E_F as a function of T

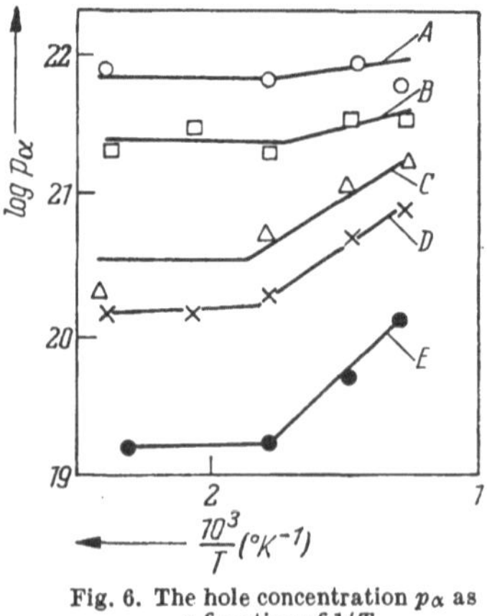

Fig. 6. The hole concentration p_α as a function of $1/T$

[4]) From the results obtained by Ksendzov et al. [14] for Li doped NiO, it can also be seen that the contribution of A_0 to the Seebeck coefficient is usually small.

here k is the Boltzmann constant and N the density of Ni^{2+} levels in the sample (following Morin [1], $N = 5.5 \times 10^{22}$ cm^{-3}). In Fig. 6 log p_α versus $1/T$ is plotted for the five samples. These curves show that, up to nearly 500 °K, p_α exhibits only a slight variation with temperature; above this temperature p_α rises with T.

Since p_α is constant up to about 500 °K and $\sigma = e\,\mu\,p$ (μ hole mobility) we may conclude that at least up to this temperature, the activation energy ε is connected with the mobility and not with the hole concentration. This conclusion is in agreement with the fact that the chemical analysis, carried out at two different temperatures, gave the same values for the concentrations of Ni^{3+} in the samples.

It seems, therefore, that in non-stoichiometric NiO practically no energy is needed to ionize the acceptor levels in the lattice and the activation energy of the conduction is connected with the activation of holes hopping between Ni^{2+} ions.[5])

But a marked disagreement exists between the p values resulting from the chemical analysis and those computed from α (Table 3), especially for the NiO samples containing a relatively high excess of oxygen. It is worth noting that similar discrepancies were also observed by Morin [1] and Parravano [15] for Li doped NiO samples, prepared in air.

Table 3

Sample type	p_α (cm^{-3})	p (cm^{-3})	p_α/p
A	80.0×10^{20}	1.05×10^{20}	76
B	2.6×10^{20}	0.40×10^{20}	64
C	2.7×10^{20}	0.27×10^{20}	10
D	0.7×10^{20}	0.08×10^{20}	9
E	0.16×10^{20}	0.02×10^{20}	8

In order to explain these discrepancies and to give a correct interpretation of the experimental results, it is necessary to take into account grain boundary conduction in polycristalline nickel oxide. Gossel [8] has shown that in NiO prepared by the decomposition of a nickel salt in air at not too high temperatures, the Ni^{3+} ions are mainly distributed in a thin layer at the surface of the NiO grains. This is in agreement with a) the conclusions of Mitoff [16] that at not too high temperatures the diffusion of oxygen, chemisorbed on the surface, into the interior of a NiO crystal is very slow and b) the fact that the values for the surface concentration of Ni^{3+} are of the same order of magnitude for all samples (see Table 1, column 6), while the mean Ni^{3+} concentration is nearly 50 times greater in the A-type than in the E-type sample.

It may, therefore, be concluded that in the NiO grains the charge carriers are not uniformly distributed throughout the grain volume, but are mainly concentrated in a thin layer on the grain surface. The conductivity and thermoelectric power of such a grain, with a non-uniform distribution of charge carriers, may be easily obtained in a simplified model. We shall assume that the grains are of cubic shape, and that each grain comprises an internal domain l where the hole

[5]) It must, however, be mentioned that, due to the high excess oxygen content of sample A (see Table 3), an "impurity"-band conduction could occur in this sample.

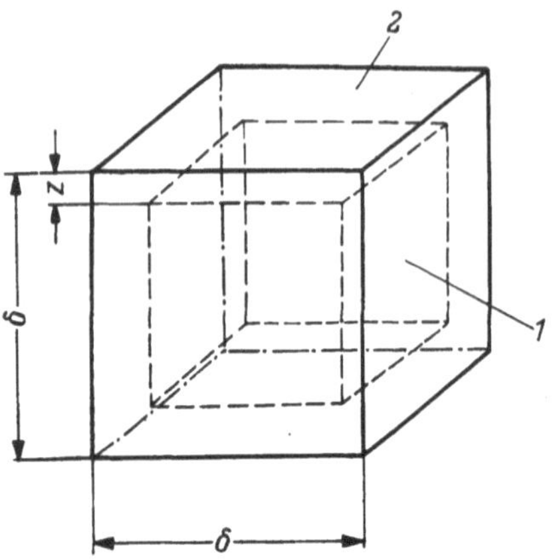

Fig. 7. Cubic shape of the grain

concentration is p_1, surrounded by a layer of thickness Z (domain 2) where the hole concentration is p_2, with $p_2 > p_1$ (Fig. 7).

The mean electrical conductivity σ_m of such a cubic grain, for currents normal to one of the faces, is given by

$$\sigma_m = \frac{4\sigma_2 Z(\delta - Z)}{\delta^2} + \frac{\sigma_1 \sigma_2 (\delta - 2Z)^2}{2\sigma_1 Z \delta + \sigma_2 \delta (\delta - 2Z)},$$

where δ is the side of the cube, and σ_1 and σ_2 are the conductivities of the domains 1 and 2, respectively. If essentially grain-boundary conduction prevails, or more exactly, when $\sigma_2/\sigma_1 \gg \delta/Z$ then

$$\sigma_m = \frac{4\sigma_2 Z(\delta - Z)}{\delta^2} = \frac{4 e \mu_2 p_2 Z(\delta - Z)}{\delta^2},$$

where μ_2 is the carrier mobility in the surface layer (domain 2). When p is the mean hole concentration (the hole concentration is averaged over the whole vo-

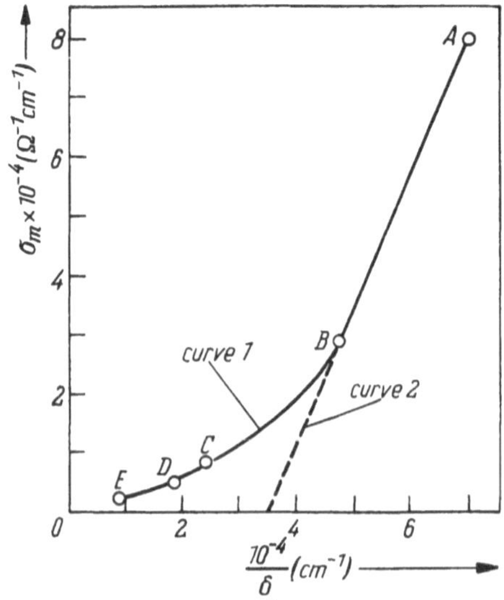

Fig. 8. σ_m as a function of $1/\delta$

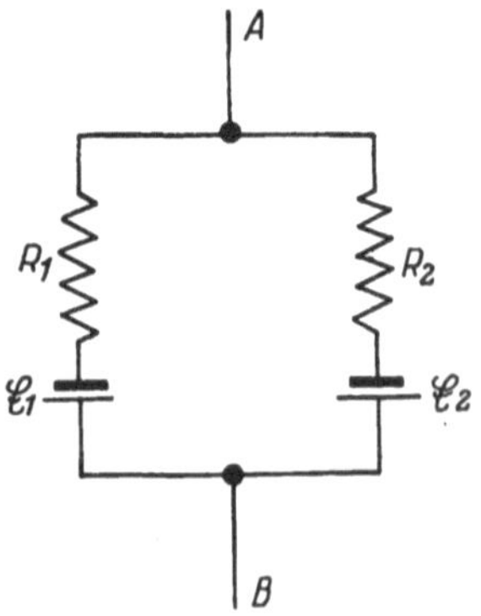

Fig. 9. Equivalent circuit of the grain

lume of the grain), and N the total number of Ni^{3+} ions in the grain, we have

$$\sigma_m = \frac{4\, e\, \mu_2\, p_2\, Z\, (\delta - Z)}{\delta^2} = \frac{e\, \mu_2\, N}{\delta^3} = e\, \mu_2\, p\,,$$

where e is the charge of an electron. Thus, under the assumptions that essentially grain-boundary conduction takes place and that the mobile charge carriers are holes on the Ni^{2+} levels, the mean electrical conductivity of a grain is determined by the mobility μ_2 of the charge carriers in the surface layer and their concentration averaged over the whole volume of the grain. We see also that the thickness of the surface layer does not enter explicitely into the expression for σ_m.

On the other hand, we see from the expression for σ_m that when $\sigma_2/\sigma_1 \gg \delta/Z$ and $Z = \text{const}$, we should have

$$\sigma_m \sim \frac{1}{\delta}$$

(because usually $Z \ll \delta$). In Fig. 8 the experimentally found dependence of σ_m on $1/\delta$ (curve 1) is shown. We see that actually no linear relationship between σ_m and $1/\delta$ holds and at least for the samples C, D, and E the values of σ_m are higher than those which one would obtain by a linear extrapolation (curve 2) of the data of samples A and B. This shows that Z increases with δ and, consequently, that at least for the samples C, D, and E a non-negligible diffusion of chemisorbed oxygen exists from the surface into the interior of the grains.

If we assume that to a first approximation the expression $\sigma_m = e\, \mu_2\, p$ is valid for all sample types we can obtain the values of μ_2 at 300 °K, the temperature at which the iodometric determinations were carried out (Table 4).

Table 4

Sample type	μ_2 (cm^2/V s)
A	4.7×10^{-5}
B	4.9×10^{-5}
C	1.8×10^{-5}
D	1.6×10^{-5}
E	2.2×10^{-5}

We see that, in spite of the approximations adopted, the values of μ_2 are in satisfactory agreement with those given by Van Houten [2].

In order to obtain the expression of the Seebeck coefficient for our model, we assume a temperature difference ΔT between two opposite faces of the grain. We may then regard the grain as a system of two parallel connected elements, with internal resistances R_1 and R_2 which give the thermoelectric voltages $\mathcal{E}_1$ and $\mathcal{E}_2$ (Fig. 9). The subscripts 1 and 2 correspond to the domains 1 and 2 in the grain. Between the two opposite faces A and B of the grain therefore a potential difference

$$V_A - V_B = \mathcal{E}_1 \frac{\sigma_1}{\dfrac{4\, Z\, (\delta - Z)}{(\delta - 2\, Z)^2}\, \sigma_2 + \sigma_1} + \mathcal{E}_2 \frac{\sigma_2}{\sigma_2 + \dfrac{(\delta - 2\, Z)^2}{4\, Z\, (\delta - Z)}\, \sigma_1}$$

exists.

The mean Seebeck coefficient is given by

$$\alpha = \alpha_1 \frac{\sigma_1}{\dfrac{4\,Z\,(\delta - Z)}{(\delta - 2\,Z)^2}\,\sigma_2 + \sigma_1} + \alpha_2 \frac{\sigma_2}{\sigma_2 + \dfrac{(\delta - 2\,Z)^2}{4\,Z\,(\delta - Z)}\,\sigma_1},$$

where α_1 and α_2 are the Seebeck coefficients for the domains 1 and 2 of the grain, respectively. Under the assumption that

$$\frac{\alpha_2\,\sigma_2}{\alpha_1\,\sigma_1} \gg \frac{\delta}{Z}$$

we have

$$\alpha \approx \alpha_2 .$$

Thus, when domain 1 has only few charge carriers, which occurs if the chemisorbed oxygen could not diffuse deep into the grain, and δ/Z is not too large $\left(\text{for } \frac{\sigma_2}{\sigma_1} > 10^3,\ \delta/Z \text{ should not exceed } 10^2\right)$, the experimentally found values for α represent the Seebeck coefficient α_2 of the surface layer.

In other words, for the samples for which the condition

$$\frac{\alpha_2\,\sigma_2}{\alpha_1\,\sigma_1} \gg \frac{\delta}{Z}$$

holds, the hole concentration p_α determined from α will be the carrier concentration p_2 in the surface layer of the grain and not the concentration averaged over the whole volume of the grain. Therefore, for samples in which boundary-conduction predominates p_α will be greater than the carrier concentration p determined by chemical analysis (Table 3) and this explains the discrepancies between p_α and p. The data given in Table 3 show that for most of the samples the carriers are actually concentrated mainly in a surface layer. With increasing preparation temperature of the samples, p_α/p becomes smaller, that is Z/δ becomes larger; from the values of p_α/p we can now obtain the thickness of the layer with enhanced electrical conduction on the grain boundary. We have

$$p = \frac{4\,p_2\,Z\,(\delta - Z)}{\delta^2} \qquad \text{and} \qquad p_2 = p_\alpha$$

so that Z may be easily computed (Table 5). Because of the simplifications and approximations which we have made, these Z values can be taken only for an order-of-magnitude estimate of the thickness of the layer with enhanced Ni^{3+} density.

Table 5

Sample type	Z (cm)
A	1×10^{-7}
B	2×10^{-7}
C	1×10^{-6}
D	1×10^{-6}
E	4×10^{-6}

The data given in Table 5 confirm the conclusion drawn from conductivity measurements, that for the samples C, D, and E, prepared by decomposition of the nickel salt at higher temperatures, a marked diffusion of oxygen into the lattice takes place.

4. Conclusions

The data presented here confirm the fact established also by other investigators that the electrical properties of non-stoichiometric NiO depend on the excess oxygen in the sample. The oxygen content varies with the decomposition temperature of the nickel salt. For NiO samples prepared at lower decomposition temperatures, electrical conduction is present mainly in a thin layer on the surface of the grain and this explains the discrepancies between the carrier concentration values determined by chemical analysis (that is the concentration averaged over the whole volume of the grain) and those determined from the Seebeck coefficient. For samples prepared at higher temperatures a non-negligible amount of oxygen, diffused into the interior of the NiO grains, leads to a more uniform distribution of charge carriers in the grains.

The temperature dependence of the hole concentration p_a determined from the temperature dependence of the Seebeck coefficient shows that p_a is constant up to the Néel temperature. Thus the exponential dependence of the conductivity in NiO on temperature must be attributed to the temperature dependence of the mobility. It seems, therefore, that at least up to this temperature, non-stoichiometric NiO exhibits a similar conduction mechanism as Li-doped nickel oxide, with holes hopping between Ni^{2+} ions in the lattice.

Acknowledgements

The authors wish to express their thanks to their colleagues T. Braun and I. Maxim for making the chemical analysis and to A. Cristofan and M. Avramescu for their technical assistance.

References

[1] F. J. Morin, Phys. Rev. **93**, 1199 (1954); Bell Syst. Techn. J. **37**, 1047 (1958).
[2] S. van Houten, J. Phys. Chem. Solids **17**, 7 (1960).
[3] R. R. Heikes and W. D. Johnston, J. chem. Phys. **26**, 582 (1957).
[4] E. J. W. Verwey, P. W. Haaijman, F. C. Romeijn, and G. W. van Oostrhout, Philips Res. Rep. **5**, 173 (1950).
[5] H. H. V. Baumbach and C. Wagner, Z. phys. Chem. **24**, 59 (1934).
[6] Y. Iida, K. Shimada, and S. Ozaki, Bull. Chem. Soc. Japan **33**, 1372 (1960).
[7] J. Deren, J. Haber, and J. Sloczynski, Bull. Acad. Sci. Polon. **9**, 245 (1961).
[8] H. Gossel, Z. Elektrochem. **65**, 98 (1961).
[9] I. Maxim and T. Braun, J. Phys. Chem. Solids **24**, 537 (1963).
[10] M. Nachman and L. Cojocaru, Radiation Damage in Solids, Vol. 3, International Atomic Energy Agency, Vienna 1963 (p. 15).
[11] I. Maxim and L. Rîbco, Studii şi Cercetări de Fizică (Rumania) **2**, 289 (1962).
[12] R. R. Heikes, A. A. Maradudin, and R. C. Miller, Ann. Phys. (France) **8**, 733 (1963).
[13] J. T. Richardson and W. O. Milligan, Phys. Rev. **102**, 1289 (1956).
[14] I. M. Ksendzov, L. N. Anselm, L. L. Vasilieva, and V. M. Latysheva, Fiz. tverd. Tela (U.R.S.S.) **5**, 1537 (1963).
[15] G. Parravano, J. chem. Phys. **5**, 23 (1955).
[16] S. P. Mitoff, J. chem. Phys. **35**, 882 (1961).

(Received December 23, 1964)

phys. stat. sol. 8, 785 (1965)

Solid State Physics Department, S.C.K.-C.E.N., Mol

Stage III Recovery in Cold-Worked Niobium

By

L. STALS and J. NIHOUL

Stage III recovery in niobium, cold-worked at room temperature, is investigated by means of electrical resistivity measurements. The associated activation energy is found to be constant in the temperature range 80 to 180 °C and equal to 1.20 ± 0.04 eV. Furthermore, the results indicate that the recovery process can be described as a diffusion-controlled bimolecular reaction which follows Waite kinetics. It is concluded that this recovery is caused by recombination of intrinsic point defects.

On the basis of the presently available data the possibility of this recovery being caused by migration of vacancies to trapped interstitials cannot be excluded. In view, however, of the relatively small activation energy found for this stage, and of the direct analogy with stage III recovery in the f.c.c. metals, it is tentatively proposed that stage III recovery in b.c.c. metals is due to interstitial migration to vacancies.

Die Erholungsstufe III in kaltverformtem Niob wurde durch Messung des elektrischen Widerstandes untersucht. Es zeigte sich, daß die zugehörige Aktivierungsenergie im Temperaturbereich von 80 bis 180 °C konstant und gleich 1,20 ± 0,04 eV ist. Weiterhin ergab sich, daß der Erholungsprozeß als eine diffusionsbestimmte bimolekulare Reaktion beschrieben werden kann, die nach der sich aus der Waiteschen Theorie ergebenden Kinetik abläuft. Es wird angenommen, daß diese Reaktion eine Rekombination von Zwischengitteratomen und Leerstellen ist.

Auf Grund der gegenwärtig verfügbaren Daten kann nicht ausgeschlossen werden, daß diese Erholung durch die Wanderung von Leerstellen zu gebundenen Zwischengitteratomen verursacht wird. Wegen der verhältnismäßig kleinen Aktivierungsenergie und der großen Ähnlichkeit zur Erholungsstufe III in k.f.z. Metallen erscheint es jedoch wahrscheinlicher, daß die Erholungsstufe III in k.r.z. Metallen durch die Wanderung von Zwischengitteratomen zu Leerstellen zustande kommt.

1. Introduction

In recent years there has been a steadily increasing interest in the annealing behaviour of damaged b.c.c. metals after cold work as well as after irradiation [1 to 19]. Only a few experiments on quenching of b.c.c. metals are found in the literature [11, 16, 20]. In analogy with the f.c.c. metals, the recovery spectrum for the electrical resistivity of b.c.c. metals shows five main annealing stages, occurring at $\frac{T}{T_m} < 0.02$ (stage I), $0.02 < \frac{T}{T_m} < 0.10$ (stage II), $0.10 < \frac{T}{T_m} < 0.20$ (stage III), $0.20 < \frac{T}{T_m} < 0.30$ (stage IV) and $0.30 < \frac{T}{T_m} < 0.35$ (stage V). T and T_m represent the mean temperature of the stage involved and the melting temperature in °K, respectively.

Tungsten and molybdenum have been studied most extensively [1 to 19], whereas for niobium and tantalum [8, 12] only stage III has been investigated. The nature of the recovery in these different stages is still quite uncertain.

The aim of the present paper is to describe stage III recovery in cold-worked niobium as followed by electrical resistivity measurements. Special attention will be paid to the annealing kinetics in the light of Waite's theory [24]. It will be examined up to where the present results are consistent with the various models proposed in the literature for stage III recovery in b.c.c. metals.

2. Experimental Method

A high purity, zone-refined niobium rod (4 ppm. C, 5 ppm. N_2, 2 ppm. O_2, 0.3 ppm. H_2, 3 ppm. Mo) of 5 mm diameter, has been reduced to 0.1 mm diameter wire after swaging and drawing at room temperature.[1])

Specimens of about 30 cm length were cut from this wire and spiraled before mounting for electrical resistivity measurements. These measurements were done at liquid nitrogen temperature and the recovery experiments were carried out following the same method as described in [11]. Temperature fluctuations in the liquid nitrogen bath were corrected by means of a reference specimen, mounted in the same bath.

3. Experimental Results

The relative concentration c/c_0 of point defects which anneal out in a well-defined recovery stage, is supposed to be proportional to the relative value of the electrical resistance, i. e.

$$\frac{c}{c_0} = \frac{R - R_e}{R_0 - R_e}, \tag{1}$$

where c_0 is the concentration of these defects at the beginning of the recovery stage, R_0 the corresponding resistance of the sample and R_e the resistance at the end of this stage. R represents the resistance of the sample after each annealing treatment, measured at the same reference temperature (in this case 77.3 °K).

Both isochronal and isothermal annealing curves have been measured (Fig. 1).

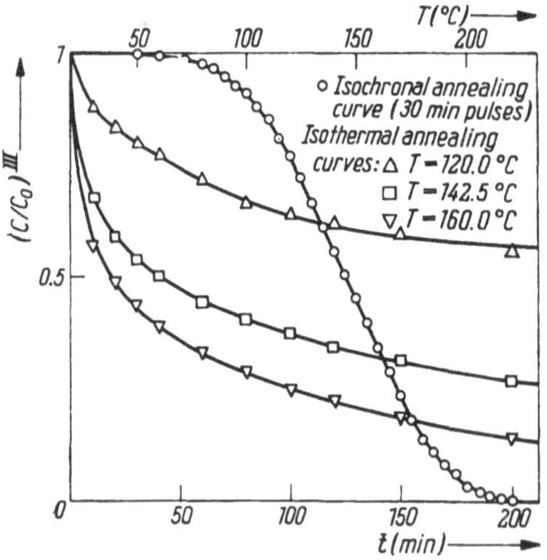

Fig. 1. Isochronal and isothermal annealing curves of cold-worked niobium in the temperature range of stage III

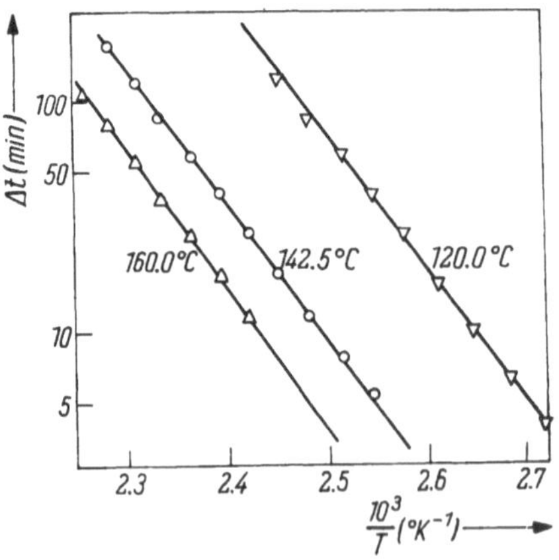

Fig. 2. Meechan-Brinkman [21] plots obtained by combining the isothermal and isochronal data of Fig. 1. The nearly parallel straight lines indicate that the recovery process is governed by a single activation energy of 1.20 ± 0.04 eV

[1]) Thanks are due to Prof. J. D. Fast and coworkers of the N. V. Philips Labs. (Eindhoven) for carrying out this work.

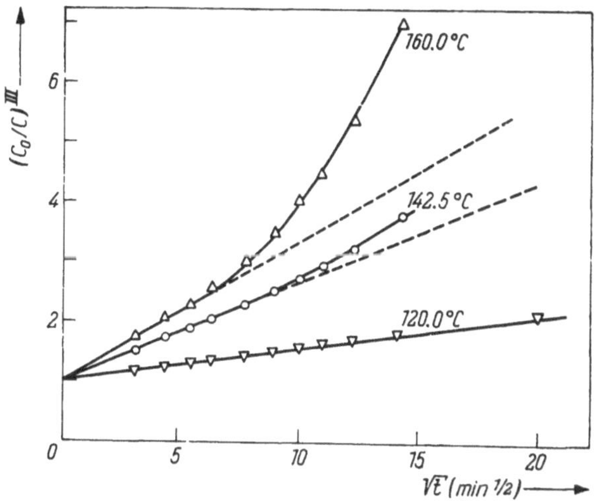

Fig. 3. Isothermal recovery of cold-worked niobium presented in a $\left(\frac{c_0}{c}\right)^{III}$ versus $\sqrt{t}$ plot. It appears that $\sqrt{t}$-kinetics is followed in the initial stage of the recovery process

In the isochronal recovery experiments equal annealing pulses of 30 minutes were used with 5 °C intervals. These results show that an annealing peak occurs at about 140 °C. Using the Meechan-Brinkman method [21], we calculated the associated activation energy to be 1.20 ± 0.04 eV (Fig. 2), in good agreement with Peacock and Johnson's result [12]. This value is also confirmed by applying Simson and Sizmann's method [22] on the isochronal data only (see section 4, Fig. 5). The results obtained by applying Overhauser's method [23] on several specimens showed a non-systematic spreading between 1.15 and 1.35 eV and are therefore not reproduced here.

4. Reaction Kinetics

As can be deduced from Fig. 3 and Fig. 4, the isothermal recovery follows a $\sqrt{t}$-law at the beginning of the recovery process: $\left[\left(\frac{c_0}{c}\right)^{III} \sim \sqrt{t}\right]$ and, subsequently as recovery proceeds, second-order kinetics are found $\left[\left(\frac{c_0}{c}\right)^{III} \sim t\right]$.

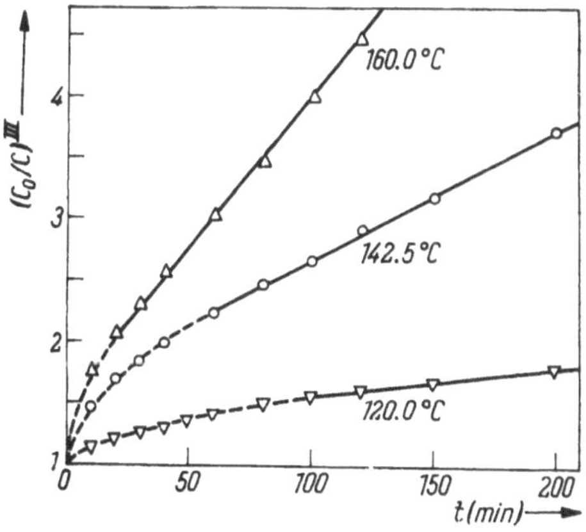

Fig. 4. Isothermal recovery of cold-worked niobium in a $\left(\frac{c_0}{c}\right)^{III}$ versus t plot. It appears that the reaction turns into second-order kinetics after a transition stage

According to Waite [24], a diffusion-limited bimolecular recovery process can be described by the equation

$$\frac{c_0}{c} - 1 = 4\,\pi\,\alpha\,r_0\,c_0\,D\left[1 + \frac{2\,r_0}{(\pi\,D t)^{1/2}}\right] t \tag{2}$$

for the case of equal initial concentrations of both reacting defects and uniform initial distribution of one kind of defect with respect to the other. D is the sum of the diffusion coefficients for both defects; r_0 stands for the capture radius, i.e., the distance to within which these defects have to approach each other to react chemically (Smoluchowski's boundary condition [25]); α is the number of atoms per cm^3; c and c_0 are atomic concentrations. If the mobility of one defect can be neglected, equation (2) can be transformed into

$$\frac{c_0}{c} - 1 = k_0\,e^{-E_m/k\,T}\,t + k_1\,e^{-E_m/2\,k\,T}\,t^{1/2}, \tag{3}$$

where $k_0 \equiv 4\,\pi\,\alpha\,r_0\,c_0\,D_0$, $k_1 \equiv 8\,\pi^{1/2}\,\alpha\,r_0^2\,c_0\,D_0^{1/2}$, E_m the migration energy and D_0 the diffusion constant of the mobile defect.

For short recovery times the second term at the right of equation (3) predominates, so that the reaction follows a $\sqrt{t}$-law, whereas for longer recovery times it is the first term that becomes predominant, and the reaction passes into pure second order. It can be deduced from equation (2) that for higher annealing temperatures the transition from $\sqrt{t}$-kinetics into second-order kinetics occurs at shorter recovery times.

Similar transitions have also been observed by Schultz for cold-worked tungsten [14] and by Simson and Sizmann for cold-worked Ni [22].

In order to apply equation (3) for isochronal data, it should be transformed as follows (method of Simson and Sizmann [22]):

$$\left(\frac{c_0}{c}\right)_j - \left(\frac{c_0}{c}\right)_{j-1} = k_0\,e^{-E_m/k\,T_j}\,\Delta t + k_1\,e^{-E_m/2\,k\,T_j}\,(\Delta t)^{1/2}, \tag{4}$$

where T_j the temperature of the j-th annealing pulse, $\left(\frac{c_0}{c}\right)_j$ the inverse relative defect concentration after the j-th pulse. For equal pulse times, $\Delta t =$ constant. The second term at the right is predominant at low temperatures, whereas at high

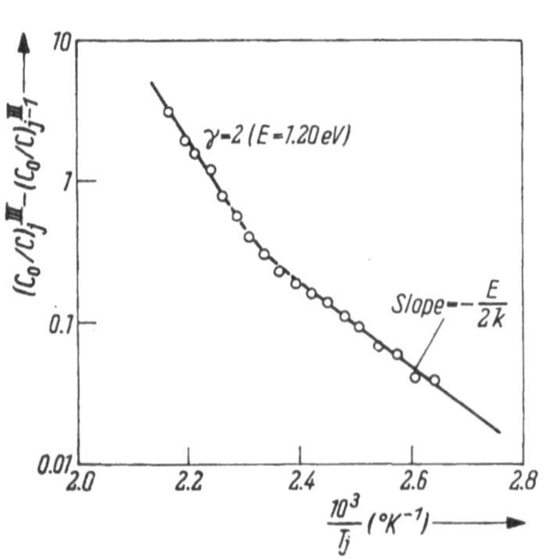

Fig. 5. Simson-Sizmann [22] plot of the isochronal recovery data of Fig. 1. The results can be fitted by a straight line with slope $- E_m/2\,k$ at the lower temperatures and a straight line with slope $- E_m/k$ at the higher temperatures, taking everytime $E_m = 1.20$ eV, in agreement with the result obtained by the Meechan-Brinkman plot (Fig. 2). (In the figure read "E_m" instead of "E".)

temperatures the first one predominates. Plotting $\ln\left[\left(\frac{c_0}{c}\right)_j - \left(\frac{c_0}{c}\right)_{j-1}\right]$ as a function of the inverse absolute temperature $\frac{1}{T}$, one obtains a straight line with slope $-E_m/2k$ at the lower temperatures and a straight line with slope $-E_m/k$ at the higher temperatures (see Fig. 5).

In an attempt to obtain a quantitative correspondence between formulae (3) and (4) on the one hand, and the experimental data of Fig. 3, 4 and 5 on the other hand, we could obtain good agreement by using r_0 and D_0 values in the ranges $20\ \text{Å} < r_0 < 43\ \text{Å}$ and $3\times10^{-4}\ \text{cm}^2/\text{s} < D_0 < 1.5\times10^{-3}\ \text{cm}^2/\text{s}$, except for small defect concentrations $\left[\left(\frac{c}{c_0}\right)^{\text{III}} < 0.25\right]$. These deviations can possibly be explained by a slightly erroneous estimation of R_e^{III}. For the initial concentration and the activation energy for migration the following values have been used: $c_0^{\text{III}} = 10^{-4}$ (see section 5) and $E_m^{\text{III}} = 1.20$ eV.

5. Discussion

The present measurements have shown that stage III recovery in cold-worked niobium can be described as a diffusion-controlled bimolecular reaction. This has also been observed for cold-worked [14] and neutron-irradiated tungsten [15] and for neutron-irradiated molybdenum [11]. Re-analysis of Peacock and Johnson's results [12] also reveals second-order kinetics for neutron-irradiated niobium [13].

For the case of cold-worked tungsten, Martinet [26] found at least two substages in stage III with distinct activation energies, whereas Schultz [14] reported a progressive increase of the activation energy as the recovery proceeds. Both authors explained these effects by the interaction of interstitials with impurity atoms. The present measurements do not yield sufficient evidence for any such effect. Instead the whole range between 80 and 180 °C is characterized by a constant activation energy.

For a review of stage III recovery in b.c.c. metals we refer to the literature [14, 15]. The authors generally agree that the recovery in this stage is connected with migration and possible interaction of point defects such as impurities, vacancies and interstitials.

As to impurities, special attention has to be paid to interstitial oxygen and carbon, since it is known that these defects are mobile in niobium at temperatures corresponding to the stage III range. As shown in Table 1, nitrogen can be excluded on this basis.

Table 1

Interstitial impurity	Temperature of migration (°C)	Activation energy for migration (eV)	References
O	150	1.20	[27, 28]
C	150	1.15	[27, 28]
N	300	1.65	[27, 28]

Although direct values are not available for hydrogen, a comparison with other b.c.c. metals e.g. Fe [30], Ta [28] and V [29] allows us to assume that hydrogen is mobile in niobium far below room temperature.

The second-order kinetics as found in stage III could be explained by assuming that carbon atoms migrate to vacancies as proposed by Sell (private communication to Schultz [14]). Gregory's results obtained on cold-worked niobium [31] might in fact indicate that recovery in the temperature range of stage III is considerably influenced by interstitial impurities. For a specimen containing 800 ppm. interstitial impurities (100 weight ppm.) Gregory finds a very small recovery peak at about 190 °C. For 4800 and 8000 ppm. this peak shifts towards about 170 °C and becomes more pronounced. Unfortunately no value for the activation energy nor the reaction order are reported. An analysis of the isochronal data suggests that the activation energy is not higher than 0.9 eV. This value is considerably smaller than the activation energy for stage III resulting from the present work and from earlier measurements [12, 32], so that it becomes questionable whether the effect reported by Gregory can be identified with stage III recovery.

In the present measurements the resistivity decrease in stage III amounted to 0.106 $\mu\Omega$ cm. This would correspond to an initial carbon or oxygen concentration of about 10^{-4} if one adopts the rule of thumb proposed by Lucasson and Walker [10], i.e. 1 at.% of impurity-vacancy pairs should yield a resistivity increase of 12 $\mu\Omega$ cm in niobium, if one assumes the resistivity of intrinsic and extrinsic interstitials to be the same.

Internal friction measurements, however, carried out in this laboratory by De Batist and Turner [33] on wires drawn from the same rod, revealed a very small Snoek peak in the temperature region of about 150 °C. From the magnitude of this peak it could be concluded that the total concentration of carbon or oxygen is of the order of 5 ppm., in agreement with the manufacturer's analysis. This concentration is too small to explain the measured electrical resistivity decrease in the stage III range.

Furthermore it should be pointed out that recombination of interstitial impurities and vacancies only gives rise to second-order kinetics if the initial concentrations of both defects are approximately the same. If the concentration of one kind of defect becomes much larger than the other one, the reaction tends towards first-order kinetics [24].

As mentioned above, the experiments carried out until now on b.c.c. metals point to second-order kinetics in stage III. If one wants to explain this stage by a recombination of vacancies and interstitial impurities one must assume that in each case the initial concentrations of both defects are approximately equal. It is hard to conceive that this condition was fulfilled in all these experiments where specimens of various degrees of impurity were subjected to various amounts of cold work and various irradiation doses.

It thus seems that, in order to account for second-order kinetics, the explanation of stage III recovery is more likely to be found in a mechanism based on a recombination of intrinsic point defects. The following possibilities can be considered here:

A) Migration of vacancies either to trapped interstitials [18] where they are annihilated, or to other vacancies where they form fast moving divacancies [14].

B) Migration of interstitials either to vacancies and consequent annihilation [11, 13, 14, 15], or to other interstitials, forming di-interstitials.

Recently Meakin, Lawley and Koo [16] reported the observation by means of electron microscopy of vacancy loops after a one hour annealing at 400 °C of molybdenum quenched from near the melting point down to about 160 °C. Annealing for one hour at 300 °C showed approximately the same results [17], but no loops could be observed in the as-quenched material. These observations seem to support vacancy migration in stage III. Taking into account, however, the very high quenching temperature (2600 °C), one cannot exclude that an appreciable number of more mobile divacancies were quenched-in. It is known, at least for f.c.c. metals, that divacancies are mobile in the temperature range of stage III, whereas vacancies migrate at higher annealing temperatures (stage IV) [44, 45].

Downey and Eyre [18] have observed interstitial dislocation loops in molybdenum foils after irradiation at 60 °C with fast neutrons up to a dose of 2×10^{20} n/cm^2. Similar loops have been observed by Higgins and Roberts [19] in molybdenum after irradiation at 80 °C with fission fragments up to doses in the range 6×10^{13} to 1×10^{15} f.f./cm^2. Downey and Eyre conclude that long range migration of interstitials occurs at 60 °C so that no free interstitials will be left over to migrate in stage III. They estimate that about 90% of the interstitials, which have survived self-annealing during irradiation, are present singly or in sub-microscopic clusters. They further assume that the single interstitials are trapped below stage III by interstitial impurities. In order to account for second-order kinetics they then conclude that stage III is caused by the migration of vacancies to these trapped interstitials. It should be pointed out, however, that stage III recovery in molybdenum is initiated at about 0 °C [5, 11] and that Downey and Eyre's irradiations were performed at 60 °C for 3 months, which corresponds to a very long mean annealing time in the stage III range. It thus seems unjustified to conclude from their experiments that long-range interstitial migration had already occurred below stage III and that this stage is caused by vacancy migration. Moreover, as shown in Table 2, the migration energies E_m^{III} are smaller than the calculated activation energies, E_m^v, for vacancy migration. These energies were calculated as the difference of the self-diffusion energy, E_{sd}, and the vacancy formation energy, E_f^v, assuming that self-diffusion in b.c.c. metals occurs by a vacancy mechanism [34].

Table 2

	E_f^v (eV)	E_{sd} (eV)	$E_m^v = E_{sd} - E_f^v$ (eV)	E_m^{III} (eV)
Nb	2.0 [35]	4.56 [36] 4.23 [34]	2.2 to 2.6	1.2 [8, 12, this work]
Mo	2.24 [37]	5.0 [38] 4.7 [39] 4.0 [40]	1.8 to 2.8	1.3 [1, 2, 11]
W	3.15 [41] 3.30 [20]	5.89 [42] 5.23 [43]	1.93 to 2.74	1.7 [2, 4, 5, 14]

If these differences between E_m^v and E_m^{III} are significant, it becomes doubtful whether stage III can be caused by vacancy migration (hypothesis A). Furthermore, it has now been widely accepted that stage III in f.c.c. metals is not caused by vacancy migration [44, 45, 46] and that it shows a marked analogy with stage III in b.c.c. metals [5, 11, 13, 14, 15].

A further argument which seems to contradict vacancy migration in stage III results from the discrepancy between the value of the diffusion constant, D_0^{III}, as estimated from the present results, and of the diffusion constant D_0^v, for vacancy migration in niobium, as derived from the self-diffusion constant, D_0^{sd}. As mentioned in section 4, the value found for D_0^{III} is of the order of 1×10^{-3} cm²/s. Taking $D_0^v = 1.66$ cm²/s [34] for the self-diffusion constant in niobium and $\exp(\Delta S_f^v/k) = 64$ [35] for the entropy change for vacancy formation one can deduce from

$$D_0^v = D_0^{sd} \exp(-\Delta S_f^v/k) \tag{5}$$

that $D_0^v \simeq 3 \times 10^{-2}$ cm²/s which is an order of magnitude larger than the value derived for D_0^{III}.

The model based on divacancy formation is questionable for additional reasons. In this case, in fact, one should expect a capture radius r_0 of only a few lattice distances, as contrasted with the present value (about 10 lattice distances). Moreover, since for divacancy formation one should use $2\,D_0$ instead of D_0 in equation (2), the above-mentioned discrepancy becomes still more important.

Hypothesis B (interstitial migration) remainsito be discussed. The formation of di-interstitials after single interstitial migration in stage III can be excluded on the basis of the presumably large formation energy (about 1 eV) for this configuration in niobium (cf. 1.08 eV for b.c.c. iron [47]). This would imply a migration energy for free interstitials of about 0.2 eV, resulting in an unreasonably large number of jumps to explain stage III recovery.

Migration of interstitials to vacancies seems to be the most probable model to explain the present results. This mechanism is consistent with second-order kinetics and the recovery in lattice parameter [48]. As shown above, the electron microscopic observations do not exclude interstitial migration in stage III. The small growth of the interstitial loops in size and in number in the temperature range of stage III is not only consistent with the picture that the majority of migrating interstitials recombine with vacancies, but it is even required to explain second-order kinetics in this model.

To conclude it may be stated that, although there seem to be more arguments in favour of interstitial migration to vacancies for explaining stage III in b.c.c. metals, a recovery mechanism based on vacancy migration to trapped interstitials cannot entirely be excluded. Decisive arguments can possibly result from electron microscopic observations on specimens damaged at a temperature below stage III or from recovery experiments after quenching. These experiments are now being prepared in this laboratory.

Acknowledgements

The authors like to thank Prof. Dr. S. Amelinckx for his continuous interest and for reading the manuscript. They are also grateful to Dr. R. Gevers, R. De Batist, R. Siems and M. Verschueren for many helpful discussions. Mr. L. Verreyt and Mr. M. George assisted with the experimental work.

References

[1] D. G. MARTIN, Acta metall. **5**, 371 (1957).
[2] G. H. KINCHIN and M. W. THOMPSON, J. nuclear Energy **6**, 275 (1958).
[3] H. SCHULTZ, Z. Naturf. **14a**, 361 (1959).
[4] L. A. NEIMARK and R. A. SWALIN, Trans. AIME **218**, 82 (1960).
[5] M. W. THOMPSON, Phil. Mag. **5**, 278 (1960).
[6] P. LUCASSON, J. nuclear Energy **11**, 108 (1960).
[7] J. BLONS, P. IMBERT, G. PERRIOT, and G. TOURAND, Coll. Principes Effects Irr. sur les Cristaux, CEA, Saclay 1960.

[8] D. E. PEACOCK, Report UKAEA (1961).
[9] P. G. LUCASSON and R. M. WALKER, Phys. Rev. **127**, 485 (1962).
[10] P. G. LUCASSON and R. M. WALKER, Phys. Rev. **127**, 1130 (1962).
[11] J. NIHOUL, phys. stat. sol. **2**, 308 (1962), Radiation Damage in Solids, IAEA, Vienna, I, 309 (1962).
[12] D. E. PEACOCK and A. A. JOHNSON, Phil. Mag. **8**, 563 (1963).
[13] J. NIHOUL, Phil. Mag. **9**, 167 (1964).
[14] H. SCHULTZ, Acta metall. **12**, 649 (1964).
[15] J. MOTEFF and J. P. SMITH, ASTM Symp. on flow and fracture behaviour of metals and alloys in nuclear environments, June 22–24 (1964).
[16] J. D. MEAKIN, A. LAWLEY, and R. C. KOO, Appl. Phys. Letters **5**, 133 (1964).
[17] J. D. MEAKIN, private communication.
[18] M. E. DOWNEY and B. L. EYRE, AERE-R4632 (1964).
[19] P. R. B. HIGGINS and A. C. ROBERTS, J. Less-Common Metals **6**, 472 (1964).
[20] H. SCHULTZ, Acta metall. **12**, 761 (1964).
[21] C. J. MEECHAN and J. A. BRINKMAN, Phys. Rev. **103**, 1193 (1956).
[22] P. SIMSON and R. SIZMANN, Z. Naturf. **17a**, 596 (1962).
[23] A. W. OVERHAUSER, Phys. Rev. **90**, 393 (1953).
[24] T. R. WAITE, Phys. Rev. **107**, 463 (1957).
[25] M. V. SMOLUCHOWSKI, Z. phys. Chem. **92**, 192 (1917).
[26] B. MARTINET, Thesis, Lausanne 1964.
[27] C. Y. ANG, Acta metall. **1**, 123 (1953).
[28] J. W. MARX, G. S. BAKER, and J. M. SIVERTSEN, Acta metall. **1**, 193 (1953).
[29] R. A. BUTERA and P. HOFSTAD, J. appl. Phys. **34**, 2172 (1963).
[30] W. R. HELLER, Acta metall. **9**, 600 (1961).
[31] D. P. GREGORY, Acta metall. **11**, 455 (1963).
[32] M. J. MAKIN and F. J. MINTER, Acta metall. **7**, 361 (1959).
[33] R. DE BATIST and T. TURNER, private communication.
[34] T. S. LUNDY, ORNL-3617 (1964).
[35] YA. A. KRAFTMAKHER, Soviet Physics Solid State **5**, 696 (1963).
[36] R. RESNICK and L. S. CASTLEMAN, Trans. AIME **218**, 307 (1960).
[37] YA. A. KRAFTMAKHER, Soviet Physics Solid State **6**, 396 (1964).
[38] E. V. BORISOV, P. L. GRUZIN, L. V. PAVLINOV, and G. B. FEDOROV, Collection: Metallurgy and Metallography of Pure Metals [in Russian], Izd. MIFI, Moscow 1959 (n° 1).
[39] W. DANNEBERG and E. KRAUTZ, Z. Naturf. **16a**, 854 (1961).
[40] J. ASKILL and D. M. TOMLIN, Phil. Mag. 8, 997 (1963).
[41] YA. A. KRAFTMAKHER and P. G. STRELKOV, Soviet Physics Solid State **4**, 1662 (1963).
[42] V. P. VASIL'EV and S. G. CHERNOMORCHENKO, Zav. Lab. **22**, 688 (1956).
[43] W. DANNEBERG, Metallwissenschaft und Technik **15**, 11 (1961).
[44] A. SEEGER, Radiation Damage in Solids, IAEA, Vienna **1**, 101 (1962).
[45] K. HOLMES, Interaction of Radiation with Matter, Proc. Intern. Summer School, Mol, North-Holland Publ. Co, Amsterdam 1963 (p. 33).
[46] A. SEEGER, Phys. Letters **8**, 296 (1964).
[47] R. A. JOHNSON, Phys. Rev. **134**, A 1329 (1964).
[48] J. ADAM and D. G. MARTIN, Phil. Mag. **3**, 1329 (1958).

(Received December 23, 1964)

phys. stat. sol. 8, 795 (1965)

Institut for Experimental Physics, Loránd Eötvös University, Budapest

Plastic Properties of Polycrystalline f.c.c. Metals

By

I. KOVÁCS and E. NAGY

For polycrystalline f. c. c. metals deformed by torsion and extension simultaneously it is shown that the change in specific resistivity depends only upon the total strain, (as given by [4]), but is independent of the ratio of the individual strains. Furthermore it is shown that the relation of the form $\Delta\varrho = f(\gamma)$ cannot be characteristic to a metal, since its form is very sensitive to sample history. A history-independent relation can however be established between the resistivity change and the instantaneous flow stress. Assuming the resistivity due to the point defects to be half of the total resistivity, this relationship gives the resistivity due to unit length of a dislocation directly, the value being in good agreement with previous experimental and theoretical estimates.

Für polykristalline kfz-Metalle, die gleichzeitig durch Scherung und Dehnung deformiert wurden, wird in Übereinstimmung mit den Ergebnissen von [4] gezeigt, daß die spezifische Widerstandsänderung von der Gesamtbelastung abhängt und unabhängig ist von dem Verhältnis der Einzelbelastungen. Ferner wird gezeigt, daß eine Beziehung der Form $\Delta\varrho = f(\gamma)$ für ein Metall nicht charakteristisch sein kann, da seine Form sehr erheblich von der Vorgeschichte der Probe abhängt. Es kann jedoch eine von der Vorgeschichte der Probe unabhängige Beziehung zwischen der Widerstandsänderung und der momentanen Flußspannung hergeleitet werden. Unter der Bedingung, daß der Widerstand, der von Punktdefekten herrührt, halb so groß ist wie der Gesamtwiderstand, ergibt diese Beziehung in guter Übereinstimmung mit früheren experimentellen und theoretischen Abschätzungen direkt den Widerstand einer Versetzung (pro Einheitslänge).

1. Introduction

The present paper attempts to discuss some plastic properties of polycrystalline f.c.c. metals, based on studies of the electrical resistivity change caused by simultaneous plastic torsion and extension. In connection with such an investigation the question evidently appears, what results are to be expected, if different methods of deformation are used simultaneously. Let γ_1 and γ_2 be two parameters, each proportional to the total shear for a single manner of deformation. If we assume that the change in a physical property depends at a constant temperature only on the total shear, further that the individual shears are produced in case of a composite deformation independently, than the resulting change depends only on the sum of the individual shears, $\gamma_1 + \gamma_2$, but is independent of the ratio γ_1/γ_2.

It is clear that between the dimension change and the change of defect structure (or the total shear strain) there is generally no simple relationship.

A single valued connection between the change of shape and the total shear can be expected only for polycrystalline metals, because in the single crystals the orientations of the slip systems in respect of the direction of the deformation are in general different. It means that the same dimensional changes of a single crystal can be brought about in general by different amount of shear strains. In the case of a fine grained polycrystalline body the distribution of the slip systems in the different grains is totally random, therefore the shears take place in all the possible directions.

The method of simultaneous torsion and extension is very favourable from the experimenter's point of view, since by this method large plastic strains can be produced without rupture. The experimental procedure is already described in the literature [1, 2].

2. Connection between the Specific Resistivity Change and Strain

It is well known that in the course of twisting of polycrystalline metal wires an axial elongation is invariably produced, as found first by Poynting [3]. This latter effect is enhanced even by small axial stresses ($\approx$ 1 kg/mm²), and the elongation of the wire becomes homogeneous and easily measurable [2]. To different axial stresses different axial elongations do belong, even for the same angle of twist.

The individual deformations, according to our previous reasoning, are characterized by parameters, connected with the total shear during that single manner of deformation. For a pure extension it is [4]

$$\gamma_l = 3\,\frac{\Delta l}{l_0}\,, \tag{1}$$

where l_0 is the original length of the wire, and Δl is the elongation.

For twisting the situation is more complicated. Assuming an ideal torsional deformation (there is no elongation) the shear strain at a radius r' from the axis of the wire is given by [5]

$$d\gamma_t = \frac{r'\,d\vartheta'}{l}\,, \tag{2}$$

where $d\vartheta'$ is the angle of plastic torsion, l the momentary wire length. Because of our interest in macroscopic properties, where the average effect of the defect structure is dominant, we can use instead of (2) the average shear strain for the total wire area

$$d\bar{\gamma}_t = \frac{1}{r^2\pi}\int_0^r \frac{r'\,d\vartheta'}{l}\,2\pi r'\,dr' = \frac{2}{3}\,\frac{r\,d\vartheta'}{l}\,,$$

where r is the momentary wire radius. For a finite torsion we find by integration

$$\bar{\gamma}_t = \frac{2}{3}\int_0^\vartheta \frac{r\,d\vartheta'}{l} = \frac{2}{3}\,\frac{r_0\,\theta}{l_0}\,, \tag{3}$$

where r_0, resp. l_0 denote the original wire radius, resp. length, and θ is given by

$$\theta = \int_0^\vartheta \frac{d\vartheta'}{\left[1+\frac{\Delta l(\vartheta')}{l_0}\right]^{3/2}}\,.$$

In (3) use has been made of volume constancy during the plastic deformation ($r_0^2\,l_0 = r^2\,l$). So through the measurement of ϑ and Δl the average shear strain can be determined. According to our supposition made in the introduction the

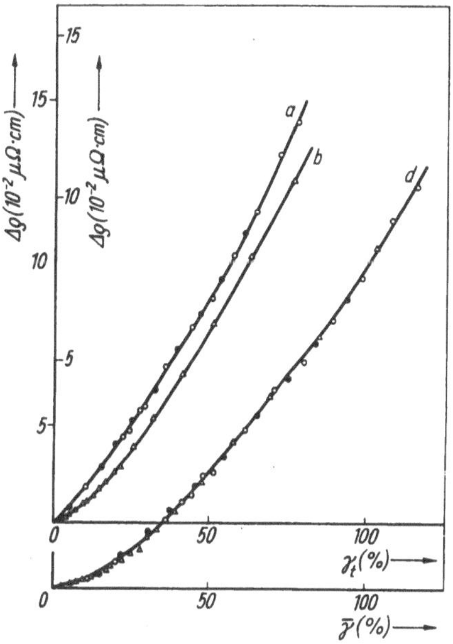

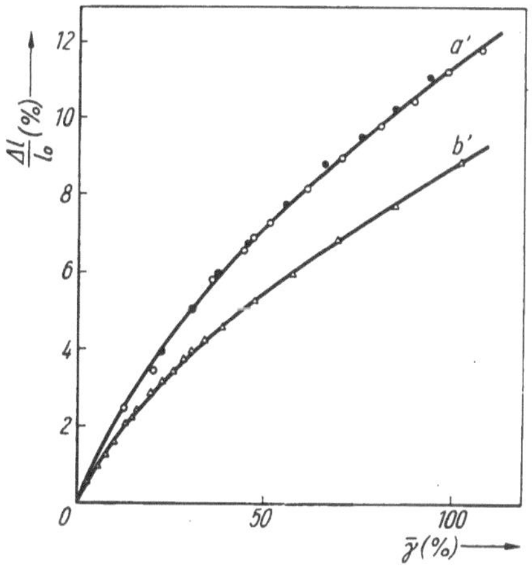

Fig. 1. Resistivity change in Au (annealed at 550 °C for 3 h) as a function of the torsional strain (curves a, b) and the total shear strain (curve d); elongations of Au wires as a function of the total shear strain (curves a′, b′)

resulting average strain is given by the following expression

$$\bar{\gamma} = \frac{2}{3}\,\frac{r_0\,\theta}{l_0} + 3\,\frac{\Delta l}{l_0}\,. \tag{4}$$

If our suppositions are indeed correct, than the change of a macroscopic property must be a single valued function of (4), and must not depend on the proportion of the individual strains.

In order to decide this issue, the change of specific resistivity of Au, Ag and Cu was measured at liquid nitrogen temperature for simultaneous torsion and extension. A typical result is shown for Au in Fig. 1. It is to be seen that if the torsional strain is only considered, than different curves are obtained (curves *a* and *b*). Similarly it can be seen that for the same torsional strain the greater specific resistivity increase corresponds to the greater elongation (curves *a′* and *b′*). If, however, the specific resistivity change is plotted in terms of the average strain, $\bar{\gamma}$, as defined by (4), a single universal curve is obtained. This result was invariably obtained for samples of different metals (Au, Ag, Cu) with different purity and previous history. Fig. 2 shows the final curves for Ag and Cu. We are fully confident that the validity of (4) is proved for all the three metals beyond doubt.

These results show that — at least for the case of twisting and extension — the resistivity change of specimens with the same previous history and purity, does not depend on the manner of the deformation, only on the total shear strain produced during deformation. To put the same in another words, for any similar samples the amount of strain and the specific resistivity change obeys a definite $\Delta\varrho = \varphi(\gamma)$ relationship. This relationship however, cannot be accepted as a final law, since the form of the function $\varphi(\gamma)$ depends on the initial condition of the specimen. So e.g. if the temperature of the annealing before the experiment is different, even for fairly high temperatures, different specific resistivity changes are found to correspond to the same strain.

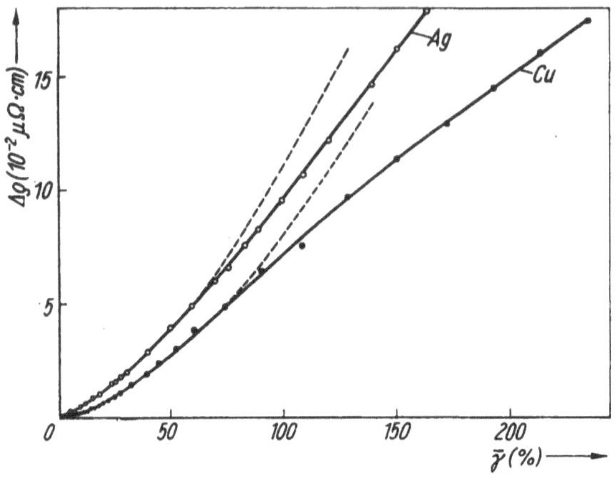

Fig. 2. Resistivity change in Ag and Cu as a function of the total shear strain

The form of the function $\varphi(\gamma)$ is greatly influenced by any previous deformation of the sample. The pioneering experiments were done by Peiffer [6]. The samples are deformed at a low temperature, than annealed at room temperature. So most of the point defects anneal out, while the dislocations immobile at room temperature remain in the sample. This leads to a resistivity decrease, measured at the low temperature. If the specimen is now further deformed the relationship now obtained is different, if the curve is regarded as commencing at the starting point of the second deformation. This procedure can be repeated, and it is invariably found that for higher previous deformation the shape of the $\Delta\varrho = \varphi(\gamma)$ curve goes over gradually from a convex shape into a concave one. Fig. 3 shows this effect for Ag and Cu deformed by torsion and extension. This result is in complete agreement with Peiffer's result, obtained for elongated Cu wires [6].

It is now evident the relationship often used in the literature,

$$\Delta\varrho = A\,\gamma^{p}\,, \tag{5}$$

is entirely uncharacteristic to the strain dependence of the specific resistivity, since A and p depend very strongly on the initial condition of the sample. Quite recently Akulov has proven in an other way [7], that a relationship of the form of (5) has no physical significance.

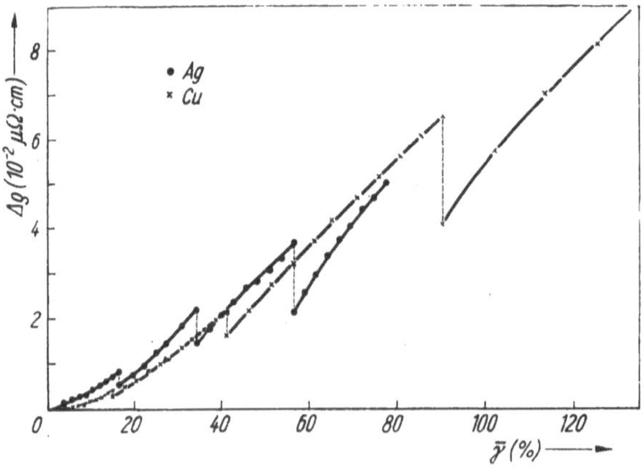

Fig. 3. The effect of prior deformation on the resistivity strain curve

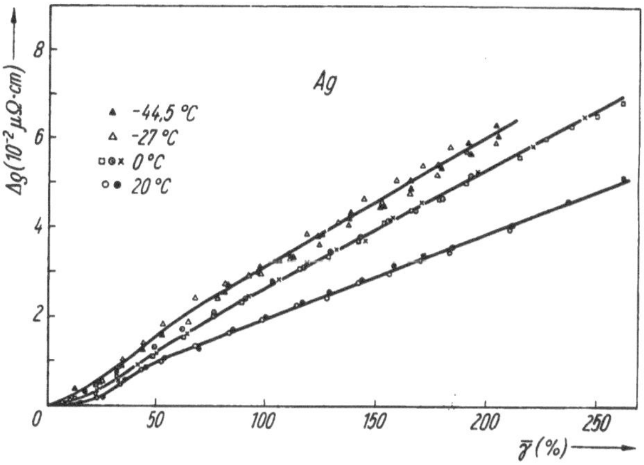

Fig. 4. Resistivity change in Ag vs. strain at different temperatures

The shape of the $\Delta\varrho = \varphi(\gamma)$ relationship depends also on the temperature. Fig. 4 shows the resistivity change of Ag wires for a few different temperatures. It is seen that although for any given temperatures the relationship $\Delta\varrho = \varphi(\gamma)$ is a unique one, nevertheless the curves themselves depend markedly on temperature.

All these facts clearly show that the connection between the amount of deformation and the specific resistivity does characterize neither — not even at a constant temperature — the material investigated, nor the physical processes occurring during the deformation. This is due to the fact that good and reproducible experimental results can be obtained for the change in specific resistivity only, this latter quantity depending, however, on the initial state of the specimen.

The initial state of the sample, however, is influenced by chemical impurities or the unavoidable straining of the sample during mounting in an apparatus, so it is no wonder that the results — mainly those obtained at different laboratories — differ substantially from each other.

3. The Connection between Stress and Strain

The defect structure of the solid determines in an unique manner the macroscopic flow stress through the mechanism of work hardening. So the momentary flow stress is always characteristic to the instantaneous defect structure of the material independently of any previous history.

It can be hoped, therefore, that the connection between the resistivity change and the flow stress is rather more inherent to the substance studied than the connection $\Delta\varrho(\gamma)$. In order to gather experimental material for checking this theoretical prediction, the stress-strain relationship for different metals was determined, for simultaneous extension and torsion. The principle of the measurement is already published [2]. Fig. 5 shows the stress-strain relationship for Au, Ag and Cu, measured at 78 °K. It is to be seen that the previous history of the sample is marked on the stress-strain curve too.

4. Resistivity Change in Dependence on the Flow Stress

Fig. 6 shows the connection between resistivity change and flow stress for Au, Ag and Cu. It can be seen that for Au and Ag the square root of the resistivity change depends linearly on the instantaneous flow stress. This linearity is inde-

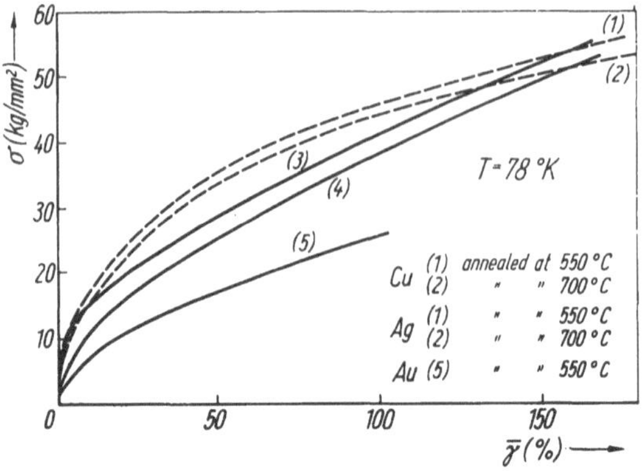

Fig. 5. Stress-strain curves of Cu, Ag and Au

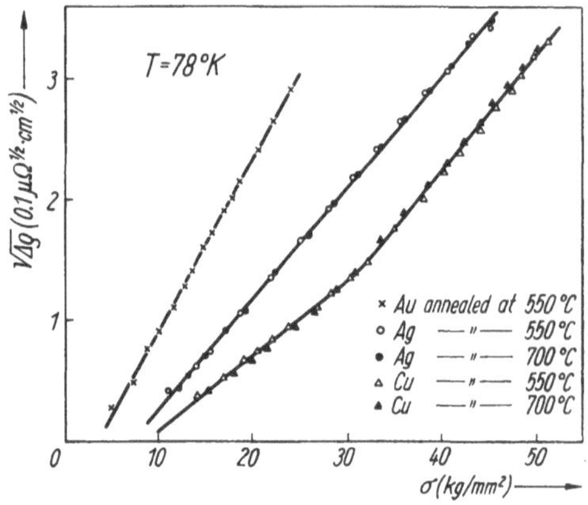

Fig. 6. Resistivity change in Cu, Ag and Au vs. flow stress

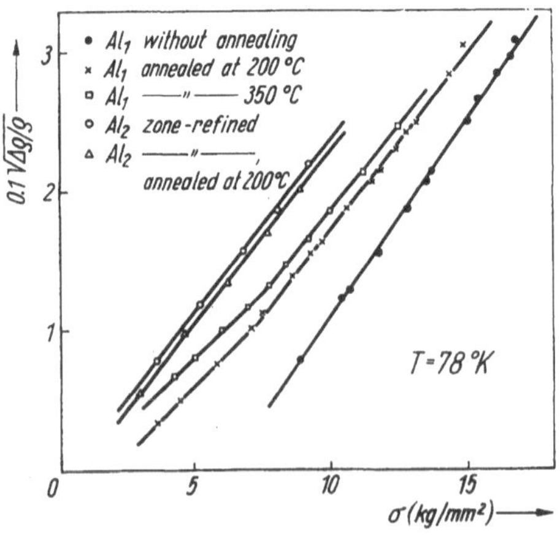

Fig. 7. Resistivity change in Al vs. flow stress (the data taken from Pistorius [9])

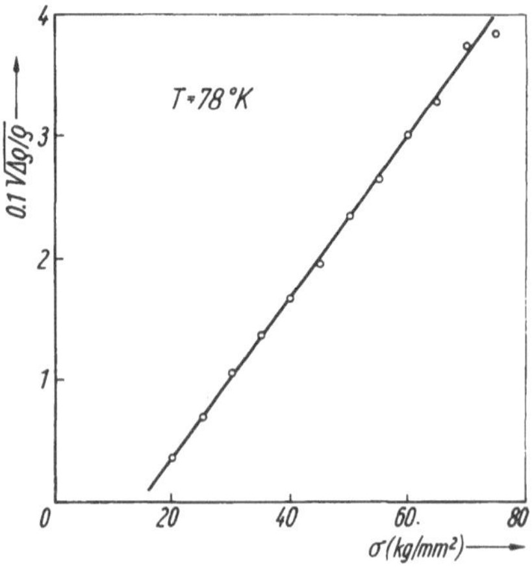

Fig. 8. Resistivity change in Rh vs. flow stress (the data taken from van Kuijk [10])

pendent of any heat treatment, further, in the case of silver it has already been shown that in a certain temperature interval, also of the temperature of deformation [8]. Our copper samples show some deviation from the case of Au and Ag, the plot consisting of two intersecting straight lines. This circumstance is probably due to impurities. The results obtained for Au (99,99%) and Ag (99.99%) seem to be special cases of a general law, substantiated by the analysis of other reliable measurements, taken from the literature and replotted by us. Pistorius' [9] data on the flow stress and resistivity change of Al samples of different purity and treatment are shown in Fig. 7. For the high purity samples a single straight line is obtained, the position of which depends, but slightly, on previous treatment (curves *a* and *b*). For the less pure materials curves similar to those found for Cu (curves *c* and *d*) were obtained. It is, however, interesting to remark that the intersection point of the straight lines is at about the same flow stress ($\approx$ 7.5 kg/mm^2). This again points to the fundamental role of the flow stress as the most characteristic parameter, further explains, why the first part is unobservable on the unannealed sample (curve *e*), since because of the large hardness, the measurement could have been started only above that point. Our conclusion for the pure materials is corroborated by the recent studies of van Kuijk [10] on Rh, shown in Fig. 8.

These findings are in complete agreement with the studies of Blewitt et al. [11] for Cu single crystals, further with those of Basinski et al. [12] for iron and potassium.

The above facts clearly indicate that the relationship between specific resistivity vs. flow stress is a universal relationship, of a particularly simple form for high purity materials, with the flow stress meeting all requirements for being a state variable.

5. Discussion

Theoretical studies of Saada [13] have shown that the number of point defects produced by plastic deformation at low temperatures is proportional to the plastic work. This result is substantiated by the findings of Blewitt et al. at 4 °K, and of Pistorius at 78 °K. Recently Dawson has proved the validity of the Saada formula for Au, Ag and Cu at 78 °K [14].

The plastic work, however, cannot be a totally representative property because of the changes in the defect structure induced by temperature dependent annealing processes. The validity of the formula was proven for a certain strain range only. In Dawson's work the shear strain was below 30%, in Pistorius' work below 60% shear, exceeding 100% only in the studies of Blewitt et al.

Recent theoretical and experimental findings indicate that the number of dislocations and point defects, produced during plastic deformation, are proportional to each other [8, 15, 16]. From this it follows that if the Saada formula is valid, than the total specific resistivity change must be also proportional to the plastic work. Our results can be used to decide this issue at 78 °K for large deformation. In Fig. 2. the dashed line gives a quantity proportional to the plastic work. An excellent agreement is seen up to a certain strain, in agreement with the studies of Pistorius and Dawson. Above a certain strain the formula does not hold any longer, a smaller resistivity change is obtained, as to be expected from the plastic work. The natural explanation is that at extremely large deformations some annealing takes place even at 78 °K. It is, however, interesting to remark that the proportionality between the plastic work and specific resistivity change loses its validity for differently pretreated samples of the same material at about the same flow stress.

The Saada formula is probably exactly valid at absolute zero temperature. At somewhat higher temperatures (e.g. that of liquid nitrogen) is valid up to not too high a strain, but becomes totally invalid at about room temperature. So the plastic work cannot be regarded as a correct state variable.

The flow stress on the other hand is in a one to one correspondence with the defect structure stabilized after the momentary increase of strain. From the cases investigated up to now, we can say with quite a good degree of certainty that for any reasonably pure f.c.c. metal the following relationship does hold

$$\Delta \varrho = \alpha \, (\sigma - \sigma_0)^2 \,, \qquad (6)$$

where the constant α is characteristic to the material.

Electron microscope investigations have shown [12] that for a number of metals the resolved shear stress is in a very simple connection with the average dislocation density, namely

$$\tau = \frac{1}{2} \mu \, b \, N^{1/2} \,, \qquad (7)$$

where μ is the shear modulus, b the Burgers vector. For some metals a similar relationship holds if the stress τ is regarded as the increment of the initial flow stress [17]. If it can be assumed, that latter case is the general one and formulas

Table 1

Material	$\alpha \left(\frac{10^{-10}\ \Omega\text{cm}}{(\text{kg/mm}^2)^2}\right)$	$\mu \left(\frac{10^{11}\ \text{dyn}}{\text{cm}^2}\right)$	b (10^{-8} cm)
Cu	0.80	4.0	2.55
Ag	0.83	3.0	2.88
Au	1.95	2.6	2.87
Al	1.21	2.7	2.86
Rh	0.22	15	2.68

Table 2

Material	$\Delta \varrho d/N$ at 78 °K (10^{-1} Ω cm³)		Resistivity of vacancies (10^{-6} Ω cm/at%)	n (10^9 cm^{-3})
	See references	Present work		
Cu	3.7 [11] 2.7 [20]	4.1	1.5 [21]	1.6
Ag	3.4 [20]	3.1	1.6 [22]	1.2
Au	3.4 [20]	5.3	1.7 [22]	1.9
Al	3.3 [20]	3.6*)	2.2 [23]	1.0
Rh	—	36*)	—	—

*) Obtained by using the data of [9, 10].

like (7) have been obtained because of the sufficiently low initial dislocation density, than the flow stress depends on the dislocation concentration, as [17, 18, 19]

$$\sigma - \sigma_0 = \mu \, b \sqrt{N} .$$

If we assume that half of the total specific resistivity change is due to point defects, the other half to dislocations, than the resistivity of a unit dislocation length is given by

$$\frac{\Delta \varrho_d}{N} = \frac{1}{2} \, \alpha \, \mu^2 \, b^2 . \tag{8}$$

The value of α can be taken from the measurements. For different metals, the data of μ and b taken from the literature and used for the calculation of $\Delta \varrho_d / N$ are compiled in Table 1. The obtained resistivities given in Table 2 are in good agreement with other results.

If one further assumes that the resistivity due to the point defects arises solely from the vacancies (this being of course a rough picture only), than the number of vacancies n corresponding to the unit length of a new dislocation can be obtained. These data are also shown in Table 2 in agreement again with other experimental and theoretical studies [11, 16].

Acknowledgement

Authors recognize the expert assistance of Mr. J. Balázs in performing a large part of the measurements.

References

[1] I. Kovács and E. Nagy, phys. stat. sol. **3**, 726 (1963).
[2] I. Kovács and P. Feltham, phys. stat. sol. **3**, 2379 (1963).
[3] J. H. Poynting, Proc. Roy. Soc. **86**, 534 (1912).
[4] P. Feltham, Phil. Mag. **8**, 989 (1963).
[5] H. G. Growe and E. Keppler, phys. stat. sol. **6**, 339 (1964).
[6] H. R. Peiffer, J. appl. Phys. **34**, 298 (1963).
[7] N. S. Akulov, Phil. Mag. **9**, 767 (1964).
[8] I. Kovács, E. Nagy, and P. Feltham, Phil. Mag. **9**, 797 (1964).
[9] C. A. Pistorius, Physica **27**, 149 (1961).
[10] J. G. M. van Kuijk, Physica **30**, 398 (1964).
[11] T. H. Blewitt, R. R. Coltman, and J. K. Redman, Report of the Conference on Defects in Crystalline Solids, The Physical Society, London 1955 (p. 369).
[12] J. S. Basinski, J. S. Dugdale, and A. Howie, Phil. Mag. **8**, 1989 (1963).
[13] V. G. Saada, Thesis, Paris 1960.
[14] H. I. Dawson, Thesis, Delft 1964.
[15] H. R. Peiffer, Acta metall. **11**, 435 (1963).
[16] P. Feltham, phys. stat. sol. **6**, 235 (1964).
[17] A. S. Keh, Direct Observations of Imperfections in Crystals, Interscience Publishers, New York 1962 (p. 213).
[18] J. E. Bailey and P. B. Hirsch, Phil. Mag. **5**, 485 (1960).
[19] J. E. Bailey, Phil. Mag. **8**, 223 (1963).
[20] L. M. Clarebrough, M. E. Hargreaves, and M. H. Loretto, Phil. Mag. **6**, 807 (1961) and **7**, 115 (1962).
[21] R. O. Simmons and R. W. Baluffi, Phys. Rev. **129**, 1535 (1963).
[22] A. Seeger, Radiation Damage in Solids, International Atomic Energy Agency, Vienna 1962 (p. 101).
[23] M. Doyama and J. S. Koehler, Phys. Rev. **134**, A 522 (1964).

(Received December 28, 1964)

phys. stat. sol. 8, 805 (1965)

Research Institute for Technical Physics of the Hungarian Academy of Sciences, Budapest

On the Theory of Radiative Recombination in High Magnetic Field in Semiconductors

By

F. BELEZNAY and G. PATAKI

The effect of a high magnetic field on the recombination between the conduction band and an acceptor is examined. It is found that at magnetic field which corresponds to the "quantum limit" for the conduction band but gives no "deformation" of the acceptor states, the transition matrix element does not depend on the magnetic field. The assumptions for the magnetic field are fulfilled for the case of laser radiation in a magnetic field. It may be said that on the basis of the hydrogen-like model, the field dependence of the threshold current must be determined by the change in the density of states.

Es wird der Einfluß eines hohen Magnetfeldes auf die Rekombination von Elektronen zwischen Leitungsband und Akzeptorniveaus untersucht. Bei Magnetfeldern, die der „Quantengrenze" für das Leitungsband entsprechen, jedoch zu keiner „Deformation" der Akzeptroniveaus führen, hängt das Übergangsmatrixelement nicht vom Magnetfeld ab. Die Annahmen bezüglich des Magnetfeldes sind für den Fall einer Laserstrahlung im Magnetfeld erfüllt. Im Gültigkeitsbereich des wasserstoffähnlichen Modells muß die Feldabhängigkeit des Schwellenstroms aus der Änderung der Zustandsdichte ermittelt werden.

1. Introduction

The GaAs laser [1] and, successively, the discovery of different injecting lasers prepared from III–V compounds, called the attention to the microscopical theory of recombination mechanism [2]. Though no definite solution is yet given, it may be supposed that the transitions occur between the injected minority electrons and the acceptor states on the p-type side [3]. The probability of radiative transitions between bands and impurities were examined by Dumke [4], who found this recombination mechanism to be sufficiently effective to exceed the others.

The hydrogen-like model of impurities in case of overdoped materials, the actual case in laser diodes, is very problematical. At high density of impurities an overlapping of wave functions takes place but it is difficult to say at which degree of overlapping the simple theory will loose its validity. The investigation of transmission and fluorescence of homogeneous GaAs still showed that at low temperatures (77 °K) the radiative transition between conduction band and acceptor atom may occur in weakly doped samples too [5].

In order to choose between various recombination mechanisms, the effect of magnetic field on the radiation of diodes was investigated by various authors. In a very high magnetic field a shift and a splitting of the line of radiation occurs. These effects can be interpreted with the aid of quantum effects taking place in high magnetic fields [7]. The application of the magnetic field changes the threshold current, needed for laser action too. The magnetic field seems to change the probability of the radiative recombination [7].

In the present paper the probability of radiative transition between conduction band and acceptor states is calculated in the case of a very high magnetic field, in quantum limit. As the hydrogen-like model is accepted to describe the acceptor state, the variation of the threshold current of lasers can be discussed qualitatively only. It was preferred by us to consider the mechanism of the radiative recombination. In the following, the effect of high magnetic field on the simple conduction band-acceptor transition is considered. Calculations relating to more complicated models are being carried on.

2. The Description of Electron States in the High Magnetic Field

For the sake of simplicity an idealized semiconducting material is assumed, having parabolic bands, the extremes of which are situated at $\boldsymbol{k} = 0$. (The effect of the more complicated valence band is considered when the acceptor atoms are treated.) The effective mass approximation is used in both the conduction and valence band. The effective masses are scalar quantities because of the simple energy-momentum relations. The wave functions have in both bands the following forms

$$\psi_i(\boldsymbol{r}) = u_{i0}(\boldsymbol{r})\, f_i(\boldsymbol{r}) \,, \tag{1}$$

where the index i refers to the bands; $u_{i0}(\boldsymbol{r})$ is the band-edge function at the point $\boldsymbol{k} = 0$, i.e. the part of Bloch's functions $e^{i\boldsymbol{k}\boldsymbol{r}}\, u_k(\boldsymbol{r})$ which has the periodicity of the lattice. Finally, the functions $f_i(\boldsymbol{r})$ are the solutions of Schrödinger's equation taking into account the perturbations (the high magnetic field and the Coulomb potential, respectively) in the effective mass approximation.

2.1 The electron states in the conduction band

It is known that in the presence of a high magnetic field the motion of the carriers in the plane perpendicular to the magnetic field should be treated quantum mechanically, i.e. it should be quantized. Let the magnetic field of strength H be of direction z. To treat the free electrons, the "Landau-measure" [8] is usually used where the vector potential $\boldsymbol{A}$ corresponding to the magnetic field H has only one component differing from zero, say the component in direction y, in detail: $A_x = A_z = 0$; $A_y = H\,x$. As the impurities have a potential of spherical symmetry, it is more suitable to use a "measure of cylindrical symmetry" and to use cylindrical coordinates.[1]) In this case the components of vector potential $\boldsymbol{A}$ are: $A_z = 0$, $A_x = -\frac{1}{2} H\,y$, $A_y = \frac{1}{2} H\,x$, or in cylindrical coordinates: $A_z = A_\varrho = 0$, $A_\varphi = \frac{1}{2} H\,\varrho$. The Hamiltonian for the electron in effective mass approximation is (the charge of the electron being negative)

$$\hat{\mathcal{H}} = \frac{1}{2\,M_e}\left(\hat{\boldsymbol{p}} + \frac{e}{c}\,\hat{\boldsymbol{A}}\right)^2, \tag{2}$$

where M_e is the effective mass of the electron. Inserting the value of the vector potential, Schrödinger's equation in cylindrical coordinates runs [9]

$$-\frac{\hbar^2}{2\,M_e}\left[\frac{1}{\varrho}\frac{\partial}{\partial\varrho}\left(\varrho\,\frac{\partial h}{\partial\varrho}\right) + \frac{\partial^2 h}{\partial z^2} + \frac{1}{\varrho^2}\frac{\partial^2 h}{\partial\varphi^2}\right] - \frac{i\,e\,\hbar}{2\,M_e c}\,H\,\frac{\partial h}{\partial\varphi} + \frac{e^2 H^2}{8\,M_e c^2}\,\varrho^2\,h = E\,h\,. \tag{3}$$

---

[1]) To calculate the matrix element of transition beyond practical purposes, it is *necessary* to use the same measure for the free electron and the bound hole state as well.

As it is well known, the solution can be factored as $h = h(z)\, h(\varrho, \varphi)$. In direction z the motion of the electron will not be perturbed by the magnetic field and the electron moves as a "free" particle described by a planewave. This motion can be characterized by the z-component of the momentum, namely the quantum number p. ($p/\hbar = k$ is the wave number.)

The part of wave function depending on z is

$$h(z) = \frac{1}{\sqrt{L_z}} \exp\left\{i \frac{p}{\hbar} z\right\}, \tag{4}$$

where L_z is the dimension of the sample in the z-direction.

The motion in the $x\,y$-plane, i.e. the solution of the "eigenvalue" problem can be described by two positive integers: n and l. The corresponding orthonormal energy-eigenfunctions are [10]

$$h(\varrho, \varphi) = (-i)^n\, (2^{l-n+1}\, \pi\, l!)^{-1/2}\, (n!)^{1/2}\, \lambda^{-1}\, (\varrho/\lambda)^{l-n} \times \\ \times \exp\{-\varrho^2/4\,\lambda^2 + i\,(n-l)\,\varphi\}\, \mathcal{L}_n^{l-n}\,(\varrho^2/2\,\lambda^2)\,, \tag{5}$$

where the generalized Laguerre polynomials $\mathcal{L}_n^{l-n}$ are defined as

$$\mathcal{L}_n^{l-n}(x) = \frac{e^x\, x^{n-l}}{n!}\, \frac{d^n}{dx^n}\,(e^{-x}\, x^l)\,. \tag{6}$$

The notation $\lambda^2 = \dfrac{\hbar\, c}{e\, H}$ was introduced. The quantity λ, the so-called magnetic length, has a dimension of length and characterizes the quantum mechanical treatment of the problem. The quantum numbers n and l have very clear meanings. If the energy of the electron is divided in two parts: into the kinetic energy of the free motion parallel to the magnetic field and into the energy of other degrees of freedom, as

$$E = E_t + \frac{p^2}{2\,M_e}\,, \tag{7}$$

then the possible values of E_t are determined by the quantum number n alone, being

$$E_t = E_n = \hbar\,\omega_c \left(n + \frac{1}{2}\right), \tag{8}$$

where $\omega_c = \dfrac{e\,H}{M_e\,c}$ is the known cyclotron frequency. At a given value of n, the quantum number l may be any positive integer: every energy eigenvalue is infinitely degenerate. The quantum number l has also a clear interpretation.

In the plane perpendicular to the magnetic field, the classical motion of the electron is a circle. In the quantum mechanical treatment an operator will be ordered to the centre of this classical circular motion. It is easy to verify that the square of this operator ($\hat{\varrho}_0^2$) is commutable with the Hamilton operator of the system. The energy eigenfunctions are, therefore, the eigenfunctions of operator $\hat{\varrho}_0^2$ too. Determining the eigenvalues, we get

$$\hat{\varrho}_0^2\, h_{n,l} = \lambda^2\,(2\,l + 1)\, h_{n,l}\,. \tag{9}$$

So the eigenfunctions of different quantum number l in the $x\,y$-plane correspond to the centres of circle which are at different distances from a given point, e.g. from the origin of the system of coordinates. The above distances have discrete values because the coordinates x_0 and y_0 of the centre of rotation do not commute with each other. The infinite degeneracy is related to the degeneracy of the corresponding classical problem, namely the energy of the electron moving on a spiral in a magnetic field does not depend on the position of the center of the circle in the $x\,y$-plane. A detailed analysis of this quantized motion is given in paper [10].

An electron state of given energy will later be needed in which the probability of finding the electron in any part of the $x\,y$-plane is constant. Such a state, as usual in the case of degeneracy, can be constructed by the linear combination of degenerate eigenfunctions.

The functions $h_{n,l}$ were the orthonormal eigenfunctions. According to equation (9) the state $l = 0$ and the other states have the weights of 1 and 2, respectively. So a possible eigenfunction is

$$f_n^{(0)} = \frac{1}{c}\left(h_{n,0} + \sum_{l\neq 0} \sqrt{2}\, h_{n,l}\right). \tag{10}$$

The value of the constant c can be obtained from the condition of normalization

$$|c|^2 = 1 + \sum_{l\neq 0} 2\,. \tag{11}$$

If the crystal has finite dimensions, say in the x and y directions it has the same lengths L, then the maximum value of $\langle \hat{\varrho}_0^2 \rangle$ is $L^2/4$, thus we get for maximum value of l

$$(2\, l_{\max} + 1) = L^2/4\,\lambda^2\,, \tag{12}$$

where equation (9) was used. We obtain from equations (11) and (12)

$$|c| = \frac{L}{2\,\lambda} \tag{13}$$

and so we get the normalized eigenfunction in the form

$$f_n^{(0)} = \frac{2\,\lambda}{L}\left[h_{n,0} + \sum_{l\neq 0} \sqrt{2}\, h_{n,l}\right]. \tag{14}$$

This is evidently not the only desired electron state. It is possible to introduce states of number $l_{\max}$ with similar properties, as follows:

$$f_n^{(i)} = \frac{2\,\lambda}{L}\left[h_{n,0} + \sum_{l\neq 0} \sqrt{2}\, e_l^{(i)}\, h_{n,l}\right], \tag{15}$$

where the complex numbers of unite absolute values $e_l^{(i)}$ serve to make the functions $f_n^{(i)}$ orthogonal.

2.2 The state of hole bound on the acceptor atom

In the effective mass approximation the Hamiltonian for the hole, moving in a Coulomb potential, is

$$\hat{\mathcal{H}} = \frac{1}{2\,M_h}\left(\hat{\boldsymbol{p}} - \frac{e}{c}\,\hat{\boldsymbol{A}}\right)^2 - \frac{e^2}{\varepsilon\, r}\,. \tag{16}$$

Here M_h is the effective mass of the hole, ε the dielectric constant. The origin of the system of coordinates is chosen to be in the centre of impurity. If the energy and the length are measured in "effective" Rydbergs and "effective" Bohrradii, respectively $\left(R\,y = \frac{M_h\, e^4}{2\,\hbar^2\,\varepsilon^2} \text{ and } a_0 = \frac{\hbar^2\,\varepsilon}{M_h\, e^2}\right)$ then the eigenvalue equation of the hole energy is

$$\left[-\nabla^2 - \gamma\, L_z + \frac{\gamma^2}{4}\,(x^2 + y^2) - \frac{2}{r}\right] g = E\, g\,, \tag{17}$$

where all operators, lengths and energies, are dimensionless quantities. L_z is for instance the dimensionless operator of the component of angular momentum operator in direction z. γ is a dimensionless parameter. It is given by

$$\gamma = \frac{\hbar\,\omega_c}{2\,R\,y} \tag{18}$$

and its value is proportional to the magnetic field, being

$$\gamma = H\left[2 \cdot 10^9 \left(\frac{M_h}{M_0\,\varepsilon}\right)^2\right]^{-1} \tag{19}$$

if H is measured in Gauss. M_0 is the mass of the free electron. As the effective mass of holes in III–V compounds is not sufficiently less than that of the "free" electrons and the dielectric constant is about 10, $\gamma \ll 1$ is obtained for magnetic fields usually applied in practice. The term in equation (17) containing γ^2 can, therefore, be omitted as the term $(x^2 + y^2)$ and the other terms have expectation values of order of unity [11]. The solution of the remaining equation can easily be obtained if we remember that the z-component of the angular momentum and the Hamiltonian for the hydrogen problem without magnetic field commute with each other. The hydrogen-like eigenfunctions of hole-states will be [12]

$$g_{n,l,m}(r, \vartheta, \varphi) = R_{nl}(r)\; Y_{lm}(\vartheta, \varphi)\,, \tag{20}$$

$$R_{n,l}(r) = -\left\{\left(\frac{2}{n\,a_0}\right)^3 \frac{(n-l-1)!}{2\,n\,[(n+l)!]^3}\right\}^{1/2} e^{-\frac{r}{n a_0}} \left(\frac{2\,r}{n\,a_0}\right)^l L_{n+l}^{2l+1}\left(\frac{2\,r}{n\,a_0}\right),$$

where $Y_{lm}(\vartheta, \varphi)$ is the normalized spherical harmonic, and

$$L_n^m(x) = \frac{d^m}{dx^m}\left\{e^x \frac{d^n}{dx^n}\,(x^n\, e^{-x})\right\}$$

is the generalized Laguerre polynomial.[2])

The energy eigenvalues in Rydbergs have the form

$$E_{n,m} = E_{n,0} - \gamma\, m \tag{21}$$

where $E_{n,0}$ denotes the n-th hydrogen-eigenvalue. As it is known, the more rigorous variational calculations for Ge and Si lead to the conclusion that the eigenfunctions (20) give a rather good description of hole states, in cases of more complicated valence bands too [13].

3. The Transition Probability

Let us examine now the probability of radiative transitions between different states of crystal electrons in a high magnetic field. The transitions are governed by the interaction with the electromagnetic field. The Hamiltonian for interaction of photon field and electrons is

$$\hat{\mathscr{H}}' = \frac{e}{n\,M_0\,c}\left[\hat{\boldsymbol{p}} + \frac{e}{c}\,\hat{\boldsymbol{A}}\right]\hat{\boldsymbol{A}}_s\,, \tag{22}$$

where the radiative part of the electromagnetic field, $\hat{\boldsymbol{A}}_s$ is considered as a perturbation and n is the refractive index of the material. If the wavelength of the radiation is large enough compared with the extension of eigenfunctions of the electron states, then the space dependence of the radiation can be neglected and for the matrix element of radiative interaction the result is

$$\mathscr{H}'_{12} = \frac{e}{n\,M_0\,c}\left(\frac{2\,\pi\,\hbar\,(N+1)}{\omega}\right)^{1/2} \langle 1|\left(\hat{\boldsymbol{p}} + \frac{e}{c}\,\hat{\boldsymbol{A}}\right)\boldsymbol{e}\,|2\rangle \tag{23a}$$

in case of photon emission, while for photon absorption

$$\mathscr{H}'_{12} = \frac{e}{n\,M_0\,c}\left(\frac{2\,\pi\,\hbar\,N}{\omega}\right)^{1/2} \langle 1|\left(\hat{\boldsymbol{p}} + \frac{e}{c}\,\hat{\boldsymbol{A}}\right)\boldsymbol{e}\,|2\rangle \tag{23b}$$

---

[2]) The generalized Laguerre polynomials, figuring in (6) and (20) have different definitions. Both forms are, however, used in applied mathematics.

is valid. Here $\boldsymbol{e}$ is the unit polarisation vector of photon, $\langle 1|$ and $\langle 2|$ are the electron states taking part in the transition. If the electron states belonging to the conduction and valence bands, respectively, are given in the effective mass approximation by (1) and $f_1(r)$, $f_2(r)$, $\boldsymbol{A}$ are slowly varying functions compared with the variations of $u_{i0}(\boldsymbol{r})$ having the periodicity of the crystal lattice, then the value of the matrix element, figuring in (23a), is

$$\mathscr{H}'_{12} = \frac{e}{n\, M_0\, c} \left(\frac{2\,\pi\,\hbar\,(N+1)}{\omega}\right)^{1/2} (\boldsymbol{p}_{12}\,\boldsymbol{e})\, I_{12} \tag{24}$$

and similarly in (23b). Here $(\boldsymbol{p}_{12}\,\boldsymbol{e})$ is the so-called momentum matrix element between bands,

$$(\boldsymbol{p}_{12}\,\boldsymbol{e}) = \int\limits_{\text{unit cell}} u^*_{10}(\boldsymbol{r})\,(\hat{\boldsymbol{p}}\,\boldsymbol{e})\,u_{20}(\boldsymbol{r})\,d\boldsymbol{r}\,, \tag{25}$$

while the integral I_{12} describes the overlapping of the envelope-functions in effective mass approximation is given by

$$I_{12} = \int\limits_V f^*_1(\boldsymbol{r})\, f_2(\boldsymbol{r})\, d\boldsymbol{r}, \tag{26}$$

where the integration is extended over the whole crystal. According to the time dependent perturbation theory the probability per unit time of transition between two states is

$$P_{12} = \left(\frac{2\,\pi}{\hbar}\right) |\mathscr{H}'_{12}|^2\, \delta\,(\hbar\,\omega + E_2 - E_1)\,. \tag{27}$$

In our case, the two states are the conduction electron in the high magnetic field and the bound hole on the acceptor atom, respectively. Let us consider the case when the ground state must be taken into account only. For integral I_{12} figuring in (24) we get

$$I_{12} = \frac{1}{L\,L_z^{1/2}}\,\frac{2^{1/2}}{\pi\,a_0^{3/2}} \int\limits_{-L_z/2}^{L_z/2} \int\limits_0^{L/2} \int\limits_0^{2\pi} d\varphi\, d\varrho\, dz\, \varrho \exp\left\{-\frac{\varrho^2}{4\,\lambda^2} + i\,\frac{p}{\hbar}\,z - \frac{r}{a_0}\right\}. \tag{28}$$

It is suitable to introduce the system of parabolic coordinates to calculate this integral. After simple calculation we get

$$I_{12} = \frac{1}{L\,L_z^{1/2}}\,\frac{2^{1/2}}{a_0^{3/2}}\,(2\,\lambda)^3 \left\{\frac{A_0}{1 + A_0\,K} - \frac{1}{A_0}\, e^{\frac{1+A_0^2K^2}{A_0^2}} \times \mathrm{Ei}\left[-\frac{1 + A_0^2\,K^2}{A_0^2}\right]\right\}, \tag{29}$$

where the following notations were introduced:

$$A_0 = \frac{a_0}{\lambda}\,, \qquad K = \frac{p}{\hbar}\,\lambda\,. \tag{30}$$

This evidently means that the length is measured in the units of the characteristic length λ. The function $\mathrm{Ei}\,(-x)$ is defined as

$$\mathrm{Ei}\,(-x) = \int\limits_{-\infty}^{-x} \frac{e^t}{t}\, dt\,. \tag{31}$$

Having the asymptotic form of $\mathrm{Ei}\,(-x)$, the value of the integral I_{12} for $x \gg 1$ is [14]

$$I_{12} \approx \frac{1}{L\,L_z^{1/2}}\,\frac{2^{1/2}}{a_0^{3/2}}\, 8\, \frac{a_0^3}{(1 + a_0^2\,k^2)^2}\,. \tag{32}$$

With the exception of a constant factor, this result is identical with the case without a magnetic field [15]. The above result is quite evident as the condition

$x \gg 1$ is equivalent with the conditions $A_0 \ll 1$ or $a_0 \ll \lambda$ and in this case the expression $e^{-\varrho^2/4\lambda^2}$ can be factored out from the integral. This means that it is impossible to distinguish the eigenfunctions in the high magnetic field and the plane wave of z-direction in the region where the hydrogen-like acceptor is extended.

4. Discussion and Application of the Results

Having the matrix element, the important physical quantities as lifetime, absorption etc. can be calculated. We have previously seen that the value of the matrix element is determined by the parameters A_0 and K. Let us estimate the possible values of A_0 and K within the frame of validity of the accepted model.

In the above calculation the lowest "Landau-level", the electron state with quantum number $n = 0$ was considered. The condition of this so-called "quantum limit" is the following

$$\hbar\,\omega_c \gg k\,T\,. \tag{33}$$

If we express the kinetic energy of the particle as $E_k \approx k\,T$, or in the effective mass approximation $E_k = 1/2\,M_e\,p^2$, the parameter K will be

$$K^2 = \frac{p^2}{\hbar^2}\lambda^2 = \frac{2\,E_k}{\hbar\,\omega_c} \ll 1\,. \tag{34}$$

It is possible to write the parameter A_0 in the form of

$$A_0^2 = \frac{a_0^2}{\lambda^2} = \frac{\hbar\,\omega_c}{2\,E_A}\cdot\frac{M_e}{M_h}\,, \tag{35}$$

where E_A means the ionization energy of the acceptor. The relation (35) can be expressed with the aid of "effective" Rydbergs introduced in (18) as follows:

$$A_0^2 = \frac{\hbar\,\omega_c}{2\,R\,y}\,\frac{M_e}{M_h} = \gamma\,\frac{M_e}{M_h}\,. \tag{36}$$

The usual eigenfunction to describe the hydrogen ground-state can be applied if condition $\gamma \ll 1$ is valid, restricting the value of the applicable magnetic field as we have seen in Section 2.1. On the other hand, in compounds III-V $M_e < M_h$ and we get $A_0^2 \ll 1$, so the variable x in the argument of the function Ei is

$$x = K^2 + \frac{1}{A_0^2} \gg 1 \tag{37}$$

which always holds if the model is valid.

To the authors knowledge the effect of the high magnetic field on the recombination was examined in case of the laser radiation only. In the following we shall discuss the behaviour of laser action on the basis of experimental data given in paper [7], assuming the validity of the hydrogen-like model. In the paper cited InSb was used at 1.7 °K at magnetic field intensity of $H = 20$ to 90 kG. As it is known, the effective masses are $M_e = 0.013\,M_0$, $M_h = 0.6\,M_0$ [16] and the dielectric constant is $\varepsilon = 17$. Using the above data, we get for the parameters of the theory:

$$K^2 = 1.64\times10^{-2}\,,\quad A_0^2 = 1.74\times10^{-4}\quad \text{in case of}\quad H = 20\,\text{kG}$$

and similarly

$$K^2 = 3.64\times10^{-3},\quad A_0^2 = 7.82\times10^{-4}\quad \text{if}\quad H = 90\,\text{kG}\,.$$

It is thus found that the model is applicable and the matrix element is independent from the magnetic field.

According to the experiments of Phelan et al. the threshold current of laser action decreases with an increasing magnetic field. The following explanation is possible to interprete the decrease of the threshold current: the number of conduction electrons in a very narrow energy band — corresponding with the line width of emission — increases, just like the density of states, with the square root of the magnetic field. So the density of the emitted photons, with unchanged transition matrix element, increases with the square root of the magnetic field.

The threshold current, however, should have a very complicated magnetic field dependence. In order to get the correct relation, the connection of the laser-mode and the other modes of different frequencies (according to the Kramer-Kronig dispersion relation) as well as the distribution of injected electrons must be taken into consideration. The usual determination of the threshold current [17], however, is very problematical, the lifetime being about 10^{-10} s [17, 4], which is comparable with the relaxation time of thermalization. So the injected electrons may be hot electrons and equilibrium statistics cannot be applied.

We have previously seen that the matrix element does not depend on the magnetic field for the accepted model, so the lifetime can be determined simply. The effects depending directly on the density of states, e.g. the line-shape of spontaneous emission etc., will show a magnetic field dependence only.

The purpose of the paper was the investigation of the microscopic mechanism of recombination. The investigation of the effect of the high magnetic field on more complicated recombination models will be done in future.

References

[1] M. I. NATHAN, W. P. DUMKE, G. BURNS, F. H. DILL, and G. J. LASHER, Appl. Phys. Letters **1**, 62 (1962).
R. N. HALL, G. E. FENNER, J. D. KINGSLEY, T. J. SOLTYS, and R. O. CARLSON, Phys. Rev. Letters **9**, 366 (1962).
J. M. QUIST, R. H. REDIKER, R. J. KEYES, W. E. KRAG, B. LAX, A. L. MCWHORTER, and H. J. ZEIGER, Appl. Phys. Letters **1**, 9 (1962).

[2] D. MEYERHOFER and R. BRAUNSTEIN, Appl. Phys. Letters **3**, 171 (1963).
T. L. LARSEN, Appl. Phys. Letters **3**, 113 (1963).
D. C. DOUSMANIS, C. W. MUELLER, and H. NELSON, Appl. Phys. Letters **3**, 133 (1963).

[3] M. I. NATHAN and G. BURNS, Appl. Phys. Letters **1**, 89 (1962).

[4] W. P. DUMKE, Phys. Rev. **132**, 1998 (1963).

[5] D. E. HILL, Phys. Rev. **133**, A 866 (1964).

[6] I. MELNGAILIS and R. H. REDIKER, Appl. Phys. Letters **2**, 202 (1963).
I. MELNGAILIS, A. J. STRAUSS, and R. H. REDIKER, Proc. IEEE **51**, 1154 (1963).
F. L. GALEENER, G. B. WRIGHT, W. E. KRAG, I. M. QUIST, and H. J. ZEIGER, Phys. Rev. Letters **10**, 472 (1963).

[7] R. J. PHELAN, A. R. CALAWA, R. H. REDIKER, R. J. KEYES, and B. LAX, Appl. Phys. Letters **3**, 143 (1963).

[8] L. LANDAU, Z. Phys. **64**, 629 (1930).

[9] Л. Д. Ландау и Е. М. Лифшиц, Квантовая механика, Москва 1963 (p. 495).

[10] M. H. JOHNSON and B. A. LIPPMANN, Phys. Rev. **76**, 828 (1949).

[11] Y. YAFET, R. W. KEYES, and E. N. ADAMS, J. Phys. Chem. Solids **1**, 137 (1956).

[12] L. I. SCHIFF, Quantum Mechanics, New York 1955 (p. 85).

[13] K. SUZUKI, M. OKAZAKI, and H. HASEGAWA, J. Phys. Soc. Japan **19**, 930 (1964).

[14] JANKE/EMDE/LÖSCH, Tafeln höherer Funktionen, Stuttgart 1960.

[15] D. M. EAGLES, J. Phys. Chem. Solids **16**, 76 (1960).

[16] C. HILSUM and A. C. ROSE-INNES, Semiconducting III-V Compounds, Pergamon Press, 1961 (p. 72).

[17] G. LASHER and F. STERN, Phys. Rev. **133**, A 553 (1964).

(Received December 28, 1964)

phys. stat. sol. 8, 813 (1965)

Institut für Theoretische Physik der Justus-Liebig-Universität Gießen

Über die irreduziblen Darstellungen der Raumgruppen und die Strahldarstellungen der kristallographischen Punktgruppen

Von

A. Kitz

Es wird ein Verfahren beschrieben, wie man aus den Strahldarstellungen der kristallographischen Punktgruppen die irreduziblen Darstellungen der Raumgruppen findet. Diese Methode läßt sich ebenso auf die doppelten Raumgruppen anwenden. Da man hierbei die Strahldarstellungen der doppelten kristallographischen Punktgruppen braucht, werden sämtliche voneinander verschiedene Strahldarstellungen dieser Gruppen aufgestellt.

A method of finding the irreducible representations of the space groups from the ray representations of the crystallographic point groups is described. This method can also be applied to the double space groups. Since the ray representations of the double crystallographic point groups are necessary, all the separate ray representations of these groups are calculated.

1. Die irreduziblen Darstellungen der Raumgruppen

Um die Symmetrieeigenschaften und Zusammenhänge der Elektronenbänder zu untersuchen, benötigt man die irreduziblen Darstellungen der Raumgruppen; denn die Wellenfunktionen zu einem Energieeigenwert spannen eine Darstellung der Symmetriegruppe des Kristalls auf. Die Symmetriegruppe ist eine der 230 Raumgruppen und besteht aus den Transformationen des Ortsvektors, gegenüber denen das Potential und damit der Hamiltonoperator invariant sind.

Die allgemeinste Gestalt einer solchen Transformation lautet

$$\boldsymbol{r}' = C\,\boldsymbol{r} + \boldsymbol{t} = C\,\boldsymbol{r} + \boldsymbol{\tau}(C) + \boldsymbol{t}_n\,. \tag{1}$$

Es ist üblich, dafür

$$\boldsymbol{r}' = \{C|\boldsymbol{\tau}(C) + \boldsymbol{t}_n\}\,\boldsymbol{r} \tag{2}$$

zu schreiben und das zugehörige Raumgruppenelement mit dem Operator $\{C|\boldsymbol{\tau}(C) + \boldsymbol{t}_n\}$ zu identifizieren. C ist eine dreidimensionale, reelle orthogonale Matrix und beschreibt bei $\det C = 1$ eine Drehung bzw. bei $\det C = -1$ eine Spiegelung oder Drehspiegelung, die zu einer Schraubung oder Gleitspiegelung werden, wenn die neben C auftretende Translation $\boldsymbol{\tau}(C) + \boldsymbol{t}_n$ eine von Null verschiedene Komponente in Richtung der Drehachse bzw. parallel zur Spiegelebene besitzt. Die einfache Translation $\boldsymbol{t}_n$ kann geschrieben werden in der Form

$$\boldsymbol{t}_n = n_1\,\boldsymbol{a}_1 + n_2\,\boldsymbol{a}_2 + n_3\,\boldsymbol{a}_3\,, \tag{3}$$

wo n_1, n_2, n_3 drei ganze Zahlen und $\boldsymbol{a}_1$, $\boldsymbol{a}_2$, $\boldsymbol{a}_3$ die drei linear unabhängigen Basisvektoren des Translationsgitters sind. Die Translationsgruppe besteht aus der Gesamtheit aller einfachen Translationen und ist ein Normalteiler der Raumgruppe.

Die Drehanteile C bilden die Punktgruppe. Wendet man die Elemente der Punktgruppe auf einen reduzierten k-Vektor der ersten Brillouinzone an, so ergibt sich ein Satz äquivalenter k-Vektoren, den man einen Stern nennt.

Für das Produkt zweier Raumgruppenelemente gilt die Regel

$$\{C_k|\boldsymbol{\tau}(C_k) + \boldsymbol{t}_{n'}\}\,\{C_i|\boldsymbol{\tau}(C_i) + \boldsymbol{t}_n\} = \{C_k\,C_i|C_k\,\boldsymbol{\tau}(C_i) + C_k\,\boldsymbol{t}_n + \boldsymbol{\tau}(C_k) + \boldsymbol{t}_{n'}\} =$$
$$= \{C_k\,C_i|\boldsymbol{\tau}(C_k\,C_i) + \boldsymbol{t}_{n''}\}\,. \qquad (4)$$

Jede irreduzible Darstellung einer Raumgruppe läßt sich durch zwei Parameter kennzeichnen. Der erste Parameter $\boldsymbol{k}$ bezieht sich auf den Stern der zu $\boldsymbol{k}$ äquivalenten k-Vektoren, und der zweite bezeichnet eine irreduzible Darstellung der Gruppe K von $\boldsymbol{k}$. K ist eine Untergruppe der Raumgruppe R und umfaßt alle Transformationen von R, deren Drehanteile $\boldsymbol{k}$ in sich überführen. Alle irreduziblen Darstellungen $D^{(\boldsymbol{k};\nu)}$ einer Raumgruppe R setzen sich aus jenen von K zusammen, wobei jeder irreduziblen Darstellung $D^{(\boldsymbol{k};\nu)}$ von R genau eine irreduzible Darstellung $D^{(\nu)}$ von K entspricht und umgekehrt. Daher sind mit den $D^{(\nu)}$ von K auch die $D^{(\boldsymbol{k};\nu)}$ von R bekannt.

Als erste sind Bouckaert, Smoluchowski und Wigner [1] zu diesen Ergebnissen gekommen. Ihre Methode zur Bestimmung der irreduziblen Darstellungen einer Raumgruppe gründet sich auf eine Arbeit von Seitz [2] und gilt nur für symmorphe Raumgruppen; das sind solche, bei denen es unter allen Gruppenelementen mit dem gleichen C stets eines ohne Translationen gibt, also $\boldsymbol{\tau}(C) = 0$ für alle C. Unabhängig voneinander haben Herring [3] und Döring und Zehler [4] am Beispiel des Diamants gezeigt, wie diese Methode auf kompliziertere Gitter auszudehnen ist. Ein anderes Verfahren zum gleichen Zweck stammt von Sugita und Yamaka [5]. Es nutzt die irreduziblen Darstellungen bekannter symmorpher Untergruppen aus. Ihm verwandt ist die Methode von Zak [6], die Streitwolf [7] in einer etwas verallgemeinerten Form dargestellt hat. Hierbei werden die Charaktere der irreduziblen Darstellungen nichtsymmorpher Raumgruppen aus denen symmorpher Untergruppen vom Index 2 oder 3 erhalten. Auch das Konstruktionsverfahren von Raghavacharyulu [8, 9], das auf der Auflösbarkeit der Raumgruppen beruht und über die sogenannten kleinen Gruppen geht, benutzt Seitzsche Ergebnisse und ist im Prinzip nicht wesentlich verschieden von den anderen.

Bethe [10] und Opechowski [11] konstruierten aus den gewöhnlichen die doppelten Punktgruppen, und in analoger Weise konnte Elliott [12] durch den Übergang zu den doppelten Raumgruppen die Spin-Bahn-Wechselwirkung bei seiner gruppentheoretischen Energiebändertheorie berücksichtigen.

Die umfassendste Arbeit über die gewöhnlichen und doppelten Raumgruppen in Form einer Zusammenfassung hat Koster [13] gegeben. Ähnliche Aussagen enthält der Artikel von Sokolov und Shirokovskii [14]. Manche Ergebnisse über die irreduziblen Darstellungen von Raumgruppen sind auch bei Johnston [15] und bei Jones [16] sowie in den Monographien [17 bis 21] über die Anwendungen der Gruppentheorie zu finden.

Ein weiteres Verfahren zur Bestimmung der irreduziblen Darstellungen der gewöhnlichen und doppelten Raumgruppen stammt von Olbrychski [22]. Bei ihm werden die irreduziblen Darstellungen zur Gruppe des k-Vektors durch Ausnutzung einer abstrakten Gruppendefinition erhalten. Nach dieser Methode sind die Darstellungen der Raumgruppen D_{4h}^{1}–D_{4h}^{16} [23] und D_{4h}^{17}–D_{4h}^{20} [24] bestimmt worden.

Einen etwas anderen Weg beschritt Döring [25]. Er erkannte seinerzeit, daß die Wellenfunktionen eines Elektrons zur gleichen Energie und demselben reduzierten k-Vektor $\boldsymbol{k}_0$ eine passend geeichte Strahldarstellung der Punktgruppe P_0 dieses k-Vektors aufspannen. P_0 ist eine Untergruppe der Punktgruppe P des Kristalls und umfaßt alle Punkttransformationen von P, die $\boldsymbol{k}_0$ in sich überführen. Aus diesem Grunde bestimmte Döring sämtliche nichtassoziierten (nicht ineinander umeichbaren), nichtäquivalenten, irreduziblen Strahldarstellungen der kristallographischen Punktgruppen.

Aus diesen Strahldarstellungen kann man nun auf sehr einfache Weise alle irreduziblen Darstellungen einer Raumgruppe ableiten. Wie das geschieht, wird im folgenden beschrieben.

Um die zu dem reduzierten k-Vektor $\boldsymbol{k}_0$ gehörigen irreduziblen Darstellungen der Raumgruppe R zu ermitteln, wende man die Elemente C der zu R gehörigen Punktgruppe P auf $\boldsymbol{k}_0$ an. Dann erhält man den Stern zu $\boldsymbol{k}_0$, der aus den verschiedenen Vektoren $\boldsymbol{k}_0, \boldsymbol{k}_1, \ldots, \boldsymbol{k}_q$ bestehen möge und deren Anzahl von $\boldsymbol{k}_0$ abhängt, sowie eine Zerlegung der Punktgruppe in eine Summe von Nebenklassen der Untergruppe P_0,

$$P = P_0 + C_1 P_0 + \cdots + C_q P_0 \,. \tag{5}$$

Dabei gilt $C \in P_0$, falls $C\,\boldsymbol{k}_0 = \boldsymbol{k}_0 + \boldsymbol{K}_n$, und $C \in C_i\,P_0$, falls $C\,\boldsymbol{k}_0 = \boldsymbol{k}_i + \boldsymbol{K}_n$ ist. $\boldsymbol{K}_n$ ist gleich dem $2\,\pi$-fachen eines Gittervektors im reziproken Gitter. Dieser Zerlegung entspricht bei der Raumgruppe R

$$R = \{E|0\}\,K_0 + \{C_1|\boldsymbol{\tau}(C_1)\}\,K_0 + \cdots + \{C_q|\boldsymbol{\tau}(C_q)\}\,K_0 \,, \tag{6}$$

wobei $\{E|0\}\,K_0 = K_0$ gilt, $\boldsymbol{\tau}(C_i)$ die mit C_i gekoppelte, reduzierte (da sie sonst nur bis auf Gittertranslationen eindeutig bestimmt ist), nichteinfache Translation ist und K_0 die Gruppe von $\boldsymbol{k}_0$ bezeichnet.

Aus jeder d-dimensionalen irreduziblen Darstellung $D_0^{(\nu)}$ von K_0 läßt sich, nachdem man sie in die Normalform (bei ihr wird $\{E|\boldsymbol{t}_n\}$ durch das $e^{i\boldsymbol{k}_0\boldsymbol{t}_n}$-fache der Einheitsmatrix dargestellt) gebracht hat, eine $(q+1)\,d$-dimensionale irreduzible Darstellung $D^{(\boldsymbol{k}_0;\nu)}$ von R in Gestalt einer $(q+1)$-dimensionalen Kästchenmatrix mit d-dimensionalen Kästchen aufbauen, und zwar lautet das Kästchen in der i-ten Zeile und k-ten Spalte von $D^{(\boldsymbol{k}_0;\nu)}\,(\{C|\boldsymbol{\tau}(C) + \boldsymbol{t}_n\})$

$$d_{ik}\,(\{C|\boldsymbol{\tau}(C) + \boldsymbol{t}_n\}) = D_0^{(\nu)}\,(\{C'|\boldsymbol{\tau}(C') + \boldsymbol{t}_m\})\,\delta_{ij} \,, \tag{7}$$

wenn

$$C\,C_k \in C_j\,P_0 \tag{8}$$

und

$$\{C_j|\boldsymbol{\tau}(C_j)\}^{-1}\,\{C|\boldsymbol{\tau}(C) + \boldsymbol{t}_n\}\,\{C_k|\boldsymbol{\tau}(C_k)\} = \{C'|\boldsymbol{\tau}(C') + \boldsymbol{t}_m\} \in K_0 \tag{9}$$

sowie δ_{ij} das Kroneckersymbol $\delta_{ij} = \begin{cases} 1 \text{ für } i = j \\ 0 \text{ für } i \neq j \end{cases}$ ist.

Ist K_0 eine symmorphe Raumgruppe (alle $\boldsymbol{\tau}(C') = 0$), so gilt

$$D_0^{(\nu)}(\{C'|\boldsymbol{t}_m\}) = e^{i\boldsymbol{k}_0\boldsymbol{t}_m}\,D^{(\nu)}(C') \tag{10}$$

oder für die Charaktere χ

$$\chi_0^{(\nu)}(\{C'|\boldsymbol{t}_m\}) = e^{i\boldsymbol{k}_0\boldsymbol{t}_m}\,\chi^{(\nu)}(C')\,. \tag{11}$$

Es gibt genau soviele irreduzible Darstellungen $D_0^{(\nu)}$ von K_0 wie irreduzible Darstellungen $D^{(\nu)}$ zur Punktgruppe P_0, und die letzteren findet man z. B. bei Koster [13, 26] tabelliert.

Ist K_0 eine nichtsymmorphe Raumgruppe und $\boldsymbol{k}_0$ ein Vektor im Innern der ersten Brillouinzone, so gilt dasselbe. Es ist

$$D_0^{(\nu)}(\{C'|\boldsymbol{\tau}(C') + \boldsymbol{t}_m\}) = e^{i\boldsymbol{k}_0\boldsymbol{t}_m}\,e^{i\boldsymbol{k}_0\boldsymbol{\tau}(C')}\,D^{(\nu)}(C')\,. \tag{12}$$

Um für k-Vektoren auf dem Rande der ersten Brillouinzone die irreduziblen Darstellungen von nichtsymmorphen Raumgruppen zu bekommen, muß man die inäquivalenten, irreduziblen Strahldarstellungen $D_{\mathrm{Str}}^{(\nu)}$ von P_0 mit der durch den k-Vektor $\boldsymbol{k}_0$ vorgeschriebenen Eichung ermitteln und dann

$$D_0^{(\nu)}(\{C'|\boldsymbol{\tau}(C') + \boldsymbol{t}_m\}) = e^{i\boldsymbol{k}_0\boldsymbol{t}_m}\,D_{\mathrm{Str}}^{(\nu)}(C') \tag{13}$$

setzen. Da alle nicht zu Vektordarstellungen assoziierten Strahldarstellungen bei Döring [25] und die Vektordarstellungen bei Koster [13, 26] tabelliert sind, ist es nur nötig, die dort aufgeführten, zu der durch $\boldsymbol{k}_0$ vorgeschriebenen Eichung assoziierten Darstellungen zu ermitteln und sie in diese umzueichen. Wie das im einzelnen geschieht, wird später an einem Beispiel erläutert.

Wendet man das beschriebene Verfahren auf jeden k-Vektor der ersten Brillouinzone an, so erhält man alle irreduziblen Darstellungen der Raumgruppe R. Da die k-Vektoren zum gleichen Stern äquivalente Darstellungen liefern, genügt es, nur für einen k-Vektor jedes Sterns diese Darstellungen zu bestimmen.

Genau die gleichen Aussagen gelten für die doppelten Raumgruppen. Man braucht dazu nur die $\{C|\boldsymbol{\tau}(C) + \boldsymbol{t}_n\}$ als Elemente der doppelten Raumgruppe R' und die C als Elemente der entsprechenden doppelten Punktgruppe P' zu betrachten, wobei aber in den Ausdrücken $C\,\boldsymbol{t}$, die als Translationsanteile bei Produkten auftreten, das C von P' durch das entsprechende C der gewöhnlichen Punktgruppe P zu ersetzen ist. Daher können nach der vorhergehenden Methode alle irreduziblen Darstellungen der doppelten Raumgruppen angegeben werden, wenn man die Vektor- und Strahldarstellungen der doppelten Punktgruppen kennt. Während die Vektordarstellungen bereits bei Koster [13, 26] tabelliert sind, werden die Strahldarstellungen der doppelten kristallographischen Punktgruppen im folgenden Abschnitt ermittelt.

2. Die Strahldarstellungen der doppelten kristallographischen Punktgruppen

Die Ermittlung sämtlicher nichtassoziierter, inäquivalenter, irreduzibler Strahldarstellungen der doppelten Punktgruppen geschieht auf die gleiche Weise wie bei Döring [25] für die gewöhnlichen Punktgruppen. Zuerst ermittelt man von allen doppelten Punktgruppen die erzeugenden Elemente sowie die Relationen, welche die Ordnungen dieser Elemente und ihre Vertauschungsregeln bestimmen. Die erzeugenden Elemente einer jeden doppelten kristallographischen Punktgruppe P' sind in der zweiten Spalte von Tabelle 1 und ihre Verknüpfungsrelationen in der dritten Spalte aufgeführt, während in der ersten Spalte derselben Zeile das Symbol von P' zu finden ist.

Tabelle 1
Die doppelten kristallographischen Punktgruppen

Gruppe	Bedeutung der erzeugenden Elemente	Verknüpfungsregeln der		Willkür	Zahlenfaktoren
		Gruppenelemente	Darstellungsmatrizen		
C_1'	$a = (1)$	$a^2 = e$	$A^2 = E$	$A: \varepsilon_2$	
C_2'	$a = (2)$	$a^4 = e$	$A^4 = E$	$A: \varepsilon_4$	
C_s'	$a = m$				
C_i'	$a = (1)$ $i =$ Inversion	$a^2 = i^2 = e$ $i\,a = a\,i$	$A^2 = I^2 = E$ $I\,A = \alpha\,A\,I$	$A: \varepsilon_2$ $I: \varepsilon_2$	$\alpha = \pm 1$
C_3'	$a = (3)$	$a^6 = e$	$A^6 = E$	$A: \varepsilon_6$	
C_4'	$a = (4)$	$a^8 = e$	$A^8 = E$	$A: \varepsilon_8$	
S_4'	$a = (\bar{4})$				
C_{2h}'	$a = (2)$ $i =$ Inversion	$a^4 = i^2 = e$ $i\,a = a\,i$	$A^4 = I^2 = E$ $I\,A = \alpha\,A\,I$	$A: \varepsilon_2$ $I: \varepsilon_2$	$\alpha = \pm 1$
C_{2v}'	$a = (2)$ $b = m \parallel a$	$a^4 = e,\ b^2 = a^2$ $b\,a = a^3\,b$	$A^4 = E,\ B^2 = A^2$ $B\,A = A^3\,B$	$A: \varepsilon_2$ $B: \varepsilon_2$	
D_2'	$a = (2)$ $b = (2) \perp a$				
C_6'	$a = (6)$	$a^{12} = e$	$A^{12} = E$	$A: \varepsilon_{12}$	
C_{3h}'	$a = (\bar{3})$				
S_6'	$a = (3)$ $i =$ Inversion	$a^6 = i^2 = e$ $i\,a = a\,i$	$A^6 = I^2 = E$ $I\,A = \alpha\,A\,I$	$A: \varepsilon_2$ $I: \varepsilon_2$	$\alpha = \pm 1$
C_{3v}'	$a = (3)$ $b = m \parallel a$	$a^6 = e,\ b^2 = a^3$ $b\,a = a^5\,b$	$A^6 = E,\ B^2 = A^3$ $B\,A = A^5\,B$	$A: \varepsilon_2$ $B: \varepsilon_2$	
D_3'	$a = (3)$ $b = (2) \perp a$				
C_{4h}'	$a = (4)$ $i =$ Inversion	$a^8 = i^2 = e$ $i\,a = a\,i$	$A^8 = I^2 = E$ $I\,A = \alpha\,A\,I$	$A: \varepsilon_2$ $I: \varepsilon_2$	$\alpha = \pm 1$
C_{4v}'	$a = (4)$ $b = m \parallel a$	$a^8 = e,\ b^2 = a^4$ $b\,a = a^7\,b$	$A^8 = E,\ B^2 = A^4$ $B\,A = A^7\,B$	$A: \varepsilon_2$ $B: \varepsilon_2$	
D_4'	$a = (4)$ $b = (2) \perp a$				
D_{2d}'	$a = (\bar{4})$ $b = (2) \perp a$				
D_{2h}'	$a = (2)$ $b = m \parallel a$ $i =$ Inversion	$a^4 = i^2 = e$ $b^2 = a^2,\ b\,a = a^3\,b$ $i\,a = a\,i,\ i\,b = b\,i$	$A^4 = I^2 = E$ $B^2 = A^2,\ BA = A^3B$ $IA = \alpha AI,\ IB = \beta BI$	$A: \varepsilon_2$ $B: \varepsilon_2$ $I: \varepsilon_2$	$\alpha = \pm 1$ $\beta = \pm 1$

Tabelle 1 (Fortsetzung)

Gruppe	Bedeutung der erzeugenden Elemente	Verknüpfungsregeln der Gruppenelemente	Verknüpfungsregeln der Darstellungsmatrizen	Willkür	Zahlenfaktoren
C'_{6h}	$a = (6)$ $i =$ Inversion	$a^{12} = i^2 = e$ $i\,a = a\,i$	$A^{12} = I^2 = E$ $I\,A = \alpha\,A\,I$	$A: \varepsilon_{12}$ $I: \varepsilon_2$	$\alpha = \pm 1$
C'_{6v}	$a = (6)$ $b = m\|\|a$	$a^{12} = e$ $b^2 = a^6$ $b\,a = a^{11}\,b$	$A^{12} = E$ $B^2 = A^6$ $B\,A = A^{11}\,B$	$A: \varepsilon_2$ $B: \varepsilon_2$	
D'_6	$a = (6)$ $b = (2) \perp a$				
D'_{3h}	$a = (\overline{3})$ $b = m\|\|a$				
D'_{3d}	$a = (3)$ $b = (2) \perp a$ $i =$ Inversion	$a^6 = i^2 = e$ $b^2 = a^3$ $b\,a = a^5\,b$ $i\,a = a\,i$ $i\,b = b\,i$	$A^6 = I^2 = E$ $B^2 = A^3$ $B\,A = A^5\,B$ $I\,A = A\,I$ $I\,B = \alpha\,B\,I$	$A: \varepsilon_2$ $B: \varepsilon_2$ $I: \varepsilon_2$	$\alpha = \pm 1$
D'_{4h}	$a = (4)$ $b = m\|\|a$ $i =$ Inversion	$a^8 = i^2 = e$ $b^2 = a^4$ $b\,a = a^7\,b$ $i\,a = a\,i,\ i\,b = b\,i$	$A^8 = I^2 = E$ $B^2 = A^4$ $B\,A = A^7\,B$ $I\,A = \alpha\,A\,I$ $I\,B = \beta\,B\,I$	$A: \varepsilon_2$ $B: \varepsilon_2$ $I: \varepsilon_2$	$\alpha = \pm 1$ $\beta = \pm 1$
D'_{6h}	$a = (6)$ $b = m\|\|a$ $i =$ Inversion	$a^{12} = i^2 = e$ $b^2 = a^6$ $b\,a = a^{11}\,b$ $i\,a = a\,i,\ i\,b = b\,i$	$A^{12} = I^2 = E$ $B^2 = A^6$ $B\,A = A^{11}\,B$ $I\,A = \alpha\,A\,I$ $I\,B = \beta\,B\,I$	$A: \varepsilon_2$ $B: \varepsilon_2$ $I: \varepsilon_2$	$\alpha = \pm 1$ $\beta = \pm 1$
T'	$a = (2)$ $b = (2) \perp a$ $c = (3)$ unt. 55° zu a und b	$a^4 = e$ $b^2 = c^3 = a^2$ $b\,a = a^3\,b$ $c\,a = b\,c$ $c\,b = a\,b\,c$	$A^4 = E$ $B^2 = C^3 = A^2$ $B\,A = A^3\,B$ $C\,A = B\,C$ $C\,B = A\,B\,C$	$C: \varepsilon_3$	
T'_h	a, b, c wie bei T' $i =$ Inversion	wie bei T', dazu $i^2 = e$ $i\,a = a\,i$ $i\,b = b\,i$ $i\,c = c\,i$	wie bei T', dazu $I^2 = E$ $I\,A = A\,I$ $I\,B = B\,I$ $I\,C = C\,I$	$C: \varepsilon_3$ $I: \varepsilon_2$	

Tabelle 1 (Fortsetzung)

<table>
<tr><th rowspan="2">Gruppe</th><th rowspan="2">Bedeutung der erzeugenden Elemente</th><th colspan="2">Verknüpfungsregeln der</th><th rowspan="2">Willkür</th><th rowspan="2">Zahlen-faktoren</th></tr>
<tr><th>Gruppenelemente</th><th>Darstellungs-matrizen</th></tr>
<tr><td>T'_d</td><td>$a = (2)$
$b = (2) \perp a$
$c = (3)$ u. 55° zu a und b
$d = m \| c$ und $\|$ der Winkel-halbierenden von a und b</td><td rowspan="2">$a^4 = e$
$b^2 = c^3 = d^2 = a^2$

$b\,a = a^3\,b$
$c\,a = b\,c$
$c\,b = a\,b\,c$
$d\,a = b^3\,d$
$d\,b = a^3\,d$
$d\,c = c^5\,d$</td><td rowspan="2">$A^4 = E$
$B^2 = C^3 = D^2 =$
$= A^2$
$B\,A = A^3\,B$
$C\,A = B\,C$
$C\,B = A\,B\,C$
$D\,A = B^3\,D$
$D\,B = A^3\,D$
$D\,C = C^5\,D$</td><td rowspan="2">D: ε_2</td><td rowspan="2"></td></tr>
<tr><td>O'</td><td>a, b, c wie bei T'_d
$d = (2) \perp c$ und $\|$ der Winkel-halbierenden von a und b</td></tr>
<tr><td>O'_h</td><td>a, b, c, d wie bei O'
$i =$ Inversion</td><td>wie bei O', dazu

$i^2 = e$
$i\,a = a\,i$
$i\,b = b\,i$
$i\,c = c\,i$
$i\,d = d\,i$</td><td>wie bei O', dazu

$I^2 = E$
$I\,A = A\,I$
$I\,B = B\,I$
$I\,C = C\,I$
$I\,D = \alpha\,D\,I$</td><td>D: ε_2
I: ε_2</td><td>$\alpha = \pm 1$</td></tr>
</table>

In einer Strahldarstellung [27] einer Gruppe H mit den Elementen a_i $(i = 1, 2, \ldots, h)$ ist jedem a_i eine nichtsinguläre Matrix $D(a_i)$ derart zugeordnet, daß die Darstellungsmatrix des Gruppenprodukts

$$a_i\,a_k = a_l \tag{14}$$

bis auf einen Zahlenfaktor r_{ik} gleich dem Produkt der Darstellungsmatrizen seiner beiden Faktoren ist,

$$D(a_i)\,D(a_k) = r_{ik}\,D\,(a_i\,a_k) = r_{ik}\,D(a_l)\,. \tag{15}$$

Durch Umeichen, d. h. durch Hinzufügen irgendwelcher von Null verschiedener Zahlenfaktoren c_i entsteht aus einer Strahldarstellung $D(a_i)$ eine zu ihr assoziierte

$$D'(a_i) = c_i\,D(a_i)\,, \tag{16}$$

so daß

$$D'(a_i)\,D'(a_k) = r'_{ik}\,D'(a_l) \tag{17}$$

mit

$$r'_{ik} = \frac{c_i\,c_k}{c_l}\,r_{ik} \tag{18}$$

gilt.

Im folgenden wird über die Eichung so verfügt, daß möglichst wenige der h^2 Koeffizienten r_{ik} von 1 verschieden sind. Das ist zweckmäßig bei der Ermittlung der Strahldarstellungen nach einer Methode, die sich auf den folgenden, von Schur [27] bewiesenen Satz gründet: Zu jeder endlichen Gruppe H gibt es eine hinreichend ergänzte Gruppe A mit einem Zentrum Z mit der Eigenschaft, daß die Faktorgruppe A/Z mit H isomorph ist und alle Strahldarstellungen von H einer der Vektordarstellungen von A assoziiert sind. Man erhält die Verknüpfungsformeln der Gruppe A, indem man in den Beziehungen zwischen den vorher geeichten Darstellungsmatrizen von H die Matrizen als Gruppenelemente von A und die Zahlenkoeffizienten als die Matrizen des Zentrums von A auffaßt.

Auf Grund der in Spalte 3 von Tabelle 1 angegebenen Relationen kann jedes Gruppenelement von P' eindeutig als Produkt

$$a^{\nu_1} b^{\nu_2} \ldots i^{\nu_g} \quad \text{mit} \quad 1 \leqq \nu_l \leqq n_l \qquad (l = 1, 2, \ldots, g) \quad (19)$$

der g erzeugenden Elemente $a, b, \ldots, i$ mit den Ordnungen $n_1, n_2, \ldots, n_g$ geschrieben werden. Um die Eichung mit möglichst wenigen $r_{ik} \neq 1$ zu erreichen, setzt man

$$D\,(a^{\nu_1} b^{\nu_2} \ldots i^{\nu_g}) = A^{\nu_1} B^{\nu_2} \ldots I^{\nu_g}\,, \qquad (20)$$

wenn $A, B, \ldots, I$ die Darstellungsmatrizen von $a, b, \ldots, i$ sind. Bei allen Matrizen außer $A, B, \ldots, I$ ist die Eichung dadurch festgelegt. Den in Spalte 3 von Tabelle 1 aufgeführten Relationen $a^{n_1} = e$, $b^{n_2} = e, \ldots$ entsprechen bei den Darstellungsmatrizen $A^{n_1} = \alpha E$, $B^{n_2} = \beta E, \ldots$, und $A, B, \ldots, I$ werden so geeicht, daß möglichst viele der Koeffizienten $\alpha, \beta, \ldots$ gleich 1 werden.

Die bei den doppelten Punktgruppen für diese Eichung entstehenden Verknüpfungsrelationen der Matrizen sind in der vierten Spalte von Tabelle 1 angegeben, die bei den Darstellungsmatrizen noch willkürlichen n-ten Einheitswurzeln ε_n in der fünften Spalte und schließlich die in den Verknüpfungsrelationen übrigbleibenden, von 1 verschiedenen Zahlenfaktoren in der letzten Spalte. Alle doppelten Punktgruppen, bei denen in der letzten Spalte kein von 1 verschiedener Faktor auftritt, besitzen nur zu Vektordarstellungen assoziierte Strahldarstellungen.

Wie Tabelle 1 zeigt, ist das nur bei den Gruppen

$$C_i' = S_2',\ C_{2h}',\ S_6',\ C_{4h}',\ D_{2h}',\ C_{6h}',\ D_{3d}',\ D_{4h}',\ D_{6h}',\ O_h'$$

nicht der Fall. Von ihnen sind aber die doppelte Punktgruppe C_i' zur gewöhnlichen Punktgruppe C_{2h}, C_{2h}' zu C_{4h} und S_6' zu C_{6h} isomorph. Da isomorphe Gruppen dieselben Darstellungen haben, sind die Strahldarstellungen von $S_2' = C_i'$, C_{2h}' und S_6' bekannt. Die nicht zu Vektordarstellungen assoziierten von ihnen sind in den Tabellen 2, 4 und 7 für C_{2h}, C_{4h} und C_{6h} bei Döring [25] zu finden. Die irreduziblen Strahldarstellungen der restlichen Gruppen C_{4h}', D_{2h}', C_{6h}', D_{3d}', D_{4h}', D_{6h}', O_h' werden aus den Vektordarstellungen der entsprechenden hinreichend ergänzten Gruppen bestimmt.

Um die Vektordarstellungen der hinreichend ergänzten Gruppen zu berechnen, wird ein Verfahren verwendet, das von Seitz [2] stammt und bei endlichen, auflösbaren Gruppen benutzt werden kann. Von solchen Gruppen läßt sich eine Kompositionsreihe aufstellen. Diese ist nach dem Satz von Jordan-Hölder eindeutig bis auf die Reihenfolge der Faktorgruppen, die bei auflösbaren Gruppen zyklische

Gruppen von Primzahlordnung sind. Bei diesem Verfahren werden die irreduziblen Darstellungen einer Gruppe sukzessive aus den irreduziblen Darstellungen ihrer Normalteiler aufgebaut. Allerdings ist es bei Seitz in keiner sehr verständlichen und praktischen Form dargestellt. Deshalb soll im folgenden kurz beschrieben werden, wie man es am zweckmäßigsten anwendet.

Zuerst stellt man von der gegebenen Gruppe H eine Kompositionsreihe auf,

$$H = N_0 \supset N_1 \supset \cdots \supset N_r = E\,. \tag{21}$$

$E = N_r$ besteht nur aus dem Einheitselement e und besitzt eine einzige irreduzible Darstellung, bei der dem Element e die einreihige Einheitsmatrix zugeordnet ist. Dann erzeugt man aus dieser Darstellung durch das folgende Verfahren, das der Allgemeinheit halber für den Übergang von $N_{\mu+1}$ zu N_μ beschrieben wird, die Darstellungen von N_{r-1} usw., bis man zu den gesuchten Darstellungen von H gelangt. Die irreduziblen Darstellungen von N_μ gewinnt man auf folgende Weise aus denjenigen von $N_{\mu+1}$: Hat die Faktorgruppe $N_\mu/N_{\mu+1}$ die Primzahlordnung p und läßt sich N_μ durch das Element $b \in (N_\mu - N_{\mu+1})$ aus $N_{\mu+1}$ erzeugen, dann bilde man die Tabelle

$$e\,|a_1 = a_{10},\, b\, a_1\, b^{-1} = a_{11}, \ldots, b^{f_1} a_1\, b^{-f_1} = a_{1f_1}|\, a_2 = a_{20}, \ldots, a_{2f_2}| \ldots |\,. \tag{22}$$

Zuerst kommt das Einheitselement e, das einen Abschnitt für sich bildet. (Abschnitte sind durch senkrechte Striche angedeutet.) Danach nimmt man ein Element a_1 von $N_{\mu+1}$, bildet $b\, a_1\, b^{-1}$, $b^2\, a_1\, b^{-2}$ usw., bis man bei $b^{f_1+1}\, a_1\, b^{-f_1-1}$ wieder a_1 erhält, und schreibt die Elemente $a_1 = a_{10}, b\, a_1\, b^{-1} = a_{11}, \ldots, b^{f_1} a_1\, b^{-f_1} = a_{1f_1}$ als zweiten Abschnitt. Von den verbleibenden Elementen von $N_{\mu+1}$ greift man wieder eines heraus (es heiße $a_2 = a_{20}$) und wendet den gleichen Prozeß an, der jetzt den dritten Abschnitt liefert. Das führt man so lange fort, bis kein Element von $N_{\mu+1}$ mehr übrig ist, das noch nicht in einem der Abschnitte vorkommt. In der entstandenen Tabelle tritt dann jedes Element von $N_{\mu+1}$ genau einmal auf und sonst keine weiteren.

Nun nimmt man der Reihe nach jede irreduzible Darstellung D von $N_{\mu+1}$ vor und schreibt die Darstellungsmatrizen unter die Gruppenelemente der in Abschnitte eingeteilten Tabelle (22). Dadurch ist auch bei den Darstellungsmatrizen eine Einteilung in Abschnitte gegeben. Man schaut nach, ob alle Matrizen eines Abschnitts dieselbe Spur, d. h. den gleichen Charakter haben.

1. Ist dies bei allen Abschnitten der Fall, so sind die durch

$$D^{(\lambda)}(a_{ij}) = D(a_{ij-\lambda}) = D\,(b^{-\lambda}\, a_{ij}\, b^{\lambda}) \tag{23}$$

für $\lambda = 0, 1, \ldots, p-1$ aus D erzeugten konjugierten Darstellungen $D^{(\lambda)}$ alle äquivalent und man gewinnt p verschiedene irreduzible Darstellungen $\Delta^{(\lambda)}$ von N_μ, indem man die Matrix M aus

$$D(a_{ij})\, M = M\, D\,(b^{-1}\, a_{ij}\, b) \tag{24}$$

und

$$M^p = D(b^p)\,, \qquad (b^p \in N_{\mu+1}) \tag{25}$$

bestimmt,

$$\Delta^{(\lambda)}(a_{ij}) = D(a_{ij})\,, \qquad \Delta^{(\lambda)}(b) = e^{\frac{2\pi i}{p}\lambda}\, M \tag{26}$$

setzt und die übrigen Matrizen durch

$$\Delta^{(\lambda)}\,(a_{ij}\, b) = \Delta^{(\lambda)}(a_{ij})\, \Delta^{(\lambda)}(b) \tag{27}$$

erhält.

2. Tritt nicht bei allen Matrizen eines jeden Abschnitts der gleiche Charakter auf, so sind die aus D erzeugten konjugierten Darstellungen $D^{(\lambda)}$ $(\lambda = 0, 1, \ldots, p-1)$ von $N_{\mu+1}$ alle inäquivalent, und man gewinnt nur eine irreduzible Darstellung von N_μ durch die Festsetzung

$$\Delta(a_{ij}) = \begin{pmatrix} D^{(0)}(a_{ij}) & & & & 0 \\ & D^{(1)}(a_{ij}) & & & \\ & & D^{(2)}(a_{ij}) & & \\ & & & \ddots & \\ 0 & & & & D^{(p-1)}(a_{ij}) \end{pmatrix}, \quad \Delta(b) = \begin{pmatrix} 0 & \cdot & \cdot & \cdot & \cdot & 0 & D^{(0)}(b^p) \\ E_n & 0 & \cdot & \cdot & \cdot & 0 & 0 \\ & E_n & 0 & \cdot & \cdot & 0 & 0 \\ & & \cdot & \cdot & 0 & \cdot & \cdot \\ & & & \cdot & \cdot & \cdot & \cdot \\ & 0 & & & \cdot & \cdot & \cdot \\ & & & & & E_n & 0 \end{pmatrix} \tag{28}$$

mit $D(e) = E_n$ und erhält die restlichen Matrizen von Δ durch

$$\Delta\,(a_{ij}\,b) = \Delta(a_{ij})\,\Delta(b)\,. \tag{29}$$

Auf diese Weise werden auch alle irreduziblen Darstellungen von N_μ erhalten. Denn nach Clifford [28] induziert umgekehrt jede irreduzible Darstellung Δ von N_μ eine Darstellung D des Normalteilers $N_{\mu+1}$, die entweder selbst irreduzibel ist, wenn alle zu ihr konjugierte nDarstellungen von $N_{\mu+1}$ äquivalent sind (Fall 1), oder die vollständig reduzibel ist und in irreduzible, inäquivalente, konjugierte Darstellungen $D^{(\lambda)}$ von $N_{\mu+1}$ zerfällt, von denen jede genau einmal auftritt (Fall 2).

Mit dem Seitzschen Verfahren wurden die Vektordarstellungen der entsprechenden hinreichend ergänzten Gruppen bestimmt und auf diese Weise die Strahldarstellungen der doppelten Punktgruppen

$$C'_{4h},\ D'_{2h},\ C'_{6h},\ D'_{3d},\ D'_{4h},\ D'_{6h},\ O'_h \tag{30}$$

gefunden. Sie sind in den entsprechenden Tabellen 2 bis 8 wiedergegeben. Die zu Vektordarstellungen assoziierten Strahldarstellungen (sie ergeben sich für $\alpha = \beta = 1$) wurden dabei, um die Tabellen nicht unnötig zu vergrößern, weggelassen. Man findet sie in dem bereits mehrfach zitierten Artikel [13] und in dem Buch [26] von Koster u. a..

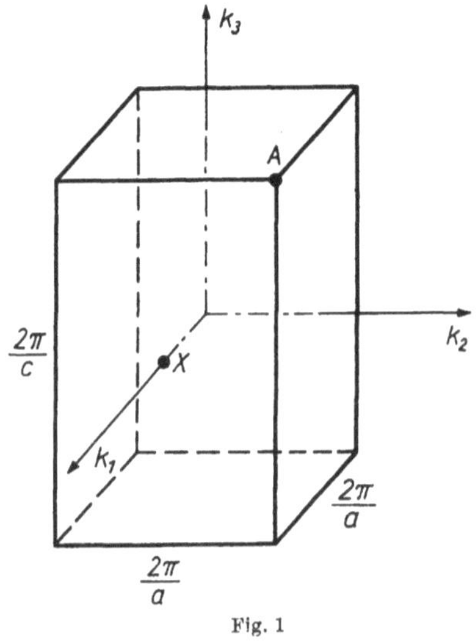

Fig. 1

Im folgenden soll am Beispiel der doppelten Raumgruppe D'^{5}_{4h} gezeigt werden, wie man in praxi Umeichungen durchführt und die Darstellungen der Gruppe zu einem bestimmten k-Vektor der Brillouinzone findet. Die Basisvektoren der Brillouinzone lauten in diesem Fall

$$\boldsymbol{K}_1 = 2\,\pi\,(1/a, 0, 0)\,,$$
$$\boldsymbol{K}_2 = 2\,\pi\,(0, 1/a, 0)\,,$$
$$\boldsymbol{K}_3 = 2\,\pi\,(0, 0, 1/c)\,.$$

Die Gestalt dieser Zelle ist in Fig. 1 wiedergegeben. Die Raumgruppe D'^{5}_{4h} ist durch die folgenden erzeugenden Elemente und Verknüpfungsrelationen definiert:

Die einfachen Translationen

$$\{E|\boldsymbol{t}_n\} = \{E|\boldsymbol{a}_1\}^{n_1}\{E|\boldsymbol{a}_2\}^{n_2}\{E|\boldsymbol{a}_3\}^{n_3} = t_1^{n_1} t_2^{n_2} t_3^{n_3} = t_n \qquad (31)$$

mit den Basistranslationen

$$\boldsymbol{a}_1 = (a, 0, 0)\,, \quad \boldsymbol{a}_2 = (0, a, 0)\,, \quad \boldsymbol{a}_3 = (0, 0, c) \qquad (32)$$

bilden einen Normalteiler von $D_{4h}'^{5}$. Die übrigen erzeugenden Elemente sind

$$\begin{aligned} a &= \{{}^zC_4|0\}\,, \quad {}^zC_4 = (4) \text{ um die } z\text{-Achse},\\ b &= \{{}^xC_2|\boldsymbol{\tau}\}\,, \quad {}^xC_2 = (2) \text{ um die } x\text{-Achse}, \qquad \boldsymbol{\tau} = (a/2, a/2, 0)\,,\\ i &= \{I|0\}\,, \qquad I = \text{Inversion}. \end{aligned} \qquad (33)$$

Sie genügen den Relationen

$$a^8 = i^2 = e\,, \quad b^2 = t_1 a^4\,, \quad b\,a = t_2 a^7 b\,, \quad i\,a = a\,i\,, \quad i\,b = t_1^{-1} t_2^{-1} b\,i\,. \qquad (34)$$

Als erstes Beispiel wird der k-Vektor $\boldsymbol{k}_0 = (\pi/a, \pi/a, \pi/c)$ zu dem Oberflächenpunkt A der Brillouinzone betrachtet. Er ist invariant gegenüber den Elementen von D_{4h}', so daß die zugehörige Gruppe von $\boldsymbol{k}_0$ mit $D_{4h}'^{5}$ identisch ist. Bei den in die Normalform gebrachten Darstellungen dieser Gruppe ist der einfachen Translation $\boldsymbol{t}_n$ das $e^{i\boldsymbol{t}_n\boldsymbol{k}_0}$-fache der Einheitsmatrix zugeordnet. Daher genügen die Darstellungsmatrizen A, B, I der Elemente a, b, i den Relationen

$$A^8 = I^2 = E\,, \quad B^2 = -A^4\,, \quad B\,A = -A^7 B\,, \quad I\,A = A\,I\,, \quad I\,B = B\,I\,. \qquad (35)$$

Um die in den Tabellen verwendete Eichung zu erhalten, ist

$$A = i\,A'\,, \quad B = i\,B'\,, \quad I = I' \qquad \left(i = \sqrt{-1}\right) \qquad (36)$$

zu setzen. Dann werden alle Zahlenfaktoren gleich 1, d. h., die Matrizen A', B', I' bilden eine Vektordarstellung von D_{4h}'. Um die gesuchten Charaktere der durch die Matrizen A, B, I gekennzeichneten Strahldarstellungen zu erhalten, sind demnach alle Vektordarstellungen von D_{4h}' mit Hilfe von (36) umzueichen.

Als zweites Beispiel seien die Darstellungen zur Gruppe des k-Vektors $\boldsymbol{k}_0 = (\pi/a, 0, 0)$ ermittelt. Dieser k-Vektor führt zum Oberflächenpunkt X der ersten Brillouinzone und ist invariant gegenüber den Elementen der Punktgruppe D_{2h}'. Die Gruppe zu $\boldsymbol{k}_0$ wird daher gebildet von dem Normalteiler der einfachen Translationen $t_n = \{E_n|\boldsymbol{t}_n\}$ sowie den erzeugenden Elementen

$$a = \{{}^zC_2|0\}\,, \quad b = \{{}^xC_2|\boldsymbol{\tau}\}\,, \quad i = \{I|0\} \qquad \left(\boldsymbol{\tau} = (a/2,\ a/2, 0)\right) \qquad (37)$$

mit

$$a^4 = i^2 = e\,, \quad b^2 = t_1 a^2\,, \quad b\,a = t_1 t_2 a^3 b\,, \quad i\,a = a\,i\,, \quad i\,b = t_1^{-1} t_2^{-1} b\,i\,. \qquad (38)$$

Wegen $e^{i\boldsymbol{t}_1\boldsymbol{k}_0} = -1$, $e^{i\boldsymbol{t}_2\boldsymbol{k}_0} = 1$ genügen die Darstellungsmatrizen A, B, I den Relationen

$$A^4 = I^2 = E\,, \quad B^2 = -A^2\,, \quad B\,A = -A^3 B\,, \quad I\,A = A\,I\,, \quad I\,B = -B\,I\,. \qquad (39)$$

Setzt man

$$A = i\,A'\,, \quad B = B'\,, \quad I = I'\,, \qquad \left(i = \sqrt{-1}\right) \qquad (40)$$

so erfüllen die Matrizen A', B', I' die geforderte Eichung und bilden eine Strahldarstellung mit den Zahlenfaktoren $\alpha = 1$, $\beta = -1$. Diese Strahldarstellungen in der Tabelle 3 für D_{2h}' sind umzueichen, indem man die Charaktere von dem dortigen A mit i multipliziert. Die Umeichung ergibt für die Punktgruppe des k-Vektors $\boldsymbol{k}_0 = (\pi/a, 0, 0)$ die folgende Charakterentafel:

E	A^2	A	A^3	A^2B B	A^3B AB	I	A^2I	A^3I AI	A^2BI BI	A^3BI ABI
2	−2	$2i$	$-2i$	0	0	0	0	0	0	0
2	−2	$-2i$	$2i$	0	0	0	0	0	0	0
2	2	0	0	0	0	0	0	−2	0	0
2	2	0	0	0	0	0	0	2	0	0

Diese beiden Beispiele mögen genügen, um das Prinzip einer Umeichung zu erläutern. In sehr vielen Fällen erübrigt sich sogar eine solche, weil die Matrizen bereits genauso wie in den Tabellen geeicht sind. Daher gehören die beiden Beispiele schon zu den komplizierteren Fällen. Sie wurden ausgewählt, weil diese Darstellungen neben weiteren in der Arbeit [23] auf eine andere Weise ermittelt und hier ihre Richtigkeit bestätigt wurden.

Aus der Kenntnis der Strahldarstellungen aller einfachen und doppelten kristallographischen Punktgruppen können nun ohne weiteres nach dem im ersten Teil beschriebenen Verfahren zu jedem k-Vektor alle irreduziblen Darstellungen der gewöhnlichen und doppelten Raumgruppen angegeben werden. Man braucht sie nicht wie bisher für jede Raumgruppe einzeln herzuleiten.

Herrn Professor Dr. W. Döring danke ich für die Anregung zu dieser Arbeit und für wertvolle Ratschläge.

Tabelle 2: C'_{4h} $(i = \sqrt{-1})$

E	A	A^2	A^3	A^4	A^5	A^6	A^7	I	AI	A^2I	A^3I	A^4I	A^5I	A^6I	A^7I	α
2	0	2	0	2	0	2	0	0	0	0	0	0	0	0	0	−1
2	0	−2	0	2	0	−2	0	0	0	0	0	0	0	0	0	−1
2	0	$2i$	0	−2	0	$-2i$	0	0	0	0	0	0	0	0	0	−1
2	0	$-2i$	0	−2	0	$2i$	0	0	0	0	0	0	0	0	0	−1

Tabelle 3: D'_{2h}

E	A^2	A^3 A	A^2B B	A^3B AB	I	A^2I	AI	A^3I	BI	A^2BI	ABI	A^3BI	α	β
2	2	2	0	0	0	0	0	0	0	0	0	0	1	−1
2	2	−2	0	0	0	0	0	0	0	0	0	0	1	−1
2	−2	0	0	0	0	0	$2i$	$-2i$	0	0	0	0	1	−1
2	−2	0	0	0	0	0	$-2i$	$2i$	0	0	0	0	1	−1
2	2	0	2	0	0	0	0	0	0	0	0	0	−1	1
2	2	0	−2	0	0	0	0	0	0	0	0	0	−1	1
2	−2	0	0	0	0	0	0	0	$2i$	$-2i$	0	0	−1	1
2	−2	0	0	0	0	0	0	0	$-2i$	$2i$	0	0	−1	1
2	2	0	0	2	0	0	0	0	0	0	0	0	−1	−1
2	2	0	0	−2	0	0	0	0	0	0	0	0	−1	−1
2	−2	0	0	0	0	0	0	0	0	0	$2i$	$-2i$	−1	−1
2	−2	0	0	0	0	0	0	0	0	0	$-2i$	$2i$	−1	−1

Tabelle 4: C'_{6h} $\left(\omega = e^{i\frac{\pi}{6}}\right)$

E	A	A^2	A^3	A^4	A^5	A^6	A^7	A^8	A^9	A^{10}	A^{11}	I	AI	A^2I	A^3I	A^4I	A^5I	A^6I	A^7I	A^8I	A^9I	$A^{10}I$	$A^{11}I$	α
2	0	2	0	2	0	2	0	2	0	2	0	0	0	0	0	0	0	0	0	0	0	0	0	−1
2	0	$-2\omega^2$	0	$2\omega^4$	0	2	0	$-2\omega^2$	0	$2\omega^4$	0	0	0	0	0	0	0	0	0	0	0	0	0	−1
2	0	$2\omega^4$	0	$-2\omega^2$	0	2	0	$2\omega^4$	0	$-2\omega^2$	0	0	0	0	0	0	0	0	0	0	0	0	0	−1
2	0	−2	0	2	0	−2	0	2	0	−2	0	0	0	0	0	0	0	0	0	0	0	0	0	−1
2	0	$2\omega^2$	0	$2\omega^4$	0	−2	0	$-2\omega^2$	0	$-2\omega^4$	0	0	0	0	0	0	0	0	0	0	0	0	0	−1
2	0	$-2\omega^4$	0	$-2\omega^2$	0	−2	0	$2\omega^4$	0	$2\omega^2$	0	0	0	0	0	0	0	0	0	0	0	0	0	−1

Tabelle 5: D'_{3d} $(i = \sqrt{-1})$

E	A^3	A^5, A	A^4, A^2	A^4B, A^2B, B	A^5B, A^3B, AB	I	A^3I	AI	A^5I	A^2I	A^4I	A^4BI, A^2BI, BI	A^5BI, A^3BI, ABI	α
2	2	2	2	0	0	0	0	0	0	0	0	0	0	−1
2	2	−1	−1	0	0	0	0	$i\sqrt{3}$	$-i\sqrt{3}$	$-i\sqrt{3}$	$i\sqrt{3}$	0	0	−1
2	2	−1	−1	0	0	0	0	$-i\sqrt{3}$	$i\sqrt{3}$	$i\sqrt{3}$	$-i\sqrt{3}$	0	0	−1
2	−2	−2	2	0	0	0	0	0	0	0	0	0	0	−1
2	−2	1	−1	0	0	0	0	$i\sqrt{3}$	$-i\sqrt{3}$	$i\sqrt{3}$	$-i\sqrt{3}$	0	0	−1
2	−2	1	−1	0	0	0	0	$-i\sqrt{3}$	$i\sqrt{3}$	$-i\sqrt{3}$	$i\sqrt{3}$	0	0	−1

Tabelle 6: D'_{4h} . $(i = \sqrt{-1})$

E	A^4	A^6 A^2	A^7 A	A^5 A^3	A^6B A^4B A^2B B	A^7B A^5B A^3B AB	I	A^4I	A^2I	A^6I	AI	A^7I	A^3I	A^5I	A^4BI BI	A^6BI A^2BI	A^5BI ABI	A^7BI A^3BI	α	β
2	2	2	2	2	0	0	0	0	0	0	0	0	0	0	0	0	0	0	1	-1
2	2	2	-2	-2	0	0	0	0	0	0	0	0	0	0	0	0	0	0	1	-1
2	2	-2	0	0	0	0	0	0	0	0	$2i$	$-2i$	$-2i$	$2i$	0	0	0	0	1	-1
2	2	-2	0	0	0	0	0	0	0	0	$-2i$	$2i$	$2i$	$-2i$	0	0	0	0	1	-1
2	-2	0	$-\sqrt{2}$	$\sqrt{2}$	0	0	0	0	$2i$	$-2i$	$-i\sqrt{2}$	$-i\sqrt{2}$	$-i\sqrt{2}$	$i\sqrt{2}$	0	0	0	0	1	-1
2	-2	0	$\sqrt{2}$	$-\sqrt{2}$	0	0	0	0	$2i$	$-2i$	$i\sqrt{2}$	$-i\sqrt{2}$	$i\sqrt{2}$	$-i\sqrt{2}$	0	0	0	0	1	-1
2	-2	0	$\sqrt{2}$	$-\sqrt{2}$	0	0	0	0	$-2i$	$2i$	$-i\sqrt{2}$	$i\sqrt{2}$	$-i\sqrt{2}$	$i\sqrt{2}$	0	0	0	0	1	-1
2	-2	0	$-\sqrt{2}$	$\sqrt{2}$	0	0	0	0	$-2i$	$2i$	$i\sqrt{2}$	$-i\sqrt{2}$	$i\sqrt{2}$	$-i\sqrt{2}$	0	0	0	0	1	-1
2	2	2	0	0	2	0	0	0	0	0	0	0	0	0	0	0	0	0	-1	1
2	2	2	0	0	-2	0	0	0	0	0	0	0	0	0	0	0	0	0	-1	1
2	2	-2	0	0	0	0	0	0	0	0	0	0	0	0	2	-2	0	0	-1	1
2	2	-2	0	0	0	0	0	0	0	0	0	0	0	0	-2	2	0	0	-1	1
4	-4	0	0	0	0	0	0	0	0	0	0	0	0	0	0	0	0	0	-1	1
2	2	2	0	0	0	2	0	0	0	0	0	0	0	0	0	0	0	0	-1	-1
2	2	2	0	0	0	-2	0	0	0	0	0	0	0	0	0	0	0	0	-1	-1
2	2	-2	0	0	0	0	0	0	0	0	0	0	0	0	0	0	2	-2	-1	-1
2	2	-2	0	0	0	0	0	0	0	0	0	0	0	0	0	0	-2	2	-1	-1
4	-4	0	0	0	0	0	0	0	0	0	0	0	0	0	0	0	0	0	-1	-1

Tabelle 7: D'_{6h} $\left(i=\sqrt{-1}\right)$

E	A^6	A^{10} A^2	A^8 A^4	A^9 A^3	A^{11} A	A^7 A^5	$A^{10}B$ A^8B A^6B A^4B A^2B B	$A^{11}B$ A^9B A^7B A^5B A^3B AB	I	A^6I	A^2I	$A^{10}I$	A^4I	A^8I	A^3I	A^9I	AI	$A^{11}I$	A^5I	A^7I	A^8BI A^4BI BI	$A^{10}BI$ A^6BI A^2BI	A^9BI A^5BI ABI	$A^{11}BI$ A^7BI A^3BI	α	β
2	2	2	2	2	2	2	0	0	0	0	0	0	0	0	0	0	0	0	0	0	0	0	0	0	1	-1
2	2	2	2	-2	-2	-2	0	0	0	0	0	0	0	0	0	0	0	0	0	0	0	0	0	0	1	-1
2	2	-1	-1	-2	1	1	0	0	0	0	$i\sqrt{3}$	$-i\sqrt{3}$	$-i\sqrt{3}$	$i\sqrt{3}$	0	0	$i\sqrt{3}$	$-i\sqrt{3}$	$-i\sqrt{3}$	$i\sqrt{3}$	0	0	0	0	1	-1
2	2	-1	-1	2	-1	-1	0	0	0	0	$i\sqrt{3}$	$-i\sqrt{3}$	$-i\sqrt{3}$	$i\sqrt{3}$	0	0	$-i\sqrt{3}$	$i\sqrt{3}$	$i\sqrt{3}$	$-i\sqrt{3}$	0	0	0	0	1	-1
2	2	-1	-1	2	-1	-1	0	0	0	0	$-i\sqrt{3}$	$i\sqrt{3}$	$i\sqrt{3}$	$-i\sqrt{3}$	0	0	$i\sqrt{3}$	$-i\sqrt{3}$	$-i\sqrt{3}$	$i\sqrt{3}$	0	0	0	0	1	-1
2	2	-1	-1	-2	1	1	0	0	0	0	$-i\sqrt{3}$	$i\sqrt{3}$	$i\sqrt{3}$	$-i\sqrt{3}$	0	0	$-i\sqrt{3}$	$i\sqrt{3}$	$i\sqrt{3}$	$-i\sqrt{3}$	0	0	0	0	1	-1
2	-2	-2	2	0	0	0	0	0	0	0	0	0	0	0	$2i$	$-2i$	$-2i$	$2i$	$-2i$	$2i$	0	0	0	0	1	-1
2	-2	-2	2	0	0	0	0	0	0	0	0	0	0	0	$-2i$	$2i$	$2i$	$-2i$	$2i$	$-2i$	0	0	0	0	1	-1
2	-2	1	-1	0	$\sqrt{3}$	$-\sqrt{3}$	0	0	0	0	$i\sqrt{3}$	$-i\sqrt{3}$	$i\sqrt{3}$	$-i\sqrt{3}$	$2i$	$-2i$	i	$-i$	i	$-i$	0	0	0	0	1	-1
2	-2	1	-1	0	$-\sqrt{3}$	$\sqrt{3}$	0	0	0	0	$i\sqrt{3}$	$-i\sqrt{3}$	$i\sqrt{3}$	$-i\sqrt{3}$	$-2i$	$2i$	$-i$	i	$-i$	i	0	0	0	0	1	-1
2	-2	1	-1	0	$-\sqrt{3}$	$\sqrt{3}$	0	0	0	0	$-i\sqrt{3}$	$i\sqrt{3}$	$-i\sqrt{3}$	$i\sqrt{3}$	$2i$	$-2i$	i	$-i$	i	$-i$	0	0	0	0	1	-1
2	-2	1	-1	0	$\sqrt{3}$	$-\sqrt{3}$	0	0	0	0	$-i\sqrt{3}$	$i\sqrt{3}$	$-i\sqrt{3}$	$i\sqrt{3}$	$-2i$	$2i$	$-i$	i	$-i$	i	0	0	0	0	1	-1
2	2	2	2	0	0	0	2	0	0	0	0	0	0	0	0	0	0	0	0	0	0	0	0	0	-1	1
2	2	2	2	0	0	0	-2	0	0	0	0	0	0	0	0	0	0	0	0	0	0	0	0	0	-1	1
2	-2	-2	2	0	0	0	0	0	0	0	0	0	0	0	0	0	0	0	0	0	$2i$	$-2i$	0	0	-1	1
2	-2	-2	2	0	0	0	0	0	0	0	0	0	0	0	0	0	0	0	0	0	$-2i$	$2i$	0	0	-1	1
4	4	-2	-2	0	0	0	0	0	0	0	0	0	0	0	0	0	0	0	0	0	0	0	0	0	-1	1
4	-4	2	-2	0	0	0	0	0	0	0	0	0	0	0	0	0	0	0	0	0	0	0	0	0	-1	1
2	2	2	2	0	0	0	0	2	0	0	0	0	0	0	0	0	0	0	0	0	0	0	0	0	-1	-1
2	2	2	2	0	0	0	0	-2	0	0	0	0	0	0	0	0	0	0	0	0	0	0	0	0	-1	-1
2	-2	-2	2	0	0	0	0	0	0	0	0	0	0	0	0	0	0	0	0	0	0	0	$2i$	$-2i$	-1	-1
2	-2	-2	2	0	0	0	0	0	0	0	0	0	0	0	0	0	0	0	0	0	0	0	$-2i$	$2i$	-1	-1
4	4	-2	-2	0	0	0	0	0	0	0	0	0	0	0	0	0	0	0	0	0	0	0	0	0	-1	-1
4	-4	2	-2	0	0	0	0	0	0	0	0	0	0	0	0	0	0	0	0	0	0	0	0	0	-1	-1

Tabelle 8: O_h' $\left(i = \sqrt{-1}\right)$

E	A^2	A^3B A^2B AB B A^3 A	A^3BC^2 A^2BC^2 A^3C^2 A^2C^2 A^3BC A^2BC A^3C C	ABC^2 BC^2 AC^2 C^2 ABC BC AC A^2C	A^3C^2D A^2C^2D AC^2D C^2D BCD A^2BCD A^2CD CD A^3BD ABD A^2D D	ACD A^3BC^2D A^3BCD BC^2D A^2BD AD	$ABCD$ A^2BC^2D A^3CD ABC^2D A^3D BD	I	A^2I	A^3BI A^2BI ABI BI A^3I AI	A^3BCI A^2BCI A^3CI CI	A^3BC^2I A^3C^2I A^2BC^2I A^2C^2I	$ABCI$ BCI ACI A^2CI	ABC^2I BC^2I AC^2I C^2I	A^3C^2DI A^2C^2DI AC^2DI C^2DI $BCDI$ A^2BCDI A^2CDI CDI A^3BDI $ABDI$ A^2DI DI	$ACDI$ A^3BC^2DI A^3BCDI BC^2DI A^2BDI ADI	$ABCDI$ A^2BC^2DI A^3CDI ABC^2DI A^3DI BDI	α
2	2	2	2	2	0	0	0	0	0	0	0	0	0	0	0	0	0	–1
2	2	2	–1	–1	0	0	0	0	0	0	$i\sqrt{3}$	$-i\sqrt{3}$	$i\sqrt{3}$	$-i\sqrt{3}$	0	0	0	–1
2	2	2	–1	–1	0	0	0	0	0	0	$-i\sqrt{3}$	$i\sqrt{3}$	$-i\sqrt{3}$	$i\sqrt{3}$	0	0	0	–1
6	6	–2	0	0	0	0	0	0	0	0	0	0	0	0	0	0	0	–1
4	–4	0	2	–2	0	0	0	0	0	0	0	0	0	0	0	0	0	–1
4	–4	0	–1	1	0	0	0	0	0	0	$i\sqrt{3}$	$-i\sqrt{3}$	$-i\sqrt{3}$	$i\sqrt{3}$	0	0	0	–1
4	–4	0	–1	1	0	0	0	0	0	0	$-i\sqrt{3}$	$i\sqrt{3}$	$i\sqrt{3}$	$-i\sqrt{3}$	0	0	0	–1

Literatur

[1] L. P. BOUCKAERT, R. SMOLUCHOWSKI und E. WIGNER, Phys. Rev. **50**, 58 (1936).
[2] F. SEITZ, Ann. Math. **37**, 17 (1936).
[3] C. HERRING, J. Franklin Inst. **233**, 525 (1942).
[4] W. DÖRING und V. ZEHLER, Ann. Phys. (Germany) **13**, 215 (1953).
[5] T. SUGITA und E. YAMAKA, Rep. ECL, NTT **2**, 24 (1954).
[6] J. ZAK, J. math. Phys. **1**, 165 (1960).
[7] H. W. STREITWOLF, Zur Bestimmung der Charaktere von irreduziblen Darstellungen nichtsymmorpher Raumgruppen, Dtsch. Akad. Wiss. Berlin, Veröff. Phys.-Techn. Inst. Nr. 2 (1962).
[8] I. V. V. RAGHAVACHARYULU, Canad. J. Phys. **39**, 830 (1961).
[9] I. V. V. RAGHAVACHARYULU und I. BHAVANACHARYULU, Canad. J. Phys. **40**, 1490 (1962).
[10] H. BETHE, Ann. Phys. (Germany) **3**, 133 (1929).
[11] W. OPECHOWSKI, Physica **7**, 552 (1940).
[12] R. J. ELLIOTT, Phys. Rev. **96**, 280 (1954).
[13] G. F. KOSTER, Solid State Physics **5**, 173 (1957).
[14] A. V. SOKOLOV und V. P. SHIROKOVSKII, Soviet Physics Uspekhi **3**, 551 (1961).
[15] F. D. JOHNSTON, Group Theory in Solid State Physics, in: Reports on Progress in Physics, vol. XXIII, 1960 (S. 66).
[16] H. JONES, The Theory of Brillouin Zones and Electronic States in Crystals, North Holland, Amsterdam 1960.
[17] J. S. LOMONT, Applications of Finite Groups, Academic Press, New York 1959.
[18] V. HEINE, Group Theory in Quantum Mechanics, Pergamon Press, London 1960.
[19] G. J. LJUBARSKI, Anwendungen der Gruppentheorie in der Physik, Deutscher Verlag der Wissenschaften, Berlin 1962.
[20] P. H. E. MEIJER und E. BAUER, Group Theory, North Holland, Amsterdam 1962.
[21] M. TINKHAM, Group Theory and Quantum Mechanics, McGraw-Hill, New York 1964.
[22] K. OLBRYCHSKI, phys. stat. sol. **3**, 1868 (1963).
[23] K. OLBRYCHSKI, phys. stat. sol. **3**, 2143 (1963).
[24] Z. SEK, phys. stat. sol. **3**, 2155 (1963).
[25] W. DÖRING, Z. Naturf. **14a**, 343 (1959).
[26] G. F. KOSTER, J. O. DIMMOCK, R. G. WHEELER und H. STATZ, Properties of the Thirty-two Point Groups, The M.I.T. Press, Cambridge (Mass.) 1963.
[27] J. SCHUR, J. reine u. angew. Math. **127**, 20 (1904); **132**, 85 (1907); **139**, 155 (1911).
[28] A. H. CLIFFORD, Ann. Math. **38**, 533 (1937).

(Received December 28, 1964)

Note added in proof:

Es sei ein neues Buch genannt, das u. a. die zitierten Arbeiten [1], [2], [3], [11] und [12] als Reprints enthält:
R. S. KNOX, A. GOLD, Symmetry in the Solid State, W. A. Benjamin, Inc., New York 1964.

phys. stat. sol. 8, 831 (1965)

Institut für angewandte Physik der Technischen Hochschule Karlsruhe

Elektrische und optische Eigenschaften von amorphem Tellur

Von
H. KELLER und J. STUKE

An aufgedampften amorphen Te-Schichten wurde die Temperaturabhängigkeit der spezifischen Leitfähigkeit gemessen. Bei hohen Temperaturen (> 200 °K) ist die Eigenleitung mit $\Delta E \approx$ 0,87 eV vorherrschend. Bei tiefen Temperaturen ergibt sich eine Störleitung, die sich durch Erwärmung auf relativ niedrige Temperaturen (~ 150 °K) irreversibel vermindern läßt. Mit dem Übergang kristallin-amorph ist eine Verschiebung der Absorptionskante um 0,40 eV zu höherer Energie verbunden, die um 0,15 eV kleiner ist als die Zunahme des elektrisch gemessenen Bandabstandes. Amorphes Te zeigt Photoleitung, deren Abhängigkeit von der Beleuchtungsstärke und von der Temperatur sowie deren spektrale Verteilung untersucht wurde. Die Eigenschaften von amorphem Te stimmen qualitativ mit denen von amorphem Se und von amorphem Ge überein.

The temperature dependence of the specific conductivity is measured for evaporated amorphous Te-layers. At high temperatures (>200 °K) the intrinsic conductivity with $\Delta E \approx$ 0.87 eV is dominant. At low temperatures impurity conduction occurs which can be decreased irreversibly by heating to temperatures which are relatively low (~ 150 °K). The transformation from the crystalline to the amorphous state implies a shift to higher energies of the absorption edge .40 eV which is lower than the increase of the electrically measured band gap by 0.15 eV. The photoconductivity of amorphous Te is measured as a function of light intensity, temperature and wavelength. The properties of amorphous Te are qualitatively similar to those of amorphous Se and amorphous Ge.

1. Einleitung

Die amorphen Stoffe haben in den letzten Jahren stark an Interesse gewonnen, da ihre Bandstruktur und ihr Leitungsmechanismus von grundsätzlicher Bedeutung für die Festkörperphysik sind [1, 2, 3]. Bei den Elementhalbleitern ist neben dem seit langer Zeit bekannten amorphen Se, das geradezu als Prototyp der amorphen Halbleiter gelten kann, vor einigen Jahren auch das amorphe Te gefunden worden [4, 5, 6], welches sich bei Aufdampfen auf kalte Unterlage ergibt. Seine Umwandlungstemperatur ist mit etwa +10 °C [5] deutlich niedriger als bei amorphem Se, wo sie — je nach Reinheit und Umwandlungsbedingungen — bei 80 bis 120 °C liegt [7]. Dieser Unterschied ist auf die bei Te stärkere Bindung zwischen den Ketten zurückzuführen.

Bei amorphem Te ist außer der Tatsache, daß die elektrische Leitfähigkeit beim Umwandlungspunkt sprunghaft um etwa drei Zehnerpotenzen zunimmt, wenig bekannt. Nach den orientierenden Messungen von Shih Tuan und Regel [5] hängt die Leitfähigkeit nur schwach von der Temperatur ab. Aus einer zusätzlich durchgeführten Thermokraftmessung geht hervor, daß es sich dabei um Löcherleitung handelt und daß die Beweglichkeit etwa 10^{-2} cm²/Vs beträgt. Über die optischen Eigenschaften von amorphem Te ist bisher nichts bekannt geworden.

Es war deshalb das Ziel dieser Arbeit, die wichtigsten optischen und elektrischen Eigenschaften von amorphem Te zu bestimmen. Dazu wurde die Temperaturabhängigkeit der Leitfähigkeit, die Absorptionskonstante und die spektrale Verteilung der Photoleitung sowie deren Abhängigkeit von der Beleuchtungsstärke und von der Temperatur gemessen.

2. Meßmethode

In einem Kryostaten wurde bei einem Vakuum von etwa 10^{-4} Torr aus einem seitlich angebrachten Ofen zonengereinigtes Te relativ langsam auf ein Quarzplättchen aufgedampft, das zur Messung der elektrischen Leitfähgkeit mit aufgedampften Au-Elektroden versehen war. Besondere Aufmerksamkeit wurde darauf verwendet, eine Bedampfung der Fenster des Kryostaten zu vermeiden. Das wurde durch ein mehrfaches Blendensystem und durch eine Abdeckung der Fenster während der Bedampfung erreicht. Zur Messung der optischen und lichtelektrischen Eigenschaften konnte der Kühlfinger des Kryostaten bei tiefer Temperatur in den Strahlengang des Monochromators gedreht und außerdem um einen Winkel von etwa 20° geschwenkt werden.

Die optischen Messungen wurden mit der gleichen Anordnung durchgeführt, die in einer früheren Arbeit [8] beschrieben ist. Die Bestimmung der Schichtdicke erfolgte durch Wägung. Für die Dichte von amorphem Te wurde dabei der Wert $\varrho = 5{,}93\ \mathrm{g/cm^3}$ [9] benutzt. Zur Messung der sich über etwa 13 Zehnerpotenzen erstreckenden elektrischen Leitfähigkeit diente ein Keithley-Röhrenvoltmeter. Die Spektralverteilung der Photoleitung wurde mit einem energiegleichen Spektrum gemessen; die Variation der Bestrahlungsstärke erfolgte durch Graufilter.

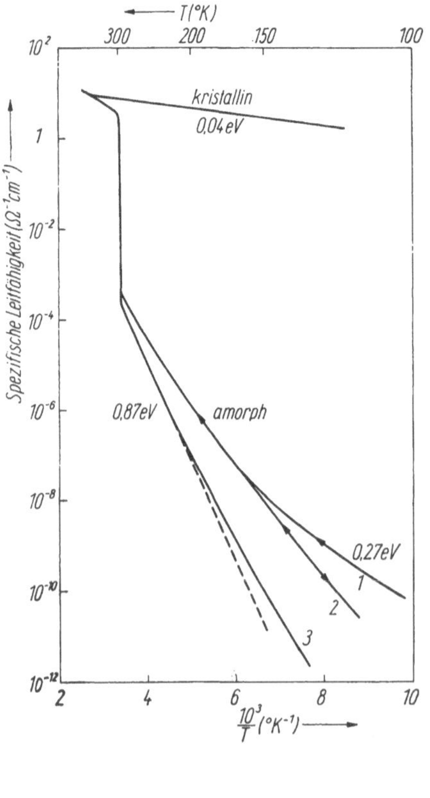

Fig. 1. Temperaturabhängigkeit der spezifischen Leitfähigkeit von amorphem und polykristallinem Te.
Kurve 1: Nach Aufdampfen bei 100 °K mit steigender Temperatur gemessen
Kurve 2: Messung derselben Schicht nach Erwärmung auf 180 °K
Kurve 3: Nach Aufdampfen bei 250 °K

3. Meßergebnisse

Für die Messung der elektrischen Leitfähigkeit war zunächst die Frage zu klären, ob es sich bei dem Kontakt zwischen amorphem Te und Au um einen Ohmschen Kontakt handelt. Strom-Spannungs-Kennlinien zeigen, daß bei dem für die Messung der Dunkel-Leitfähigkeit verwendeten Elektrodenabstand $L = 1$ mm das Ohmsche Gesetz auch bei tiefen Temperaturen bis zu einer Spannung von etwa 100 V erfüllt ist. Bei der normalerweise benutzten Spannung von 10 V ist der Au-Kontakt demnach zur Messung der spezifischen Leitfähigkeit gut geeignet. Bei kleinen Elektroden-Abständen ($L = 50\ \mu$m) treten im Gebiet tiefer Temperaturen dagegen überproportionale Kennlinien auf, die wahrscheinlich durch raumladungsbegrenzte Ströme zustande kommen; darüber soll später gesondert berichtet werden.

Für die elektrische Leitfähigkeit von amorphem Te im Dunkeln wurde im Gegensatz zu der Messung von Shih Tuan und Regel [5] eine starke Temperaturabhängigkeit gefunden, welche bei hohen Temperaturen (200 °K $< T <$ 280 °K) relativ gut der Beziehung $\sigma = \sigma_0 e^{-\frac{\Delta E}{2kT}}$ entspricht (Fig. 1). Bei tiefen Temperaturen weicht der Kurvenverlauf von dieser Formel ab, hier wird die Neigung deutlich geringer. Die Größe dieser Neigungsänderung hängt von den Aufdampfbedingungen ab. Die spezifische Leitfähigkeit im Gebiet tiefer Temperaturen von Proben, welche bei tiefer Temperatur ($\sim$ 100 °K) aufgedampft wurden (Kurve 1), ist nämlich beträchtlich größer als von den bei hoher Temperatur ($\sim$ 250 °K) hergestellten Schichten (Kurve 3). Steigert man nach Aufdampfen bei 100 °K die Temperatur auf etwa 180 °K (Kurve 1), so ergibt sich eine irreversible Leitfähigkeitsabnahme (Kurve 2), die mit einer Zunahme der Neigung verbunden ist. Hierbei kann es sich also nicht um die Eigenleitung von amorphem Te handeln, sondern dafür müssen „Gitter"-Störungen verantwortlich sein, die schon bei relativ niedriger Temperatur ausheilen. Dieses Ergebnis entspricht genau dem Verhalten von amorphem Ge [10], nur daß sich dort im Störleitungsgebiet eine definierte Aktivierungsenergie ergibt. Ob diese hier bei amorphem Te bis 100 °K schon erreicht ist, läßt sich nicht sicher entscheiden; der kleinste bei tiefer Temperatur gemessene ΔE-Wert betrug 0,27 eV.

Anders verhalten sich dagegen amorphe Te-Schichten, die bei etwa 250 °K aufgedampft wurden. Sie zeigen — besonders dann, wenn man sie nach dem Aufdampfen 15 bis 20 min bei dieser Temperatur gehalten hat — eine reversible Temperaturabhängigkeit der Leitfähigkeit im ganzen untersuchten Temperaturgebiet. Die bei hoher Temperatur gemessene Aktivierungsenergie war bei verschiedenen Proben nicht genau gleich. Der höchste gemessene Wert betrug $\Delta E = 0{,}87$ eV (Fig. 1, Kurve 3).

Wahrscheinlich handelt es sich hierbei nahezu um die Aktivierungsenergie der Eigenleitung. Dafür spricht einmal die Tatsache, daß die Neigung dieser Kurve über etwa drei Zehnerpotenzen der Leitfähigkeit konstant ist, während sich bei kleineren Werten von ΔE auch bei hohen Temperaturen meistens noch eine schwache Krümmung der Kurve bemerkbar macht. Weiterhin nähern sich hier die anderen Kurven in ihrem Absolutwert deutlich dieser Kurve an. Ein wichtiges Argument ist ferner, daß die Photoleitung (siehe Fig. 4) bei etwa der gleichen Energie im Spektrum einsetzt. Der Bandabstand von amorphem Te liegt also nahe bei 0,87 eV und ist demnach um mindestens 0,55 eV größer als bei kristallinem Te, wo er 0,32 eV [11] beträgt.

Die Umwandlung in kristallines Te setzt bei 280 °K relativ plötzlich ein; dabei nimmt die Leitfähigkeit um mehr als vier Zehnerpotenzen zu. Bei Wiederabkühlung ergibt sich für die Störleitung eine Aktivierungsenergie von 0,04 eV, die an aufgedampften kristallinen Schichten auch von Sakurai und Munesue [4] gefunden wurde.

Mit dem Übergang von kristallinem zu amorphem Te ist ebenfalls eine starke Änderung der Lage der Absorptionskante verbunden (Fig. 2). Der starke Anstieg der Absorptionskonstante K verläuft bei amorphem Te annähernd parallel zu dem von kristallinem Te, jedoch um 0,40 eV zu höheren Energiewerten verschoben. Der spektrale Verlauf der Absorptionskonstante unterscheidet sich außerdem noch dadurch, daß die bei kristallinem Te vorhandene Struktur in dem untersuchten Bereich bei amorphem Te fehlt. Die Kurve steigt hier mit einer stetig abnehmenden Neigung an und hat bei etwa 4,5 eV ein flaches Maximum. Dieser qualitative Unterschied tritt in dem Verlauf des Absorptionsindex $k = \frac{K\lambda}{4\pi}$ noch deutlicher hervor (Fig. 3). Während bei kristallinem Te in dem hier untersuchten Bereich zwei Maxima vorhanden sind, ergibt sich bei amorphem Te ein einziges breites Maximum. Dieses Verhalten ist ganz analog wie bei Se; auch hier zeigen sich bei der kristallinen Phase zwei Maxima, bei der amorphen jedoch nur ein breites Maximum [12]. Die Temperaturabhängigkeiten der Absorptionskonstante bei kristallinem Se und Te stimmen ebenfalls überein. In beiden Fällen werden die

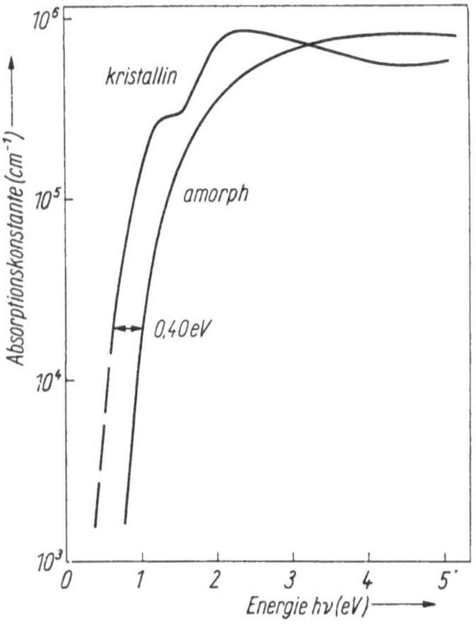

Fig. 2. Absorptionskonstante von amorphem und polykristallinem Te

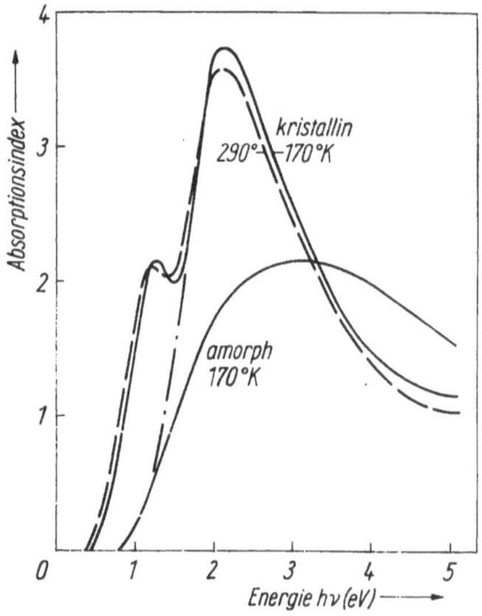

Fig. 3. Absorptionsindex k von amorphem und polykristallinem Te

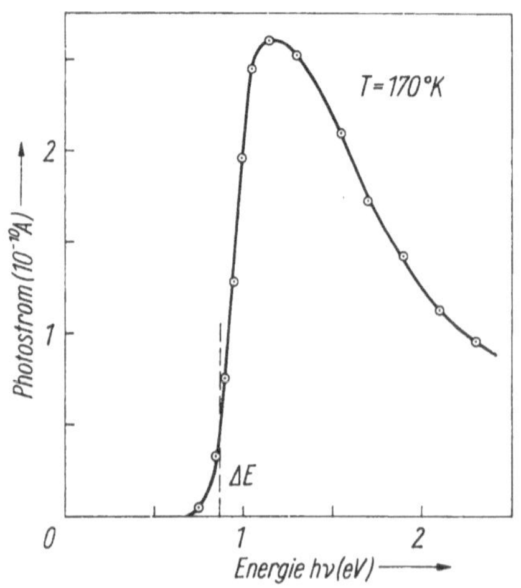

Fig. 4. Spektralverteilung der Photoleitung von amorphem Te. $B = 5 \cdot 10^{-13}\,\mathrm{cm^{-2}\,s^{-1}}$, $U = 15$ V

Fig. 5a. Abhängigkeit des Photostroms von der Best.ahlungsstärke, $h\nu = 1{,}2$ eV, $U = 15$ V

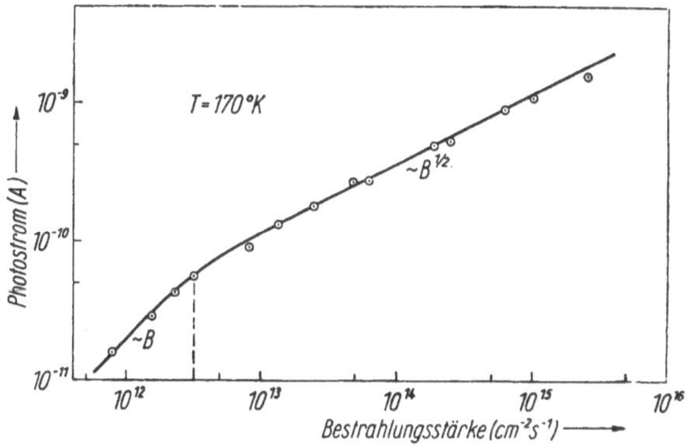

Maxima mit abnehmender Temperatur schärfer. Die wesentliche Änderung in dem Verlauf des Absorptionsindex k bzw. der Absorptionskonstante K beim Übergang von der geordneten zur ungeordneten Phase besteht demnach bei Te ebenso wie bei Se darin, daß die Absorptionskante eine Verschiebung zu höherer Energie erfährt. Wie aus Fig. 3 zu ersehen ist, kommt diese Verschiebung durch die starke Abschwächung des vorgelagerten ersten Maximums zustande. Die Kante des zweiten Hauptmaximums wird dagegen nicht wesentlich geändert (strichpunktierte Kurve). Hier wird aber das bei kristallinem Te vorhandene relativ scharfe Maximum stark verschmiert.

Der niedrige Wert der Leitfähigkeit ließ vermuten, daß amorphes Te auch ein Photoleiter ist. Tatsächlich ergab sich bei tiefen Temperaturen eine schwache Photoleitung mit einem Maximum bei etwa 1,2 eV und einer langwelligen Grenze, die recht gut mit dem Bandabstand übereinstimmt (Fig. 4). Das Maximum liegt ähnlich wie bei der Photoleitung von amorphem Se [13] relativ weit in der Eigenabsorption bei einem Wert der Absorptionskonstante $K_{max} = 6 \cdot 10^4$ cm^{-1}. Die Abhängigkeit des Photostroms J_{ph} von der Beleuchtungsstärke B gehorcht bei

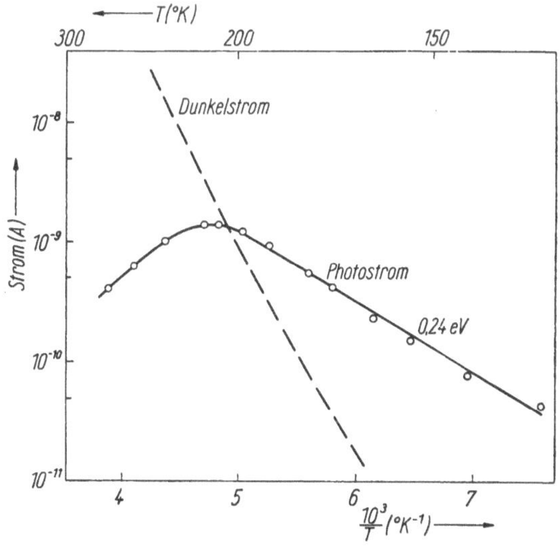

Fig. 5b. Photostrom in Abhängigkeit von der Temperatur, $h\nu = 1{,}2$ eV, $B = 8 \cdot 10^{14}$ cm^{-2} s^{-1}, $U = 10$ V

Einstrahlung im Maximum bei 1,2 eV für hohe Bestrahlungsstärken der einfachen Beziehung: $J_{\text{ph}} \sim B^{1/2}$ (Fig. 5a), die bei einem bimolekularen Rekombinationsprozeß zu erwarten ist. Bei kleinen Bestrahlungsstärken wird die Abhängigkeit stärker; hier ist annähernd $J_{\text{ph}} \sim B$.

Der Absolutwert des Photostroms ist relativ klein; er nimmt ausgehend von 100 °K mit zunehmender Temperatur zunächst annähernd exponentiell zu (Fig. 5b), mit einer Aktivierungsenergie von 0,24 eV, welche praktisch mit der in der Störleitung gemessenen übereinstimmt. Bei weiter steigender Temperatur fällt dann der Photostrom nach einem Maximum bei etwa 200 °K wieder ab und wird in der Nähe der Umwandlungstemperatur so klein, daß er kaum noch meßbar ist. Amorphes Te ist demnach wie amorphes Se ein relativ schlechter Photoleiter, der nur bei tiefen Temperaturen einen nennenswerten Photostrom zeigt.

4. Diskussion

Das Energiespektrum von amorphen Stoffen unterscheidet sich, wie Gubanov [14], Makinson und Roberts [15] gezeigt haben, von dem von Kristallen hauptsächlich dadurch, daß in der ungeordneten Phase eine große Anzahl von lokalisierten Energie-Zuständen auftritt. Beim Übergang kristallin–amorph werden die in den Kristallen in der Nähe der Bandränder vorhandenen nichtlokalisierten Energieniveaus infolge der in den amorphen Stoffen fehlenden Ordnung erheblich gestört und gehen in mehr oder weniger stark lokalisierte Zustände über. Falls die Zustandsdichte und die Übergangswahrscheinlichkeit zwischen den Zuständen bei der Umwandlung kristallin–amorph nicht wesentlich geändert wird, bleiben die optischen Eigenschaften der Kristalle in erster Näherung auch für die amorphe Phase erhalten. Die Lage der Absorptionskante wird sich beispielsweise nur wenig ändern. Bei amorphen Halbleitern ist aber im Gegensatz zu Kristallen die Lage der Absorptionskante nicht mehr durch den Bandabstand gegeben. Die Bandkante der amorphen Phase, d. h. die Grenze der nichtlokalisierten Zustände wird nämlich zu Energiewerten verschoben, die bei Kristallen relativ weit innerhalb des betreffenden Bandes liegen [16]. Die elektrischen Eigenschaften von amorphen Halbleitern müssen sich darum stark von denen der Kristalle unterscheiden, insbesondere nimmt der elektrisch gemessene Bandabstand beim Übergang kristallin–amorph beträchtlich zu.

Unter den Elementhalbleitern findet man bei Ge diese Vorstellung gut bestätigt. Hier stimmt nach Messungen von Tauc und Mitarbeitern [17] die Lage der Absorptionskante bei kristallinem und amorphem Ge weitgehend überein. Ein Unterschied in den optischen Eigenschaften besteht im wesentlichen nur darin, daß die bei Ge-Kristallen im Spektrum der optischen Konstanten vorhandenen Maxima bei amorphem Ge stark verschmiert sind. Dieses unterschiedliche Verhalten läßt sich nach Tauc und Mitarbeitern durch die Annahme erklären, daß in der amorphen Phase bei den Band-Band-Übergängen nur noch der Energiesatz, im Gegensatz zu den Kristallen aber nicht mehr der Impulssatz erfüllt sein muß. Die elektrischen Eigenschaften von amorphem Ge unterscheiden sich dagegen beträchtlich von denen der Kristalle. Nach Messungen von Grigorovici, Croitoru, Devenyi und Teleman [10] nimmt der Bandabstand bei der Umwandlung von 0,72 eV bei kristallinem Ge auf 1,10 eV bei amorphem Ge zu (Tabelle 1). Die Absorptionskonstante von amorphem Ge beträgt bei der Energie des Bandabstandes $K_{\Delta E} \approx 3 \cdot 10^4\ \text{cm}^{-1}$ [17]. Dieser Wert ist viel größer als der entsprechende bei Ge-Kristallen, welcher $< 10^2\ \text{cm}^{-1}$ ist [18]. Die Bandkante liegt also bei

Tabelle 1

Substanz		ΔE (eV)	$K_{\Delta E}$ (cm^{-1})	ε (eV)	δ (eV)	Δ (eV)
Se	hexagonal	~1,75 [19]	~10^3 [21]	0,56	0,25 [12]	0,31
	amorph	2,31 [20]	$1 . 10^5$ [12]			
Te	hexagonal	0,32	<10^2 [22]	0,55	0,40	0,15
	amorph	0,87	$7 . 10^3$			
Ge	Diamantgitter	0,72	<10^2	0,38	~0	0,38
	amorph	1,10	$3 . 10^4$			

ε Zunahme des Bandabstandes aus Leitfähigkeitsmessungen[1]) beim Übergang kristallin—amorph
δ Verschiebung der Absorptionskante beim Übergang kristallin—amorph
$\Delta = \varepsilon - \delta$

amorphem Ge relativ weit im entsprechenden Band des kristallinen Ge. Die vor der Kante liegenden Zustände haben den Charakter von Haftstellen. Die Beweglichkeit bei Zimmertemperatur ist nämlich mit $\mu = 10^{-3}$ cm²/Vs sehr klein, und sie nimmt mit zunehmender Temperatur in exponentieller Form mit einer Aktivierungsenergie von 0,18 eV zu, weil die in den Haftstellen befindlichen Ladungsträger — es sind Löcher — in zunehmendem Maße ins Band gelangen können.

Bei Te und Se wächst nun aber beim Übergang kristallin–amorph nicht nur der Bandabstand, sondern es verschiebt sich auch die Absorptionskante zu höherer Energie (Tabelle 1). Dieser Unterschied zum Ge ist auf die verschiedene Bindung zurückzuführen. Bei Ge wird die energetische Lage und Dichte der Zustände durch die starke homöopolare Bindung zwischen den benachbarten Ge-Atomen bestimmt, welche auch in der amorphen Phase nicht wesentlich geändert ist. Bei Se und Te dagegen hat neben der homöopolaren Bindung innerhalb der Ketten auch die relativ schwache van der Waalssche Bindung zwischen den Ketten für die Bandstruktur eine erhebliche Bedeutung. Die Absorptionskante von polykristallinem Se und Te ist durch relativ schmale Bänder gegeben, welche von der schwachen Wechselwirkung zwischen den Ketten herrühren [8]. Da in der amorphen Phase die Ordnung der Ketten aber ganz wesentlich gestört ist, werden diese Bänder eine starke Veränderung erfahren im Gegensatz zu den von der starken Bindung innerhalb der Ketten herrührenden breiten Bändern. Dementsprechend nimmt der Beitrag der schmalen Bänder zur Absorption in der amorphen Phase stark ab, und genau das zeigt der Verlauf des Absorptionsindex k in Fig. 3. Die Tatsache, daß bei Te die Verschiebung der Absorptionskante δ deutlich größer ist als bei Se (Tabelle 1), hängt offenbar damit zusammen, daß wegen der gegenüber Se stärkeren Bindung zwischen den Ketten die vorgelagerten Bänder bei Te breiter sind als bei Se [8].

Berücksichtigt man diesen Unterschied in der Bindung zwischen Se und Te einerseits und Ge andererseits, so ergeben sich wieder ähnliche Verhältnisse. Beim

[1]) Da bei hexagonalem Se derartige Messungen nicht vorliegen, wurde hier ein aus optischen Messungen bestimmter Wert benutzt.

Übergang kristallin–amorph ist nämlich die Zunahme des Bandabstandes ε bei Se und Te deutlich größer als die zugehörige Verschiebung der Absorptionskante δ (Tabelle 1). Bei Se beträgt die Differenz dieser Änderungen $\Delta = \varepsilon - \delta = 0{,}31$ eV. Dieser Wert stimmt gut mit dem von Ge überein, wo analog $\Delta \approx \varepsilon = 0{,}38$ eV beträgt; hier ist ja δ klein. Auch bei amorphem Se sind demnach in einem relativ breiten Energiebereich vor der Band-Kante zahlreiche lokalisierte Niveaus vorhanden. Es handelt sich ebenfalls um Haftstellen, da Lanyon [23] mit einer exponentiellen Verteilung dieser Zustände neben dem exponentiellen Abfall der Absorptionskonstante an der Kante auch die im Falle raumladungsbegrenzter Ströme auftretenden stark überproportionalen Strom-Spannungs-Kennlinien einheitlich erklären konnte. Diese Haftstellen haben eine niedrige Drift-Beweglichkeit der Löcher zur Folge; sie steigt mit zunehmender Temperatur in exponentieller Form mit einer Aktivierungsenergie von 0,14 eV an [24]; bei Zimmertemperatur ist die wesentlich größer als bei amorphem Ge, nämlich etwa 10^{-1} cm^{-2}/Vs.

Bei Te ist die Differenz zwischen der Zunahme des Bandabstandes und der Absorptionskanten-Verschiebung mit $\Delta = 0{,}15$ eV zwar geringer als bei Se und Ge, aber ohne Zweifel vorhanden. Da es zudem nicht ausgeschlossen ist, daß der Bandabstand von amorphem Te noch geringfügig größer ist als $\Delta E = 0{,}87$ eV, hat man diesen Wert vorerst als untere Grenze anzusehen. Auf die Anwesenheit von zahlreichen Haftstellen vor der Bandkante deutet auch hier der niedrige Wert der Beweglichkeit und außerdem die schwache Störleitung bei tiefen Temperaturen hin. Die Störleitung von amorphem Te ist ja viel weniger ausgeprägt als bei der kristallinen Phase (Fig. 1). In der Nähe der Umwandlungstemperatur ist bei den amorphen Schichten schon die Eigenleitung vorherrschend, während kristallines Te trotz eines um fast den Faktor 3 kleineren Bandabstandes sich noch in der Störleitung befindet.

Dieser starke Unterschied ist für den Übergang kristallin–amorph charakteristisch. Lange Zeit war es sogar fraglich, ob amorphe Stoffe überhaupt eine richtige Störleitung haben. Die Ergebnisse an amorphem Ge [10] und die vorliegenden Messungen an amorphem Te zeigen nun, daß dies mit Sicherheit der Fall ist, allerdings ist in beiden Fällen die Störleitung gegenüber der von Kristallen stark abgeschwächt. Dafür sind wahrscheinlich die in der amorphen Phase vor der Bandkante liegenden zahlreichen lokalisierten Niveaus verantwortlich. Die z.B. aus flachen Akzeptoren A stammenden Löcher werden nämlich fast vollständig von diesen Haftstellen eingefangen, wenn deren Tiefe E_H und Konzentration H wesentlich größer ist als die der Akzeptoren. Für die Temperaturabhängigkeit der Leitfähigkeit ist dann also die Energieverteilung der Haftstellen, für den Absolutwert der Leitfähigkeit aber das Verhältnis $r = A/H$ maßgebend [25]. Die irreversible Abnahme der Störleitung bei Temperung kann demnach durch eine Abnahme der Akzeptorkonzentration oder durch eine Zunahme der Haftstellen-Konzentration erklärt werden. Da aber ein Anstieg von H bei einer Temperung schwer verständlich ist, ist wahrscheinlich für diese Abnahme der Störleitung eine irreversible Verminderung der Akzeptor-Konzentration A verantwortlich. Dieses Ergebnis ist plausibel, wenn man annimmt, daß bei tiefer Temperatur die aufgedampften Kettenstücke sich nur teilweise zu größeren Ketten bzw. Ringen zusammenschließen können. Es bleibt dann eine große Anzahl von freien Kettenenden übrig, welche wegen ihrer freien Valenz als Akzeptoren wirken. Bei Steigerung der Temperatur wird die Konzentration dieser Akzeptoren abnehmen, da sich mehr und mehr Ketten zu größeren Gebilden zusammenfinden. Die Haftstellen-Niveaus vor der Bandkante haben dagegen nichts mit freien Valenzen zu tun, sondern kommen durch die Unordnung selbst zustande, die erst bei der Umwandlungstemperatur größtenteils verschwindet. Die vorliegenden Ergebnisse reichen noch nicht aus, um über die Konzentration und Energieverteilung der Haftstellen Auskunft zu geben. Zur Klärung dieser Zusammenhänge sind weitere Untersuchungen im Gange, speziell Messungen der Temperaturabhängigkeit von Thermokraft und raumladungsbegrenzten Strömen.

Auch die lichtelektrischen Eigenschaften von amorphen Stoffen lassen sich durch die lokalisierten Niveaus vor der Bandkante einfach erklären. Das Maximum der Photoleitung liegt bei amorphem Te nämlich ebenso wie bei amorphem Se bei einem relativ hohen Wert der Absorptionskonstante. Die entsprechenden Werte an kristallinen Schichten vergleichbarer Dicke sind größenordnungsmäßig kleiner [26]. Dieser Unterschied, auf den Moss [26] bei amorphem Se besonders hingewiesen hat, kommt dadurch zustande, daß bei den amorphen Stoffen die Absorptionskonstante bei der Energie des Bandabstandes wegen der vorhandenen lokalisierten Zustände wesentlich höhere Werte hat als bei Kristallen (Tabelle 1). Eine Anregung in die vor der Bandkante liegenden lokalisierten Niveaus hat noch keine Photoleitung zur Folge. An amorphen Halbleitern tritt in diesem Spektralgebiet also auch dann, wenn das Licht in der betreffenden Schicht vollständig absorbiert wird, kein Photostrom auf. Diese Tatsache zeigt noch einmal besonders deutlich den Unterschied zwischen Absorptionskante und Bandabstand. Die zahlreichen Haftstellen-Niveaus haben außerdem zur Folge, daß der Photostrom auch bei Anregung ins Band relativ klein ist. Mit abnehmender Temperatur wird ein immer größerer Anteil der durch Licht zusätzlich erzeugten Löcher in diese lokalisierten Zustände eingefangen.

Herrn Prof. Dr. F. Stöckmann danken wir für die Durchsicht des Manuskripts, Herrn cand. phys. G. Zimmerer sind wir für die Durchführung von Messungen zu Dank verpflichtet.

Literatur

[1] A. E. Joffe und A. R. Regel, Progress in Semiconductors, Vol. 4, Heywood, London 1960 (S. 237).
[2] N. E. Cusack, Rep. Progr. Phys. **26**, 361 (1963).
[3] B. T. Kolomiets, phys. stat. sol. **7**, 359 und 713 (1964).
[4] T. Sakurai und S. Munesue, Phys. Rev. **85**, 921 (1952).
[5] Y. Shih-Tuan und A. R. Regel, Soviet Physics — Solid State **3**, 2627 (1962).
[6] C. G. Oancea, J. Teodorescu und P. Cristea, Kristallografiya **7**, 560 (1962).
[7] Siehe z. B. Gmelin, Handbuch der anorg. Chemie, 8. Aufl. Nr. 10, Teil A, Verlag Chemie, Weinheim 1953.
[8] J. Stuke und H. Keller, phys. stat. sol. **7**, 189 (1964).
[9] Gmelin, Handbuch der anorgan. Chemie, 8. Aufl. Nr. 11, Verlag Chemie, Berlin 1940 (S. 173).
[10] R. Grigorovici, N. Croitoru, A. Devenyi und E. Teleman, Proc. Int. Conf. Semicond. Phys., Dunod, Paris 1964 (S. 423).
[11] A. Nussbaum, Phys. Rev. **94**, 337 (1954).
[12] J. Stuke, Z. Phys. **134**, 194 (1953).
[13] P. K. Weimer und A. D. Cope, RCA Rev. **12**, 314 (1951).
[14] A. I. Gubanov, Fiz. tverd. Tela **4**, 2873 (1962).
[15] R. E. B. Makinson und A. P. Roberts, Proc. Phys. Soc. **79**, 630 (1962).
[16] L. Banyai, Proc. Int. Conf. Semicond. Phys., Dunod, Paris 1964 (S. 417).
[17] J. Tauc, A. Abraham, L. Pajasova, R. Grigorovici und A. Vancu, Proc. Int. Conf. Physics of Non-Crystalline Solids, Delft 1964 (im Druck).

[18] W. C. DASH and R. NEWMAN, Phys. Rev. **99**, 1151 (1955).
[19] F. ECKART und W. HENRION, phys. stat. sol. **2**, 841 (1962).
[20] H. W. HENCKELS und J. MACZUK, J. appl. Phys. **24**, 1056 (1953).
[21] H. GOBRECHT und A. TAUSEND, Z. Phys. **161**, 205 (1961).
[22] J. S. BLAKEMORE und K. C. NOMURA, Phys. Rev. **127**, 1024 (1962).
[23] H. P. D. LANYON, Phys. Rev. **130**, 134 (1963).
[24] J. L. HARTKE, Phys. Rev. **125**, 1177 (1962).
[25] M. POLKE, J. STUKE und E. VINARICKY, phys. stat. sol. **3**, 1885 (1963).
[26] T. S. MOSS, Optical Properties of Semiconductors, Butterworth, London 1959.

(Received December 30, 1964)

phys. stat. sol. 8, 841 (1965)

Laboratorium voor Kristallografie en Studie van Vaste Stoffen, Gent

Interpretation of the Yellow Colour of Heat Treated ZnO Powder

By

F. VAN CRAEYNEST, W. MAENHOUT-VAN DER VORST, and W. DEKEYSER

Two batches of analytical grade ZnO powder (ZnO_N) are heated to 920 °C and one, (ZnO_A), quenched to room temperature, and the other, (ZnO_B), cooled slowly. Samples of both ZnO_A and ZnO_B are afterwards heated for one hour at temperatures between 100 and 800 °C and quenched. From all these samples the amount of free zinc is determined and the reflection and luminescence spectra measured. The reflection spectra with ZnO_N as blank indicate the presence of a supplementary absorption band having an intensity which depends upon the previous treatment. From this variation and correlation with the amount of free zinc the yellow colour of the samples can be explained and new models for the centres deduced.

Zwei Teile analysenreines ZnO-Pulver (ZnO_N) wurden bis 920 °C erhitzt. Der eine Teil wurde auf Zimmertemperatur abgeschreckt (ZnO_A), der zweite Teil langsam abgekühlt (ZnO_B). ZnO_A- und ZnO_B-Proben wurden danach für eine Stunde bei Temperaturen zwischen 100 und 800° C erhitzt und abgeschreckt. Bei allen Proben wurden der Gehalt an freiem Zink bestimmt und die Reflexions- und Lumineszenzspektren untersucht. Die Reflexions-Spektren bei ZnO_N gestatten neben der b-Bande eine zusätzliche Absorptionsbande festzustellen, deren Intensität von der Vorbehandlung abhängt. Aus dieser Änderung und ihrer Beziehung zur Menge freien Zinks konnte die gelbe Farbe der Proben gedeutet und ein neues Modell für diese Zentren hergeleitet werden.

1. Introduction

Although the optical properties of ZnO powders have been extensively studied, both by transmission and reflection measurements, many features remain unexplained. It is well known that ZnO contains an amount of excess Zn which varies with temperature. When under the form of neutral zinc, this amount can be determined by chemical means, and is found to reach a maximum at 400 °C [1]. The experiments reported in this paper were undertaken in order to find a correlation between the amount of excess Zn, determined by chemical analysis, and the optical properties of pretreated specimens. For the sake of clarity we summarize first briefly the essentials of earlier observations which are necessary for what follows.

a) White ZnO powder exhibits, at room temperature, a steep absorption edge at 385 nm. This is the fundamental absorption band, due to band to band transitions (3.2 eV).

b) A shift to longer wavelengths, and a less steep rise of this edge is observed when powders, quenched from temperatures above 400 °C to room temperature, are examined. Such powders are coloured (yellow to orange) and bleach slightly when kept for days at room temperature in air.

c) When the specimens have been heated in Zn vapour the same unresolved peak near the absorption edge is more pronounced. This shoulder is attributed, by some authors [2], to the absorption of Zn^+ ion, others [3] consider that it is due to unspecified lattice defects resulting from the diffusion of atomic Zn.

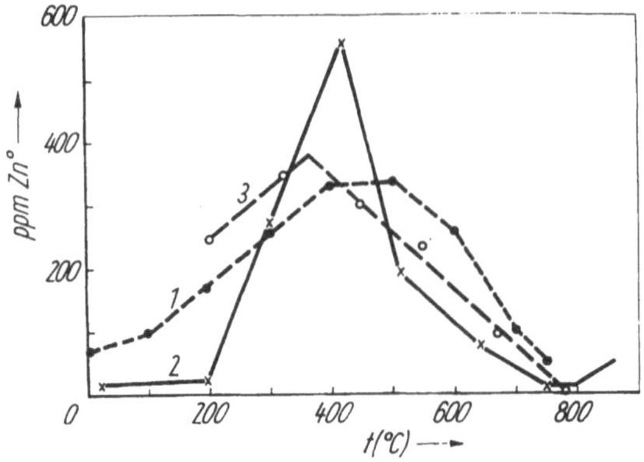

Fig. 1. Amount of free zinc in ZnO as a function of temperature.
Curve 1: Allsopp (100×) [1]
Curve 2: ZnO_B immediately after heat treatment.
Curve 3: ZnO_B exposed to air at room temperature for 60 h after heat treatment (10×)

2. Experimental

In order to work with samples of known history, we started from white, analytical grade ZnO powder. Two equal amounts of this ZnO_N were heated at 920 °C for 1 h. The first batch was afterwards quenched to room temperature and is labelled ZnO_A or type A. The second amount was cooled to room temperature at a rate of 50 °C/h, this is type B or ZnO_B.

Samples, both from ZnO_A and ZnO_B, were afterwards heated for one hour at temperatures situated between 100 and 800 °C, quenched to room temperature, and the amount of free zinc was determined by a colorimetric bichromate method. For type B we found a curve similar to that published by Allsopp [1]. This means that the excess of zinc reaches a maximum at 400 °C (Fig. 1). For type A the excess of zinc decreases with increasing temperature. For optical measurements, thin layers of these different powder samples were used. They were obtained by sedimentation, on thin brass sheets, of the crystalline powder dispersed in water. The layers obtained in this way are granulous after drying. As they are too thick (0.5 mm) for absorption measurements, reflection spectra were made by means of a Beckman D.U. spectrophotometer. Wood's light was used for the excitation of the luminescence and these spectra were recorded with a Perkin Elmer spectrophotometer.

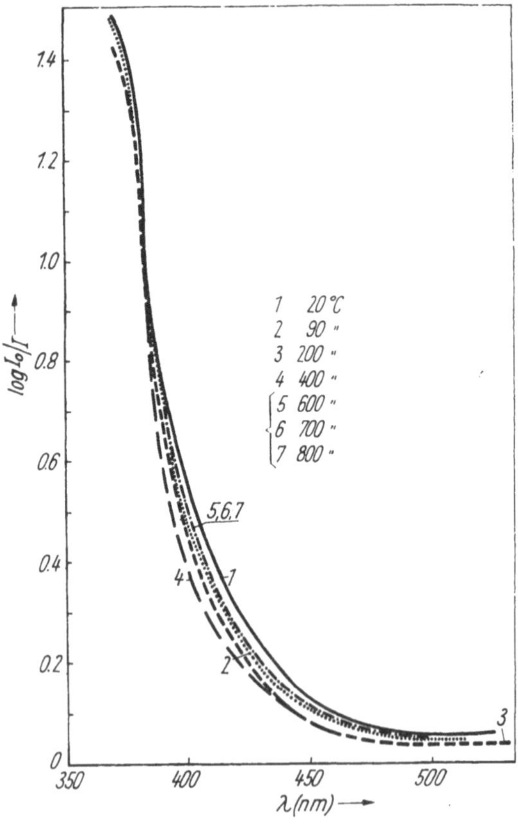

Fig. 2. Reflection spectra of ZnO_A with MgO as blank

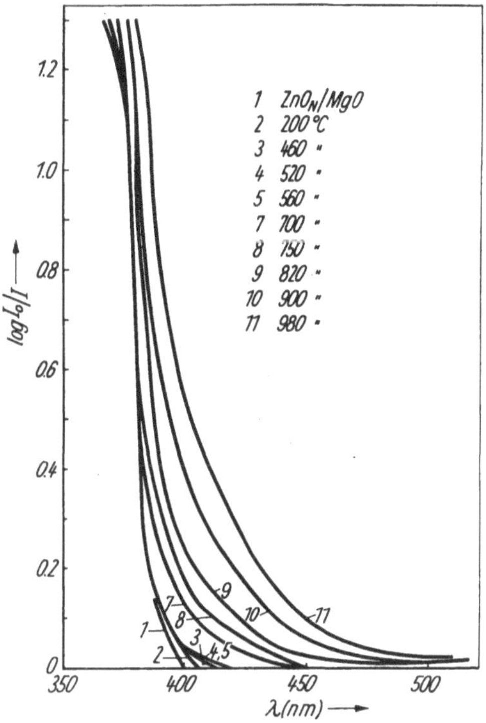

Fig. 3. Reflection spectra of ZnO_B with MgO as blank

3. Experimental Results

3.1 Reflection spectra

A series of reflection spectra were made, first with respect to MgO. The absorption curves are shown in Fig. 2 for specimens derived from type A material and in Fig. 3 for type B ones.

As a result of the heat treatment a shoulder at the fundamental absorption edge, the so called b-band is formed. Its intensity depends upon the temperature at which the ZnO powder was treated.

For type A this b-band is already present in the starting material and its intensity decreases for samples heated up to 400 °C, from 600 °C onwards it increases again.

For type B a real shoulder is only observed for samples heated at temperatures higher than 600 °C.

When the reflection curves of the same samples are now taken with ZnO_N as blank, we obtain the results shown in Fig. 4 and 5.

In these relative reflection spectra we have now, for A type material (Fig. 4), a resolved band at 380 nm as well as the b-band. The 380 nm band increases with

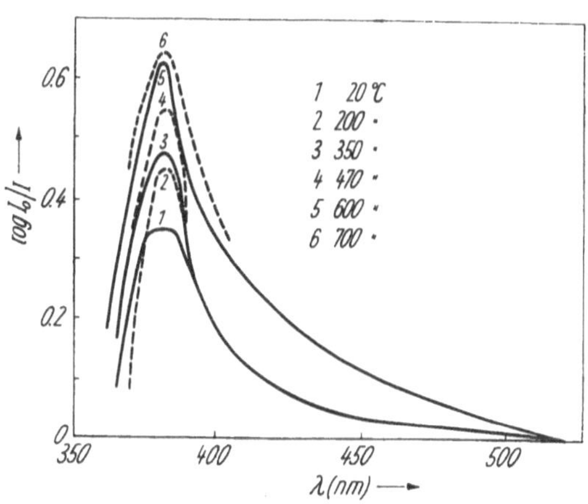

Fig. 4. Reflection spectra of ZnO_A with ZnO_N as blank

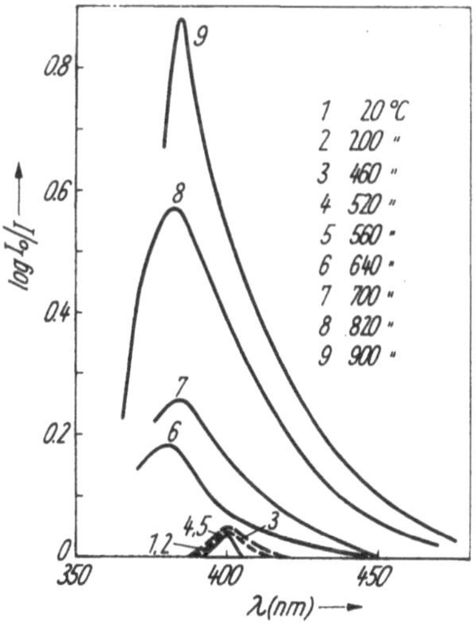

Fig. 5. Reflection spectra of ZnO_B with ZnO_N as blank

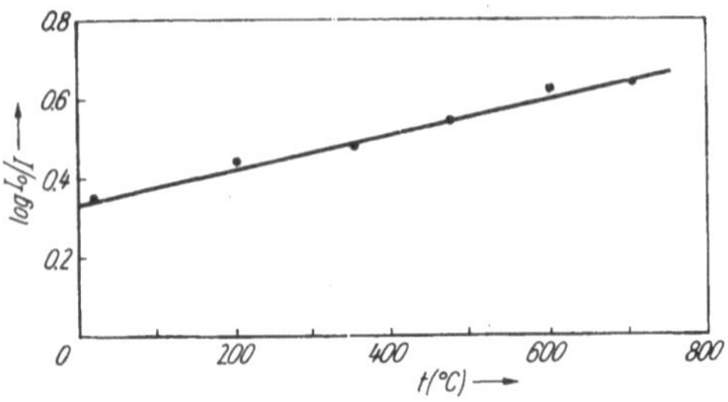

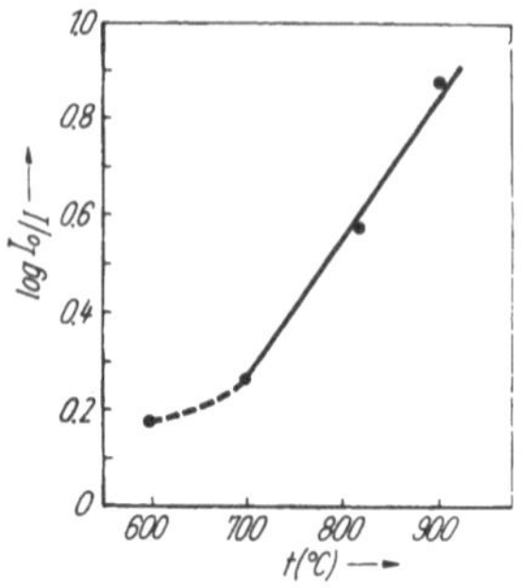

Fig. 6. Intensity of the absorption peak at 380 nm in function of temperature of ZnO_A

Fig. 7. Intensity of the absorption peak at 380 nm in function of temperature of ZnO_B ▶

increasing temperatures and reaches a saturation value for temperatures higher than 700 °C. The shoulder remains nearly constant until 600 °C, for higher temperatures its intensity increases and shifts to longer wavelengths.

In Fig. 5 the evolution of B type material is plotted. Here, a very small band about 400 nm is observed. At 600 °C this 400 nm band disappears and a band at 380 nm as well as the shoulder are formed; they both increase with increasing temperature.

If we plot the absorption constant at $\lambda = 380$ nm in function of temperature for the two types of ZnO, we find a linear relation between log I_0/I and T for type A up to 800 °C and for type B this linearity is only available at temperatures exceeding 600 °C (Fig. 6 and Fig. 7).

3.2 Luminescence spectra

The same specimens described above were excited by ultra violet light and the intensity of the luminescence was measured. We only were interested in the green luminescence, which has its maximum at a wavelength of about 510 nm.

We observed that the intensity and wavelength of the peak does not change very much for type A specimens. The intensive luminescence, resulting from the initial heating and subsequent quenching seems to be insensitive to another short temperature treatment below 920 °C.

For specimens of type B however, we found, on the contrary, a temperature dependent effect, again two temperature ranges can be considered (Fig. 8). The initial luminescence of ZnO_B decreases in intensity after a second treatment at temperatures ranging from 20 to 500 °C. From 500 to 900 °C the intensity

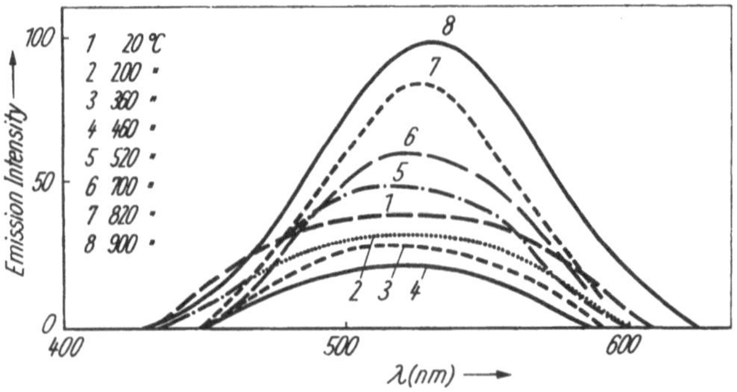

Fig. 8. Green luminescence of ZnO_B

of the luminescence increases strongly with temperature and exceeds the initial ZnO_B value. The maximum of the peak shifts to longer wavelengths with increasing temperature.

3.3 Discussion of the results

To explain the behaviour of the optical absorption in our experiments, we first summarize different features which were taken into account in our interpretation.

a) From the dark conductivity, it was deduced that the electrons, responsible for it result from the dissociation of ZnO in zinc and oxygen

$$ZnO \rightarrow Zn^\circ + \frac{1}{2} O_2 . \tag{1}$$

This neutral zinc, will on its turn dissociate. The two following reactions are then possible:

$$Zn^\circ \rightleftarrows Zn^+ + e , \tag{2}$$

$$Zn^\circ \rightleftarrows Zn^{2+} + 2\,e . \tag{3}$$

It is accepted that relation (2) is valid at temperatures lower than 750 °C and that reaction (3) dominates at higher temperatures.

b) Besides this, the Allsopp curve indicates that the amount of neutral zinc increases with increasing temperature up to 400 °C and decreases again at higher temperatures.

c) The electrical conductivity is highly dependent on the pretreatment of the specimens up to 450 °C. Above this temperature, the σ/T curves are the same for all kinds of specimens [5].

d) This effect of pretreatment is also found in our experiments, i.e. the different optical behaviour of type A and type B ZnO up to 600 °C.

e) As the absorption coefficient of ZnO is very high at the considered wavelength, we may accept that our reflection-measurements describe only the behaviour of a thin surface layer, so that we may suppose that the oxygen adsorption plays an important role in these properties.

f) From the oxygen pressure dependency of the conductivity it is generally accepted that, below 600 °C, we have monovalent ions such as O^- which are chemisorbed [4]. Above 600 °C the chemisorption of O^{2-} is more plausible.

g) An electron paramagnetic resonance study of our specimens proved that no signal corresponding to Zn^+ could be detected. Instead, another signal, corresponding to a $g = 1.96$ is observed. For specimens treated at temperatures higher than 450 °C this signal increases with temperature.

h) The reflection curves of ZnO powder heated in a nitrogen or argon atmosphere, with ZnO_N as blank, do not show the 380 nm band. This indicates that oxygen is needed for the formation of this center.

All these different facts induced us to interpret our results in following way.

At temperatures exceeding 400 °C the neutral zinc formed by reaction (1) dissociates according to relation (2) or (3) for temperatures above 750 °C.

The center which is responsible for the b-band is not a pure F-center [6] because it does not bleach by illumination. It must be a more complex center with an unpaired electron.

Therefore we propose an F-center associated to a Zn^+ ion (see Fig. 9a). At temperatures exceeding 750 °C the Zn^+ looses a second electron which is supposed to be captured by the F-center and forms an F′-center associated to a Zn^{2+} (Fig. 9b).

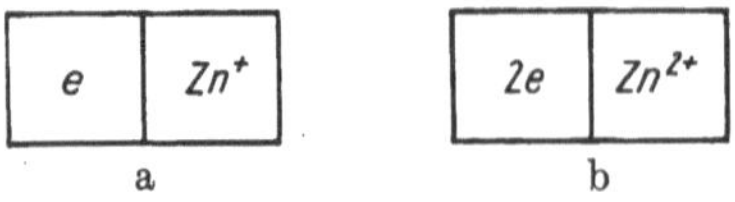

Fig. 9. Models for the b-band absorption

This effect will explain why the wavelength of the b-band shifts to longer wavelengths at higher temperatures. These two centers may capture oxygen at the surface and oxydize to ZnO_S as follows:

center (a) $[Zn^+/\square^{2+} \S\, e/\square^{2-}] + 1/2\ O_2 \rightarrow$

$$[Zn^+/\square^{2+} \S\, O^-/\square^{2-}] \rightarrow Zn^+O^- \rightarrow Zn^{2+}O^{2-} \rightarrow ZnO_S\,, \quad (4)$$

center (b) $[Zn^{2+}/\square^{2+} \S\, 2\,e/\square^{2-}] + 1/2\ O_2 \rightarrow$

$$[Zn^{2+}/\square^{2+} \S\, O^{2-}/\square^{2-}] \rightarrow Zn^{2+}O^{2-} \rightarrow ZnO_S\,. \quad (5)$$

Reaction (4) will happen at temperatures below 600 °C as O^- is then chemisorbed. Reaction (5) occurs at temperatures exceeding 600 °C which corresponds with the chemisorption of O^{2-}.

In the two cases we find finally that the proposed b-centers transform to ZnO_S (S surface) i.e. a layer of highly imperfect ZnO.

By comparing the normal ZnO_N with the heat treated ZnO, we finally find a supplementary absorption peak at 380 nm. We ascribe this peak to the formation of the ZnO_S layer. In the case of type A we observed already, at 20 °C, both the b-band and the peak at 380 nm. In type B however this peak is only observable at temperatures exceeding 600 °C. In type A the Zn_2O centers and a ZnO_S layer are already formed during the pretreatment: i.e. heating at 920 °C and quenching.

These b-centers are first oxydized and the ZnO_S layer is partly dissociated during the second heat treatment from 20 to 800 °C, and this in a different degree. At temperatures below 450 °C Zn° will be formed again. This agrees with the decrease of the b-absorption with increasing temperature from 20 to 600 °C and the corresponding increase of the 380 nm peak. For temperatures exceeding 600 °C, the b-center is formed again and the b-absorption increases with temperature while the 380 nm tends slowly to a constant intensity. This last fact we interprete as if the surface layer is saturated by ZnO_S and has reached a critical thickness.

Acknowledgements

This work is part of a research program sponsored by the I.R.S.I.A. (Brussels) and Gevaert-Agfa N.V. (Antwerp) to which we are greatly indebted.

References

[1] H. J. ALLSOPP, Analyst 82, 474 (1957).
[2] E. SCHAROWSKY, Z. Phys. **135**, 318 (1953).
[3] G. HEILAND, E. MOLLWO, and Z. STÖCKMANN, Solid State Phys. 8, 191 (1959).
[4] T. I. BARRY and F. S. STONE, Proc. Roy. Soc. **A 255**, 124 (1960).
[5] E. E. HAHN, J. appl. Phys. **22**, 855 (1951).
[6] W. EHRET and A. GREENSTONE, J. Amer. Chem. Soc. **65**, 872 (1943).

(Received January 4, 1965)

phys. stat. sol. 8, 847 (1965)

Physikalisch-Technisches Institut der Deutschen Akademie der Wissenschaften zu Berlin, Abteilung Theoretische Physik

Theory of Electron-Phonon Interaction in Metals in the Presence of an External Magnetic Field

By

J. MERTSCHING and H. STOLZ

Semiclassical treatments of the interaction of conduction electrons in metals with ultrasonic waves in the presence of a magnetic field use two coordinate systems: the laboratory system (LS) and a moving system (MS) which shares the motion of the ionic lattice. In the present work the relationship between these two formulations is studied by a quantum-mechanical treatment of the problem. It is shown that in the MS the semiclassical representation of the charge and current densities by means of an electron distribution function is essentially correct. In the LS, however, the off-diagonal elements of the electron density matrix are important, and give rise to the Holstein-effect, i. e. a component of electron current exactly in phase with the lattice current and originating from the motion of the periodic potential. Generally the formulation of the theory for Bloch electrons is more complex in the LS than in the MS. Only for the free electron model the basic epuations obtain their simplest form in the LS.

In den früheren halbklassischen Behandlungen der Wechselwirkung von Leitungselektronen in Metallen mit Ultraschallwellen bei Anwesenheit eines Magnetfeldes wurden zwei Koordinatensysteme benutzt: das Laboratoriumsystem (LS) und das sich mitbewegende System (MS), das an der Bewegung des Ionengitters teilnimmt, so daß die Ionen von diesem System aus gesehen ruhen. In dieser Arbeit wird das Problem quantenmechanisch behandelt und die Beziehung zwischen den beiden Formulierungen im LS und MS untersucht. Es wird gezeigt, daß die halbklassische Betrachtungsweise, in der die Ladungs- und Stromdichten mit Hilfe einer Verteilungsfunktion für Elektronen dargestellt werden, für das MS eine korrekte Approximation ist. Im LS sind jedoch die Nicht-Diagonalelemente der Elektronendichtematrix von Bedeutung, die zu dem sogenannten Holstein-Effekt Anlaß geben, d. h. zu einem Beitrag zum Elektronenstrom, der streng in Phase mit dem Gitterstrom ist und von der Bewegung des periodischen Potentials herrührt. Allgemein ist die Formulierung der Theorie für Blochelektronen im LS komplexer als im MS, nur für das Modell freier Elektronen erhalten die grundlegenden Gleichungen im LS ihre einfachste Gestalt.

1. Introduction

The interaction of the conduction electrons in metals with ultrasonic waves in the presence of an external magnetic field was first studied quasiclassically in the "jellium model" (free electrons in a positively charged continuum [1 to 5]). For Bloch electrons the problem was treated in two different coordinate systems [6 to 18]: in the laboratory system (LS) and in a moving system (MS) fixed to the oscillating lattice. It was shown by Holstein [6] that in the LS the off-diagonal elements of the interaction Hamiltonian give rise to a contribution to the electron current which is strictly in phase with the lattice current. In [18] an attempt was made to take into account this effect in the Boltzmannian framework using a properly defined equivalent Hamiltonian in the LS. By a number of authors [7, 9, 10, 12 to 17] the MS was used to calculate the electron current density from a distribution function obeying Boltzmann's equation, especially in order to derive the equations of motion for the elastic continuum [15 to 17]. It is not clear, whether or not the off-diagonal elements of the density matrix are important in the MS too and why the Holstein effect appears explicitly only in the LS.

In the present work the problem of the interaction of Bloch-electrons with ultrasonic waves is treated quantum mechanically. At first the Hamiltonian in the LS is transformed to the MS by means of an unitary transformation [8, 19 to 21]. The equation of motion of the electron density matrix is solved in the MS, first without, later in the presence of an external magnetic field. Collisions with impurities are taken into account, the collision terms being averaged over the position of the impurities [22]. From the density matrix the electron charge and current densities are derived which enter into Maxwell's equations which must be solved to obtain the selfconsistent electromagnetic field produced by the ultrasonic waves [1]. The equations of motion of the lattice which is suitably approximated by an elastic continuum are derived from the Hamiltonian in the MS by averaging over the electron variables. It is shown that in the MS the off-diagonal elements of the electron density matrix are unimportant and the validity of the results of the semiclassical treatment [17] on the basis of Boltzmann's equation is confirmed. From the equations of motion of the lattice the ultrasonic attenuation coefficient is derived and in an appendix its derivations from the electron entropy which are presented in the literature, are criticized.

In order to understand the roles played by the "collision drag"-effect and by the Holstein-effect, the electron density matrix is transformed from the MS to the LS. The result is that both effects being implicitly included in the formulation in the MS now become apparent in the LS [1, 6]. In addition it is seen that for Bloch electrons it is more convenient to use the MS since the calculations in this case are generally simpler in the MS than in the LS. For the free electron model, however, the expressions for the ultrasonic attenuation and the sound velocity (e.g.) can be easily obtained using the LS.

2. The Hamiltonian in the Laboratory System and the Transformation to the Moving System

We start from the following Hamiltonian in the LS:

$$\mathcal{H}' = \int \left\{\frac{1}{2\,\mu}\,\boldsymbol{\Pi}^2(\boldsymbol{r}) + \frac{1}{2}\sum_{ij}\theta_{ij}(\boldsymbol{r})\,s_{ij}(\boldsymbol{r}) + \varepsilon\,\mu\,\varphi(\boldsymbol{r})\right\} d^3\boldsymbol{r} + \\ + \frac{1}{2\,m}\,\boldsymbol{P}^2 + \sum_{\boldsymbol{R}} v_i\left(\boldsymbol{r}_e - \left(\boldsymbol{R} + \mathbf{s}(\boldsymbol{R})\right)\right) + e\,\varphi(\boldsymbol{r}_e) + \Phi'(\boldsymbol{r}_e)\,. \qquad (2.1)$$

The meaning of the symbols is as follows:

$\boldsymbol{\Pi}(\boldsymbol{r})$	lattice kinetic momentum density,
$\theta_{ij}(\boldsymbol{r})$	strain tensor of the lattice of the bare ions,
$\mathbf{s}(\boldsymbol{r})$	displacement vector,
$s_{ij} = \dfrac{\partial s_i}{\partial x_j}$	deformation tensor,
ε, μ	lattice specific charge and mass density, respectively, ($\varepsilon > 0$),
e, m	electron charge and mass, respectively, ($e < 0$),

$\varphi(\boldsymbol{r})$	selfconsistent scalar electric potential produced by the electrons,
$\boldsymbol{r}_e$, $\boldsymbol{P}$	electron position and kinetic momentum, respectively,
$v_i\,(\boldsymbol{r} - (\boldsymbol{R} + \mathbf{s}(\boldsymbol{R}))$	potential at $\boldsymbol{r}$ of a bare ion displaced by $\mathbf{s}(\boldsymbol{R})$ from its equilibrium position $\boldsymbol{R}$,
$\Phi'(\boldsymbol{r})$	potential of the impurities in the deformed lattice.

The corresponding commutation relations are

$$[\boldsymbol{P}, \boldsymbol{r}_e] = \frac{\hbar}{i}\,1\,, \qquad [\boldsymbol{P}, \boldsymbol{P}] = \frac{e\,\hbar}{i\,c}\,\boldsymbol{B}\times\,, \tag{2.2}$$

$$[\boldsymbol{\Pi}(\boldsymbol{r}), \mathbf{s}(\boldsymbol{r}')] = \frac{\hbar}{i}\,1\,\delta\,(\boldsymbol{r} - \boldsymbol{r}')\,, \qquad [\boldsymbol{\Pi}(\boldsymbol{r}), \boldsymbol{\Pi}(\boldsymbol{r}')] = \frac{\varepsilon\,\mu\,\hbar}{i\,c}\,\delta\,(\boldsymbol{r} - \boldsymbol{r}')\,\boldsymbol{B}\times\,, \tag{2.3}$$

where

$$\boldsymbol{B} = \operatorname{rot}\vec{\mathcal{A}} \tag{2.4}$$

and

$$\vec{\mathcal{A}} = \boldsymbol{A}_0 + \boldsymbol{A} \tag{2.5}$$

is the sum of the external vector potential $\boldsymbol{A}_0$ and the selfconsistent vector potential $\boldsymbol{A}$ produced by the total (electron + ion) electric current.

The mass density of the lattice is related to the displacement by

$$\mu(\boldsymbol{r}) = \mu_0 + \mu_1(\boldsymbol{r}) = \mu_0\,(1 - \operatorname{div}\mathbf{s})\,, \tag{2.6}$$

where μ_0 is the equilibrium density.

Obviously, we have replaced the lattice by a continuum only with respect to its "bare" energy and its interaction with the external and the selfconsistent field, but not with respect to the electron–bare ion interaction.

The next step is to eliminate the potential v_i of the bare ions by a procedure first introduced by Bardeen [23]. We write the periodic lattice potential $V_0(\boldsymbol{r})$ in the form

$$V_0(\boldsymbol{r}) = \sum_{\boldsymbol{R}} v_a\,(\boldsymbol{r} - \boldsymbol{R})\,, \tag{2.7}$$

where $v_a\,(\boldsymbol{r} - \boldsymbol{R})$ vanishes everywhere except in the Wigner-Seitz-cell around $\boldsymbol{R}$ and is supposed to be the sum of $v_i\,(\boldsymbol{r} - \boldsymbol{R})$ and of the potential of a negative charge distribution compensating the ionic charge so that the Wigner-Seitz-cell is neutral as a whole. Extending this picture to the deformed lattice, we write

$$\sum_{\boldsymbol{R}} v_i\,(\boldsymbol{r}_e - (\boldsymbol{R} + \mathbf{s}(\boldsymbol{R})) = \sum_{\boldsymbol{R}} v_a\,(\boldsymbol{r}_e - (\boldsymbol{R} + \mathbf{s}(\boldsymbol{R})) - e\int \frac{\varepsilon\,\mu_0\,\operatorname{div}\mathbf{s}(\boldsymbol{r})}{|\boldsymbol{r}_e - \boldsymbol{r}|}\,d^3\boldsymbol{r}\,.\,^{1)} \tag{2.8}$$

Now we subject $\mathcal{H}'$ to the unitary transformation

$$\left.\begin{aligned} \mathcal{H} &= e^{iS}\,\mathcal{H}'\,e^{-iS}\,,\\ S &= \frac{1}{2\,\hbar}\,[\boldsymbol{P}\cdot\mathbf{s}(\boldsymbol{r}_e)]_+\,. \end{aligned}\right\} \tag{2.9}$$

---

1) A zero order term was dropped because it is cancelled by the zero order term of $e\,\varphi$ in (2.1).

We perform the transformation up to the second order in S dropping all terms of the kinds $s_i s_{jk}$.[2]) Using (2.7) to (2.9) we obtain

$$\mathcal{H} = \mathcal{H}_{\text{ion}} + H + \Phi(\boldsymbol{r}_e) + \frac{1}{2M} \boldsymbol{P}^2 , \tag{2.10}$$

$$\mathcal{H}_{\text{ion}} = \int \left\{ \frac{1}{2\mu_0} \boldsymbol{\Pi}^2(\boldsymbol{r}) + \frac{1}{2} \sum_{ij} \theta_{ij}^{(0)}(\boldsymbol{r})\, s_{ij}(\boldsymbol{r}) - \varepsilon \mu_0 \operatorname{div} \mathbf{s}(\boldsymbol{r})\, \varphi(\boldsymbol{r}) \right\} d^3\boldsymbol{r} , \tag{2.11}$$

$$H = \frac{1}{2m} \boldsymbol{P}^2 + V_0(\boldsymbol{r}) + e\,\varphi(\boldsymbol{r}) + \sum_{ij} \left\{ \Xi_{ij} \frac{\partial s_j}{\partial x_i} - \frac{1}{4m} \left[P_j, \left[P_i, \frac{\partial s_j}{\partial x_i} \right]_+ \right]_+ \right\} -$$
$$- \frac{1}{2\mu_0} [\boldsymbol{P} \cdot \boldsymbol{\Pi}]_+ + \frac{e}{2mc} [\boldsymbol{P} \cdot [\mathbf{s} \times \boldsymbol{B}]]_+ + \frac{e^2}{mc^2} [\mathbf{s} \times \boldsymbol{B}]^2 - \frac{e}{4\mu_0 c} [\boldsymbol{\Pi} \cdot [\mathbf{s} \times \boldsymbol{B}]]_+ . \tag{2.12}$$

Here M means the ion mass and the quantity $\boldsymbol{\Xi}$ is defined by

$$\Xi_{ij} = \sum_{\boldsymbol{R}} (x_i - X_i) \frac{\partial v_a(\boldsymbol{r} - \boldsymbol{R})}{\partial x_j} \tag{2.13}$$

and $e\varphi$ has been redefined to include the second term of (2.8). Accordingly the long range Coulomb-interaction of the ions has been extracted from the second term of (2.11) and is now comprised by the third term of this expression. Thus $\varphi(\boldsymbol{r})$ is to be calculated from

$$\Box \varphi = -4\pi\varrho , \tag{2.14}$$

ϱ being the first order total electric charge density of the system. Accordingly the selfconsistent vector potential $\boldsymbol{A}$ is given by

$$\Box \boldsymbol{A} = -\frac{4\pi}{c} \boldsymbol{j} , \tag{2.15}$$

where $\boldsymbol{j}$ is the total electric current density.

3. The Electron Density Matrix for $\boldsymbol{B}_0 = 0$ in the Moving System

In order to obtain the selfconsistent field from (2.14) and (2.15) it is necessary to calculate ϱ and $\boldsymbol{j}$ from the electron density matrix W. For a given state of the lattice W obeys the equation of motion

$$\frac{dW}{dt} = \frac{i}{\hbar} [W, H + \Phi] = \frac{i}{\hbar} [W, H] + \left(\frac{\partial W}{\partial t} \right)_{\text{coll}} . \tag{3.1}$$

We write

$$W = f(H + \Phi) + W_1 , \tag{3.2}$$

f being the Fermi distribution of the electrons of the density n_0, and linearize (3.1) since it is sufficient to know W up to the first order in the deformation:

$$\frac{dW}{dt} = \frac{i}{\hbar} [W_1, H_0 + \Phi] . \tag{3.3}$$

[2]) It can be shown that these terms lead to a renormalization of the elastic tensor. So we assume them to be absorbed into $\theta_{ij}^{(0)}(\boldsymbol{r})$.

Here H_0 is the zero order part of H (2.12). Putting

$$H_1 = \sum_{\omega} H_1(\omega)\, e^{-i\omega t}\,, \qquad W_1 = \sum_{\omega} W_1(\omega)\, e^{-i\omega t} \tag{3.4}$$

and using the Bloch respresentation

$$\langle \boldsymbol{k}\, \nu\, | H_0 |\, \boldsymbol{k}'\, \nu' \rangle = \varepsilon_{\boldsymbol{k}\nu}\, \delta_{\boldsymbol{k}\boldsymbol{k}'}\, \delta_{\nu\nu'} \qquad (\boldsymbol{k} \text{ reduced, } \nu \text{ band index}) \tag{3.5}$$

we obtain (3.3) up to the first order in the form

$$\frac{i}{\hbar} (\varepsilon_{\boldsymbol{k}\nu} - \varepsilon_{\boldsymbol{k}'\nu'} - \hbar\,\omega)\, \langle \boldsymbol{k}\, \nu\, | W_1(\omega) |\, \boldsymbol{k}'\, \nu' \rangle - i\,\omega\, \langle \boldsymbol{k}\, \nu\, | H_0(\omega) |\, \boldsymbol{k}'\, \nu' \rangle \frac{f(\varepsilon_{\boldsymbol{k}\nu}) - f(\varepsilon_{\boldsymbol{k}'\nu'})}{\varepsilon_{\boldsymbol{k}\nu} - \varepsilon_{\boldsymbol{k}'\nu'}} =$$

$$= -\frac{i}{\hbar} \sum_{\boldsymbol{k}''\nu''} \{ \langle \boldsymbol{k}\, \nu\, | \Phi |\, \boldsymbol{k}''\, \nu'' \rangle \langle \boldsymbol{k}''\, \nu''\, | W_1(\omega) |\, \boldsymbol{k}'\, \nu' \rangle -$$

$$- \langle \boldsymbol{k}''\, \nu''\, | \Phi |\, \boldsymbol{k}'\, \nu' \rangle \langle \boldsymbol{k}\, \nu\, | W_1(\omega) |\, \boldsymbol{k}''\, \nu'' \rangle \}\,. \tag{3.6}$$

According to the existence of the dispersion relation $\omega = \omega(\boldsymbol{q})$ the only non-vanishing matrix elements of $H_1(\omega)$ are $\langle \boldsymbol{k}\, \nu\, | H_0(\omega) |\, \boldsymbol{k} - \boldsymbol{q} - \boldsymbol{K}\, \nu \rangle$, $\boldsymbol{K}$ being a reciprocal lattice vector and $\boldsymbol{q}$ designing the ultrasonic wave vector. Thus the matrix elements $\langle \boldsymbol{k}\, \nu\, | W_1(\omega) |\, \boldsymbol{k}'\, \nu' \rangle$ for $\boldsymbol{k}' \neq \boldsymbol{k} - \boldsymbol{q} + \boldsymbol{K}$ are one order of magnitude smaller with respect to Φ than for $\boldsymbol{k}' = \boldsymbol{k} - \boldsymbol{q} + \boldsymbol{K}$ [22]. The interband elements $\langle \boldsymbol{k} + \boldsymbol{q}/2\, \nu\, | W_1(\omega) |\, \boldsymbol{k} - \boldsymbol{q}/2\, \nu' \rangle$[3]) with $\nu' \neq \nu$ are much smaller than the intraband elements with $\nu' = \nu$ and may be neglected if the conditions $|\varepsilon_{\boldsymbol{k}+\boldsymbol{q}/2\,\nu} - \varepsilon_{\boldsymbol{k}-\boldsymbol{q}/2\,\nu}| \gg \hbar\omega\,, \frac{\hbar}{\tau}$ hold. Introducing the abbreviations

$$\langle \boldsymbol{k} + \boldsymbol{q}/2\, \nu\, | W_1(\omega) |\, \boldsymbol{k} - \boldsymbol{q}/2\, \nu \rangle = -\chi_\nu^{\boldsymbol{k}} \frac{\partial f_{\boldsymbol{k}\nu}}{\partial \varepsilon}\,, \qquad (f_{\boldsymbol{k}\nu} \equiv f(\varepsilon_{\boldsymbol{k}\nu}))\,, \tag{3.7}$$

$$\langle \boldsymbol{k} + \boldsymbol{q}/2\, \nu\, | H_1(\omega) |\, \boldsymbol{k} - \boldsymbol{q}/2\, \nu \rangle = K_\nu^{\boldsymbol{k}}\,, \tag{3.8}$$

and using the approximation

$$\frac{1}{\hbar} (\varepsilon_{\boldsymbol{k}+\boldsymbol{q}/2\,\nu} - \varepsilon_{\boldsymbol{k}-\boldsymbol{q}/2\,\nu}) \approx \boldsymbol{q} \cdot \frac{1}{\hbar} \frac{\partial \varepsilon_{\boldsymbol{k}\nu}}{\partial \boldsymbol{k}} = \boldsymbol{q} \cdot \boldsymbol{v}_{\boldsymbol{k}\nu}$$

we obtain the drift term on the left side of (3.6)

$$-\frac{\partial \chi_\nu^{\boldsymbol{k}}}{\partial t} = i\, (\boldsymbol{q} \cdot \boldsymbol{v}_{\boldsymbol{k}\nu} - \omega)\, \chi_\nu^{\boldsymbol{k}} + K_\nu^{\boldsymbol{k}}\,. \tag{3.9}$$

In close analogy to the work of Kohn and Luttinger [22] the collision term on the right side of (3.6) is simplified by a relaxation time approximation. To achieve this the matrix elements $\langle \boldsymbol{k}\, \nu\, | W_1(\omega) |\, \boldsymbol{k}'\, \nu' \rangle$ for $\boldsymbol{k} - \boldsymbol{k}' \neq \boldsymbol{q} + \boldsymbol{K}$ are calculated from (3.6) to first order with respect to Φ:

$$\langle \boldsymbol{k}\, \nu\, | W_1(\omega) |\, \boldsymbol{k}'\, \nu' \rangle = \frac{\langle \boldsymbol{k}\, \nu\, | \Phi |\, \boldsymbol{k}'\, \nu' \rangle}{\varepsilon_{\boldsymbol{k}\nu} - \varepsilon_{\boldsymbol{k}'\nu'} - \hbar\,\omega} \left(\chi_{\nu'}^{\boldsymbol{k}'} \frac{\partial f_{\boldsymbol{k}'\nu'}}{\partial \varepsilon} - \chi_\nu^{\boldsymbol{k}} \frac{\partial f_{\boldsymbol{k}\nu}}{\partial \varepsilon} \right). \tag{3.10}$$

---

[3]) If $\boldsymbol{k} \pm \frac{\boldsymbol{q}}{2}$ is not reduced, this quantity is easily defined with the help of the periodicity property $\psi_{\boldsymbol{k}+\boldsymbol{K}\,\nu}(\boldsymbol{r}) = \psi_{\boldsymbol{k}\nu}(\boldsymbol{r})$ of the Bloch functions.

Inserting this into the collision term of (3.6) we obtain

$$\left(\frac{\partial \chi_\nu^{\boldsymbol{k}}}{\partial t}\right)_{\text{coll}} \frac{\partial f_{\boldsymbol{k}\nu}}{\partial \varepsilon} = \frac{i}{\hbar} \sum_{\boldsymbol{k}'\nu'} |\langle \boldsymbol{k}\,\nu\,|\Phi|\,\boldsymbol{k}'\,\nu'\rangle|^2 \left(\chi_\nu^{\boldsymbol{k}} \frac{\partial f_{\boldsymbol{k}\nu}}{\partial \varepsilon} - \chi_{\nu'}^{\boldsymbol{k}'} \frac{\partial f_{\boldsymbol{k}'\nu'}}{\partial \varepsilon}\right) \times$$
$$\times \left[\frac{1}{\varepsilon_{\boldsymbol{k}'\nu'} - \varepsilon_{\boldsymbol{k}\nu} - \hbar\,\omega} + \frac{1}{\varepsilon_{\boldsymbol{k}\nu} - \varepsilon_{\boldsymbol{k}'\nu'} - \hbar\,\omega}\right]. \tag{3.11}$$

Averaging of (3.11) over the positions of the N impurities, each of which contribute the potential Φ_s to Φ, leads to the relaxation time approximation

$$\left(\frac{\partial \chi_\nu^{\boldsymbol{k}}}{\partial t}\right)_{\text{coll}} = -\frac{\chi_\nu^{\boldsymbol{k}} - \overline{\chi_\nu^{\boldsymbol{k}}}}{\tau} \tag{3.12}$$

with

$$\frac{1}{\tau} \sim \frac{2\,\pi}{\hbar} N_s \sum_{\boldsymbol{k}'\nu'} |\langle \boldsymbol{k}\,\nu\,|\Phi_s|\,\boldsymbol{k}'\,\nu'\rangle|^2\, \delta\,(\varepsilon_{\boldsymbol{k}\nu} - \varepsilon_{\boldsymbol{k}'\nu'}) \tag{3.13}$$

and the bar designing the average over the Fermi surface.

Since after (3.2), (3.4) and (3.7) $\chi_\nu^{\boldsymbol{k}} \frac{\partial f_{\boldsymbol{k}\nu}}{\partial \varepsilon}$ is the matrix element of the Fourier component of $W_1 = W - f\,(H + \Phi)$, we see from (3.12) that W relaxes into the equilibrium distribution

$$W_{eq} = f\,(H + \Phi - \overline{\chi_\nu^{\boldsymbol{k}}})\,, \tag{3.14}$$

where $H + \Phi$ is the Hamiltonian of the electrons in the MS and $\overline{\chi_\nu^{\boldsymbol{k}}}$ represents a renormalization of the Fermi level. The form of (3.14) is referred to as the "collision drag" effect [1, 6].

It must be kept in mind however, that the averaging process over the positions of the impurities performed already on the stage of equation (3.11) is open to criticism. Correctly it would be the final expressions as ultrasonic attenuation or sound velocity e.g. which might be averaged over the impurity positions [24], not to speak of the problem of the existence of a relaxation time at all. The collision drag problem appears to be very difficult already in the free electron model and seems to be still unsolved for the case of impurity scattering [24].

4. The Electron Density Matrix for $\boldsymbol{B}_0 \neq 0$ in the Moving System

In order to obtain the electron density matrix in the presence of an external magnetic field it is convenient to use a non-orthogonal representation introduced by Stinchcombe [25]. We assume $\boldsymbol{B}_0 = \{0, 0, B_0\}$ and $\boldsymbol{A}_0 = B_0\,\hat{A}\cdot\boldsymbol{r}$, where $\hat{A}$ is a constant gauge matrix. Then the Stinchcombe-representation of an arbitrary operator C is defined by

$$\{\boldsymbol{k}\,\nu\,|C|\,\boldsymbol{k}'\,\nu'\} = \int e^{-\frac{i e B_0}{2\hbar c}(\boldsymbol{r}-\boldsymbol{r}')\cdot\hat{A}\cdot(\boldsymbol{r}+\boldsymbol{r}')} \langle \boldsymbol{k}\,\nu\,|\,\boldsymbol{r}\rangle\,\langle \boldsymbol{r}\,|C|\,\boldsymbol{r}'\rangle\,\langle \boldsymbol{r}'\,|\,\boldsymbol{k}'\,\nu'\rangle\, d^3\boldsymbol{r}\, d^3\boldsymbol{r}'\,, \tag{4.1}$$

$\langle \boldsymbol{r}\,|\,\boldsymbol{k}\,\nu\rangle$ being the Bloch functions. The advantage presented by the Stinchcombe representation becomes apparent in the relation

$$\{\boldsymbol{k}\,\nu\,|C(\boldsymbol{P}, \boldsymbol{r})|\,\boldsymbol{k}'\,\nu'\} = \langle \boldsymbol{k}\,\nu\,|C(\boldsymbol{p}, \boldsymbol{r})|\,\boldsymbol{k}'\,\nu'\rangle\,, \tag{4.2}$$

where

$$\boldsymbol{P} = \boldsymbol{p} - \frac{e}{c}\boldsymbol{A}_0 \tag{4.3}$$

is the electron kinetic momentum in the field $\boldsymbol{B}_0$. This relation holds exactly for a rather wide class of operators $C(\boldsymbol{P}, \boldsymbol{r})$ as will be discussed in appendix I. For any other operator $C(\boldsymbol{P}, \boldsymbol{r})$ (4.2) is correct apart from quantum corrections.

Making use of (4.2) it follows that

$$\begin{aligned} \operatorname{Tr}\{C(\boldsymbol{P}, \boldsymbol{r})\, W\} &= \sum_{\boldsymbol{k}\nu} \sum_{\boldsymbol{k}'\nu'} \{\boldsymbol{k}'\,\nu'\, |C(\boldsymbol{P}, \boldsymbol{r})|\, \boldsymbol{k}\,\nu\}\, \{\boldsymbol{k}\,\nu\, |W|\, \boldsymbol{k}'\,\nu'\} \\ &= \sum_{\boldsymbol{k}\nu} \sum_{\boldsymbol{k}'\nu'} \langle \boldsymbol{k}'\,\nu'\, |C(\boldsymbol{p}, \boldsymbol{r})|\, \boldsymbol{k}\,\nu\rangle\, \{\boldsymbol{k}\,\nu\, |W|\, \boldsymbol{k}'\,\nu'\}\,. \end{aligned} \tag{4.4}$$

The first line of (4.4) is by no means trivial, since the Stinchcombe representation is not orthogonal. From (4.4) it is seen that the expectation value of the operator C in the presence of a magnetic field can be calculated from the Bloch matrix elements of $C(\boldsymbol{p}, \boldsymbol{r})$ and the Stinchcombe matrix elements of W.

Apparently the next we must do is to solve the equation of motion (3.1) of W in the Stinchcombe representation. The corresponding calculations for the case $\boldsymbol{q} = 0$, $\omega = 0$ were carried out by Stinchcombe [25] in analogy to the work of Kohn and Luttinger [22] for zero magnetic field.

Turning to the Stinchcombe representation we write instead of (3.7) and (3.8)

$$\{\boldsymbol{k} + \boldsymbol{q}/2\,\nu\, |W_1(\omega)|\, \boldsymbol{k} - \boldsymbol{q}/2\,\nu\} = -\,\chi_\nu^{\boldsymbol{k}}\, \frac{\partial f_{\boldsymbol{k}\nu}}{\partial \varepsilon}\,, \tag{4.5}$$

$$\{\boldsymbol{k} + \boldsymbol{q}/2\,\nu\, |H_1(\omega)|\, \boldsymbol{k} - \boldsymbol{q}/2\,\nu\} = K_\nu^{\boldsymbol{k}} \tag{4.6}$$

(this is a new definition of $\chi_\nu^{\boldsymbol{k}}$ and $K_\nu^{\boldsymbol{k}}$). Then along the lines of the work of Stinchcombe [25] the following results are obtained: The off-diagonal elements $\{\boldsymbol{k}\,\nu\, |W_1|\, \boldsymbol{k}'\,\nu'\}$ ($\nu \neq \nu'$) may be neglected as for $\boldsymbol{B}_0 = 0$, if in addition the condition $|\varepsilon_{\boldsymbol{k}'\nu'} - \varepsilon_{\boldsymbol{k}\nu}| \gg \hbar\,\omega_c$ is fulfilled, ω_c being the cyclotron frequency of the electrons. Neglecting the influence of the magnetic field on the collisions, the expression (3.12) for the collision term is retained. Quantum effects may be neglected if $\hbar\,\omega_c \ll k\,T$, ε_F and $\hbar/\tau \ll \varepsilon_F$ (ε_F Fermi energy), so the only change in the drift term (3.9) of $\chi_\nu^{\boldsymbol{k}}$ is introduced by the appearance of the Lorentz force, thus we have instead of (3.9)

$$\left\{i\,(\boldsymbol{q}\cdot\boldsymbol{v}_{\boldsymbol{k}\nu} - \omega) + \frac{e}{\hbar\,c}\,[\boldsymbol{v}_{\boldsymbol{k}\nu} \times \boldsymbol{B}_0]\cdot\frac{\partial}{\partial \boldsymbol{k}}\right\}\chi_\nu^{\boldsymbol{k}} + i\,\omega\,K_\nu^{\boldsymbol{k}} = -\,\frac{\chi_\nu^{\boldsymbol{k}} - \overline{\chi_\nu^{\boldsymbol{k}}}}{\tau}\,. \tag{4.7}$$

In order to obtain an expression for $K_\nu^{\boldsymbol{k}}$ we eliminate $\boldsymbol{\Pi}(\boldsymbol{r}, t)$ from (2.12) by means of

$$\begin{aligned} \dot{\mathbf{s}}(\boldsymbol{r}, t) &= \frac{i}{\hbar}\,[\mathcal{H}, \mathbf{s}(\boldsymbol{r}, t)] = \frac{1}{\mu_0}\,\boldsymbol{\Pi}(\boldsymbol{r}, t) - \\ &- \frac{1}{2\,\mu_0}\,[\delta\,(\boldsymbol{r} - \boldsymbol{r}_e),\, \boldsymbol{P}]_+ - \frac{e}{4\,\mu_0\,c}\,[\delta\,(\boldsymbol{r} - \boldsymbol{r}_e),\, [\mathbf{s}(\boldsymbol{r}, t) \times \boldsymbol{B}]]_+ \end{aligned} \tag{4.8}$$

and thus obtain for the first order term H_1 of $H = H_0 + H_1 + H_2$

$$\begin{aligned} H_1 &= e\,\varphi(\boldsymbol{r}, t) + \frac{e}{2\,m\,c}\,\{[\boldsymbol{P}\cdot[\mathbf{s}(\boldsymbol{r}, t) \times \boldsymbol{B}_0]]_+ - [\boldsymbol{P}\cdot\boldsymbol{A}(\boldsymbol{r}, t)]_+\} + \\ &+ \sum_{ij}\left\{\Xi_{ij}\,\frac{\partial s_j}{\partial x_i} - \frac{1}{4\,m}\left[P_j, \left[P_i, \frac{\partial s_j}{\partial x_i}\right]_+\right]_+\right\} - \frac{1}{2}\,[\boldsymbol{P}\cdot\dot{\mathbf{s}}(\boldsymbol{r}, t)] \end{aligned} \tag{4.9}$$

neglecting terms of the order m/M.

Now we introduce the Fourier representations [4])

$$\mathbf{s}(\boldsymbol{r}, t) = \sum_{\boldsymbol{q}} \mathbf{s}(\boldsymbol{q}, \omega)\, e^{i(\boldsymbol{q}\cdot\boldsymbol{r}-\omega t)}, \qquad \mathbf{s}(\boldsymbol{r}, t) = \sum_{\boldsymbol{q}} \boldsymbol{u}(\boldsymbol{q}, \omega)\, e^{i(\boldsymbol{q}\cdot\boldsymbol{r}-\omega t)} \tag{4.10}$$

and write for brevity $\mathbf{s} \equiv \mathbf{s}(\boldsymbol{q}, \omega)$, $\boldsymbol{u} \equiv \boldsymbol{u}(\boldsymbol{q}, \omega) = -i\,\omega\,\mathbf{s}$. Similar representations [4]) are inserted for $\boldsymbol{A}(\boldsymbol{r}, t)$ and $\varphi(\boldsymbol{r}, t)$. Making use of the abbreviations

$$\boldsymbol{\Lambda}_{\boldsymbol{k}\nu} = \left\langle \boldsymbol{k}\,\nu \left| \boldsymbol{\Xi} - \frac{1}{m}\,\boldsymbol{p}\,\boldsymbol{p} \right| \boldsymbol{k}\,\nu \right\rangle = \boldsymbol{\Xi}'_{\boldsymbol{k}\nu} - m\,\boldsymbol{v}_{\boldsymbol{k}\nu}\,\boldsymbol{v}_{\boldsymbol{k}\nu}, \tag{4.11}$$

$$\boldsymbol{\Xi}'_{\boldsymbol{k}\nu} = \boldsymbol{\Xi}_{\boldsymbol{k}\nu} - \frac{1}{m} \sum_{\nu' \neq \nu} \langle \boldsymbol{k}\,\nu\,|\boldsymbol{p}|\,\boldsymbol{k}'\,\nu'\rangle \langle \boldsymbol{k}'\,\nu'\,|\boldsymbol{p}|\,\boldsymbol{k}\,\nu\rangle, \tag{4.12}$$

$$\boldsymbol{\Xi}_{\boldsymbol{k}\nu} = \langle \boldsymbol{k}\,\nu\,|\boldsymbol{\Xi}|\,\boldsymbol{k}\,\nu\rangle \tag{4.13}$$

we obtain from (4.9) by means of (4.6) and (4.3) to the lowest order in $\boldsymbol{q}$

$$K^{\boldsymbol{k}}_{\nu} = i\,\boldsymbol{q}\cdot\boldsymbol{\Lambda}_{\boldsymbol{k}\nu}\cdot\mathbf{s} + e\,\varphi - \frac{e}{c}\,\boldsymbol{v}_{\boldsymbol{k}\nu}\cdot\{\boldsymbol{A} - \mathbf{s}\times\boldsymbol{B}_0 + m\,\boldsymbol{u}\}. \tag{4.14}$$

The solution of the kinetic equation (4.7) is given by

$$\chi^{\boldsymbol{k}}_{\nu} = -\,i\,\omega\,R_{\boldsymbol{k}\nu}\left(K^{\boldsymbol{k}}_{\nu} + \frac{\overline{R_{\boldsymbol{k}\nu}\,K^{\boldsymbol{k}}_{\nu}}}{\tau - \overline{R}_{\boldsymbol{k}\nu}}\right), \tag{4.15}$$

with (2.9) and eliminate from the result the quantity $\boldsymbol{\Pi}$ by means of equation (4.8). Thus we find
$R_{\boldsymbol{k}\nu}$ being the reciprocal (integral-)operator of the differential operator

$$D_{\boldsymbol{k}\nu} = \frac{1}{\tau} + i\,(\boldsymbol{q}\cdot\boldsymbol{v}_{\boldsymbol{k}\nu} - \omega) + \frac{e}{\hbar\,c}\,[\boldsymbol{v}_{\boldsymbol{k}\nu}\times\boldsymbol{B}_0]\cdot\frac{\partial}{\partial\boldsymbol{k}} \tag{4.16}$$

obeying $D_{\boldsymbol{k}\nu}\,R_{\boldsymbol{k}\nu} = 1$.

5. Calculation of the Electron Charge and Current Densities in the Moving System

Now we calculate the Fourier components ϱ and $\boldsymbol{j}$ of the electron charge and current density, respectively, in the MS. Since in this coordinate system the ions are at rest ϱ and $\boldsymbol{j}$ are equal to the total charge and current density, respectively. Using the abbreviation

$$-\frac{2}{\Omega}\sum_{\boldsymbol{k}\nu}\frac{\partial f_{\boldsymbol{k}\nu}}{\partial\varepsilon} = \mathcal{N} \tag{5.1}$$

for the density of states on the Fermi surface we obtain from the density matrix (3.2) to the lowest order in $\boldsymbol{q}$

$$\varrho = \frac{e}{\Omega}\,\mathrm{Tr}\,\{e^{-i(\boldsymbol{q}\cdot\boldsymbol{r}-\omega t)}\,W\} = \frac{2\,e}{\Omega}\sum_{\boldsymbol{k}\nu}(K^{\boldsymbol{k}}_{\nu} - \chi^{\boldsymbol{k}}_{\nu})\,\frac{\partial f_{\boldsymbol{k}\nu}}{\partial\varepsilon} = \mathcal{N}\,e\,(\overline{\chi^{\boldsymbol{k}}_{\nu}} - \overline{K^{\boldsymbol{k}}_{\nu}}), \tag{5.2}$$

$$\begin{aligned}\boldsymbol{j} &= \frac{e}{\Omega}\,\mathrm{Tr}\left\{\frac{1}{2}\,[\dot{\boldsymbol{r}}, e^{-i(\boldsymbol{q}\cdot\boldsymbol{r}-\omega t)}]_+\,W\right\} = \\ &= \frac{e}{\Omega}\,\mathrm{Tr}\left\{\frac{i}{2\,\hbar}\,[[H,\boldsymbol{r}], e^{-i(\boldsymbol{q}\cdot\boldsymbol{r}-\omega t)}]_+\,f(H)\right\} + \frac{e}{\Omega}\,\mathrm{Tr}\left\{\frac{1}{2\,m}\,[\boldsymbol{P}, e^{-i(\boldsymbol{q}\cdot\boldsymbol{r}-\omega t)}]_+\,W_1\right\} \approx \\ &\approx \mathcal{N}\,e\,\overline{\boldsymbol{v}_{\boldsymbol{k}\nu}\,\chi^{\boldsymbol{k}}_{\nu}}\end{aligned} \tag{5.3}$$

---

[4]) It is in the spirit of the continuum approximation to confine $\boldsymbol{q}$ to the first Brillouin zone.

since the contribution of the first term is of higher order with respect to $\boldsymbol{q}$.

Evidently ϱ and $\boldsymbol{j}$ fulfill the equation of continuity

$$i\,\boldsymbol{q}\cdot\boldsymbol{j} - i\,\omega\,\varrho = 0 \tag{5.4}$$

as follows immediately averaging (4.7) over the Fermi surface. In addition $\boldsymbol{j}$ and ϱ are gauge invariant: After (4.14) $K_\nu^{\boldsymbol{k}}$ is augmented by a term proportional to $i\,(\boldsymbol{q}\cdot\boldsymbol{v}_{\boldsymbol{k}\nu} - \omega) \equiv D_{\boldsymbol{k}\nu} - \frac{1}{\tau}$ as a consequence of the gauge transformation

$$\varphi \to \varphi + \frac{i\,\omega}{c}\,F\,, \qquad \boldsymbol{A} \to \boldsymbol{A} + i\,\boldsymbol{q}\,\boldsymbol{F}\,.$$

According to (4.15) and

$$R_{\boldsymbol{k}\nu}\left(D_{\boldsymbol{k}\nu} - \frac{1}{\tau} + \frac{\overline{R_{\boldsymbol{k}\nu}\left(D_{\boldsymbol{k}\nu} - \frac{1}{\tau}\right)}}{\tau - \overline{R_{\boldsymbol{k}\nu}}}\right) = 1 \tag{5.5}$$

this additional term of $K_\nu^{\boldsymbol{k}}$ leads to a constant additive term in $\chi_\nu^{\boldsymbol{k}}$ which, as is seen from (5.3) and (5.4), makes no contribution to $\boldsymbol{j}$ and to ϱ neither. For convenience we choose the gauge $\varphi \equiv 0$ in the following sections.

6. The Equation of Motion of the Ions in the Continuum Approximation

In order to obtain the equations of motion of the ionic continuum we calculate

$$\dot{\boldsymbol{\Pi}} = \frac{i}{\hbar}\,[\mathcal{H}, \boldsymbol{\Pi}] - \frac{\varepsilon\,\mu_0}{c}\frac{\partial \vec{\mathcal{A}}}{\partial t}\,, \tag{6.1}$$

$$\begin{aligned}
\mu_0\,\ddot{\mathbf{s}} + \operatorname{Div}\boldsymbol{\theta}^{(0)} = {} & \varepsilon\,\mu_0\left\{-\frac{1}{c}\frac{\partial\vec{\mathcal{A}}}{\partial t} + \frac{1}{c}\,[\dot{\mathbf{s}}\times\boldsymbol{B}_0]\right\} + \operatorname{grad}_{\boldsymbol{r}}\delta\,(\boldsymbol{r}-\boldsymbol{r}_e)\,\boldsymbol{\Xi} - \\
& - \frac{1}{4\,m}\sum_i\left[P_i, \left[\boldsymbol{P}, \frac{\partial}{\partial x_i}\,\delta\,(\boldsymbol{r}-\boldsymbol{r}_e)\right]_+\right]_+ + \\
& + \frac{e}{2\,m\,c}\{[\boldsymbol{P}\times\boldsymbol{B}_0]\,\delta\,(\boldsymbol{r}-\boldsymbol{r}_e) - \delta\,(\boldsymbol{r}-\boldsymbol{r}_e)\,[\boldsymbol{B}_0\times\boldsymbol{P}]\} - \\
& - \frac{1}{2}\frac{d}{dt}\,[\boldsymbol{P}, \delta\,(\boldsymbol{r}-\boldsymbol{r}_e)]_+ + \frac{e^2}{m\,c^2}\,[\mathbf{s}\times\boldsymbol{B}_0]\times\boldsymbol{B}_0\,\delta\,(\boldsymbol{r}-\boldsymbol{r}_e) - \\
& - \frac{e}{2\,c}\,\{[\dot{\mathbf{s}}\times\boldsymbol{B}]\,\delta\,(\boldsymbol{r}-\boldsymbol{r}_e) - \delta\,(\boldsymbol{r}-\boldsymbol{r}_e)\,[\boldsymbol{B}\times\dot{\mathbf{s}}]\}\,.
\end{aligned} \tag{6.2}$$

Now the Fourier component of this equation is taken and averaged over the electron distribution W. If we eliminate $\boldsymbol{P}$ by means of the relation

$$\boldsymbol{V} = \frac{i}{\hbar}\,[H_0 + H_1, \boldsymbol{r}_e] = \frac{1}{m}\,\boldsymbol{P} - \frac{e}{m\,c}\{\boldsymbol{A} - [\mathbf{s}\times\boldsymbol{B}_0]\} - \frac{1}{2\,m}\sum_i\left[\left(\frac{\partial\mathbf{s}}{\partial x_i} + \operatorname{grad} s_i\right), P_i\right]_+ \tag{6.3}$$

the result turns out to be

$$-(\mu_0 + n_0 m)\,\omega^2 \mathbf{s} + i\,\boldsymbol{q}\cdot\boldsymbol{\theta}^{(0)} = \mathcal{N}\, i\,\boldsymbol{q}\cdot\overline{\Lambda_{\boldsymbol{k}\nu}\,\chi_\nu^{\boldsymbol{k}}} + \frac{1}{c}[\boldsymbol{j}\times\boldsymbol{B}_0] + \frac{m}{e}\, i\,\omega\,\boldsymbol{j}\,, \tag{6.4}$$

some terms which only give rise to a renormalization of $\boldsymbol{\theta}^{(0)}$ being discarded.

The equation of motion (6.4) was first derived in the framework of a semiclassical theory by Skobov and Kaner [17].

7. Ultrasonic Attenuation

From (6.4) it is easily derived that the product of the attenuation coefficient α and the intensity I of an ultrasonic wave with frequency ω and wave vector $\boldsymbol{q}$ is given by the expression

$$\alpha\, I = -\left\langle \boldsymbol{u}\cdot\left(\mathcal{N}\, i\,\boldsymbol{q}\cdot\overline{\Lambda_{\boldsymbol{k}\nu}\,\chi_\nu^{\boldsymbol{k}}} + \frac{1}{c}[\boldsymbol{j}\times\boldsymbol{B}_0] + \frac{m}{e}\, i\,\omega\,\boldsymbol{j}\right)\right\rangle, \tag{7.1}$$

where the brackets $\langle\,\rangle$ design the time average over the period $T = \dfrac{2\pi}{\omega}$. With the aid of (4.9) and (5.3) we find

$$-\langle \mathcal{N}\, i\,\boldsymbol{q}\cdot\overline{\Lambda_{\boldsymbol{k}\nu}\,\chi_\nu^{\boldsymbol{k}}}\cdot\boldsymbol{u}\rangle = \left\langle \mathrm{Tr}\left\{\sum_{ij} \Xi_{ij}\frac{\partial s_j}{\partial x_i} - \frac{1}{4m}\left[P_i,\left[P_j,\frac{\partial \dot{s}_j}{\partial x_i}\right]_+\right]_+\right\} W\right\rangle =$$

$$= \mathrm{Tr}\left\langle \frac{\partial H_1}{\partial t} W_1\right\rangle + \frac{1}{c}[\boldsymbol{j}\times\boldsymbol{B}_0]\cdot\boldsymbol{u} - \langle \boldsymbol{j}\cdot\boldsymbol{E}\rangle + \frac{m}{e}\langle \boldsymbol{j}\cdot\dot{\boldsymbol{u}}\rangle\,. \tag{7.2}$$

Inserting (7.2) into (7.1) and using $\langle \boldsymbol{j}\cdot\boldsymbol{E}\rangle \equiv 0$ which follows from Maxwell's equations we obtain

$$\alpha\, I = \mathrm{Tr}\left\langle \frac{\partial H_1}{\partial t} W_1\right\rangle = \mathrm{Tr}\left\langle \frac{\partial H}{\partial t} W\right\rangle = -\,\mathrm{Tr}\left\langle H\frac{\partial W}{\partial t}\right\rangle = -\,\mathrm{Tr}\left\langle H\left(\frac{\partial W}{\partial t}\right)_{\mathrm{coll}}\right\rangle. \tag{7.3}$$

Now looking at (4.7) we realize

$$\mathrm{Tr}\left\langle \frac{\partial H_1}{\partial t} W_1\right\rangle = \frac{1}{2}\,\mathrm{Re}\; i\,\omega\;\mathrm{Tr}\,\langle H_1^*(\omega)\, W_1(\omega)\rangle =$$

$$= -\frac{1}{2}\,\mathrm{Re}\; i\,\omega\,\frac{2}{\Omega}\sum_{\boldsymbol{k}\nu} K_\nu^{\boldsymbol{k}*}\,\chi_\nu^{\boldsymbol{k}}\,\frac{\partial f_{\boldsymbol{k}\nu}}{\partial\varepsilon} =$$

$$= -\frac{1}{2}\,\mathrm{Re}\,\frac{2}{\Omega}\sum_{\boldsymbol{k}\nu}\frac{(\chi_\nu^{\boldsymbol{k}} - \overline{\chi_\nu^{\boldsymbol{k}}})^*}{\tau}\,\chi_\nu^{\boldsymbol{k}}\,\frac{\partial f_{\boldsymbol{k}\nu}}{\partial\varepsilon} = \frac{1}{2}\,\frac{\mathcal{N}}{\tau}\,|\tilde{\chi}_\nu^{\boldsymbol{k}}|^2\,, \tag{7.4}$$

thus

$$\alpha\, I = \frac{1}{2}\,\frac{\mathcal{N}}{\tau}\,|\tilde{\chi}_\nu^{\boldsymbol{k}}|^2\,, \tag{7.5}$$

where

$$\tilde{\chi}_\nu^{\boldsymbol{k}} = \chi_\nu^{\boldsymbol{k}} - \bar{\chi}_\nu^{\boldsymbol{k}}\,. \tag{7.6}$$

This expression was used by many authors [7, 9, 10, 12 to 14, 18] as a starting point for the discussion of ultrasonic attenuation and especially of the magneto-acoustic effects. The derivation of (7.5) given in the literature which starts from the statistical expression of the electron entropy is not correct as we shall discuss in appendix II.

Introducing the abbreviation

$$\tilde{\Lambda}_{k\nu} = \Lambda_{k\nu} - \bar{\Lambda}_{k\nu} \tag{7.7}$$

we obtain from (4.7), (4.14) and (5.2) the following equation for the function $\tilde{\chi}_\nu^k$

$$\left\{\frac{1}{\tau} + i\,(\boldsymbol{q}\cdot\boldsymbol{v}_{k\nu} - \omega) + \frac{e}{\hbar\,c}[\boldsymbol{v}_{k\nu}\times\boldsymbol{B}_0]\cdot\frac{\partial}{\partial\boldsymbol{k}}\right\}\tilde{\chi}_\nu^k =$$

$$= \boldsymbol{v}_{k\nu}\cdot\left\{e\,\boldsymbol{E} - i\,\boldsymbol{q}\,i\,\boldsymbol{q}\cdot\bar{\Lambda}_{k\nu}\cdot\boldsymbol{s} + \frac{e}{c}[\boldsymbol{u}\times\boldsymbol{B}_0] - m\,\dot{\boldsymbol{u}}\right\} - i\,\boldsymbol{q}\cdot\Lambda_{k\nu}\cdot\boldsymbol{u} - i\,(\boldsymbol{q}\cdot\boldsymbol{v}_{k\nu} - \omega)\frac{\varrho}{\mathcal{N}e}. \tag{7.8}$$

If the ultrasonic frequency ω is much smaller compared with the electron plasma frequency $\omega_p = \left(\frac{4\,\pi\,n_0\,e^2}{m}\right)^{1/2}$ the system can be assumed to be electrically neutral, i.e. $\varrho \approx 0$. From this assumption the longitudinal part of $e\,\boldsymbol{E} - i\,\boldsymbol{q}\,i\,\boldsymbol{q}\cdot\bar{\Lambda}_{k\nu}\cdot\boldsymbol{s}$ can be determined. The resulting $\tilde{\chi}_\nu^k$ then only depends on $\tilde{\Lambda}_{k\nu}$ (but not on $\bar{\Lambda}_{k\nu}$). For this case of charge neutrality the equation (7.8) was set up by Skobov and Kaner [17]. The deviation $\tilde{\chi}_\nu^k$ of the electron distribution from the equilibrium distribution [3.14] in the case $\varrho \approx 0$ may be interpreted as the result of the simultaneous action of the real force

$$\boldsymbol{F}_{\text{real}} = e\,\boldsymbol{E} + \frac{e}{c}[\boldsymbol{u}\times\boldsymbol{B}_0] + i\,\omega\,m\,\boldsymbol{u} - i\,\boldsymbol{q}\,i\,\boldsymbol{q}\cdot\bar{\Lambda}_{k\nu}\cdot\boldsymbol{s} \tag{7.9}$$

and the fictitious force

$$\boldsymbol{F}_{\text{fict}} = -\frac{1}{v^2}\,\boldsymbol{v}\,i\,\boldsymbol{q}\cdot\tilde{\Lambda}_{k\nu}\cdot\boldsymbol{u}\,. \tag{7.10}$$

On these lines the problem of the interaction of the conduction electrons with ultrasonic waves was first studied by Pippard [11] making use of semiclassical kinetic principles.

The transverse current density and the transverse electric field are obtained from (5.3)

$$\boldsymbol{j} = \mathcal{N}\,e\,\overline{\boldsymbol{v}_{k\nu}\,\tilde{\chi}_\nu^k} \tag{7.11}$$

and the Maxwell equation

$$i\,\boldsymbol{q}\times[i\,\boldsymbol{q}\times\boldsymbol{E}] = \frac{4\,\pi}{c^2}\,i\,\omega\,\boldsymbol{j}\,. \tag{7.12}$$

8. Formulation of the Theory in the Laboratory System

In the preceding sections the theory of the interaction of Bloch electrons with ultrasonic waves has been developed in the moving coordinate system (MS). We now turn to the formulation of the theory in the laboratory system, in which the "collision drag" effect and the Holstein effect will appear explicitly.

At first we transform the density matrix (3.2) into the LS according to (2.9). Thus we obtain to the first order

$$W' = e^{-iS}\,W\,e^{iS} = f\,(e^{-iS}\,H_0\,e^{iS} + H_1 + \Phi') + W_1\,. \tag{8.1}$$

It is

$$e^{-iS} H_0 e^{iS} \approx H_0 + \{V_0(\boldsymbol{r} - \mathbf{s}(\boldsymbol{r})) - V_0(\boldsymbol{r})\} + \sum_{ij} \Xi_{ij} \frac{\partial s_j}{\partial x_i} - \frac{e}{2mc} [\boldsymbol{P} \cdot \boldsymbol{A}]_+ - \\ - \frac{1}{2} [\boldsymbol{P} \cdot \dot{\mathbf{s}}] + \Phi' . \tag{8.2}$$

If we now write down W' in the form

$$W' = f\left(H_0 + \sum_{ij} \Xi_{ij} \frac{\partial s_j}{\partial x_i} - \frac{e}{2mc} [\boldsymbol{P} \cdot \boldsymbol{A}]_+\right) + W_1'' \tag{8.3}$$

the matrix elements of the Fourier component of W_1'' turn out to be

$$\{\boldsymbol{k} + \boldsymbol{q}/2 \ \ \nu \,|W_1''|\, \boldsymbol{k} - \boldsymbol{q}/2 \ \ \nu\} = -\eta_\nu^{\boldsymbol{k}} \frac{\partial f_{\boldsymbol{k}\nu}}{\partial \varepsilon} \tag{8.4}$$

with

$$\eta_\nu^{\boldsymbol{k}} = \chi_\nu^{\boldsymbol{k}} - m\, \boldsymbol{v}_{\boldsymbol{k}\nu} \cdot \boldsymbol{u} \tag{8.5}$$

if we put $\{\boldsymbol{k} + \boldsymbol{q}/2\, \nu \,|\{V_0(\boldsymbol{r} - \mathbf{s}(\boldsymbol{r})) - V_0(\boldsymbol{r})\}|\, \boldsymbol{k} - \boldsymbol{q}/2\, \nu\} \approx 0$.

Introducing the abbreviation

$$\varkappa_{\boldsymbol{k}\nu} = \frac{m}{\hbar} \frac{\partial}{\partial \boldsymbol{k}} \boldsymbol{v}_{\boldsymbol{k}\nu} = \frac{m}{\hbar^2} \frac{\partial^2 \varepsilon_{\boldsymbol{k}\nu}}{\partial \boldsymbol{k}\, \partial \boldsymbol{k}} \tag{8.6}$$

and making use of the relation

$$-\left\{i\, (\boldsymbol{q} \cdot \boldsymbol{v}_{\boldsymbol{k}\nu} - \omega) + \frac{e}{\hbar c} [\boldsymbol{v}_{\boldsymbol{k}\nu} \times \boldsymbol{B}_0] \cdot \frac{\partial}{\partial \boldsymbol{k}}\right. m\, \boldsymbol{v}_{\boldsymbol{k}\nu} \cdot \boldsymbol{u} + i\, \omega\, K_\nu^{\boldsymbol{k}} = \\ = i\, \boldsymbol{q} \cdot \boldsymbol{\Xi}'_{\boldsymbol{k}\nu} \cdot \boldsymbol{u} - e\, \boldsymbol{v}_{\boldsymbol{k}\nu} \cdot \boldsymbol{E} + \frac{e}{c} [\boldsymbol{v}_{\boldsymbol{k}\nu} \times \boldsymbol{B}_0] \cdot (1 - \varkappa_{\boldsymbol{k}\nu}) \cdot \boldsymbol{u} , \tag{8.7}$$

which follows from (4.11), (4.12) and (4.14) we find for $\eta_\nu^{\boldsymbol{k}}$ the equation

$$\left\{\frac{1}{\tau} + i\, (\boldsymbol{q} \cdot \boldsymbol{v}_{\boldsymbol{k}\nu} - \omega) + \frac{e}{c\hbar} [\boldsymbol{v}_{\boldsymbol{k}\nu} \times \boldsymbol{B}_0] \cdot \frac{\partial}{\partial \boldsymbol{k}}\right\} \eta_\nu^{\boldsymbol{k}} = \\ = \boldsymbol{v}_{\boldsymbol{k}\nu} \cdot \left\{e\, \boldsymbol{E} + \frac{m\, \boldsymbol{u}}{\tau} + \frac{e}{c} [(1 - \varkappa_{\boldsymbol{k}\nu}) \cdot \boldsymbol{u} \times \boldsymbol{B}_0]\right\} - i\, \boldsymbol{q} \cdot \boldsymbol{\Xi}'_{\boldsymbol{k}\nu} \cdot \boldsymbol{u} + \frac{1}{\tau} \overline{\eta_\nu^{\boldsymbol{k}}} . \tag{8.8}$$

The electron charge density ϱ' in the LS is the difference of the total charge density ϱ and the ion charge density $n_0\, e\, i\, \boldsymbol{q} \cdot \mathbf{s}$. Accordingly the electron current density $\boldsymbol{j}'$ is expressed by the difference $\boldsymbol{j} - (-n_0\, e\, \boldsymbol{u})$. With the aid of the formula

$$\mathcal{N}\, m\, \overline{\boldsymbol{v}_{\boldsymbol{k}\nu}\, \boldsymbol{v}_{\boldsymbol{k}\nu}} = -\sum_{\boldsymbol{k}\nu} m\, \boldsymbol{v}_{\boldsymbol{k}\nu}\, \boldsymbol{v}_{\boldsymbol{k}\nu} \frac{\partial f_{\boldsymbol{k}\nu}}{\partial \varepsilon} = \sum_{\boldsymbol{k}\nu} \varkappa_{\boldsymbol{k}\nu}\, f_{\boldsymbol{k}\nu} \tag{8.9}$$

we obtain ϱ' and $\boldsymbol{j}'$ from (5.2) and (5.3) inserting into these equations the quantity $\chi_\nu^{\boldsymbol{k}}$ according to (8.5):

$$\varrho' = \varrho - n_0\, e\, i\, \boldsymbol{q} \cdot \mathbf{s} = \mathcal{N} e\, (\eta_\nu^{\boldsymbol{k}} - i\, \boldsymbol{q} \cdot \boldsymbol{\Xi}'_{\boldsymbol{k}\nu} \cdot \mathbf{s}) - e\, i\, \boldsymbol{q} \cdot \sum_{\boldsymbol{k}\nu} (1 - \varkappa_{\boldsymbol{k}\nu})\, f_{\boldsymbol{k}\nu} \cdot \mathbf{s}, \tag{8.10}$$

$$\boldsymbol{j}' = \boldsymbol{j} + n_0\, e\, \boldsymbol{u} = \mathcal{N} e\, \overline{\boldsymbol{v}_{\boldsymbol{k}\nu}\, \eta_\nu^{\boldsymbol{k}}} + e \sum_{\boldsymbol{k}\nu} (1 - \varkappa_{\boldsymbol{k}\nu})\, f \cdot \boldsymbol{u} . \tag{8.11}$$

Now it is interesting to express the deformation potential $\Lambda_{\boldsymbol{k}\nu}$ in the MS which enters into the equation of motion (6.4) of the continuum, by the deformation potential $\Xi'_{\boldsymbol{k}\nu}$ in the LS. By the aid of the relation

$$\mathcal{N} i \boldsymbol{q} \cdot m \overline{\boldsymbol{v}_{\boldsymbol{k}\nu} \boldsymbol{v}_{\boldsymbol{k}\nu} \chi_\nu^{\boldsymbol{k}}} = \mathcal{N} m \overline{\boldsymbol{v}_{\boldsymbol{k}\nu} \left(i \omega \chi_\nu^{\boldsymbol{k}} - \frac{e}{\hbar c} [\boldsymbol{v}_{\boldsymbol{k}\nu} \times \boldsymbol{B}_0] \cdot \frac{\partial \chi_\nu^{\boldsymbol{k}}}{\partial \boldsymbol{k}} - i \omega K_\nu^{\boldsymbol{k}} - \frac{\chi_\nu^{\boldsymbol{k}} - \overline{\chi_\nu^{\boldsymbol{k}}}}{\tau} \right)} =$$

$$= \frac{m}{e} i \omega \boldsymbol{j} + \mathcal{N} \overline{\varkappa_{\boldsymbol{k}\nu} \cdot \frac{e}{c} [\boldsymbol{v}_{\boldsymbol{k}\nu} \times \boldsymbol{B}_0] \chi_\nu^{\boldsymbol{k}}} +$$

$$+ \left\{ e \left(\boldsymbol{E} + \frac{\boldsymbol{u}}{c} \times \boldsymbol{B}_0 \right) + i \omega m \boldsymbol{u} \right\} \cdot \sum_{\boldsymbol{k}\nu} \varkappa_{\boldsymbol{k}\nu} f_{\boldsymbol{k}\nu} - \frac{m}{e \tau} \boldsymbol{j} \tag{8.12}$$

the equation of motion then takes the form

$$- \left(\mu_0 + \sum_{\boldsymbol{k}\nu} (1 - \varkappa_{\boldsymbol{k}\nu}) f_{\boldsymbol{k}\nu} m \right) \omega^2 \cdot \mathbf{s} + i \boldsymbol{q} \cdot \boldsymbol{\theta}^{(0)} = \mathcal{N} i \boldsymbol{q} \cdot \overline{\Xi'_{\boldsymbol{k}\nu} \chi_\nu^{\boldsymbol{k}}} -$$

$$- e \left(\boldsymbol{E} + \frac{\boldsymbol{u}}{c} \times \boldsymbol{B}_0 \right) \cdot \sum_{\boldsymbol{k}\nu} \varkappa_{\boldsymbol{k}\nu} f_{\boldsymbol{k}\nu} + \mathcal{N} \overline{(1 - \varkappa_{\boldsymbol{k}\nu}) \cdot \frac{e}{c} (\boldsymbol{v}_{\boldsymbol{k}\nu} \times \boldsymbol{B}_0) \chi_\nu^{\boldsymbol{k}}} + \frac{m}{e \tau} \boldsymbol{j} \, . \tag{8.13}$$

Let us now discuss the basic equations (8.8), (8.10), (8.11) and (8.13) of the theory as expressed in the LS. The "collision drag" effect contributes explicitly the force $\frac{m \boldsymbol{u}}{\tau}$ to the right side of (8.8). As a consequence of the Holstein effect the electrons are dragged in phase with the ions to the amount $(1 - \varkappa_{\boldsymbol{k}\nu}) \cdot \boldsymbol{u}$ so we can think of them according to the laws of electrodynamics in moving media [18]. We thus understand the induction term $\left[(1 - \varkappa_{\boldsymbol{k}\nu}) \cdot \frac{\boldsymbol{u}}{c} \times \boldsymbol{B}_0 \right]$ of (8.8) which appears in addition to the electric field $\boldsymbol{E}$. The collision term in (8.8) is

$$\left(\frac{\partial \eta_\nu^{\boldsymbol{k}}}{\partial t} \right)_{\text{coll}} = - \frac{\eta_\nu^{\boldsymbol{k}} - \overline{\eta_\nu^{\boldsymbol{k}}} - m \boldsymbol{v}_{\boldsymbol{k}\nu} \cdot \boldsymbol{u}}{\tau} \tag{8.14}$$

and has the form which was at first derived by Holstein [6]. From (8.14) we see that in the LS the electron distribution relaxes into

$$\left\{ \boldsymbol{k} + \boldsymbol{q}/2 \, \nu \left| f \left(H_0 + \sum_{ij} \Xi_{ij} \frac{\partial s_j}{\partial x_i} - \frac{e}{2 m c} [\boldsymbol{P} \cdot \boldsymbol{A}]_+ - \overline{\chi_\nu^{\boldsymbol{k}}} \right) \right| \boldsymbol{k} - \boldsymbol{q}/2 \, \nu \right\} - m \boldsymbol{v}_{\boldsymbol{k}\nu} \cdot \boldsymbol{u} \frac{\partial f_{\boldsymbol{k}\nu}}{\partial \varepsilon} \approx$$

$$\approx f \left(\varepsilon_\nu \left(\boldsymbol{k} - \frac{e}{\hbar c} \boldsymbol{A} - \frac{m \boldsymbol{u}}{\hbar} \right) + \Xi_{ij} \frac{\partial s_j}{\partial x_i} - \overline{\chi_\nu^{\boldsymbol{k}}} \right) \tag{8.15}$$

i.e. a displaced Fermi distribution.

The Holstein contributions proportional to $1 - \varkappa_{\boldsymbol{k}\nu}$ to $\boldsymbol{j}'$ and ϱ' arise from off-diagonal elements of W' in the LS. Of course the Holstein effect is taken into account correctly also in the expressions (5.2) and (5.3) of the total current and charge densities in the MS but it does not enter explicitly into them in a manner comparable with its appearing in (8.10) and (8.11). This is because of that there are no relevant off-diagonal elements of W in the MS.

The Holstein effect manifests itself in the equation of motion (8.13) in the term

$$e\left\{\boldsymbol{E}+\frac{\boldsymbol{u}}{c}\times\boldsymbol{B}_0\right\}\cdot\sum_{\boldsymbol{k}\nu}(1-\varkappa_{\boldsymbol{k}\nu})\,f_{\boldsymbol{k}\nu}+\overline{\mathcal{N}(1-\varkappa_{\boldsymbol{k}\nu})\cdot\frac{e}{c}[\boldsymbol{v}_{\boldsymbol{k}\nu}\times\boldsymbol{B}]\,\chi_\nu^{\boldsymbol{k}}} \tag{8.16}$$

which is to be added to the electromagnetic force $-n_0\left(e\,\boldsymbol{E}-\frac{e}{c}[\boldsymbol{u}\times\boldsymbol{B}_0]\right)$, the deformation force $\mathcal{N}\,i\,\boldsymbol{q}\cdot\overline{\Xi'_{\boldsymbol{k}\nu}\,\chi_\nu^{\boldsymbol{k}}}$ and the collision drag force $\frac{m}{e\tau}\boldsymbol{j}$ acting on the ions.

Since in the free electron model $\Xi'_{\boldsymbol{k}\nu}=0$, $\varkappa_{\boldsymbol{k}\nu}=1$ the Holstein effect and the deformation force vanish in this case and the equations (8.8), (8.10), (8.11) and (8.13) take their simple forms in which they establish the fundamentals of the work of Cohen, Harrison and Harrison [2]. For Bloch electrons, however, the equations in the LS are considerably complicated so that the MS appears to be more useful.

Acknowledgement

We wish to express our gratitude to H. Schulz for valuable remarks and suggestions on the generalization of the results of the work of Stinchcombe [25].

Appendix I

Derivation of equation (4.2)

In order to derive the relation (4.2)

$$\{\boldsymbol{k}\,\nu\,|C(\boldsymbol{P},\boldsymbol{r})|\,\boldsymbol{k}'\,\nu'\}=\langle\boldsymbol{k}\,\nu\,|C(\boldsymbol{p},\boldsymbol{r})|\,\boldsymbol{k}'\,\nu'\rangle \tag{A I.1}$$

we introduce the Fourier representation of the operator $C(\boldsymbol{P},\boldsymbol{r})$:

$$C(\boldsymbol{P},\boldsymbol{r})=\int c(\boldsymbol{R},\boldsymbol{Q})\,e^{i\left(\frac{1}{\hbar}\boldsymbol{P}\cdot\boldsymbol{R}+\boldsymbol{Q}\cdot\boldsymbol{r}\right)}\,d^3\boldsymbol{R}\,d^3\boldsymbol{Q}=$$

$$=\int c(\boldsymbol{R},\boldsymbol{Q})\,e^{\frac{i}{2}\left(\boldsymbol{Q}-\frac{eB_0}{\hbar c}\boldsymbol{R}\cdot\hat{A}\right)\cdot\boldsymbol{R}}\,e^{i\left(\boldsymbol{Q}-\frac{eB_0}{\hbar c}\boldsymbol{R}\cdot\hat{A}\right)\cdot\boldsymbol{r}}\,e^{\frac{i}{\hbar}\boldsymbol{p}\cdot\boldsymbol{R}}\,d^3\boldsymbol{R}\,d^3\boldsymbol{Q}\,. \tag{A I.2}$$

This Fourier representation always exists, since the commutators of the components of $\boldsymbol{P}$ and $\boldsymbol{r}$ are numbers. According to (4.1) we obtain

$$\{\boldsymbol{k}\,\nu\,|C(\boldsymbol{P},\boldsymbol{r})|\,\boldsymbol{k}'\,\nu'\}=$$

$$=\int e^{-\frac{ieB_0}{2\hbar c}(\boldsymbol{r}-\boldsymbol{r}')\cdot\hat{A}\cdot(\boldsymbol{r}+\boldsymbol{r}')}\,\langle\boldsymbol{k}\,\nu\,|\,\boldsymbol{r}\rangle\,c(\boldsymbol{R},\boldsymbol{Q})\,e^{\frac{i}{2}\left(\boldsymbol{Q}-\frac{eB_0}{\hbar c}\boldsymbol{R}\cdot\hat{A}\right)\cdot\boldsymbol{R}}\times$$

$$\times\,e^{i\left(\boldsymbol{Q}-\frac{eB_0}{\hbar c}\boldsymbol{R}\cdot\hat{A}\right)\cdot\boldsymbol{r}}\,\delta(\boldsymbol{r}-\boldsymbol{r}'+\boldsymbol{R})\,\langle\boldsymbol{r}'\,|\,\boldsymbol{k}'\,\nu'\rangle\,d^3\boldsymbol{R}\,d^3\boldsymbol{Q}\,d^3\boldsymbol{r}\,d^3\boldsymbol{r}'=$$

$$=\int\langle\boldsymbol{k}\,\nu\,|\,\boldsymbol{r}\rangle\,c(\boldsymbol{R},\boldsymbol{Q})\,e^{\frac{i}{2}\boldsymbol{Q}\cdot\boldsymbol{R}}\,e^{i\boldsymbol{Q}\cdot\boldsymbol{r}}\,\delta(\boldsymbol{r}-\boldsymbol{r}'+\boldsymbol{R})\,\langle\boldsymbol{r}'\,|\,\boldsymbol{k}'\,\nu'\rangle\,d^3\boldsymbol{R}\,d^3\boldsymbol{Q}\,d^3\boldsymbol{r}\,d^3\boldsymbol{r}'=$$

$$=\int\langle\boldsymbol{k}\,\nu\,|\,\boldsymbol{r}\rangle\,c(\boldsymbol{R},\boldsymbol{Q})\left\langle\boldsymbol{r}\left|e^{i\left(\frac{1}{\hbar}\boldsymbol{p}\cdot\boldsymbol{R}+\boldsymbol{Q}\cdot\boldsymbol{r}\right)}\right|\boldsymbol{r}'\right\rangle\langle\boldsymbol{r}'\,|\,\boldsymbol{k}'\,\nu'\rangle\,d^3\boldsymbol{R}\,d^3\boldsymbol{Q}\,d^3\boldsymbol{r}\,d^3\boldsymbol{r}'=$$

$$=\left\langle\boldsymbol{k}\,\nu\left|\int c(\boldsymbol{R},\boldsymbol{Q})\,e^{i\left(\frac{1}{\hbar}\boldsymbol{p}\cdot\boldsymbol{R}+\boldsymbol{Q}\cdot\boldsymbol{r}\right)}\,d^3\boldsymbol{R}\,d^3\boldsymbol{Q}\right|\boldsymbol{k}'\,\nu'\right\rangle. \tag{A I.3}$$

From (A I.3) it follows, that $c(\boldsymbol{P}, \boldsymbol{r})$ obeys the relation (A I.1) if $C(\boldsymbol{p}, \boldsymbol{r})$ has the same Fourier coefficients $c(\boldsymbol{R}, \boldsymbol{Q})$ as $C(\boldsymbol{P}, \boldsymbol{r})$, i.e. if

$$C(\boldsymbol{p}, \boldsymbol{r}) = \int c(\boldsymbol{R}, \boldsymbol{Q})\, e^{i\left(\frac{1}{\hbar}\boldsymbol{p}\cdot\boldsymbol{R}+\boldsymbol{Q}\cdot\boldsymbol{r}\right)}\, d^3\boldsymbol{R}\, d^3\boldsymbol{Q}\,. \tag{A I.4}$$

With the exception of $f(H_0)$ this condition is fulfilled by all operators appearing in the text, so that the application of (A I.1) is justified. Generally the Fourier coefficients of $C(\boldsymbol{P}, \boldsymbol{r})$ and $C(\boldsymbol{p}, \boldsymbol{r})$ are differing by quantum correction terms which arise from the non-commutability of $\boldsymbol{P}_x$ and $\boldsymbol{P}_y$ and are neglected in the quasiclassical approximation. Thus generally (A I.1) is correct apart from these corrections.

Appendix II

In order to justify the use of the expression (7.5) by several authors [7, 9, 10, 12] the argument was presented that it would follow from the rate of change of the entropy by means of

$$\alpha\, I = \left\langle T \left(\frac{\partial S}{\partial t}\right)_{\text{coll}} \right\rangle. \tag{A II.1}$$

$\left(\frac{\partial S}{\partial t}\right)_{\text{coll}}$ being given by

$$\left(\frac{\partial S}{\partial t}\right)_{\text{coll}} = -\frac{1}{\Omega}\, k\, \text{Tr}\left\{[\log W - \log(1 - W)] \left(\frac{\partial W}{\partial t}\right)_{\text{coll}}\right\}. \tag{A II.2}$$

Eliminating $\left(\frac{\partial W}{\partial t}\right)_{\text{coll}}$ with the aid of (3.1) we see however at once, that

$$\left\langle \frac{\partial S}{\partial t} \right\rangle_{\text{coll}} = \left\langle -k\, \text{Tr}\left\{[\log W - \log(1 - W)]\, \frac{\partial W}{\partial t}\right\}\right\rangle = \left\langle \frac{\partial S}{\partial t} \right\rangle = 0\,.$$

It was by a second order error in the evaluation of (A II.1), that from (A II.1) the correct form (7.5) of the attenuation was obtained. On account of the irreversibility of the dissipation processes the heat $-\,dQ$ delivered by the system is related to the change dS in entropy by

$$dQ = T\, dS - T(dS)_{\text{irr}}\,.$$

$T(dS)_{\text{irr}}$ being the entropy produced by the irreversible processes. Instead of (A II.1) we have

$$\alpha\, I = -\left\langle \frac{dQ}{dt} \right\rangle = T \left\langle \frac{dS}{dt} \right\rangle_{\text{irr}} \tag{A II.3}$$

with

$$\left(\frac{dS}{dt}\right)_{\text{irr}} = -\frac{k}{\Omega}\, \text{Tr}\left\{\left[\log \frac{W}{1 - W} - \log \frac{W_{\text{eq}}}{1 - W_{\text{eq}}}\right] \frac{\partial W}{\partial t}\right\},$$

W_{eq} being the solution of

$$\left(\frac{\partial W_{\text{eq}}}{\partial t}\right)_{\text{coll}} = 0\,.$$

From (A I.3) and (A I.4) the correct expression (7.5) of the attenuation is obtained.

References

[1] A. B. PIPPARD, Phil. Mag. **46**, 1104 (1955).
[2] M. H. COHEN, M. J. HARRISON, and W. A. HARRISON, Phys. Rev. **117**, 937 (1960).
[3] M. J. HARRISON, Phys. Rev. Letters **9**, 299 (1962).
[4] S. RODRIGUEZ, Phys. Rev. **130**, 1778 (1963).
[5] S. RODRIGUEZ, Phys. Rev. **132**, 535 (1963).
[6] T. HOLSTEIN, Phys. Rev. **113**, 479 (1959).
[7] A. J. ACHIEZER, M. J. KAGANOV, and J. LIUBARSKI, Zh. eksper. teor. Fiz. **32**, 837 (1957).
[8] E. J. BLOUNT, Phys. Rev. **114**, 418 (1959).
[9] V. L. GUREVICH, Zh. eksper. teor. Fiz. **37**, 71 (1959).
[10] V. L. GUREVICH, Zh. eksper. teor. Fiz. **37**, 1680 (1959).
[11] A. B. PIPPARD, Proc. Roy. Soc. **A 257**, 165 (1960).
[12] E. A. KANER, Zh. eksper. teor. Fiz. **38**, 212 (1960).
[13] V. P. SILIN, Zh. eksper. teor. Fiz. **38**, 977 (1960).
[14] E. A. KANER, V. G. PESCHANSKY, and J. A. PRIVOROTSKY, Zh. eksper. teor. Fiz. **40**, 214 (1961).
[15] V. M. KANTOROVICH, Zh. eksper. teor. Fiz. **45**, 1638 (1963).
[16] K. B. VLASOV and B. N. FILIPPOV, Zh. eksper. teor. Fiz. **46**, 223 (1964).
[17] V. G. SKOBOV and E. A. KANER, Zh. eksper. teor. Fiz. **46**, 273 (1964).
[18] H. STOLZ, phys. stat. sol. **3**, 1153 (1963).
[19] T. TSUNETO, Phys. Rev. **121**, 402 (1961).
[20] G. D. WHITFIELD, Phys. Rev. **121**, 720 (1961).
[21] H. STOLZ, Ann. Phys. (Germany) **7**, 353 (1961).
[22] W. KOHN and J. M. LUTTINGER, Phys. Rev. **108**, 590 (1957).
[23] J. BARDEEN, Phys. Rev. **52**, 688 (1937).
[24] Y. NAGAOKA, Progr. theor. Phys. **26**, 589 (1961).
[25] R. B. STINCHCOMBE, Proc. Roy. Soc. **78**, 275 (1961).

(Received January 4, 1965)

phys. stat. sol. 8, 863 (1965)

Max-Planck-Institut für Metallforschung, Stuttgart, und Institut für theoretische und angewandte Physik der Technischen Hochschule Stuttgart

Die Erholung des elektrischen Widerstandes in abgeschrecktem Kupfer

Von

F. RAMSTEINER, G. LAMPERT, A. SEEGER und W. SCHÜLE[1])

Es wird das Ausheilen des elektrischen Widerstandes in abgeschrecktem Kupfer untersucht, wobei je eine Erholungsstufe im Bereich 0 bis 100 °C und 100 bis 200 °C gefunden wird. Die obere Stufe wird durch die Wanderung der Einfachleerstellen erklärt, wobei sich (in Einklang mit Selbstdiffusions- und Hochtemperaturdaten) als Aktivierungsenergie der Wanderung $W_1 = 1{,}06$ eV ergibt. Die untere Stufe kann als Ausheilen von Doppelleerstellen, die mit Einfachleerstellen im lokalen Gleichgewicht stehen, gedeutet werden. Als Wanderungs- und Bindungsenergie der Doppelleerstellen werden $W_2 = 0{,}67$ eV bzw. $B_2 = 0{,}12$ eV gefunden. Der Einfluß der Vorbehandlung und des Reinheitsgrades der Proben auf das Erholungsverhalten wird untersucht und diskutiert.

The electrical resistance of quenched copper was found to anneal out in two stages between 0 to 100 °C, and 100 to 200 °C. The upper stage is explained by the migration of a single vacancy, giving an activation energy of migration of $W_1 = 1.06$ eV (which is in agreement with the high-temperature and diffusion data). The lower stage can be interpreted in terms of the annealing of di-vacancies which are in local equilibrium with single vacancies. The migration and binding energies of the di-vacancies are $W_1 = 0.67$ eV and $B_2 = 0.12$ eV respectively. The effects of pre-treatment and purity on the recovery are investigated and discussed.

Die vorliegende Arbeit schließt eine in ihren wichtigsten Ergebnissen schon früher [1] umrissene Untersuchungsreihe über atomare Fehlstellen in Ni, Cu, Ag und Au vorläufig ab. Bei den hier zu behandelnden Abschreckexperimenten an Kupfer — die Verformungsexperimente wurden an anderer Stelle dargestellt [2] — ist wegen der besonderen mit dem Abschrecken von Cu verknüpften experimentellen Schwierigkeiten neben dem früher [1, 3 bis 7] benützten Abschreckverfahren (I) ein weiteres (II) herangezogen worden, um den Einfluß von Vorbehandlung (verschiedene Schutzgase) und Verunreinigungen genauer zu studieren. Wie bei den entsprechenden früheren Untersuchungen haben wir uns dabei durchweg auf das quantitative Studium der *Erholungs*vorgänge beschränkt, haben also nicht versucht, aus Abschreckexperimenten die Bildungsenergie der Leerstellen zu ermitteln. Wir werden jedoch unsere Ergebnisse mit den in der Literatur vorliegenden Messungen dieser Art vergleichen.

1. Versuchsdurchführung

1.1 Abschreckmethode I

Zum Abschrecken wurden etwa 0,1 mm dicke Kupferfolien verwendet, die vor dem Abschrecken etwa 30 min in einem evakuierten Quarzrohr bei etwa 1000 °C angelassen wurden. Nach Einstellung der Abschrecktemperatur wurde etwa zwei

[1]) Jetzt CCR EURATOM, Ispra (Italien).

Minuten vor dem Abschrecken Argon in das Quarzrohr eingeleitet und danach die Probe in Wasser von etwa 0 °C bzw. 20 °C abgeschreckt. Die so abgeschreckten Folien konnten in zwei Meßproben zerschnitten werden, die gleiche Vorgeschichte (Abschrecktemperatur, Abschreckgeschwindigkeit sowie sonstige Vorgänge während des Abschreckens) hatten. Das Erholungsverhalten der Proben wurde anhand von elektrischen Widerstandsmessungen bei der Temperatur des flüssigen Sauerstoffs zwischen 0 °C und 250 °C untersucht. Eine Schwäche des Verfahrens ist die schlechte Reproduzierbarkeit der Abschreckbedingungen.

1.2 Abschreckmethode II

Diese Abschreckmethode, die im wesentlichen auf Bauerle und Koehler [8] zurückgeht, wurde — von einigen geringfügigen Änderungen [9] abgesehen — bei früheren Experimenten über Kleinwinkelstreuung [10] und bei elektronenmikroskopischen Untersuchungen [11, 12] angewandt. Etwa 0,07 mm dicke Folien wurden in einem Schutzgas (wahlweise gereinigtes Argon, Argon mit Alkoholdampf, Formiergas [92% N_2 + 8% H_2]) über einer Wasseroberfläche durch Wechselstrom auf etwa 1000 °C aufgeheizt (Temperaturbestimmung erfolgte mit einem optischen Pyrometer), bei dieser Temperatur etwa 10 min geglüht und danach in Wasser von 0 °C abgeschreckt. Bei einigen Versuchen wurde die Wasseroberfläche mit einer dünnen Ölschicht abgedeckt, um das Eindringen von Wasserdampf in das Schutzgas zu verhindern. Etwa 3 s nach dem Eintauchen in Wasser wurde die Probe auf etwa —50 °C (in gekühltem Aceton) weiter abgekühlt und anschließend das Erholungsverhalten zwischen —20 °C und 250 °C untersucht. Die elektrischen Widerstandsmessungen wurden meistens bei der Temperatur des flüssigen Stickstoffs ausgeführt.

Zur Untersuchung des Einflusses der Vorbehandlung wurden Proben verschiedener Reinheitsgrade bei 1000 °C geglüht und durch Abschalten des Heizstromes auf Raumtemperatur abgekühlt. Der spezifische Widerstand ϱ wurde aus dem Verhältnis des Gesamtwiderstandes R der Probe bei 295 °K und 77 °K entsprechend

$$\varrho = \frac{R_{77}}{R_{295}} \varrho_{295} \approx \varrho_{77}^{ideal} + \varrho_R \tag{1}$$

bestimmt (Vernachlässigung des Restwiderstandes ϱ_R bei Raumtemperatur, $\varrho_{295} = 1{,}69 \cdot 10^{-6}\ \Omega\text{cm}$).

Tabelle 1

Erholung in Stufe IV nach Abschrecken mit Methode I

Abschreck-temperatur (°C)	Reinheits-grad (%)	Widerstandsänderung in Isochronen (%)	Widerstandsänderung in Isothermen (%)	Anlaßtemperatur der Isothermen (°C)	Aktivierungs-energie (eV)
700	99,99	0,7	0,7	162	1,08 ± 0,03
800	99,99	2,1	2,45	177	1,08 ± 0,03
830	99,999	1,25	—	—	—
900	99,99	1,05	1,9	182	1,06 ± 0,03
1000	99,999	1,5	1,4	178	1,08 ± 0,05

2. Versuchsergebnisse

2.1 Ergebnisse mit Abschreckmethode I

Die Abschreckexperimente wurden mit den gleichen Kupfersorten (Reinheitsgrade: 99,999%, 99,99%) wie die Verformungsexperimente [2] durchgeführt; die Ergebnisse (Tabelle 1) weisen im untersuchten Temperaturgebiet keine wesentliche, reproduzierbare Abhängigkeit vom Reinheitsgrad auf.

Wie aus Fig. 1 am Beispiel der von 900 °C (99,99%) bzw. 830 °C (99,999%) in Wasser (0 °C) abgeschreckten und gleich danach in flüssigem Sauerstoff abgekühlten Proben ersichtlich, treten im untersuchten Temperaturgebiet zwei Erholungsstufen auf. Zwischen −20 °C und 50 °C wird eine Tieftemperaturerholung beobachtet, die mit diesem Abschreckverfahren nicht näher untersucht werden konnte, da der Betrag des ausheilenden Widerstandes hierfür zu gering war. Wir haben daher die meisten Proben in Wasser von etwa 20 °C abgeschreckt und die Erholung unterhalb 50 °C mit diesem Verfahren nicht mehr untersucht.

Eine zweite Stufe (in der üblichen Nomenklatur [13] als Stufe IV bezeichnet) wird zwischen 100 °C und 250 °C beobachtet. Die Erholung erfolgt, nahezu unabhängig vom Verunreinigungsgrad und von der Abschrecktemperatur, ungefähr immer im gleichen Temperaturgebiet. Die prozentualen Gesamtwiderstandsänderungen stehen nicht in einfachem Zusammenhang mit der Abschrecktemperatur, was bei der schlechten Reproduzierbarkeit der Abschreckbedingungen auch nicht erwartet werden kann. In äußerlich oxydierten Proben war im betrachteten Temperaturgebiet meistens keine Widerstandsänderung festzustellen. Die den oben beschriebenen Isochronen entsprechenden (d. h. an Proben gleicher Vorgeschichte gewonnenen) Isothermen zeigt Fig. 2. Das Stufenende ϱ_{IV} dieser Isothermen wurde durch Anlassen bei ca. 240 °C bestimmt; bei der von 1000 °C abgeschreckten Probe (99,999% rein) erfolgte die Auslagerung bei 280 °C. Nach dem Meechan-

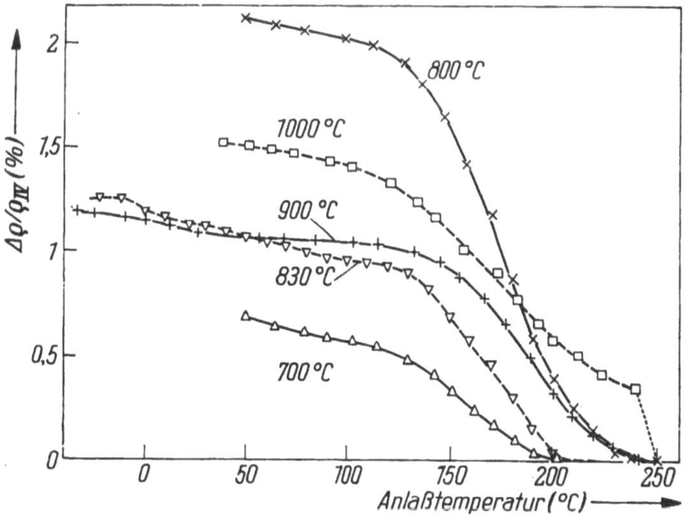

Fig. 1. Isochrone Erholung des elektrischen Widerstandes nach Abschrecken (Argonatmosphäre) von den angegebenen Temperaturen.
——— Cu (99,99%) Anlaßzeit 15 min je Anlaßtemperatur,
– – – Cu (99,999%) Anlaßzeit 10 min je Anlaßtemperatur.
Die Widerstandswerte sind auf den Endwiderstand ϱ_{IV} der Stufe IV bezogen

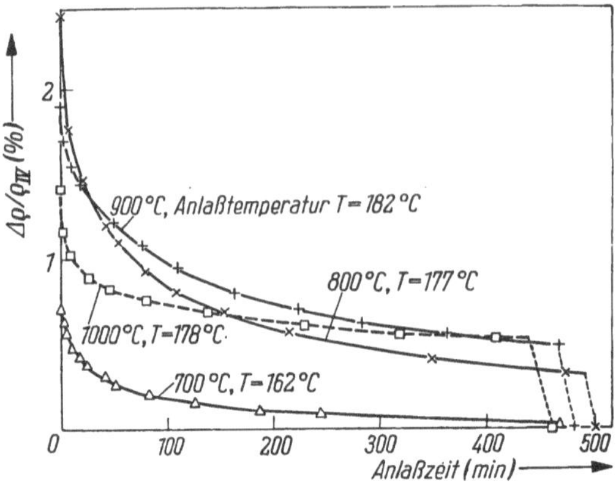

Fig. 2. Isotherme Erholung in Stufe IV nach Abschrecken (gleiche Vorgeschichte der Proben wie in Fig. 1)

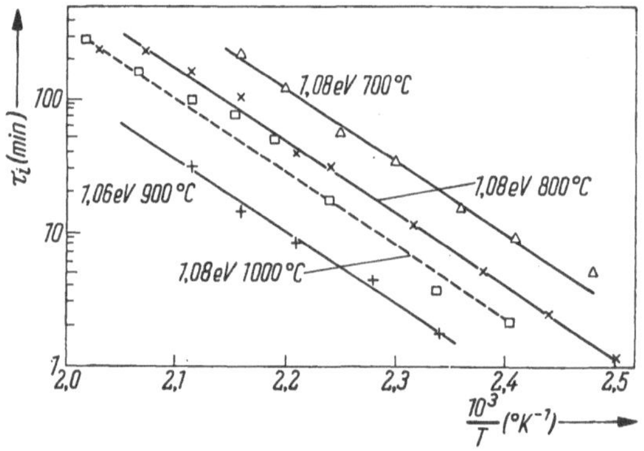

Fig. 3. Bestimmung der Aktivierungsenergie des Erholungsprozesses von Stufe IV nach dem Meechan-Brinkman-Verfahren (aus Fig. 1 und 2)

Brinkman-Verfahren [14] erhält man durch Vergleich dieser Isothermen mit den entsprechenden Isochronen für den Erholungsprozeß der Stufe IV als Aktivierungsenergie $Q^{IV} = 1{,}08 \pm 0{,}03$ eV (Tabelle 1, Fig. 3). In Übereinstimmung hiermit ergibt sich unter Verwendung des etwas ungenaueren Temperaturwechselverfahrens $Q^{IV} = (1{,}1 \pm 0{,}1)$ eV (Fig. 4).

Die für den Erholungsprozeß der Stufe IV charakteristische Ausheilkinetik ist bei verschiedenen Proben am Anfang etwas uneinheitlich; das Ende des Erholungsprozesses kann jedoch bei allen Proben durch eine Reaktionsordnung 1 (Fig. 5) beschrieben werden.[2])

2.2 Ergebnisse mit Abschreckmethode II

Der Gaseinfluß auf den elektrischen Widerstand von Kupfer verschiedener Reinheitsgrade nach Glühen bei 1000 °C ist aus Fig. 6 ersichtlich. Nach einer anfänglichen raschen Widerstandsabnahme als Folge der Entfestigung in den gewalzten Proben stellt sich bei Glühung in Formiergas oder Argon und Alkoholdampf ein größerer Wert für den spezifischen Widerstand (reduzierter Zustand) ein als bei Glühung in Argon (innerlich oxydierter Zustand). Durch Änderung des Schutzgases läßt

---

[2]) Am besten ist der gesamte Erholungsverlauf empirisch durch

$$\Delta\varrho = A\,(1/\tau + a\,e^{-\tau}) \qquad (2)$$

beschreibbar, mit $\tau = b\,t + M$ und den von der Isothermentemperatur unabhängigen Konstanten A, a und M.

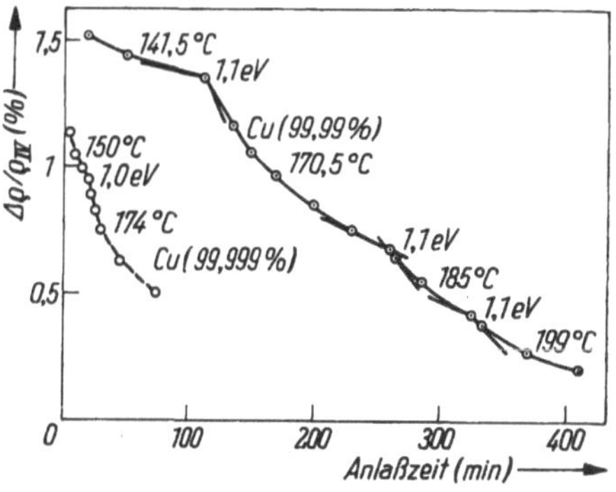

Fig. 4. Bestimmung der Aktivierungsenergie des Erholungsprozesses von Stufe IV nach dem Temperaturwechselverfahren

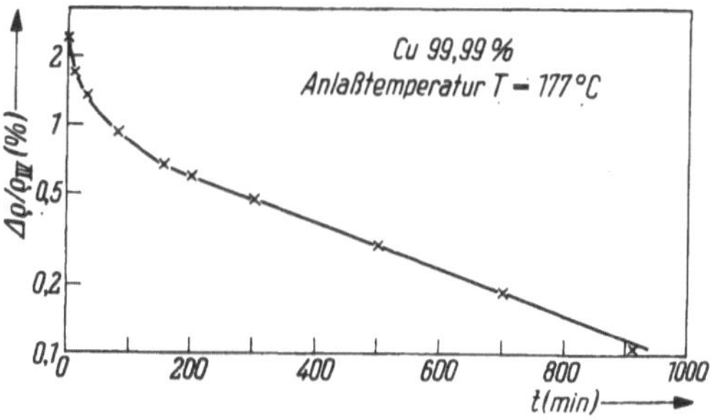

Fig. 5. Prüfung auf Reaktionsordnung 1 für den Ausheilvorgang der Stufe IV (Abschrecktemperatur 800 °C)

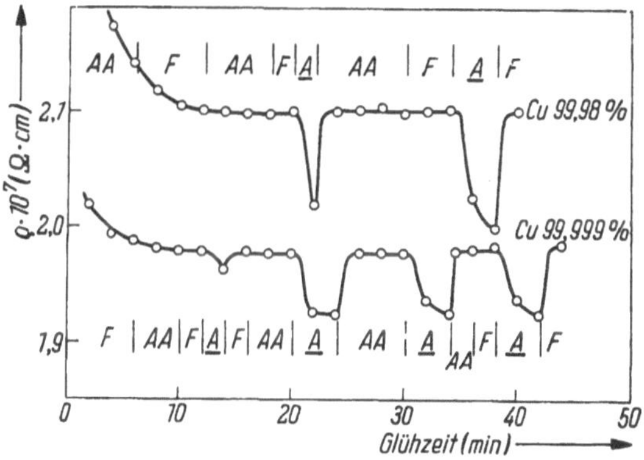

Fig. 6. Spezifischer elektrischer Widerstand (Meßtemperatur 77 °K) von Kupfer nach Glühen bei 1000 °C in verschiedenen Gasen. F: Formiergas, A: gereinigtes Argon, AA: Argon und Alkoholdampf

sich der oxydierte in den reduzierten Zustand und umgekehrt überführen. Die Unterschiede zwischen den beiden Zuständen sind in reinen Proben kleiner als in unreinen.

Die Erholungsvorgänge nach Abschrecken in *Formiergas* lassen sich — von Einzelheiten zunächst abgesehen — in einem Zweistufenschema zusammenfassen (Fig. 7). Zwischen 0° und 100 °C wird eine Tieftemperaturstufe beobachtet, für deren Erholungsprozeß durch Temperaturwechsel Aktivierungsenergien von 0,5 bis 0,7 eV (Fig. 8) bestimmt werden; für den Erholungsprozeß der zwischen 100° und 200 °C gelegenen Hochtemperaturstufe werden Aktivierungsenergien von $(1{,}1 \pm 0{,}1)$ eV (Fig. 8) gemessen. Im einzelnen werden vier Typen von Isochronen beobachtet.

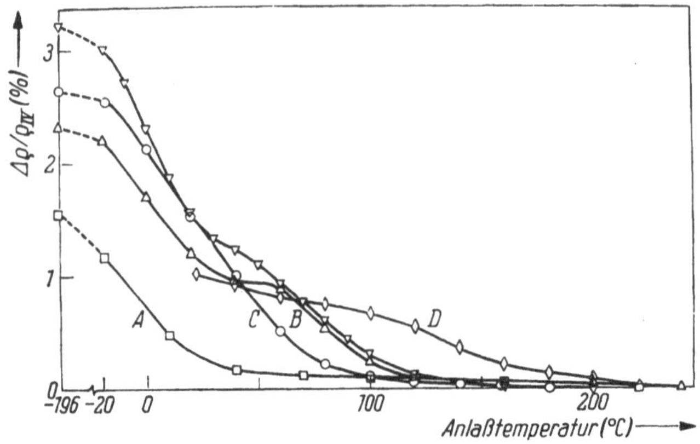

Fig. 7. Isochrone Erholung des elektrischen Widerstandes nach Abschrecken von 1000 °C in Formiergasatmosphäre für verschiedene Proben. ◇ ○ Cu (99,98%) Anlaßzeit 10 min je Anlaßtemperatur. □ △ Cu (99,999%) Anlaßzeit 10 min je Anlaßtemperatur. ▽ Cu (99,999%) Anlaßzeit 5 min je Anlaßtemperatur

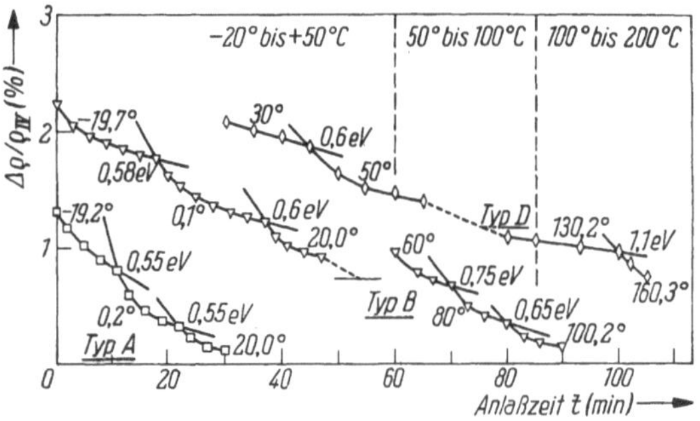

Fig. 8. Bestimmung der Aktivierungsenergien für die Erholungsprozesse zwischen −20 und 200 °C mit Hilfe des Temperaturwechselverfahrens (Formiergasatmosphäre, Anlaßtemperatur in °C). □ ▽ Cu (99,999%) ◇ Cu (99,98%)

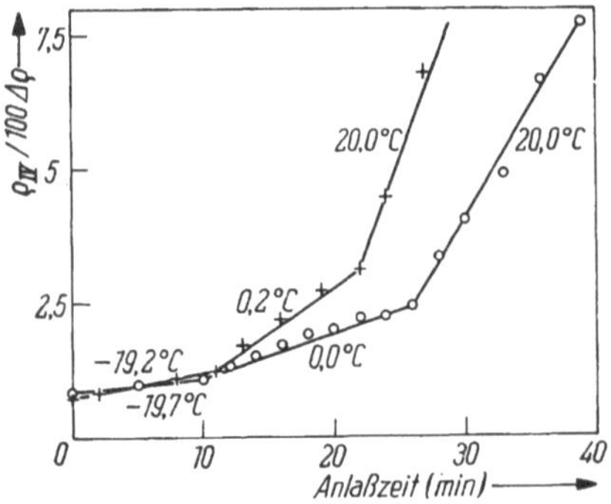

Fig. 9. Prüfung auf Reaktionsordnung 2 für den Erholungsprozeß bei etwa 0 °C (Cu 99,999%)

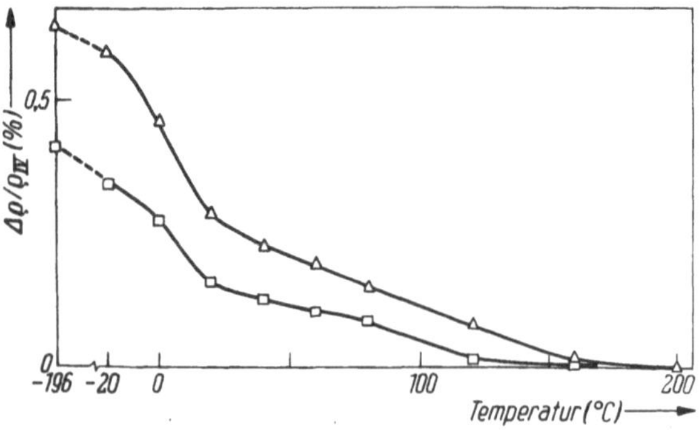

Fig. 10. Isochrone Erholung des elektrischen Widerstandes von zwei verschiedenen Proben (Cu 99,999%) nach Abschrecken von 1000 °C in Argonatmosphäre (Anlaßzeit 10 min je Anlaßtemperatur)

Die meisten Proben zeigen bei 0 °C eine deutliche Erholungsstufe (Aktivierungsenergie: $(0{,}55 \pm 0{,}05)$ eV, Reaktionsordnung ungefähr 2 (Fig. 9)), danach wird bis 200 °C keine weitere nennenswerte Erholung beobachtet (Typ A). Neben dieser 0°C-Stufe wird bei einigen Proben noch eine Erholung bei 70 °C festgestellt (Aktivierungsenergie: 0,7 eV) (Typ B). Anstatt in zwei Stufen wie bei Typ B kann die Erholung zwischen 0 und 100 °C auch in einer Stufe erfolgen (Aktivierungsenergie: 0,5 eV bis 0,7 eV)[3]) (Typ C). Zusätzlich zum Erholungsprozeß von Typ C wird an manchen Proben ein weiterer zwischen 100 und 200 °C beobachtet (Aktivierungsenergie: $(1{,}1 \pm 0{,}1)$ eV), der bezüglich Aktivierungsenergie und Temperaturlage dem im vorigen Abschnitt der Stufe IV zugeordneten entspricht (Typ D).

Im allgemeinen wird in den Proben höherer Reinheit häufiger Typ A, in den etwas unreineren Proben mehr Typ B, C oder D beobachtet, jedoch ist der Reinheitsgrad nicht allein ausschlaggebend für den Typ des Erholungsvorganges.

Nach Abschrecken in gereinigtem *Argon* beobachtet man (Fig. 10) im allgemeinen eine Stufe bei 0 °C (Aktivierungsenergie: 0,4 bis 0,7 eV), an die sich eine nahezu kontinuierliche — bei 130 °C etwas stärker ausgeprägte — Erholung bis etwa 180 °C anschließt. Die Erholung des elektrischen Widerstandes ist jedoch zu klein, um weitere, quantitative Aussagen über Einzelheiten der Erholungsvorgänge machen zu können. Der Reinheitsgrad der Proben hat wenig Einfluß auf das Erholungsspektrum im untersuchten Temperaturgebiet.

Die isochronen Erholungsvorgänge nach Abschrecken in *Argon* und *Alkoholdampf* zeigen im wesentlichen das gleiche Erholungsverhalten wie nach Abschrecken in Formiergas, weshalb wir hier auf die Wiedergabe der entsprechenden Ergebnisse verzichten.

Eine Ölschicht auf der Wasseroberfläche beim Abschrecken bewirkt, daß die Gesamterholung des elektrischen Widerstandes im untersuchten Temperaturgebiet kleiner ist als nach Abschrecken in reinem Wasser; Einflüsse auf das Erholungsspektrum konnten dagegen nicht nachgewiesen werden.

3. Diskussion

3.1 Vorbehandlung, Erholungsstufen nach Abschrecken

Der spezifische Widerstand geglühten Kupfers ist — wie bereits Coltman und Mitarbeiter [15] festgestellt haben — vom verwendeten Schutzgas abhängig. Da die Unterschiede in den Werten für den spezifischen Widerstand des oxydierten bzw. reduzierten Zustandes in reinem Kupfer geringer als in unreinem sind, spielen Verunreinigungen offenbar eine wesentliche Rolle. Die Verunreinigungen agglomerieren wahrscheinlich während der Glühung in Argon, bleiben dagegen dispers verteilt bei Anwesenheit von Wasserstoff oder Alkoholdampf.

In Zusammenfassung unserer in 2. beschriebenen Erholungsdaten ergibt sich nach Abschrecken — von Einzelheiten zunächst einmal abgesehen — eine Tieftemperaturstufe zwischen 0 und 100 °C, deren Erholungsprozeß, in Übereinstimmung mit früheren Messungen [10, 16], durch Aktivierungsenergien zwischen

---

[3]) An anderer Stelle mitgeteilte Experimente [16] ergaben als Kinetik ungefähr Reaktionsordnung 2, wie sie (durch Neuauftragung) für den Beginn des Erholungsprozesses auch aus Kleinwinkelstreuexperimenten [10] gefunden wird.

0,5 und 0,7 eV charakterisiert ist, sowie eine Hochtemperaturstufe bei etwa 150 °C (Stufe IV), deren Erholungsprozeß durch eine Aktivierungsenergie von etwa 1,08 eV gekennzeichnet ist. Die untere Stufe, die mit der Bildung und dem Wachstum von Leerstellenagglomeraten verbunden ist [10, 12, 17] wurde auch von zahlreichen anderen Autoren beobachtet [18 bis 23]. Clarebrough und Mitarbeiter [18] berichten nach Abschrecken in Kohlenmonoxydatmosphäre über einen Erholungsprozeß zwischen —50 und 100 °C (hinsichtlich der Ausheiltemperatur unserer Stufe bei 0 °C entsprechend). Nach Abschrecken in Argon folgt diesem Prozeß eine weitere Erholungsstufe bei 80 °C, welche hinsichtlich der Ausheiltemperatur unserem Erholungstyp B nach Abschrecken in Formiergas entspricht. Nähere Angaben über Kinetik oder Aktivierungsenergien machen die Autoren nicht.

Es verdient festgehalten zu werden, daß die Erholungsprozesse zwischen 0 und 200 °C nach Abschrecken nahezu unabhängig von Reinheitsgrad, Schutzgas und Abschreckmethode — wenn auch unterschiedlich deutlich und in Einzelheiten etwas modifiziert — auftreten. Wir dürfen deshalb erwarten, daß ihnen im wesentlichen wohldefinierte Ausheilvorgänge von Einfach- oder Mehrfachleerstellen zugrunde liegen. Auf Einzelheiten werden wir in 3.3 eingehen.

Außerhalb des von uns etwas genauer untersuchten Temperaturgebiets sind zwei weitere Erholungsstufen beobachtet worden. Clarebrough und Mitarbeiter [18] und Budin und Mitarbeiter [19, 20] berichten nach Abschrecken in Argon- bzw. Wasserstoffatmosphäre über eine Erholungsstufe bei etwa —100 °C, deren Erholungsprozeß durch eine Aktivierungsenergie von etwa 0,4 eV gekennzeichnet ist. Wegen der Übereinstimmung dieses Wertes mit der Wanderungsenergie von Wasserstoff in Kupfer [24] ordnen Budin und Mitarbeiter diese Stufe der Wasserstoffdiffusion zu. Ferner wurde von Clarebrough und Mitarbeitern sowie Airoldi, Bacchella und Germagnoli [25] auch zwischen 250 und 400 °C ein Erholungsprozeß beobachtet, dessen Aktivierungsenergie zu etwa 1,3 bis 1,6 eV angegeben wird [25]. Kleinwinkelstreuexperimente [10] ergaben, daß in Kupfer oberhalb 200 °C Leerstellenagglomerate ausheilen oder sich umordnen; in Kupferlegierungen erfolgt bei diesen Temperaturen Ordnungseinstellung über Selbstdiffusion [26, 27, 28]. Man kann daher annehmen, daß die Erholungsprozesse oberhalb 200 °C, abgesehen von Verunreinigungseinflüssen, mit dem Auflösen von Leerstellenagglomeraten verbunden sind, die in abgeschrecktem Kupfer während des Anlassens bei tieferen Temperaturen entstanden waren.

Unter den nach Abschrecken beobachteten verschiedenen Erholungsprozessen wird bei den Edelmetallen üblicherweise derjenige der Einfachleerstellenwanderung zugeordnet, dessen Aktivierungsenergie mit der über

$$W_1 = Q^D - U_1 \tag{3}$$

berechneten Wanderungsenergie W_1 für Leerstellen übereinstimmt (U_1 Leerstellen-Bildungsenergie, Q^D Selbstdiffusionsenergie). Bei Kupfer fehlen leider für Q^D und U_1 allgemein anerkannte genaue Werte, weshalb die Zuordnung von Leerstellen zu einem der beobachteten Erholungsprozesse nach Abschrecken einer sorgfältigen Diskussion bedarf.

3.2 Bildungsenergie für Leerstellen und Selbstdiffusionsenergie

Bei der folgenden Auswertung der Hochtemperatur- und Selbstdiffusionsdaten beschränken wir uns auf die Betrachtung von Einfach- und Doppelleerstellen, vernachlässigen also größere Leerstellenagglomerate.

3.2.1 Bildungsenergie der Leerstellen

Zwischen der Gleichgewichtskonzentration c_1 und der Bildungsenergie U_1 der Einfachleerstellen besteht der Zusammenhang

$$c_1(T) = \exp(S_1^B/k)\exp(-U_1/k\,T) \tag{4}$$

(S_1^B Leerstellen-Bildungsentropie). U_1 kann entweder

a) aus Hochtemperaturmessungen unter Gleichgewichtsbedingungen oder
b) aus Abschreckversuchen ermittelt werden.

a) Die Konzentration der leeren Gitterplätze c im *thermischen Gleichgewicht* ist durch

$$c = c_1 + 2\,c_2 = c_1 + 12\exp(\varDelta S/k)\exp(B_2/k\,T)\,c_1^2 \tag{5}$$

gegeben, wobei das zweite Glied Doppelleerstellen berücksichtigt ($\varDelta S$ und B_2 bedeuten Bindungsentropie bzw. -energie). Simmons und Balluffi [29] finden bei 1000 °C bzw. 1075 °C die Konzentrationen $c = (0{,}9 \pm 0{,}5)\,10^{-4}$ bzw. $c = (1{,}9 \pm 0{,}5)\,10^{-4}$. Aus diesen Daten kann die Bildungsenergie für Einfachleerstellen über (4, 5) nur ermittelt werden, wenn Werte für die Bildungsentropie der Einfachleerstellen sowie für die Bindungsenergie und -entropie der Doppelleerstellen angenommen werden. Wie aus Tabelle 2 ersichtlich, haben Unsicherheiten in den (sich auf die Doppelleerstellen beziehenden) $\varDelta S$- und B_2-Werten — im Gegensatz zur Bildungsentropie S_1^B der Einfachleerstellen — auf den Zahlenwert von U_1 nur geringen Einfluß. Simmons und Balluffi erhalten mit der Annahme $S_1^B = (1{,}5 \pm 0{,}5)\,k$ den Wert $U_1 = (1{,}17 \pm 0{,}11)$ eV, während sich mit den theoretischen Werten $S_1^B = 0{,}5\,k$, $\varDelta S = 1{,}8\,k$ [31] und der experimentell bestimmten Bindungsenergie $B_2 = 0{,}13$ eV [10] $U_1 = (1{,}06 \pm 0{,}05)$ eV errechnet [30, 31].

In Fig. 11 sind die auf Grund dieser beiden Auffassungen zu erwartenden Leerstellenkonzentrationen als Funktion der Temperatur aufgetragen. Starke, der experimentellen Nachprüfung zugängliche Unterschiede sind demnach erst bei verhältnismäßig tiefen Temperaturen zu erwarten. Im thermodynamischen Gleichgewicht ist der (in Fig. 11 nicht berücksichtigte) Bruchteil der von Doppelleerstellen eingenommenen leeren Gitterplätze nicht sehr groß. Mit den oben angegebenen Werten [10, 30, 31] für $\varDelta S$ und B_2 ergibt er sich am Schmelzpunkt zu 4,5% der Einfachleerstellenkonzentration. Es erscheint in der Tat gerechtfertigt, größere Leerstellenkomplexe zu vernachlässigen.

Tabelle 2

Bildungsenergie U_1 (in eV) der Einfachleerstellen*)

	$10^4\,c(T_s)$	$(2{,}0 \pm 0{,}5)$	$(2{,}0 \mp 0{,}5)$	$(2{,}0 \pm 0{,}5)$	$(2{,}0 \mp 0{,}5)$
	S_1^B/k	$(0{,}5 \pm 0{,}0)$	$(0{,}5 \pm 0{,}2)$	$(1{,}5 \pm 0{,}0)$	$(1{,}5 \pm 0{,}5)$
$\varDelta S$	B_2				
0	0	$1{,}05 \mp 0{,}03$	$1{,}05 \pm 0{,}06$	$1{,}17 \mp 0{,}03$	$1{,}17 \pm 0{,}09$
$1{,}8\,k$	0,13 eV	$1{,}06 \mp 0{,}03$	$1{,}06 \pm 0{,}06$	$1{,}18 \mp 0{,}03$	$1{,}18 \pm 0{,}09$

*) Berechnet über Gleichung (4) und (5) aus Gleichgewichtsmessungen [29] am Schmelzpunkt unter verschiedenen Annahmen für Bindungsentropie $\varDelta S$, Bindungsenergie B_2 und Bildungsentropie S_1^B.

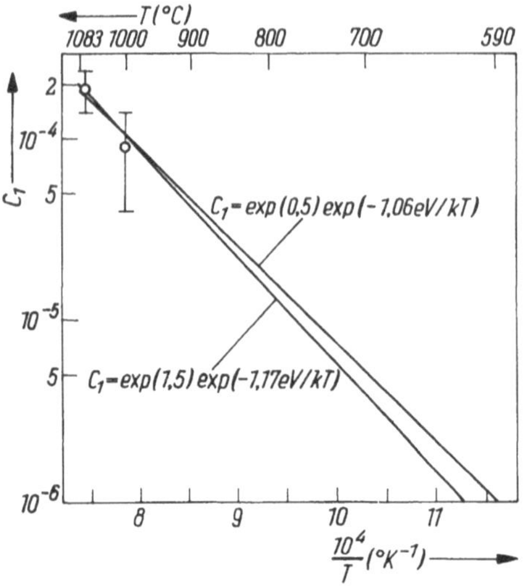

Fig. 11. Temperaturabhängigkeit der Leerstellenkonzentration, aus den Gleichgewichtsmessungen von Simmons und Balluffi [29] auf Grund der beiden in 3.2.1a erörterten Hypothesen errechnet

b) *Abschreckexperimente* zeigen recht unterschiedliche und z. T. sich widersprechende Ergebnisse. Budin und Mitarbeiter [19, 20] berichten über Abschreckexperimente von Kupfer in Wasserstoffatmosphäre, wobei die Autoren den der eingefrorenen Leerstellenkonzentration proportionalen Zusatzwiderstand aus der Differenz des gesamten eingefrorenen Widerstandes und des vom gelösten Wasserstoff herrührenden Widerstands abschätzen. Eine explizite Temperaturabhängigkeit von c wird nicht angegeben. Die Autoren berichten über eine gute Übereinstimmung ihrer Ergebnisse mit denen von Simmons und Balluffi, wenn sie als Leerstellenwiderstand $\Delta\varrho_L = 1{,}1\ \mu\Omega\,\mathrm{cm}/\%$ Leerstelle annehmen. Die Experimente von Budin und Mitarbeitern scheinen uns jedoch keinesfalls im Widerspruch mit dem von uns bevorzugten Wertepaar $S_1^B = 0{,}5\,k$, $U_1 = 1{,}06$ eV zu stehen. Airoldi und Mitarbeiter [25], die als erste Abschreckexperimente an Kupfer beschreiben, geben als Bildungsenergie aus der Temperaturabhängigkeit (entsprechend (4)) des eingefrorenen elektrischen Zusatzwiderstandes $U_1 = (1{,}0 \pm 0{,}1)$ eV an. Der gemessene Zusatzwiderstand, etwa ausgedrückt in der für den Schmelzpunkt berechneten Leerstellenkonzentration von $c = 1{,}25 \cdot 10^{-3}$ ($\Delta\varrho_L = 1{,}6\ \mu\Omega\,\mathrm{cm}/\%$ Leerstelle), ist jedoch im Vergleich zu den Messungen von Simmons und Balluffi um den Faktor 6 zu groß. Der von Budin und Mitarbeitern beobachtete starke Einfluß des Wasserstoffs auf die eingefrorenen Zusatzwiderstände läßt es denkbar erscheinen, daß auch in [25] Gaseinflüsse vorhanden sind. Ebenfalls aus der Temperaturabhängigkeit des eingefrorenen Zusatzwiderstandes ermittelten Hasiguti, Nakao und Kimura [32] in neuerer Zeit als Bildungsenergie $U_1 = (1{,}05 \pm \pm 0{,}05)$ eV, was in guter Übereinstimmung mit dem von uns angegebenen Wert ist.[4])

[4]) Es sei noch erwähnt, daß durch eine (vom theoretischen Standpunkt aus nicht ganz unbedenkliche) Auswertung des Temperaturkoeffizienten des elektrischen Widerstandes Meechan und Eggleston [33] $U_1 = (0{,}9 \pm 0{,}05)$ eV, Gertsriken und Slyusar [34] $U_1 = 0{,}65$ eV erhalten haben. Als Leerstellenkonzentration am Schmelzpunkt geben Meechan und Eggleston $2 \cdot 10^{-3}$ ($\Delta\varrho_L \approx 1{,}4\ \mu\Omega\,\mathrm{cm}/\%$ Leerstelle) an, was zwar ungefähr mit dem von Airoldi und Mitarbeitern, nicht jedoch mit dem von Simmons und Balluffi angegebenen Wert übereinstimmen würde. (S. Anmerkung bei der Korrektur S. 879.)

Messungen der Ordnungseinstellung in Cu_3Au nach Abschrecken ergaben etwa $U_1 = 1$ eV [35]. Allerdings braucht der Wert für die Legierung nicht zwingenderweise mit dem für reines Kupfer übereinzustimmen.

3.2.2 *Selbstdiffusion*

Der mit radioaktiven Isotopen gemessene Selbstdiffusionskoeffizient ist gegeben durch (die Indizes 1 und 2 beziehen sich auf Einfach- bzw. Doppelleerstellen)

$$D \equiv D_0 \exp(-Q^D/k\,T) = D_1^D + D_2^D\,, \tag{6}$$

wobei

$$D_1^D = f_1\, a^2\, \nu_1^0 \exp\left[(S_1^B + S_1^W)/k\right] \exp\left[-Q_1^D/k\,T\right], \tag{7}$$

$$D_2^D = f_2\, 4\, a^2\, \nu_2^0 \exp\left[(2\,S_1^B + \Delta S + S_2^W)/k\right] \exp\left[-Q_2^D/k\,T\right] \tag{8}$$

mit $Q_1^D = U_1 + W_1$ und $Q_2^D = 2\,U_1 - B_2 + W_2$ ist (W_i Wanderungsenergien, S_i^W Wanderungsentropien, ν_i^0 Frequenzfaktoren, Korrelationsfaktoren $f_1 = 0{,}78$, $f_2 = 0{,}54$ [36], a Kantenlänge des Elementarwürfels). Mercer und Shuttleworth [37] bestimmten im Temperaturgebiet von 850 °C bis 1060 °C für die Selbstdiffusionsenergie $\left(\text{definiert durch } Q^D \equiv -\frac{d \ln D}{d\,(1/k\,T)}\right)$ $Q^D = (2{,}15 \pm 0{,}013)$ eV (präexponentieller Faktor $D_0 = 0{,}62\ \mathrm{cm^2\,s^{-1}}$). Einen ähnlichen Wert ($Q^D = 2{,}13$ eV) erhielten auch Maier und Nelson [38] an Einkristallen (Temperaturgebiet 840° bis 980 °C). Der von Kuper und Mitarbeitern [39] angegebene Wert $Q^D = (2{,}05 \pm 0{,}02)$ eV (Temperaturgebiet 640 bis 1060 °C) ist kleiner als die beiden oben aufgeführten. Slifkin [40] und Le Claire [41] weisen darauf hin, daß Q^D wegen des Einflusses der Versetzungen zu niedrig ausfallen kann, wenn, wie in [39], bei verhältnismäßig tiefen Temperaturen gewonnene D-Werte mitverwendet werden. Unter Beachtung dieses Gesichtspunktes hat Le Claire aus den Ergebnissen von Kuper und Mitarbeitern den höheren Wert $Q^D = 2{,}10$ eV erhalten.

Der Beitrag der Doppelleerstellen zu Q^D ist nicht sehr groß und läßt sich z. B. unter Verwendung der in Tabelle 3 aufgeführten Werte zu $Q^D - Q_1^D \approx 0{,}025$ eV abschätzen [30]. Wir werden im folgenden mit $Q_1^D = 2{,}12$ eV rechnen, womit sich aus dem experimentellen Wert von D_0 als Frequenzfaktor für die Leerstellenwanderung $\nu_1^0 \exp(S_1^W/k) \approx 2 \cdot 10^{14}\ \mathrm{s^{-1}}$ ergibt [30].

3.3 *Ausheilen von Einfach- und Doppelleerstellen*

3.3.1 *Einfachleerstellen*

Aus den in 3.1 und 3.2 erhaltenen Werten $U_1 + W_1 = 2{,}12$ eV und $U_1 = 1{,}06$ eV folgt als Wanderungsenergie der Einfachleerstellen $W_1 = 1{,}06$ eV. Dieser Wert stimmt gut mit der im Temperaturgebiet der Erholungsstufe IV von uns beob-

Tabelle 3
Zusammenstellung der charakteristischen Werte für Einfach- und Doppelleerstellen

Bildungsenergie der Einfachleerstellen U_1	1,06 eV
Bildungsentropie der Einfachleerstellen S_1^B	0,5 k
Wanderungsenergie der Einfachleerstellen W_1	1,06 eV
Wanderungsenergie der Doppelleerstellen W_2	0,67 eV
Bindungsenergie der Doppelleerstellen B_2	0,12 eV
Bindungsentropie der Doppelleerstellen ΔS	1,8 k
Frequenzfaktor der Einfachleerstellenwanderung $\nu_1^0 \exp(S_1^W/k)$	$2 \,.\, 10^{14}\ \mathrm{s^{-1}}$
Frequenzfaktor der Doppelleerstellenwanderung $\nu_2^0 \exp(S_2^W/k)$	$5 \,.\, 10^{13}\ \mathrm{s^{-1}}$

achteten *Aktivierungsenergie* überein. Wir sind deshalb der Ansicht, daß für diesen Erholungsprozeß die Wanderung von Einfachleerstellen maßgebend ist. Experimentelle Ergebnisse über die *Erholungskinetik* in Stufe IV liegen an denjenigen Proben vor, die keine nennenswerte Erholung zwischen 0 °C und 100 °C zeigen (2.1). Während am Anfang des Erholungsvorganges Unterschiede von Probe zu Probe beobachtet werden, findet man gegen Ende des Erholungsprozesses einheitlich angenähert Reaktionsordnung 1. Die aus der Halbwertszeit $\tau_{1/2}$ der Widerstandsänderung gemäß

$$n_{1/2} = 12\, \nu_1^0 \exp(S_1^W/k) \exp(-W_1/k\,T)\, \tau_{1/2} \tag{9}$$

errechnete mittlere Sprungzahl ergibt sich zu $n_{1/2} \approx 5 \cdot 10^6$, wenn die gesamte Widerstandsänderung der Stufe IV betrachtet wird, und zu $n_{1/2} \approx 3 \cdot 10^7$, wenn nur der mit Reaktionsordnung 1 ausheilende Anteil berücksichtigt wird. Die beobachtete Reaktionsordnung und die Größenordnung der Sprungzahlen, die beide mit einem Ausheilen an Versetzungen verträglich sind, sprechen dafür, daß während des Anlassens in Stufe IV kaum neue Senken, etwa in der Form von Leerstellenagglomeraten, gebildet werden.

In abgeschreckten Kupferlegierungen [42, 43, 27, 28] wurden im Temperaturgebiet der Stufe IV Platzwechselvorgänge beobachtet, die von einem Teil der Autoren mit Ordnungseinstellung infolge von Leerstellenwanderung erklärt werden. Als Aktivierungsenergie wurde von Li und Nowick [43] in CuAl (17%) mit Hilfe des Zener-Orientierungseffektes (1,08 ± 0,08) eV gefunden. Die gute Übereinstimmung mit dem oben angegebenen W_1 darf allerdings nicht überbewertet werden, da die Wanderungsenergien von Fehlstellen in Legierungen und in reinen Metallen nicht genau gleich zu sein brauchen und außerdem zur Ordnungseinstellung auch die Bewegung von Mehrfachleerstellen beizutragen vermag.

3.3.2 Doppelleerstellen

Es liegt nahe, die Erholung zwischen 0 °C und 100 °C, mit der sich die Mehrzahl der bisherigen Veröffentlichungen [10, 16 bis 23] befaßt (und die nach Ausweis röntgenographischer und elektronenmikroskopischer Untersuchungen mit der Bildung und dem Wachstum von Leerstellenagglomeraten verbunden ist), wegen der beobachteten verhältnismäßig kleinen Aktivierungsenergie von etwa 0,55 eV mit der Wanderung von Doppelleerstellen in Zusammenhang zu bringen. Wie wir jedoch sogleich zeigen werden, kann

a) diese Erholung mit dem Ausheilen von Doppelleerstellen *allein* nicht erklärt werden. Dagegen ist

b) eine Deutung der Ergebnisse auf Grund der nächstliegenden Verallgemeinerung möglich, bei der das Ausheilen von miteinander in *lokalem Gleichgewicht* befindlichen Einfach- und Doppelleerstellen betrachtet wird.

a) Versucht man aus den Erholungsdaten für Typ A — der die am besten definierten Aktivierungsenergien aufweist — die Einfangkonstante α für das einfache Zusammenlagern (Reaktionsordnung 2) von zwei Doppelleerstellen aus

$$\frac{dc_2}{dt} = -\alpha \frac{D_2}{a^2} c_2^2 \tag{10}$$

$(D_2 = 1/6\, a^2 \nu_2^0 \exp(S_2^W/k) \exp(-W_2/k\,T))$ zu bestimmen, so ergibt sich $\alpha \approx 0{,}1$ bis 0,5 (mit $\nu_2^0 \exp(S_2^W/k) = 5 \cdot 10^{13}\ \mathrm{s}^{-1}$, $W_2 = 0{,}55$ eV, $\Delta\varrho_{DL} = 2\,\Delta\varrho_L = 3{,}2\ \mu\Omega\mathrm{cm}/\%\ DL$). Die mittleren Sprungzahlen für diesen Erholungsprozeß ergeben sich mit

$$n_{1/2} = 8\, \nu_2^0 \exp(S_2^W/k) \exp(-W_2/k\,T)\, \tau_{1/2} \tag{11}$$

zu $n_{1/2} \approx 3 \cdot 10^7$. Die so berechnete Einfangkonstante α ist jedoch um mindestens eine Größenordnung kleiner als man theoretisch ($\alpha \approx 10$ [44]) für das Zusammenlagern von Doppelleerstellen erwarten würde; die errechneten Sprungzahlen sind bei den in unseren Versuchen vorhandenen Konzentrationen an leeren Gitterplätzen von ungefähr $2 \cdot 10^{-5}$ um mehrere Größenordnungen zu groß. Dieselben Schlüsse ergeben sich für den Erholungsprozeß bei 60 °C ($\alpha \approx 0{,}4$, $n_{1/2} \approx 3 \cdot 10^7$) und insbesondere den Erholungstyp C ($\alpha \approx 0{,}003$, $n_{1/2} \approx 2 \cdot 10^9$ [16]).[5]) Wegen der Ähnlichkeit (kleine Werte für α, ungefähr gleiche Aktivierungsenergien) der zwischen 0 °C und 100 °C beobachteten Erholungsstufen (Typ A, B, C und wohl auch untere Stufe von Typ D) liegt die Annahme nahe, daß es sich bei diesen Stufen im Prinzip durchweg um den gleichen Ausheilmechanismus handelt, der in den Einzelheiten vielleicht durch Verunreinigungen oder unterschiedliche Versetzungsdichten etwas modifiziert wird.

b) Im folgenden betrachten wir den Fall, daß neben den Doppelleerstellen auch Einfachleerstellen am Ausheilvorgang beteiligt sind und beide Fehlstellenarten miteinander in lokalem Gleichgewicht stehen [45, 46]. Unter diesen Voraussetzungen ergibt sich als effektive Diffusionskonstante für leere Gitterplätze [47]

$$D(c) = \frac{c_1 D_1 + 4 c_2 D_2}{c_1 + 4 c_2} \tag{12}$$

$\left(D_1 = a^2 \nu_1^0 \exp (S_1^W/k) \exp (-W_1/k\,T),\ D_2 \text{ siehe Gleichung (10)}\right)$. Die durch (12) gemäß

$$W_{\text{eff}} \equiv -\,\partial \ln D(c)/\partial(1/k\,T) \tag{13}$$

definierte effektive Aktivierungsenergie ist im allgemeinen temperatur- und konzentrationsabhängig. Neben den trivialen Grenzfällen $W_{\text{eff}} = W_1$ und $W_{\text{eff}} = W_2$ ist die einzige ausgezeichnete Aktivierungsenergie nach (12)

$$W_{\text{eff}} = W_2 - B_2\,, \tag{13a}$$

die für $c_1 \gg c_2$, $c_1 D_1 \ll 2 c_2 D_2$ (Überwiegen der Einfachleerstellen, jedoch Diffusionsstrom hauptsächlich durch Doppelleerstellen getragen) erhalten wird [46]. Es liegt deshalb nahe, die bei unseren Experimenten häufig auftretende und für den Erholungstyp A charakteristische Aktivierungsenergie von 0,55 eV mit (13a) zu identifizieren, also

$$W_2 - B_2 = 0{,}55\ \text{eV} \tag{13b}$$

zu setzen. Die gelegentlich beobachteten Abweichungen von diesem Wert können dadurch erklärt werden, daß entweder die Voraussetzungen für die Gültigkeit von (12) nicht zutreffen oder daß die zu (13a) führende Vereinfachung nicht zulässig ist. So kann z. B. das nahezu kontinuierliche Spektrum der Aktivierungsenergien beim Typ C durch W_{eff}-Werte außerhalb des Gültigkeitsbereichs von (13b) verstanden werden. Ist die Leerstellenkonzentration, gemessen an der Senkenkonzentration, zu niedrig, so heilen die Doppelleerstellen aus, bevor sich das lokale Gleichgewicht einstellen kann, und man beobachtet anschließend W_1 (Typ D). Beim Typ B ist denkbar, daß entweder Verunreinigungen oder in (12) vernachlässigte Fehlstellen wie Dreifachleerstellen eine wesentliche Rolle spielen.

---

[5]) Macht man die (realistische) Annahme, daß beim Zusammenlagern von zwei Doppelleerstellen zu einer (praktisch als unbeweglich anzusehenden) Vierfachleerstelle nur ein Bruchteil von $\Delta\varrho_{DL}$ ausheilt, so werden die oben beschriebenen Diskrepanzen noch verstärkt.

De Jong [48] hat gegen die vorstehende Deutung (die im wesentlichen mit einer früher skizzierten [10, 16] übereinstimmt) Bedenken erhoben, weil die Beweglichkeit der Einfachleerstellen zu einer Einstellung des Gleichgewichts innerhalb der Versuchszeiten nicht ausreiche. Letzteres ist zwar richtig, aber gänzlich belanglos, solange sich das Gleichgewicht auf irgendeine Weise einstellt. Ein mögliches Beispiel hierfür ist: Doppelleerstelle wandert zu Einfachleerstelle, die entstehende Dreifachleerstelle wandert zu Einfachleerstelle und bildet zwei Doppelleerstellen.

Die Bestimmung der Einzelwerte für W_2 und B_2 macht etwas größere Schwierigkeiten, jedoch lassen sich folgende Abschätzungen angeben:

α) Die zur Ableitung der Gleichung (13a) gemachte Annahme $c_1 \gg c_2$ bedeutet

$$1 \gg 6 \exp (\Delta S/k) \exp (B_2/k\,T)\, c_1 \,, \tag{14}$$

was bei einer Leerstellenkonzentration von $2 \cdot 10^{-5}$, Anlaßtemperatur $T = 0$ °C (bzw. 30 °C) und $\Delta S = 1{,}8\, k$ zu

$$B_2 < 0{,}17 \text{ eV (bzw. } 0{,}19 \text{ eV)}$$

führt.[6]) Wäre diese Bedingung nicht erfüllt, so gäbe der vorgeschlagene Erholungsmechanismus keinen ausgezeichneten Wert für die effektive Aktivierungsenergie.

β) Die gemessene Erholungskinetik ist in erster Näherung durch eine Reaktionsordnung 2 gekennzeichnet. Berechnet man die Erholungsisothermen durch Integration der Reaktionsgleichung

$$\frac{dc}{dt} = -\lambda_0^2\, D(c)\, c, \tag{15}$$

so ergibt sich mit der erlaubten Vereinfachung $c_1 D_1 \ll 4\, c_2 D_2$

$$\frac{3}{2} \ln (x-1) + \frac{1}{2} \ln (x+1) - \frac{1}{x-1} = -A\,t + B\,, \tag{16}$$

wo $x = \sqrt{1 + 48 \exp (\Delta S/k) \exp (B_2/k\,T)\, c}$, $A = \lambda_0^2 D_2$ und B eine Integrationskonstante sind. λ_0^2 ist dabei eine die Geometrie der Senken charakterisierende Größe, die sich im Fall von kugelförmigen Senken mit Radien r_0 der Senke und R_0 des Einzugsgebiets näherungsweise ($R_0 \gg r_0$) zu

$$\lambda_0^2 = \frac{3\, r_0}{R_0^3} \tag{17}$$

ergibt [49]. Die durch (16) dargestellte Ausheilkurve ist mit der von uns für Typ A gemessenen Reaktionsordnung in Einklang zu bringen, falls

$$B_2 \lesssim 0{,}13 \text{ eV}\,.$$

Eine Auswertung [10, 16] der Erholungskinetik (Typ C) hinsichtlich der Abhängigkeit der Ausheilgeschwindigkeit von der Fehlstellenkonzentration gemäß (15) ergab

$$B_2 = (0{,}13 \pm 0{,}04) \text{ eV}\,,$$

unabhängig von der Wahl für $W_2 - B_2$.

---

[6]) Die zweite in diesem Zusammenhang gemachte Annahme $c_1 D_1 \ll 2\, c_2 D_2$ ergibt mit den aus Tabelle 3 ersichtlichen Werten ($T = 0$ °C) $W_2 - B_2 < 0{,}84$ eV, was für unseren experimentellen Wert gut erfüllt ist.

γ) Von $W_2 - B_2 = 0{,}55$ eV ausgehend, läßt sich aus den Messungen von Seeger, Gerold und Rühle [10] durch Vergleich der bei verschiedenen Temperaturen gemessenen Isothermen (Typ C)

$$0{,}03 \text{ eV} < B_2 < 0{,}14 \text{ eV}$$

abschätzen.

Als Wertepaar, das mit diesen Abschätzungen und mit den übrigen quantitativen experimentellen Daten [10, 12, 16] innerhalb deren Meßgenauigkeit am besten übereinstimmt, ergibt sich $B_2 = 0{,}12$ eV und $W_2 = 0{,}67$ eV. Die geringfügige Abänderung von B_2 gegenüber dem in 3.2 verwendeten vorläufigen Wert ist ohne Einfluß auf die dort gezogenen quantitativen Folgerungen.

Die in (15) eingeführte Größe λ_0^2 läßt sich aus der oben benützten „Einfangkonstante" α (Gleichung (10)) umrechnen und ergibt sich gemäß

$$\lambda_0^2 = \frac{\alpha \, \Delta\varrho_L(1 + 36 \exp(\Delta S/k) \exp(B_2/k\,T)\, c)}{4\, a^2 \, \Delta\varrho_{DL} \, 6 \exp(\Delta S/k) \exp(B_2/k\,T)} \tag{18}$$

zu $\lambda_0^2 \approx 5 \cdot 10^{11}$ cm^{-2} (Typ A, $W_2 = 0{,}67$ eV). Es überrascht nicht, daß dieser Wert größer ist als der direkt aus Kleinwinkelstreuexperimenten bestimmte Zahlenwert $\lambda_0^2 = 5 \cdot 10^9$ cm^{-2} [10], da bei der Versuchsführung in [10] besonders große Streueffekte (d. h. bei etwa konstanter Gesamtleerstellenkonzentration große Einzugsradien R_0 und somit kleine λ_0^2) angestrebt worden waren.[7])

Die Ergebnisse anderer Autoren über den Erholungsprozeß zwischen 0 und 100 °C lassen sich, soweit quantitative Vergleiche überhaupt möglich sind, weitgehend in das oben entworfene Schema einordnen. So stellten Kimura und Mitarbeiter [23] über die Messung der Verfestigung in abgeschrecktem Kupfer eine Erholung fest, die durch Aktivierungsenergien zwischen 0,6 und 0,7 eV gekennzeichnet ist. Lediglich Budin und Mitarbeiter [19, 20] berichten von einem Erholungsprozeß bei 0 °C, der durch eine Aktivierungsenergie von $0{,}80 \pm 0{,}02$ eV und eine Reaktionsordnung 1 gekennzeichnet sei. Die Autoren deuten diesen Prozeß durch Einfachleerstellenwanderung, weil Simmons und Balluffi [29] auf Grund ihrer Bildungsenergie von 1,17 eV über Gleichung (3) mit $Q^D = 2{,}05$ eV als Wanderungsenergie der Einfachleerstellen $W_1 = 0{,}88$ eV errechneten. Die aus diesen Erholungsdaten mit Hilfe von (9) berechneten Sprungzahlen (Fig. 3 in [19]) von $n_{1/2} = 2 \cdot 10^4$ (0 °C) sind jedoch zu klein um — entsprechend der mitgeteilten Reaktionsordnung 1 — mit dem Ausheilen von Leerstellen an Versetzungen oder Leerstellenagglomeraten erklärt werden zu können. Bei Reaktionen mit Verunreinigungen hingegen dürfte nicht der gesamte elektrische Widerstand der Leerstellen ausheilen. Selbst unter der extremen Annahme, daß die Leerstellen unmittelbar nach Bildung einer Doppelleerstelle ausgeheilt sind — wofür man allerdings Reaktionsordnung 2 messen müßte — würden mindestens 10^6 Sprünge erforderlich sein, entsprechend der Leerstellenkonzentration von weniger als 10^{-6} (Fig. 11) bei der von Budin und Mitarbeitern angegebenen Abschrecktemperatur von 590 °C.

---

[7]) Verwendet man $W_2 = 0{,}67$ eV und $B_2 = 0{,}12$ eV zur Auswertung der Kleinwinkelstreuexperimente [10], so ergibt sich als Frequenzfaktor für die Doppelleerstellenwanderung der in der vorliegenden Arbeit benützte Wert $\nu_0^2 \exp(S_2^W/k) = 5 \cdot 10^{13}$ s^{-1}, in Abänderung des in [10] angegebenen, im Vergleich zum Frequenzfaktor für Einfachleerstellenwanderung etwas zu niedrig erscheinenden Wertes $\nu_2^0 \exp(S_2^W/k) = 4 \cdot 10^{12}$ s^{-1}. Die richtige Größenordnung des aus den Experimenten folgenden Frequenzfaktors stellt eine wichtige Kontrolle der Konsistenz des hier vorgeschlagenen Mechanismus dar, vergleiche die Diskussion in [10].

Änderte man die Aktivierungsenergie von 0,80 eV im Sinne von Simmons und Balluffi auf 0,88 eV, so würden die Diskrepanzen noch um den Faktor 20 größer werden. Wir glauben daher, daß für den von Budin und Mitarbeitern beobachteten Erholungsprozeß nicht allein Einfachleerstellen, sondern, ähnlich wie bei unseren Messungen, auch Doppelleerstellen eine wesentliche Rolle spielen, zumal der Verlauf der Isochrone (Fig. 2 in [19]) mit unserem Erholungstyp C übereinstimmt.

3.3.3 Schlußbemerkungen

Zusammenfassend kann gesagt werden, daß die in der vorliegenden Arbeit mitgeteilten Abschreckexperimente, gemeinsam mit der Analyse der Hochtemperatur- und Selbstdiffusionsdaten in 3.2 betrachtet, für eine Wanderungsenergie der *Einfachleerstellen* von $W_1 = 1{,}06$ eV sprechen. Für den von Simmons und Balluffi [29] angegebenen Wert $W_1 = 0{,}88$ eV konnten dagegen keinerlei experimentelle Stützen gefunden werden. Dies gilt auch für Verformungs- und Bestrahlungsexperimente, die ebenfalls auf eine Wanderungsenergie von etwa 1,05 eV hinweisen (siehe die Zusammenstellung und Diskussion in [2]). Auch die für die *Doppelleerstellen* charakteristischen Größen konnten in der vorliegenden Arbeit bestimmt werden.

Diejenigen Werte, die die zur Zeit an Kupfer vorliegenden Experimente nach unserer Auffassung am besten beschreiben, sind in Tabelle 3 zusammengestellt. Abschließend sei noch erwähnt, daß die Bildungs- und Bindungsenergien in guter Übereinstimmung mit der Theorie stehen (vergleiche [6]).

Die Autoren danken der Deutschen Forschungsgemeinschaft für die Förderung dieser Arbeit. Herr Professor Hasiguti hat uns dankenswerter Weise die Arbeit [32] als Übersetzung aus dem Japanischen zugänglich gemacht. Den Herren Prof. Dr. L. Slifkin und Dr. A. D. Le Claire haben wir für wertvolle Korrespondenz zu danken.

Literatur

[1] W. Schüle, A. Seeger, F. Ramsteiner, D. Schumacher und K. King, Z. Naturf. **16a**, 323 (1961).
[2] F. Ramsteiner, W. Schüle und A. Seeger, phys. stat. sol. **7**, 937 (1964).
[3] D. Schumacher, W. Schüle und A. Seeger, Z. Naturf. **17a**, 228 (1962).
[4] W. Schüle, A. Seeger, D. Schumacher und K. King, phys. stat. sol. **2**, 1199 (1962).
[5] F. Ramsteiner, W. Schüle und A. Seeger, phys. stat. sol. **2**, 1005 (1962).
[6] A. Seeger und D. Schumacher, Lattice Defects in Quenched Metals, Int. Conf. Argonne Nat. Lab. 1964, Academic Press, New York–London 1965.
[7] W. Schüle, J. Phys. Soc. Japan **18**, Suppl. III, 292 (1963).
[8] J. E. Bauerle und J. S. Koehler, Phys. Rev. **107**, 1493 (1957).
[9] H. Mughrabi, Diplomarbeit, Stuttgart 1964.
[10] A. Seeger, V. Gerold und M. Rühle, Z. Metallk. **54**, 493 (1963).
[11] S. Mader, A. Seeger und E. Simsch, Z. Metallk. **52**, 785 (1961).
[12] K. P. Chik, A. Seeger und M. Rühle, V. Int. Conf. for Electron Microscopy, Philadelphia 1962, J 11, Academic Press, New York–London 1962.
[13] H. G. v. Bueren, Z. Metallk. **46**, 272 (1955).
[14] C. J. Meechan und J. A. Brinkman, Phys. Rev. **103**, 1193 (1956).
[15] R. R. Coltman, T. H. Blewitt und S. T. Sekula, Nouvelles Propriétés Physiques et Chimiques des Métaux de très Haute Pureté, Colloques Internationaux du Centre National de la Recherche Scientifique XC, Paris 1959, Centre National de la Recherche Scientifique, Paris 1960 (S. 73).
[16] A. Seeger, V. Gerold, K. P. Chik und M. Rühle, Phys. Letters (Netherlands) **5**, 107 (1963).

[17] J. Galligan und J. Washburn, Phil. Mag. 8, 1455 (1963).
[18] L. M. Clarebrough, R. L. Segall, M. H. Loretto und M. E. Hargreaves, Phil. Mag. 9, 377 (1964).
[19] C. Budin, F. Denayrou, A. Lucasson und P. Lucasson, C. R. Acad. Sci. (France) 256, 1518 (1963).
[20] P. Lucasson, A. Lucasson, C. Budin und F. Denayrou, J. Phys. Radium 24, 508 (1963).
[21] M. J. Makin, Acta metall 6, 305 (1958).
[22] D. G. Martin, Phil. Mag. 7, 803 (1962).
[23] H. Kimura, R. Maddin und D. Kuhlmann-Wilsdorf, Acta metall. 7, 154 (1959).
[24] W. Eichenauer und A. Pebler, Z. Metallk. 48, 373 (1957).
[25] G. Airoldi, G. L. Bacchella und E. Germagnoli, Phys. Rev. Letters 2, 145 (1959).
[26] R. E. Feder, M. Mooney und A. S. Nowick, Acta metall. 6, 266 (1958).
[27] S. Matsuo und L. M. Clarebrough, Acta metall. 11, 1195 (1963).
[28] M. Hirabayashi, S. Nagasaki und H. Kono, J. appl. Phys. 28, 1070 (1957).
[29] R. O. Simmons und R. W. Balluffi, Phys. Rev. 129, 1533 (1963).
[30] A. Seeger, Phys. Letters (Netherlands) 12, 176 (1964).
[31] G. Schottky, A. Seeger und G. Schmid, phys. stat. sol. 4, 439 (1964).
[32] R. R. Hasiguti, Y. Nakao und H. Kimura, private Mitteilung, s. a. 18th Annual Meeting of Physical Society Japan, April 1963 (part 4, p. 23).
[33] C. J. Meechan und R. R. Eggleston, Acta metall. 2, 680 (1954).
[34] S. D. Gertsriken und B. F. Slyusar, Phys. Metals Metallography 6, 103 (1958).
[35] R. A. Dugdale, Phil. Mag. 1, 537 (1956).
[36] G. Schottky, Phys. Letters (Netherlands) 12, 95 (1964).
[37] W. L. Mercer und R. Shuttleworth, zitiert von D. Turnbull und R. E. Hoffmann, Acta metall. 7, 407 (1959), und A. D. Le Claire, Phil. Mag. 7, 141 (1962).
[38] M. S. Maier und H. R. Nelson, Trans. AIME 147, 39 (1942).
[39] A. Kuper, H. Letaw, jr., L. Slifkin, E. Sonder und C. T. Tomizuka, Phys. Rev. 98, 1870 (1955).
[40] L. Slifkin, private Mitteilung, s. a. Diskussion zu [6], Lattice Defects in Quenched Metals, Int. Conf. Argonne Nat. Lab. 1964, Academic Press, New York–London 1965.
[41] A. D. Le Claire, private Mitteilung, s. a. Phil. Mag. 10, 641 (1964).
[42] I. A. Brinkman, C. E. Dixon und C. J. Meechan, Acta metall. 2, 38 (1954).
[43] C. Y. Li und A. S. Nowick, Phys. Rev. 103, 294 (1956).
[44] T. R. Waite, J. chem. Phys. 28, 103 (1958).
[45] J. S. Koehler, F. Seitz und J. E. Bauerle, Phys. Rev. 107, 1499 (1957).
[46] G. Schottky, Z. Phys. 160, 16 (1960).
[47] M. De Jong und J. S. Koehler, Phys. Rev. 129, 40 (1963).
[48] M. De Jong, Phys. Letters (Netherlands) 7, 302 (1963).
[49] F. S. Ham, J. Phys. Chem. Solids 6, 335 (1958).

(Received January 6, 1965)

Note added on proof:
In einer etwas ausführlicheren, uns erst jetzt zugegangenen Arbeit geben Hasiguti, Nakao und Kimura (J. Phys. Soc. Japan, demnächst) $U_1 = (1{,}06 \pm 0{,}07)$ eV an.

phys. stat. sol. 8, 881 (1965)

Physikalisch-Technisches Institut der Deutschen Akademie der Wissenschaften zu Berlin, Bereich elektrischer Durchschlag, und IV. Physikalisches Institut der Humboldt-Universität zu Berlin

Thermisch bedingte nichtmonotone Strom–Spannungs-Charakteristiken

Von

E. NEBAUER und E. JAHNE

Es wird die Strom–Spannungs-Charakteristik eines durch Joulesche Erwärmung aufgeheizten Halbleiters untersucht. Für den Fall, daß bei jeder angelegten Spannung die erwärmte Probe eine einheitliche Temperatur hat, die jedoch vom Wert der angelegten Spannung abhängt, wird eine Parameterdarstellung für die Strom–Spannungs-Charakteristik angegeben. Bei nichtmotononer Temperaturabhängigkeit der elektrischen Leitfähigkeit können sowohl strom- als auch spannungskontrollierte negative differentielle Widerstände auftreten. Für die bei Halb- und Photoleitern zu erwartenden Charakteristiken werden Beispiele angeführt. Bei inhomogener Verteilung von Temperatur und Leitfähigkeit werden die Strom–Spannungs-Charakteristiken mit Hilfe eines Differenzenverfahrens ermittelt.

The current-voltage characteristic of a semiconductor which is warmed by Joule heating is investigated. If, for every applied voltage, the sample has a uniform temperature, a parametric representation can be given for the current–voltage characteristic. For the case in which the temperature dependence of the electrical conductivity is non-monotonic, current- and voltage controlled negative differential resistances can occur. Examples are given of the characteristics expected with semiconductors and photoconductors. For an inhomogeneous distribution of temperature and conductivity, the current–voltage characteristics are obtained by a method involving difference equations.

1. Einleitung

In den letzten Jahren ist das Auftreten von negativen differentiellen Widerständen in gewissen Abschnitten stationärer Strom(I)–Spannungs(U)-Charakteristiken von Halb- und Photoleitern sowie Dielektrika verstärkt theoretisch diskutiert und auch experimentell oft gefunden worden, wobei zahlreiche Anwendungsgebiete erschlossen werden konnten. Grundsätzlich hat man zwei Typen von negativen differentiellen Widerständen zu unterscheiden: Beim spannungskontrollierten Typ („N-Typ“) nimmt der Strom nach dem Durchlaufen eines Strommaximums (hier gilt $dI/dU = 0$) mit wachsender Spannung ab. Beim stromkontrollierten Typ („S-Typ“) nimmt der Strom nach dem Durchlaufen eines Spannungsmaximums (Kippunkt, hier gilt $dU/dI = 0$) bei fallender Spannung zu. Über die unterschiedlichen Stabilisierungsbedingungen dieser beiden Typen ist schon an anderer Stelle berichtet worden [1]. Die physikalischen Ursachen für das Auftreten von Bereichen, in denen der differentielle Widerstand negativ ist, sind sehr verschiedener Natur. Wir wollen im folgenden dafür einige Beispiele anführen.

Der N-Typ (Fig. 1a) kommt vor

a) bei der Esakidiode. Hier ist das Auftreten eines $dU/dI < 0$-Bereiches durch die Tunnelprozesse der Majoritätsträger erklärbar [2].

b) im Bereich „heißer" Elektronen. Eine Ursache für den negativen differentiellen Widerstand ist der Störstellenbarriereneffekt [3]; es existieren jedoch auch andere Vorstellungen [4].

c) bei den Charakteristiken dünner Metall-Halbleiter-Schichten. Hier kann zur Zeit noch keine befriedigende Erklärung für den Effekt gegeben werden, z. B. [5].

d) bei der Feldtilgung, insbesondere beim Vorliegen inhomogener Feldstärkeverteilungen (sogenannte Ringe oder Domänengebiete) [1, 6].

e) bei Charakteristiken an Photoleitern. Ursache des negativen differentiellen Widerstandes ist die Joulesche Erwärmung in Verbindung mit einer mit wachsender Temperatur abnehmender elektrischen Leitfähigkeit [7].

Der S-Typ (Fig. 1b) kommt vor

a) bei der Doppelinjektion. Der $dU/dI < 0$-Bereich ist durch die im Falle der Haftstellenerschöpfung stark anwachsende Löcherlebensdauer zu erklären [8].

b) beim Tieftemperaturdurchschlag. Hier stellt die Stoßionisation flacher Störstellen den verantwortlichen Prozeß dar [9].

c) bei Charakteristiken an Halbleitern und Isolatoren. Der $dU/dI < 0$-Abschnitt läßt sich durch die starke Zunahme der elektrischen Leitfähigkeit mit der Temperatur und das begrenzte Vermögen des Prüflings, Wärme an die Umgebung abzugeben, verstehen [10, 11, 12].

In dieser Arbeit soll in Weiterführung früherer Untersuchungen, insbesondere von Böer und Kümmel [7, 12], näher erläutert werden, unter welchen Umständen thermisch bedingte negative differentielle Widerstände vom N-Typ und solche vom S-Typ auftreten und wann bei nichtmonotoner Temperaturabhängigkeit der elektrischen Leitfähigkeit in ein und derselben Charakteristik Bereiche negativen differentiellen Widerstandes von beiden Typen auftreten (Charakteristiken vom N–S-Typ).

Die beiden wichtigsten Beispiele für nichtmonotone Temperaturabhängigkeit der elektrischen Leitfähigkeit $\sigma(T)$ findet man

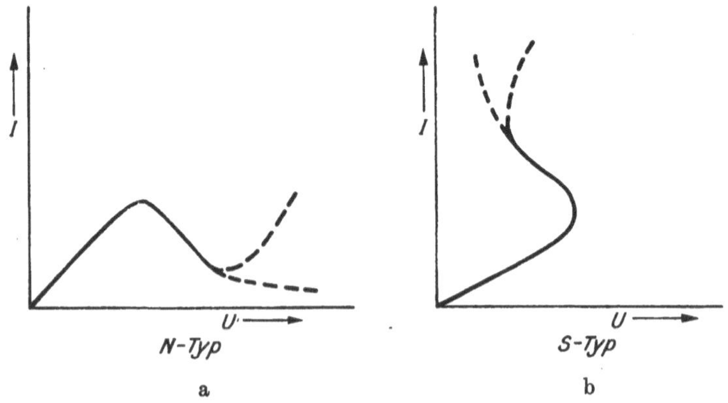

Fig. 1. Strom–Spannungs-Charakteristik mit verschiedenen Typen negativen differentiellen Widerstandes (schematisch)
a) N-Typ (spannungskontrollierte $dU/dI < 0$-Bereiche)
b) S-Typ (stromkontrollierte $dU/dI < 0$-Bereiche)

a) bei Photoleitern: Hier kommt die Nichtmonotonie dadurch zustande, daß die Leitfähigkeit bei niedrigen Temperaturen durch die Einstrahlung des Lichtes bestimmt wird (Photoleitung) und mit wachsender Temperatur abnimmt (Temperaturtilgung durch Abnahme der Lebensdauer infolge des sich mit der Temperatur ändernden Rekombinationsmechanismus), daß jedoch bei höheren Temperaturen die mit der Temperatur exponentiell anwachsende Dunkelleitung dominiert. Es existieren in der Literatur zahlreiche Arbeiten, die über derartige $\sigma(T)$-Verläufe von II–VI-, III–VI- und anderen Verbindungen, insbesondere in Abhängigkeit von der eingestrahlten Lichtintensität und der Dotierung, berichten [13].

b) bei Halbleitern: Bei dotierten Halbleitern ergibt sich eine $\sigma(T)$-Nichtmonotonie dadurch, daß im Störstellenerschöpfungsgebiet die Leitfähigkeit infolge einer mit wachsender Temperatur abnehmenden Beweglichkeit (Gitterstreuung) ebenfalls abnimmt, während bei hohen Temperaturen die Leitfähigkeit infolge der Eigenleitung exponentiell anwächst. Neben den Elementhalbleitern Germanium und Silizium sind hauptsächlich die II–V- und III–V-Verbindungen untersucht worden, wobei die Nichtmonotonie von $\sigma(T)$ sehr unterschiedlich stark ausgeprägt sein kann [14 bis 17].

2. Berechnung der Strom-Spannungs-Charakteristiken vom *N*-*S*-Typ

2.1 *Allgemeines*

Zur Berechnung der stationären Strom–Spannungs-Charakteristik (bei Gleichspannung) sowie der zugehörigen räumlichen Temperatur- und Stromdichteverteilungen eines durch Joulesche Erwärmung aufgeheizten Mediums geht man von den Erhaltungssätzen für die Energie und Ladung aus [10, 11]:

$$\operatorname{div}(\varkappa \operatorname{grad} T) + (\boldsymbol{j}, \boldsymbol{E}) = 0\,, \tag{1a}$$

$$\operatorname{div}\boldsymbol{j} = 0 \tag{1b}$$

mit $\boldsymbol{j} = \sigma \boldsymbol{E}$. Dabei bedeuten $\varkappa$ Wärmeleitfähigkeit, $\boldsymbol{j}$ Stromdichte, $\boldsymbol{E}$ elektrische Feldstärke. Die Materialgrößen $\sigma(T)$ und $\varkappa(T)$ sowie die Randbedingungen für Temperatur und Potential sind vorzugeben. Die Lösung von (1a, b) gibt dann die Temperatur- und Feldstärkeverteilung im Dielektrikum an. Material- und Kontaktinhomogenitäten werden durch die Gleichungen (1a, b) nicht berücksichtigt. Demzufolge können eventuelle elektrodennahe Erscheinungen hier ebensowenig erfaßt werden, wie dies z. B. in der Plasmaphysik mit der die Bogensäule gut beschreibenden Elenbaas-Hellerschen Differentialgleichung möglich ist. Bei den Gleichungen (1a, b) handelt es sich um dieselben Grundgleichungen wie sie in der Theorie des Wärmedurchschlags [10, 11] verwendet werden, nur daß dort ein einfaches monotones $\sigma(T)$-Gesetz verwendet wird.

Wir wollen annehmen, daß die Probe die Gestalt eines Zylinders habe (Koordinaten r, φ, z mit $0 \leqq r \leqq R$, $-\frac{L}{2} \leqq z \leqq +\frac{L}{2}$, L Länge, $2\,R$ Durchmesser), an dessen Deckflächen die Kontakte angebracht sind. Als Randbedingung für die Temperaturverteilung wollen wir annehmen, daß die Temperatur an der Oberfläche durch das Newtonsche Abkühlungsgesetz bestimmt wird und daß die Temperatur in der Zylinderachse endlich bleibt. Die Potentialverteilung soll der Randbedingung genügen, daß die Deckflächen Äquipotentialflächen sind — die Differenz der Potentiale an den beiden Deckflächen ist gleich der angelegten

Spannung U — und die Normalkomponente der elektrischen Stromdichte an der Manteloberfläche soll verschwinden.

Der Einfachheit halber wollen wir uns auf solche geometrischen Verhältnisse von Länge zu Dicke beschränken, bei denen $\boldsymbol{j}$ nur eine z-Komponente hat. Da wir σ als skalare Größe annehmen, bedeutet dies, daß wir uns auf solche Potentialverteilungen Φ beschränken, bei denen Φ nur eine Funktion von z ist. Diese Potentialverteilungen genügen den oben geforderten Randbedingungen. Aus (1b) folgt

$$\frac{\dfrac{\partial^2\Phi}{\partial z^2}}{\dfrac{\partial\Phi}{\partial z}} = -\frac{1}{\sigma}\frac{d\sigma}{dT}\frac{\partial T}{\partial z}\,. \tag{2}$$

Diese Gleichung läßt sich für allgemeine $\sigma(T)$-Gesetze nur erfüllen, wenn entweder (Fall a)) sowohl die rechte als auch die linke Seite von (2) verschwinden oder (Fall b)), wenn auch die rechte Seite von (2) nur eine Funktion von z ist, d. h. $T = T(z)$. Der Fall a) läßt sich beim sehr langen Zylinder ($L \gg R$) verwirklichen, da dort der Temperaturgradient in axialer Richtung zu vernachlässigen ist. In diesem Falle ist $\boldsymbol{j} \perp \operatorname{grad} T$ und die Feldstärke $\boldsymbol{E}$ hängt nicht von den Koordinaten ab. Der Fall b) ist bei einer dünnen Scheibe ($L \ll R$) realisiert, da hier der Temperaturgradient in radialer Richtung zu vernachlässigen ist. In diesem Falle ist $\boldsymbol{j} \parallel \operatorname{grad} T$ und die elektrische Stromdichte $\boldsymbol{j}$ hängt nicht von den Koordinaten ab.

Die elektrische Leistung N ist gegeben durch

$$N = \int\limits_{(V)} (\boldsymbol{j}, \boldsymbol{E})\, dV\,. \tag{3}$$

V Volumen der Probe. Mit Hilfe der Energiebilanzgleichung (1a) folgt dann aus (3) unter Verwendung des Gaußschen Satzes

$$N = -\int\limits_{(O)} (\varkappa \operatorname{grad} T)_n\, dO\,. \tag{4}$$

Für den Fall a) des sehr langen Zylinders ist dieses Integral annähernd

$$N \cong -2\pi R L \varkappa \left.\frac{dT}{dr}\right|_R . \tag{5}$$

Das Newtonsche Abkühlungsgesetz lautet im Falle a)

$$-\varkappa \left.\frac{dT}{dr}\right|_R = h\left(T(R) - T_u\right), \tag{6}$$

wobei h die Wärmeübergangszahl an der Manteloberfläche und T_u die Temperatur der Umgebung ist. Bezeichnet man mit τ die Übertemperatur der Oberfläche gegenüber der Umgebung, d. h. $\tau = T(R) - T_u$, so wird durch Einsetzen von (6) in (5)

$$N = 2\pi R L h \tau\,. \tag{7}$$

Andererseits erhält man aus (3), da im Falle a) die elektrische Feldstärke vor das Integral gezogen werden kann und $|\boldsymbol{E}| = \dfrac{U}{L}$,

$$N = 2\pi \frac{U^2}{L} \int\limits_0^R \sigma\left(T(r)\right) r\, dr\,. \tag{8}$$

Für den Fall b) der dünnen Scheibe ist das Integral (4) annähernd

$$N \cong \pi R^2 \left(\varkappa \frac{dT}{dz}\bigg|_{-\frac{L}{2}} - \varkappa \frac{dT}{dz}\bigg|_{+\frac{L}{2}} \right). \tag{5'}$$

Das Newtonsche Abkühlungsgesetz lautet in diesem Falle

$$\pm \varkappa \frac{dT}{dz}\bigg|_{\mp \frac{L}{2}} = h \left(T\left(\mp \frac{L}{2}\right) - T_u \right), \tag{6'}$$

wobei h die Wärmeübergangszahl an den Kontaktflächen ist und gleich starke Wärmeabgabe an beiden Kontakten angenommen wurde. Bezeichnet man mit τ wieder die Übertemperatur, d. h. $\tau = T\left(\mp \frac{L}{2}\right) - T_u$, so ergibt sich anstelle von (7)

$$N = 2\pi R^2 h \tau . \tag{7'}$$

Anstelle von (8) erhält man, da im Falle b) die elektrische Stromdichte vor das Integral gezogen werden kann und $I = \pi R^2 |\boldsymbol{j}|$

$$N = \frac{I^2}{\pi R^2} \int\limits_{-\frac{L}{2}}^{+\frac{L}{2}} \frac{dz}{\sigma(T(z))} . \tag{8'}$$

Es ist klar, daß bei komplizierten $\sigma(T)$-Gesetzen selbst für diese einfachen geometrischen Verhältnisse eine analytische Lösung für $\sigma(T(r))$ bzw. $\sigma(T(z))$ nicht angegeben werden kann und somit auch eine explizite Darstellung für die Strom–Spannungs-Charakteristik nicht möglich ist. Um dennoch zu Aussagen über die Charakteristik zu gelangen, bestehen zwei Möglichkeiten:

1. Beschränkung auf den homogenen Fall, d. h. bei jeder gegebenen Spannung soll in der Probe (annähernd) eine einheitliche Temperatur und damit auch einheitliche Leitfähigkeit vorliegen, jedoch ändern sich diese Werte mit der Höhe der angelegten Spannung. In diesem Falle läßt sich, wie unten gezeigt wird, eine Parameterdarstellung für die Charakteristik angeben.
2. Behandlung des inhomogenen Falls durch ein geeignetes Näherungsverfahren. Im Abschnitt 2.4 wird hierzu das Differenzenverfahren herangezogen.

2.2 Der homogene Fall

Im Grenzfall, daß für die Temperaturverteilung in der Probe nicht die Wärmeleitung, sondern die Wärmeabgabe an der Oberfläche der dominierende Prozeß ist, entsteht in der Probe eine homogene Temperatur- und Leitfähigkeitsverteilung, d. h. $T(r) = T_u + \tau$ und $\sigma(r) = \sigma(R)$. Aus (7) und (8) erhält man in Verbindung mit der Beziehung $N = U \cdot I$

$$I = \pi R^2 \sqrt{\frac{2h}{R}} \sqrt{\tau\, \sigma\,(T_u + \tau)} , \tag{9}$$

$$U = L \sqrt{\frac{2h}{R}} \sqrt{\frac{\tau}{\sigma\,(T_u + \tau)}} . \tag{10}$$

Entsprechend erhält man im Fall b) der dünnen Scheibe aus den Gleichungen (7′) und (8′)

$$I = \pi R^2 \sqrt{\frac{2h}{L}} \sqrt{\tau \sigma (T_u + \tau)}, \tag{9'}$$

$$U = L \sqrt{\frac{2h}{L}} \sqrt{\frac{\tau}{\sigma (T_u + \tau)}}, \tag{10'}$$

d. h. gegenüber dem Fall a) ist nur die Größe h/R zu ersetzen durch h/L.

Diese Gleichungen sind bei gegebenen R, L, h und T_u eine Parameterdarstellung für die Strom–Spannungs-Charakteristik. Parameter ist – ebenso wie in [11] – die Übertemperatur oder wegen $\tau \sim N$ die elektrische Leistung. Man sieht, daß bei temperaturunabhängigem h die Größen R, L und h nur den Maßstab auf der Abszissen- bzw. Ordinatenachse festlegen. Die Form der Strom–Spannungs-Charakteristik wird durch das $\sigma(T)$-Gesetz und die Umgebungstemperatur T_u bestimmt.

Der differentielle Widerstand ist gegeben durch

$$\frac{dU}{dI} = W \frac{\sigma - \tau \frac{d\sigma}{d\tau}}{\sigma + \tau \frac{d\sigma}{d\tau}} \tag{11}$$

mit

$$W = \frac{L}{\pi R^2} \frac{1}{\sigma (T_u + \tau)}. \tag{12}$$

Die zu den Strommaxima bzw. Stromminima der Charakteristik gehörenden Parameterwerte τ^* sind die Wurzeln der Gleichung

$$\sigma + \tau \frac{d\sigma}{d\tau} = 0. \tag{13a}$$

Die zum Kippunkt gehörenden Parameterwerte τ^{**} sind die Wurzeln der Gleichung

$$\sigma - \tau \frac{d\sigma}{d\tau} = 0. \tag{13b}$$

Durch Einsetzen dieser Parameterwerte in (9) und (10) erhält man die zu den markanten Punkten der Charakteristik gehörenden Strom- und Spannungswerte.

Die zu den Extrema des differentiellen Widerstandes gehörigen Parameterwerte τ_w sind die Wurzeln der Gleichung

$$2\tau\sigma^2 \frac{d^2\sigma}{d\tau^2} - \tau^2 \left(\frac{d\sigma}{d\tau}\right)^3 - 2\tau\sigma \left(\frac{d\sigma}{d\tau}\right)^2 + 3\sigma^2 \frac{d\sigma}{d\tau} = 0. \tag{14}$$

Einsetzen von τ_w in (12) ergibt die extremalen Werte für den differentiellen Widerstand, mit welchen sich dann leicht die zur Stabilisation des Bereichs negativen differentiellen Widerstandes vom S-Typ erforderlichen Vorwiderstände berechnen lassen.

Für die Diskussion ist es günstig, die Charakteristik in doppeltlogarithmischer Darstellung zu betrachten. Die Steigung ist

$$\frac{d \log I}{d \log U} = \frac{1 + \frac{d \log \sigma}{d \log \tau}}{1 - \frac{d \log \sigma}{d \log \tau}}. \tag{15}$$

Man sieht, daß die Steigung weder von geometrischen Abmessungen, noch von der Wärmeübergangszahl[1]) noch vom Absolutwert der elektrischen Leitfähigkeit abhängt. $\frac{d \log I}{d \log U}$ hängt nur von der Steigung $\frac{d \log \sigma}{d \log \tau}$, also vom $\sigma(T)$-Gesetz und der Umgebungstemperatur T_u ab.

Das Vorzeichen von $\frac{d \log I}{d \log U}$ hängt vom Absolutbetrag von $\frac{d \log \sigma}{d \log \tau}$ ab, und zwar gilt

$$\frac{d \log I}{d \log U} \lesseqgtr 0 \qquad \text{für} \qquad \left|\frac{d \log \sigma}{d \log \tau}\right| \gtrless 1 .$$

Der Absolutbetrag von $\frac{d \log I}{d \log U}$ hängt vom Vorzeichen von $\frac{d\sigma}{d\tau}$ ab, und zwar gilt

$$\left|\frac{d \log I}{d \log U}\right| \lesseqgtr 1 \qquad \text{für} \qquad \frac{d\sigma}{d\tau} \lesseqgtr 0 .$$

Insbesondere ist für $\left|\frac{d \log \sigma}{d \log \tau}\right| \ll 1$ die Steigung ≈ 1 (ohmscher Anstieg) und für $\left|\frac{d \log \sigma}{d \log \tau}\right| \gg 1$ die Steigung ≈ -1 (Hyperbel gleicher Leistung).

Offensichtlich hat man zwei Arten von negativen differentiellen Widerständen zu unterscheiden, je nachdem, ob $\frac{d \log \sigma}{d \log \tau} < -1$ oder $> +1$ ist. Die erste Art ist von einem Strommaximum und einem Stromminimum $\left(\text{hier gilt } \frac{d \log \sigma}{d \log \tau} = -1\right)$ eingeschlossen. Die zweite Art ist von einem Spannungsmaximum und Spannungsminimum $\left(\text{hier gilt } \frac{d \log \sigma}{d \log \tau} = +1\right)$ eingeschlossen. Wir nennen in Übereinstimmung mit Ridley [1] einen von einem Strommaximum und Stromminimum begrenzten Bereich negativen differentiellen Widerstandes *spannungs*kontrolliert, da in diesem Bereich des negativen differentiellen Widerstandes und den benachbarten Bereichen positiven differentiellen Widerstandes die *Spannung* die *eindeutige* Größe und der Strom die mehrdeutige Größe ist. Die Form der Kennlinie in diesem Gebiet ist dann N-ähnlich. Die andere Art nennen wir *strom*kontrolliert, die Form der Kennlinie in diesem Gebiet ist dann S-ähnlich. Da die Steigung $\frac{d \log \sigma}{d \log \tau}$ sicherlich beschränkt ist, muß zwischen einem Bereich spannungskontrollierten negativen differentiellen Widerstandes und einem solchen stromkontrollierten negativen differentiellen Widerstandes stets ein — wenn auch kurzer — Bereich positiven differentiellen Widerstandes liegen. Jedenfalls können, wenn in einem Temperaturbereich die elektrische Leitfähigkeit genügend steil abfällt

[1]) Ist die Wärmeübergangszahl h temperaturabhängig, so gilt statt (15)

$$\frac{d \log I}{d \log U} = \frac{1 + \frac{d \log h}{d \log \tau} + \frac{d \log \sigma}{d \log \tau}}{1 + \frac{d \log h}{d \log \tau} - \frac{d \log \sigma}{d \log \tau}} .$$

Hieraus ist ersichtlich, daß die Temperaturabhängigkeit von h nur dann eine Rolle spielt, wenn der Exponent in der Darstellung $\log h$ vs. $\log \tau$ mit 1 vergleichbar wird.

und wieder anwächst, *rein thermisch* bedingt in einer Charakteristik sowohl negative differentielle Widerstände vom N- als auch vom S-Typ vorkommen. Eine derartige Kennlinie wollen wir als N–S-Charakteristik bezeichnen.

Fig. 2b zeigt eine I–U-Charakteristik in doppeltlogarithmischer Darstellung, wobei das in Fig. 2a dargestellte nichtmonotone $\sigma(T)$-Gesetz zugrundegelegt wurde. Nach einem positiven (anfänglich ohmschen) Bereich I weist der Bereich II einen negativen differentiellen Widerstand vom N-Typ auf; dieser Bereich wird durch die charakteristischen Punkte $I_{\max}$ und $I_{\min}$ (Strommaximum und Stromminimum) eingeschlossen. Dann folgt nochmals ein $dU/dI > 0$-Bereich (III), der am Kippunkt aufhört. Abschnitt IV stellt den schon aus [10, 11] bekannten Bereich negativen differentiellen Widerstands vom S-Typ dar.

Interessant ist noch, welche Temperatur die Probe an den markanten Punkten der I–U-Charakteristik besitzt. Bei gegebener Umgebungstemperatur hängt die Temperatur am Maximum von der Steilheit des Abfalls von σ ab: Bei größerer Steilheit wird das Maximum bei niedrigeren Temperaturen erreicht. Die Temperatur am Stromminimum (Fig. 2b) ist stets kleiner als die zum Leitfähigkeitsminimum (Fig. 2a) gehörende Temperatur, wobei die Differenz durch $\frac{d \log \sigma}{d \log \tau}$ bestimmt wird. Die zum Leitfähigkeitsminimum gehörende Temperatur wird an jenem Punkt im Bereich III erreicht, an dem ohmsches Verhalten vorliegt. Im Bereich IV ist die Temperatur der Probe stets höher als die zum Leitfähigkeitsminimum gehörende. Allgemein kann man sagen, daß der Bereich II (bzw. IV) einen um so kleineren negativen differentiellen Widerstand zeigt, je steiler der Abfall (bzw. Anstieg) von $\sigma(T)$ ist; der dabei vom Prüfling durchlaufende Temperaturbereich ist folglich umso kleiner, je steiler $\sigma(T)$ verläuft. Der Bereich III hingegen nimmt einen um so größeren Temperatur- und Spannungsbereich ein, je breiter das Minimum in der $\sigma(T)$-Kurve ist.

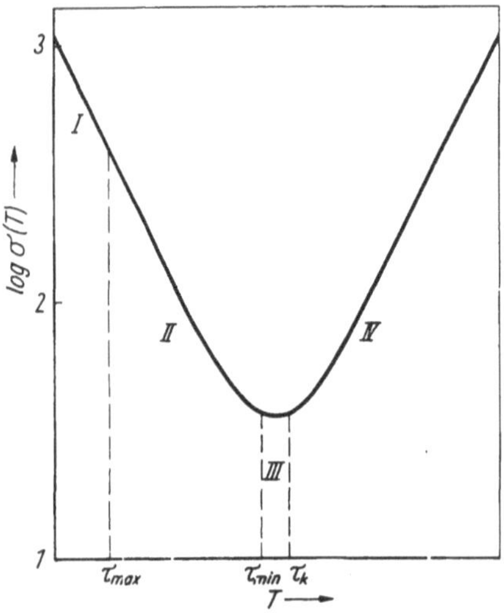

Fig. 2a. Nichtmonotoner $\sigma(T)$-Verlauf. Die Bereiche I bis IV entsprechen den in Fig. 2b gezeichneten Bereichen I bis IV

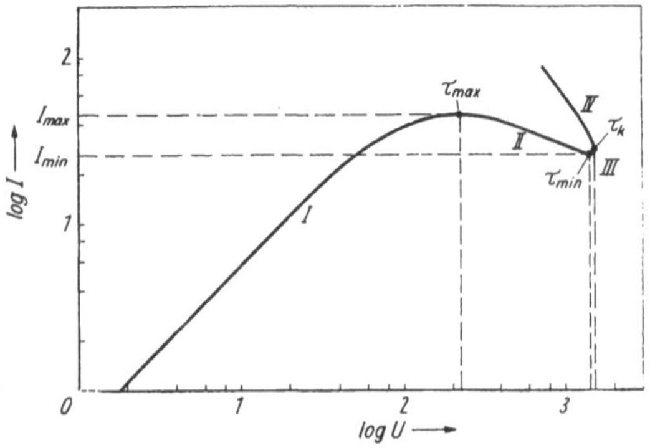

Fig. 2b. N–S-Charakteristik, ermittelt mit $\sigma(T)$ nach Fig. 2a.
Bereich I: $dU/dI > 0$,
Bereich II: $dU/dI < 0$ (N-Typ),
Bereich III: $dU/dI > 0$,
Bereich IV: $dU/dI < 0$ (S-Typ)

Da diese Resultate unter der Voraussetzung homogener σ-Verteilung erhalten wurden, ist es erforderlich, eine Bedingung für das Auftreten solcher Verteilungen anzugeben bzw. abzuschätzen, unter welchen Umständen man in praxi diese Näherung heranziehen darf. Wir entwickeln zu diesem Zweck die elektrische Leitfähigkeit σ nach dem Ort in eine Taylorreihe und brechen nach dem ersten Glied ab

$$\sigma(r) \approx \sigma(R) + (r - R) \frac{d\sigma}{dT} \frac{dT}{dr}\bigg|_R + \cdots .$$

Der Temperaturgradient läßt sich durch das Newtonsche Abkühlungsgesetz (6) ausdrücken

$$\sigma(r) \approx \sigma(R) + (R - r) \frac{d\sigma}{dT} \tau \frac{h}{\varkappa} . \tag{16}$$

Geht man mit Gleichung (16) in (8) ein, so sieht man, daß bei schwach inhomogenen Temperaturverteilungen in (9) und (10) $\sigma(T_u + \tau)$ zu ersetzen ist durch

$$\sigma(T_u + \tau) \to \sigma(T_u + \tau)\left(1 + \frac{H}{3} \frac{d \log \sigma}{d \log \tau}\right) \tag{17}$$

mit $H = \frac{h R}{\varkappa}$.[2])

Damit lassen sich die Charakteristiken auch für den Fall schwach inhomogener $\sigma(r)$-Verteilungen approximativ ermitteln. Qualitativ ersieht man aus den Gleichungen (17) und (9), (10), in welcher Richtung sich die Charakteristik verschiebt.

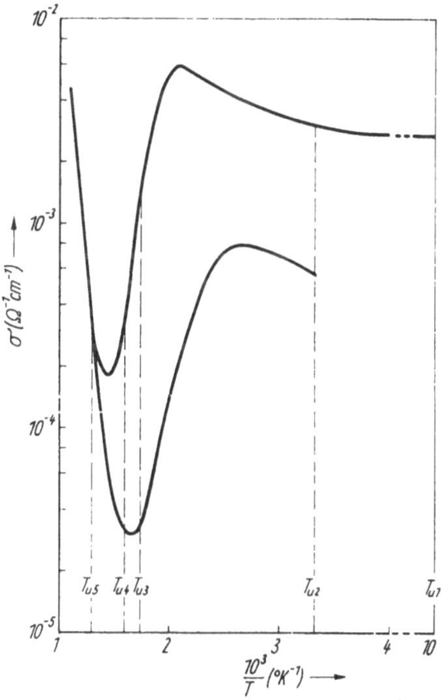

◀ Fig. 3a. $\sigma(1/T)$ an einem CdS-Einkristall bei zwei verschiedenen Lichtintensitäten (erhalten nach einer Temperung an Luft bei 550 °C) (eigene Messungen) (oben: Kurve 1, unten: Kurve 2)

Fig. 3b. Einfluß der Lichtintensität auf die I–U-Charakteristik, berechnet mit $\sigma(T)$ nach Fig. 3a und den Werten: $h = 2 \cdot 10^{-2}$ W grad^{-1} cm^{-2}, $R = 10^{-2}$ cm, $L = 1$ cm, $T_u = 300$ °K ▼

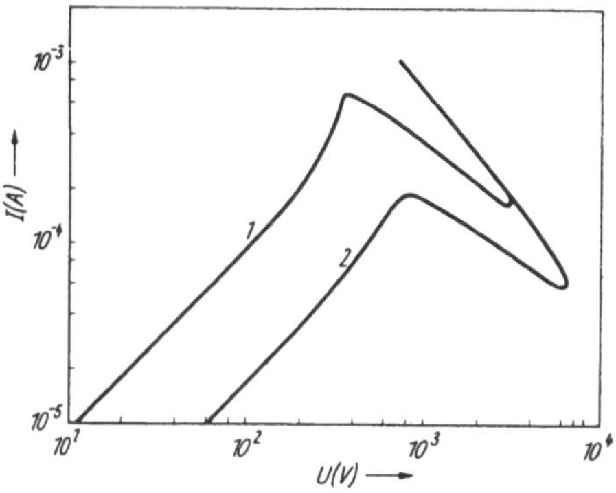

[2]) Im Falle b) erhält man eine ähnliche Korrekturformel, hier hat dann H die Bedeutung $H = \frac{h L}{\varkappa}$.

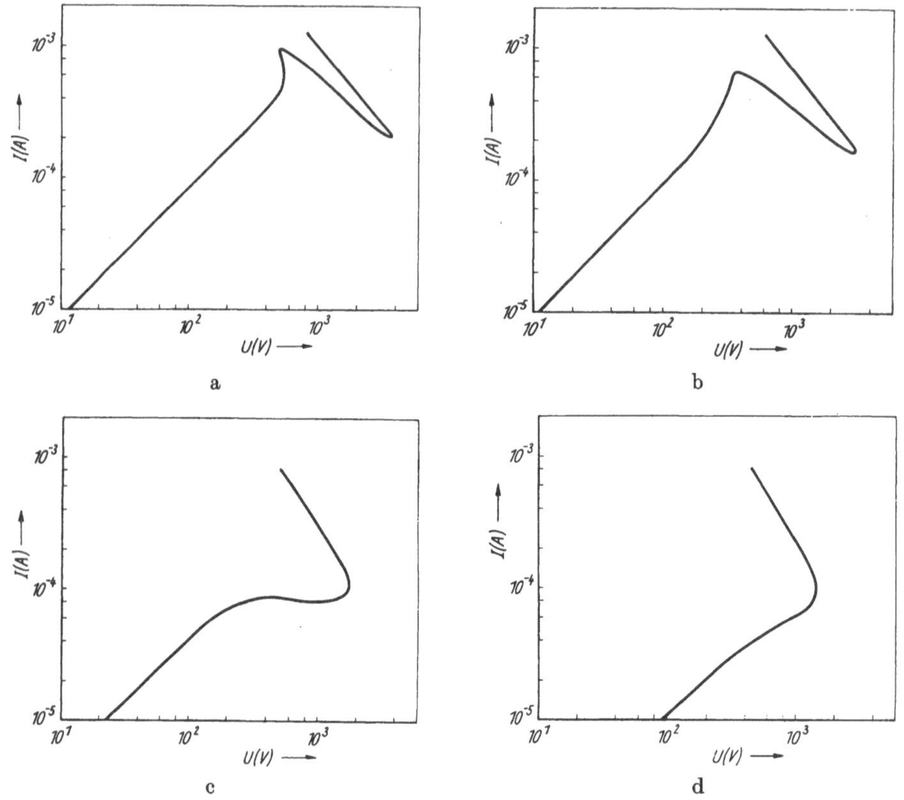

Fig. 4. Einfluß der Umgebungstemperatur auf die $I-U$-Charakteristik. Berechnet mit $\sigma(T)$ von Fig. 3a (Kurve 1) und den Werten: $h = 2 \cdot 10^{-2}$ W grad^{-1} cm^{-2}, $R = 10^{-2}$ cm, $L = 1$ cm und $T_u =$ a) 100 °K, b) 300 °K, c) 575 °K und d) 625 °K

Man darf demnach höchstens dann mit einer einheitlichen Leitfähigkeit rechnen, wenn die notwendige Bedingung $\frac{H}{3}\left|\frac{d \log \sigma}{d \log \tau}\right| \ll 1$ erfüllt ist. Da man die Größe H bei schwacher Kühlung und dünnen nadelförmigen Kristallen sehr klein gegen 1 machen kann (z. B. wird bei $h = 10^{-2}$ W grad^{-1} cm^{-2}, $R = 1 \cdot 10^{-2}$ cm und $\varkappa = 10^{-1}$ W grad^{-1} cm^{-1}: $H = 1 \cdot 10^{-3}$) und da $\frac{d \log \sigma}{d \log \tau}$ im allgemeinen zwischen 1 und 10 liegt, ist normalerweise auch $\frac{H}{3}\left|\frac{d \log \sigma}{d \log \tau}\right| \ll 1$ experimentell realisierbar. Bei nicht zu starken inhomogenen Verteilungen von Temperatur und Stromdichte bleibt der grundsätzliche Typ der Charakteristik – wie in Abschnitt 2.4 gezeigt wird – erhalten.

2.3 Beispiele zum homogenen Fall

2.3.1 Photoleiter

Wir verwenden für die $\sigma(T)$-Abhängigkeit Kurven, die auf eigenen Messungen basieren. Die in Fig. 3a gezeigten $\sigma(T)$-Verläufe wurden an getemperten CdS-Einkristallen bei verschiedenen Lichtintensitäten aufgenommen. Die mit diesen $\sigma(T)$-Verläufen berechneten Charakteristiken (Fig. 3b) zeigen mit den von Böer und Kümmel [7] an plättchenförmigen Proben gemessenen Kennlinien große Ähnlichkeit. Fig. 4 demonstriert das Verschwinden des negativen differentiellen Widerstandes vom N-Typ mit wachsender Umgebungstemperatur. Der bei $T_{u1} = 100$ °K (Fig. 4a) und $T_{u2} = 300$ °K (Fig. 4b) gut ausgeprägte Bereich II mit $dU/dI < 0$ verschwindet bei etwa $T_u = 600$ °K (Fig. 4c,), da $d\sigma/dT$ nicht mehr hinreichend groß ist. Die Charakteristik in Fig. 4a zeigt bereits vor dem Strommaximum einen Abschnitt negativen differentiellen Widerstandes vom S-Typ, der durch den σ-Anstieg oberhalb Zimmertemperatur (Fig. 3a) bedingt ist. Daß eine nichtmonotone $\sigma(T)$-Abhängigkeit nicht unbedingt zu einer N–S-Charakteristik führen muß, demonstriert Fig. 4d. Diese Kennlinie (mit $T_{u4} = 625$ °K) hat bereits große Ähnlichkeit mit einer normalen Charakteristik, die sich bei monotonem $\sigma(T)$-Gesetz ergibt. Bei den hier gezeigten N–S-Charakteristiken sind das Stromminimum und der Kippunkt stets eng benachbart. Dies ist durch die große Steilheit des Abfalls und des Anstiegs der $\sigma(T)$-Kurve in Fig. 3a bedingt.

Neben einfach nichtmonotonen $\sigma(T)$-Gesetzen treten, insbesondere bei thermisch und elektrisch noch nicht belasteten (sogenannten jungfräulichen) Kristallen,

Fig. 5b. I–U-Charakteristik, berechnet mit $\sigma(T)$ nach Fig. 6a als Beispiel einer Charakteristik mit mehreren Kippunkten. ($h = 1 \cdot 10^{-2}$ Wgrad^{-1} cm^{-2}, $R = 2 \cdot 10^{-2}$ cm, $L = 1$ cm, $T_u = 300$ °K) ▶

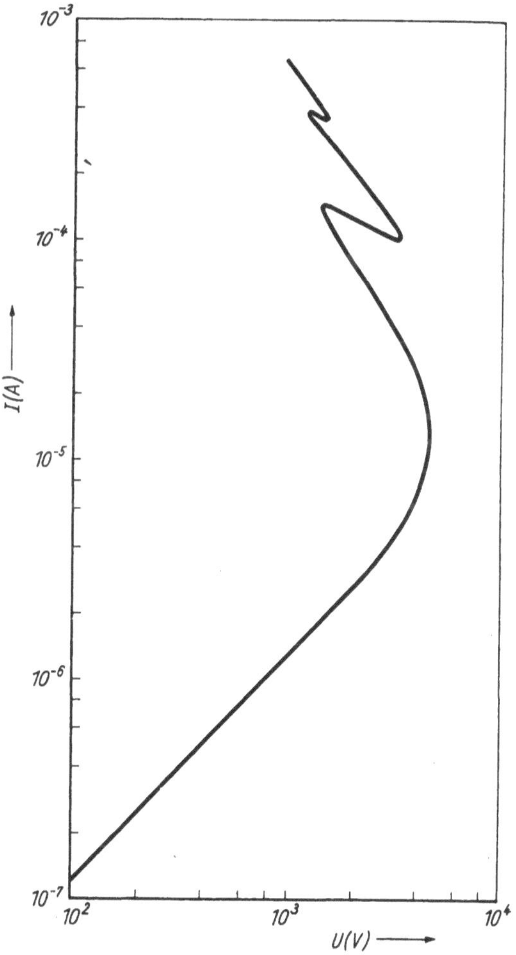

Fig. 5a. Mehrfach nichtmonotoner $\sigma(1/T)$-Verlauf eines jungfräulichen CdS-Kristalls (eigene Messungen) ▼

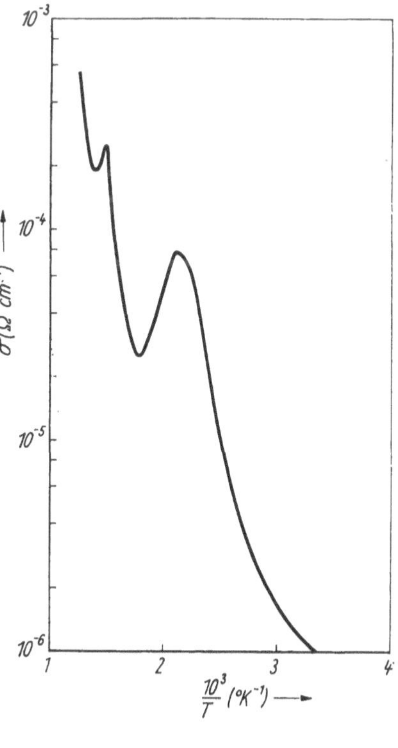

auch mehrfach nichtmonotone $\sigma(T)$-Abhängigkeiten auf. Fig. 5a zeigt eine solche $\sigma(T)$-Kurve, wie sie von uns an einem jungfräulichen CdS-Kristall gemessen wurde. Die in Fig. 5b dargestellte I–U-Charakteristik zeigt einen entsprechend komplizierten Gang: Es existieren fünf Kippunkte und je zwei Strommaxima und Stromminima, so daß es Spannungsbereiche gibt, in denen zu einem U-Wert bis zu sechs I-Werte stationär möglich sind. Es soll nochmals betont werden, daß man aus der Anzahl der relativen Minima und Maxima in solchen $\sigma(T)$-Gesetzen nicht ohne weiteres auf die Anzahl der zu erwartenden Strommaxima und Stromminima schließen kann, weil dabei stets die Größe des Abfalls und Anstiegs von σ entscheidend ist. In jedem konkreten Fall kann man sich anhand der Gleichungen (9, 10) schnell einen Überblick verschaffen.

2.3.2. *Halbleiter*

Im Gegensatz zu den Photoleitern kann man bei Halbleitern eine einfache Formel für die Abnahme der elektrischen Leitfähigkeit mit der Temperatur angeben. Im Störstellenerschöpfungsgebiet ist $\sigma \sim \mu \sim T^{-p}$ (mit p zwischen 1,5 und 2,2). Dadurch ist es auch möglich, mit Hilfe von Gleichung (15) eine übersichtliche Bedingung für das Auftreten von $dU/dI < 0$-Bereichen vom N-Typ

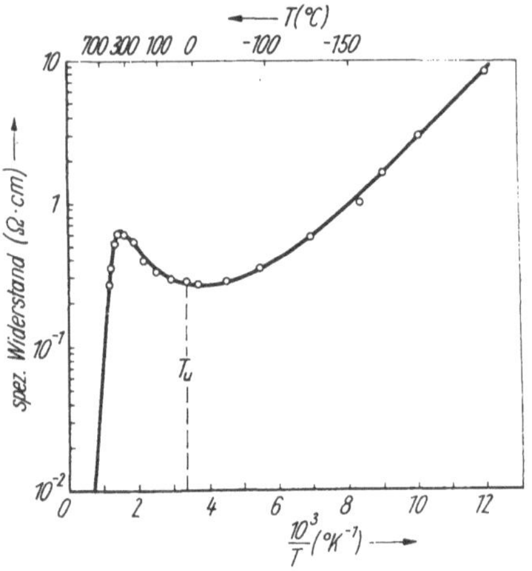

a

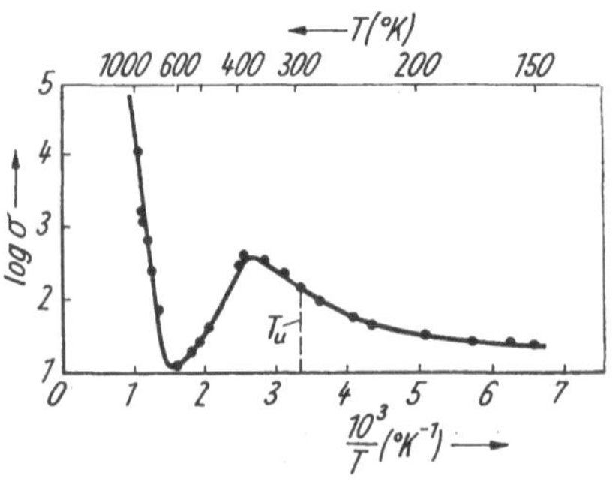

b

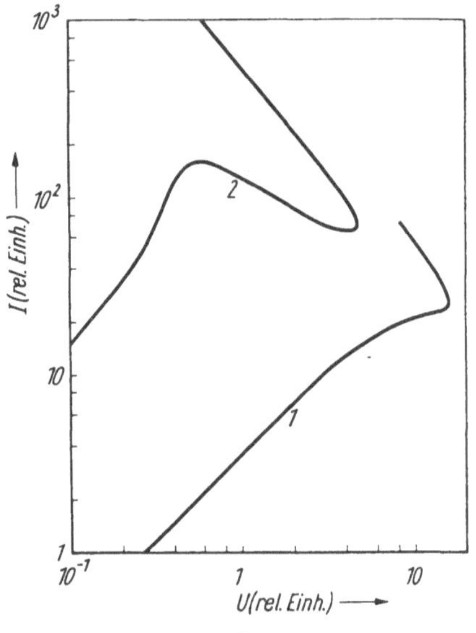

c

Fig. 6a, b. $\sigma(1/T)$ bei
a) P-dotiertem Silizium (nach [15])
b) Se-dotiertem GaSb (nach [17])

Fig. 6c. I–U-Charakteristiken, (in reduzierten Einheiten) berechnet mit $\sigma(T)$ von Fig. 6a (Kurve 1) bzw. 6b (Kurve 2); $T_u = 300\,°K$

anzugeben. Es muß nämlich wegen $1 + \frac{d \log \sigma}{d \log \tau} < 0$ gelten

$$T > \left(\frac{p}{p-1}\right) T_u . \tag{18}$$

Speziell folgt hieraus bei $p = 1{,}5$ (Streuung an akustischen Phononen), daß die Temperatur der Probe größer als die dreifache Umgebungstemperatur werden muß, wenn ein Bereich II ($dU/dI < 0$) auftreten soll.

Diese Bedingung läßt sich z. B. für die in Fig. 6a gezeigte $\sigma(T)$-Kurve (Pearson und Bardeen [15]) von dotiertem n-Silizium nicht erfüllen, so daß in der zugehörigen Charakteristik nur ein $dU/dI < 0$-Abschnitt von S-Typ auftritt (Fig. 6c, Kurve 1). Die bei Conwell [16] gezeigten $\sigma(T)$-Kurven an n-Germanium liefern hingegen zwar eine N–S-Typ-Charakteristik, allerdings nur bei sehr tiefen Umgebungstemperaturen, so daß andere Effekte (insbesondere Tieftemperaturdurchschlag) die Beobachtung einer solchen Charakteristik fraglich machen dürften. Bessere Aussichten zur Messung dieses Effektes bestehen bei einigen III–V-Verbindungen, z. B. bei GaSb (Boltaks [17]), wo die $\sigma(T)$-Abnahme erst oberhalb Zimmertemperatur auftritt (Fig. 6b). Fig. 6c, Kurve 2 zeigt die daraus berechnete Charakteristik, die einen ähnlich stark ausgeprägten Bereich II hat wie z. B. Fig. 3b.

2.4 Der inhomogene Fall

Bei Stoffen mit sehr kleiner Wärmeleitfähigkeit und steilen $\sigma(T)$-Verläufen wird (vor allem bei größeren Proben und bei starker Kühlung) der Ausdruck $\frac{H}{3}\left|\frac{d \log \sigma}{d \log \tau}\right|$ vergleichbar mit oder sogar größer als eins. Man gelangt dann weder mit der homogenen Näherung noch mit der Korrekturformel (17) zu einer annähernd richtigen I–U-Charakteristik. Für den Fall, daß starke Temperaturgradienten im Inneren der Probe auftreten, bleibt nur noch die Möglichkeit, die Bilanzgleichungen (1) mit einem Näherungsverfahren zu lösen.

An dieser Stelle beschränken wir uns auf den zylindersymmetrischen Fall bei temperaturunabhängig angenommener Wärmeleitfähigkeit $\varkappa_0$. Das Problem ist dann auf die Lösung folgender Differentialgleichung reduziert:

$$r\, T''(r) + T'(r) + r\, \sigma(T)\, E^2/\varkappa_0 = 0 . \tag{19}$$

Die Randbedingungen lauten

$$\varkappa_0 \left.\frac{dT}{dr}\right|_{r=0} = 0 , \qquad \varkappa_0 \left.\frac{dT}{dr}\right|_{r=R} = h\,[T_u - T(R)] . \tag{19a}$$

Die näherungsweise Integration von (19) erfolgt mit der Methode der Differenzengleichungen [18]. Man erhält $T(r)$ als Treppenfunktion aus folgendem, im allgemeinen nichtlinearen Gleichungssystem (bei Verwendung einfacher Differenzenquotienten):

$$\left.\begin{array}{lll} r = 0: & T_1 - T_0 + \frac{K}{4\,n^2} E^2\, \sigma(T_0) = 0 & (i = 0) \\ 0 < r < R: & \left(1 + \frac{1}{2i}\right) T_{i+1} + \left(1 - \frac{1}{2i}\right) T_{i-1} - 2\,T_i + \frac{K^2}{n^2} E^2\, \sigma(T_i) = 0 & \\ & & (i = 1 \ldots n-1) \\ r = R: & \left(\frac{H}{n} + 1\right) T_n - T_{n-1} - \frac{H}{n} T_u = 0 & (i = n) \end{array}\right\} \tag{20}$$

mit $H = h\,R/\varkappa_0$, $K = R^2/\varkappa_0$.

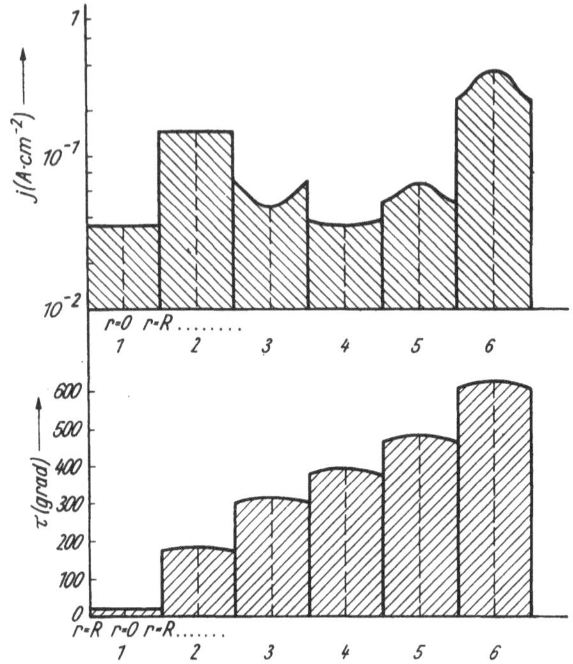

Fig. 7. Inhomogene Temperatur- und Stromdichteverteilungen an den mit Ziffern bezeichneten Punkten der I–U-Charakteristik in Fig. 8 (nach dem Differenzenverfahren ermittelt)

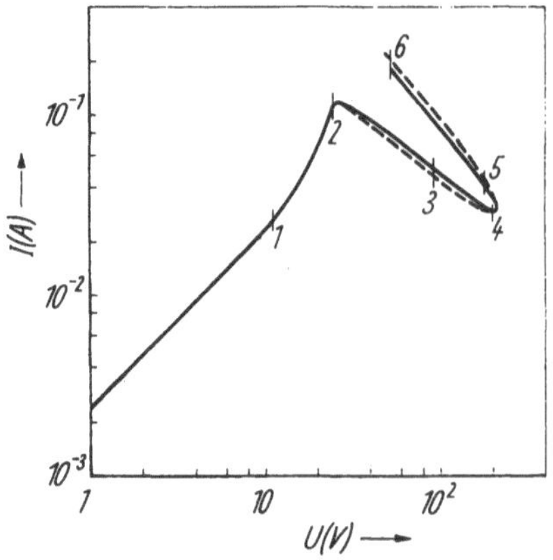

Fig. 8. Vergleich von „homogener" Näherung (———) und Näherung mit Differenzenverfahren (Maschenweite $R/3$) (- - - - -) $H = 0{,}05$ ($h = 5 \cdot 10^{-3}$ W grad^{-1} cm^{-2}, $\varkappa_0 = 5 \cdot 10^{-2}$ W grad^{-1} cm^{-1}, $R = 0{,}5$ cm), $L = 5$ cm, $T_u = 300\,°$K; $\sigma(T)$ nach Fig. 3a (Kurve 1)

Die Charakteristik ergibt sich dann approximativ aus der Stromtransportgleichung

$$I_S(U) = E\,L \sum_{i=0}^{n} \sigma(T_{is})\,F_i \quad \text{mit} \quad s = 1 \ldots S\,. \tag{21}$$

S gibt die Mehrdeutigkeit der Charakteristik an; F_i sind Kreisringe. E wird als Parameter vorgegeben.

Das Gleichungssystem (20) liefert jedoch für große Werte von H auch keine hinreichend genauen Lösungen mehr, wie man durch den Test mit der Gleichung (7) leicht nachprüfen kann. Man wird deshalb zu Gleichungen mit finiten Ausdrücken höherer Ordnung übergehen müssen, die dann für einige interessante Fälle nur noch mittels elektronischer Rechenmaschinen gelöst werden können; diesbezügliche Untersuchungen sind in Aussicht genommen.

Als Beispiel für das Differenzenverfahren wurde eine Charakteristik ermittelt, bei der $\frac{H}{3}\left|\frac{d \log \sigma}{d \log \tau}\right|$ nicht mehr sehr klein gegen eins, aber andererseits noch so klein ist, daß bereits das Gleichungssystem (20) hinreichend genaue Lösungen liefert. Während die Temperaturverteilungen stets einen monotonen Verlauf mit dem Maximum in der Achse des Zylinders zeigen (müssen), kann man (Fig. 7) bei den Stromdichten auch Verteilungen finden, die die maximalen j-Werte am Zylinderrand aufweisen; diese Verteilungen treten im Abschnitt II auf (bautförmige Stromdichteprofile wie beim Skineffekt). Fig. 8 zeigt die dazugehörigen Charakteristik, wie sie sich auf der Grundlage der Gleichungen (9, 10) der homogenen Näherung und unter Berücksichtigung der sich aus dem Differenzenverfahren ergebenden Inhomogenität von Temperatur und Stromdichte ergibt.

Für $H \gg 1$ erwartet man, daß sich in gewissen Abschnitten der Charakteristik nichtmonotone $j(r)$-Verläufe ergeben. Denn für $H \gg 1$ ist $T(r)$ bei höheren Leistungen stark inhomogen, wobei $T(R)$ nur wenig größer als T_u, die Temperatur in der Zylinderachse jedoch von einem gewissen Leistungswert ab größer als die zum Leitfähigkeitsminimum gehörende Temperatur ist, so daß im Bereich $0 < r < R$ die Stromdichte ein Minimum annehmen muß.

3. Schlußbemerkungen

Mit dieser Arbeit sollte näher ausgeführt werden, unter welchen Bedingungen die Joulesche Erwärmung zu einer nichtmonotonen Strom–Spannungs-Charakteristik führt. Dies erschien uns um so notwendiger, als trotz der grundsätzlichen Klärung dieser Frage durch die Arbeiten von Böer, Kümmel und Mitarbeitern [7, 12] auch in neuerer Zeit noch Arbeiten erscheinen, z. B. [19], bei denen der Einfluß einer Jouleschen Erwärmung außer acht gelassen wird. Gleichzeitig möchten wir darauf hinweisen, daß spontane Stromänderungen bei der Aufnahme von I–U-Charakteristiken ohne Stabilisationswiderstand ihre Ursache auch im Auftreten von thermisch bedingten $dU/dI < 0$-Bereichen haben können.

Andererseits wollten wir darauf hinweisen, daß mit einem Bereich negativen differentiellen Widerstandes vom N- bzw. S-Typ nicht notwendig — wie dies von Ridley [1] angenommen wird — eine Feld- bzw. Strominhomogenität verbunden sein muß (Domänen hohen Feldes bzw. Strom(dichte)kanäle). Vielmehr können rein thermisch bedingte N–S-Charakteristiken auch bei homogenen Feldstärke- und Stromdichteverteilungen vorkommen.

Abschließend möchten wir bemerken, daß auf Grund der Tatsache, daß der thermisch bedingte $dU/dI < 0$-Bereich vom N-Typ im Unterschied zum S-Typ beiderseits von Bereichen positiven differentiellen Widerstandes begrenzt wird, gewisse Anwendungsmöglichkeiten erwartet werden können, wenngleich die große Trägheit beachtet werden muß.

Wir danken Herrn Dr. E. Gutsche für wertvolle Diskussionen und eine kritische Durchsicht des Manuskripts sowie Herrn Dr. U. Kümmel für zahlreiche Hinweise.

Literatur

[1] B. K. Ridley, Proc. Phys. Soc. **82**, 954 (1963).
[2] L. Esaki, Phys. Rev. **109**, 603 (1958).
A. K. Jonscher, Brit. J. appl. Phys. **12**, 654 (1961).
E. O. Kane, J. appl. Phys. **32**, 83 (1961).

[3] A. BARRAUD, C. R. Acad. Sci. (France) **256**, 3632 (1963).
B. K. RIDLEY und T. B. WATKINS, J. Phys. Chem. Solids **22**, 155 (1961).
U. L. BONCH-BRUEVICH, Fiz. tverd. Tela **6**, 2047 (1964).
[4] J. ADAWI, J. appl. Phys. **32**, 1101 (1961).
E. ERLBACH, Phys. Rev. **132**, 1976 (1963).
B. K. RIDLEY und T. B. WATKINS, Proc. Phys. Soc. **78**, 293 (1961).
[5] T. W. HICKMOTT, J. appl. Phys. **33**, 2669 (1962).
[6] K. W. BÖER, J. Phys. Chem. Solids **22**, 123 (1961).
K. W. BÖER und W. E. WILHELM, phys. stat. sol. **4**, 237 (1964); **3**, 1718 (1963).
[7] K. W. BÖER, U. KÜMMEL und R. ROMPE, Z. phys. Chem. **200**, 180 (1952).
K. W. BÖER und U. KÜMMEL, Ann. Phys. (Germany) **14**, 341 (1954).
K. W. BÖER und U. KÜMMEL, Z. angew. Phys. **12**, 241 (1960); phys. stat. sol. **1**. 730 (1961).
[8] M. A. LAMBERT, J. Phys. Chem. Solids **22**, 189 (1961).
M. A. LAMBERT, Proc. IRE **50**, 1781 (1963).
[9] A. L. MCWHORTER und R. H. REDICKER, Proc. Int. Conf. on Semicond. Physics, Prag 1960 (S. 134).
J. YAMASHITA, J. Phys. Soc. Japan **16**, 720 (1961).
[10] W. FRANZ in Handbuch der Physik, Bd. **17**, Springer-Verlag, Berlin/Göttingen/Heidelberg 1956 (S. 166ff).
[11] K. W. BÖER, E. JAHNE und E. NEBAUER, phys. stat. sol. **1**, 231 (1961).
[12] K. W. BÖER, H.-J. HÄNSCH, U. KÜMMEL, H. LANGE und E. NEBAUER, phys. stat. sol. **1**, 169 (1961).
[13] W. MUSCHEID, Ann. Phys. (Germany) **13**, 305 (1953).
R. H. BUBE und E. L. LIND, Phys. Rev. **110**, 1040 (1958);**115**, 1159 (1959); **105**, 1711 (1957).
R. H. BUBE, J. Phys. Chem. Solids **1**, 234 (1957).
R. H. BUBE und L. A. BARTON, J. chem. Phys. **21**, 128 (1958).
R. H. BUBE und A. B. DREEBEN, Phys. Rev. **115**, 1578 (1959).
[14] O. MADELUNG in Handbuch der Physik, Bd. 20, Springer-Verlag, Berlin/Göttingen/Heidelberg 1957.
O. MADELUNG und H. WEISS, Z. Naturf. **9a**, 527 (1954).
[15] G. L. PEARSON und J. BARDEEN, Phys. Rev. **75**, 865 (1949).
[16] E. M. CONWELL, Proc. Inst. Radio Engrs. **40**, 1327 (1952).
[17] B. I. BOLTAKS, Diffusion in Halbleitern, FIZMATGIZ, Moskau 1961 (in russ.).
[18] L. COLLATZ, Numerische Behandlung von Differentialgleichungen, Springer-Verlag, Berlin/Göttingen/Heidelberg 1956.
[19] S. I. SHAKHOVTSOVA, Fiz. tverd. Tela **6**, 2541 (1964).

(Received January 6, 1965)

phys. stat. sol. 8, 897 (1965)

Institute of Physics, Bulgarian Academy of Sciences, Sofia

Photovoltaic Effect in CdS-CdSe Heterojunctions

By

B. KANDILAROV and R. ANDREYTCHIN

CdS—CdSe (n^+—n and n—n^+) heterojunctions are made and the photovoltaic effect investigated. Both the sign and the magnitude of the photovoltage measured depends on the doping. The spectral dependence of the short-circuit photocurrent is measured. The photovoltage is found to depend on the growth conditions.

Es wurden n^+—n und n—n^+ CdS–CdSe-Heteroübergänge hergestellt und hinsichtlich ihrer photoelektrischen Eigenschaften untersucht. Das Vorzeichen und die Größe der Photospannung hängen von der Dotierung ab. Ferner wurde die spektrale Verteilung des Kurzschluß-Photostromes gemessen. Es wurde gefunden, daß die Photospannung von den Wachstumsbedingungen abhängt.

1. Introduction

Recently a great number of studies have appeared which deal with semiconductor devices produced by the epitaxial deposition of one semiconductor on another. Until now such semiconducting structures have been produced by means of $A_{III}B_V$ semiconducting compounds and the elements Ge and Si (Ge–GaAs, Ge–Si, GaP–GaAs) [1 to 6] as well as by means of some other materials (Ge–CaF_2, ZnTe–CdS) [7, 8]. A few papers concerning the two-layer heterojunction have appeared too. Most of these works are concerned with the rectifying action of such structures and only quite few of them deal with the photovoltaic effect in the heterojunction [9 to 12]. However it is of certain interest to notice that it is possible to explain the spectral sensitivity of some types of commercial Se-photocells in terms of a p–n heterojunction photo-voltaic effect (Se–CdO, Se–CdSe) [13 to 15].

As it is well known, the number of semiconductors which allow the preparation of an epitaxial heterojunction is strongly limited. But, as a matter of fact, a heterojunction may exist not only between two epitaxially bounded crystals [16].

Therefore it is interesting to study the epitaxial deposition of one $A_{II}B_{VI}$ semiconducting compound on another as well as the photovoltaic effect in such n–n heterojunctions. We chose CdS and CdSe because these substances have the same crystal-structure with approximately equal lattice-constants (the distance between the nearest neighbours is 2.52 and 1.62 Å for CdS and CdSe, respectively). Moreover, on account of a recent tendency to relate semiconducting properties to short-range order we turned our attention to heterojunctions between crystals as well as between layers.

As it is well known, layers may have either a single-crystalline or a polycrystalline and/or amorphous structure. In the last case no epitaxy, in the true sense of this word, can exist. Although, as it is shown by our experiments, a useful photoheterojunction between two layers can be produced only if the first layer is kept at high temperatures during the deposition of the second. Only layers grown under certain conditions show both a typical n–n heterojunction rectification and a more or less perfect structure. This fact leads to the suggestion of a quasi-epitaxial growth mechanism, as a short-range order analogy of epitaxy.

This paper is the first one with which we are beginning to publish our results some of which we have already reported in brief [17, 18].

2. Sample Preparation

We have grown both a CdS film on a thin CdSe single-crystal plate and a CdSe film on a CdS single-crystal plate, making use of the vacuum evaporation method. We used the vacuum evaporation method to grow the films on an appropriately heated single crystal substrate since the technology developed by Marinace and other analogous methods require the establishment of relatively high temperatures for a relatively long time which leads to an enhancement of diffusion processes.

As one might expect, the n–n heterojunction photo-cells, fabricated as mentioned above, show a well expressed rectification action due to the heterojunction barriers. Nevertheless, the photovoltaic effect in such photoelements was relatively small (about 10 to 20 mV). The influence of a variety of surface treatments of the substrate surface, done just before the second film deposition on the photovoltage obtained from the samples has shown that in this case the photovoltaic effect in the heterojunction was quenched by interface levels caused by adsorbed atoms and other surface defects.

As it was impossible to grow a large single crystal in vacuum and immediately afterwards to grow epitaxially the second crystal on the first one in the same vacuum, we investigated layers evaporated onto each other. The layers, as it is well known, show, under certain conditions, properties similar to those of single crystals [16].

Moreover, experiments with heterojunctions between two layers enable one to verify the supposition mentioned above, that the absence of adsorption levels at the interface is a most important factor for the preparation of heterojunctions.

Our experiments have shown that the photovoltage obtained from such a two-layer heterojunction is several times larger than that obtained from some single-crystal heterojunctions.

The structure of the photoelements is shown schematically in Fig. 1. CdS layers deposited on CdSe substrates are often fragile. There is no good electrical contact between the two layers and such structures are not appropriate for photo-heterojunctions, and this is true for evaporation on a "cold" (room temperature) CdSe substrate as well as a substrate heated up to 200 °C. Therefore we have used the CdS-layer (evaporated on an appropriately heated glass substrate) as a substrate on which the CdSe film was evaporated.

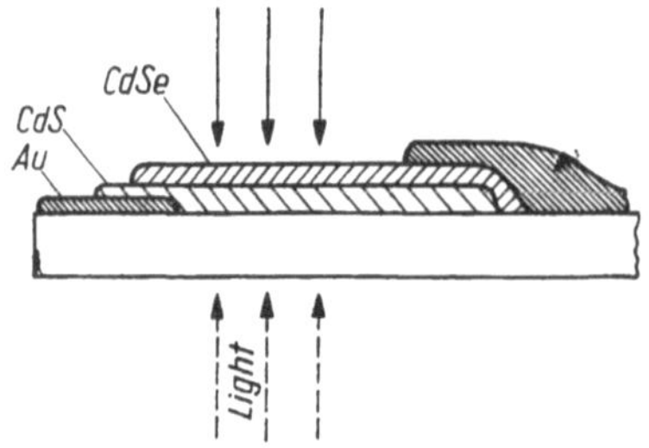

Fig. 1. Schematic view of the construction of a CdS-CdSe photoelement

A perfect CdSe layer having a thickness of 1 to 3 μm and an intimate mechanical and electrical contact with the CdS substrate, may be obtained only if the CdSe is evaporated at an appropriate evaporation rate (20 to 25 mg per min) on a CdS substrate heated up to a suitable temperature (about 160 °C).

It is of some interest to notice that we have obtained the best results when the evaporation rate was not constant during the evaporation. At the beginning the evaporation rate was low and after a fixed period of time (about 3 min) we increased it. This can be explained if we take into account that the formation energies of a CdSe nucleus on a "foreign" (CdS) and a "native" substrate (CdSe) are different [18, 19].

The experiments have shown also that the optimum evaporation rate depends both on the vacuum in the evaporation vessel and on the substrate temperature. It is not easy to explain the relation between these three factors: vacuum, substrate temperature, and evaporation rate. Further experimental as well as theoretical work is required before the answer to this question can be given.

3. Photovoltaic Effect

The energy-band diagram for a CdS–CdSe heterojunction is given in Fig. 2. The various variants (Fig. 2a, b) correspond to differently doped CdS and CdSe. The influence of the interface levels on the band bending is also shown (Fig. 2c, d).

It is well known that abrupt epitaxial heterojunctions between two semiconducting crystals can exist only if the lattice mismatch is smaller than 7 percent. But if the lattice mismatch is larger than 0.05 percent, local levels do occur at the interface [20]. Taking into account the lattice mismatch for the materials under investigation it is clear that local levels could exist even at the best epitaxial contact between CdS and CdSe.

Obviously, the presence of these interface levels may alter the band profile. The main result is their influence on the discontinuity of the band edges as well as on the width of the transition region. Such an energy band diagram, with a certain limitation, can be used in the discussion of the photoheterojunction

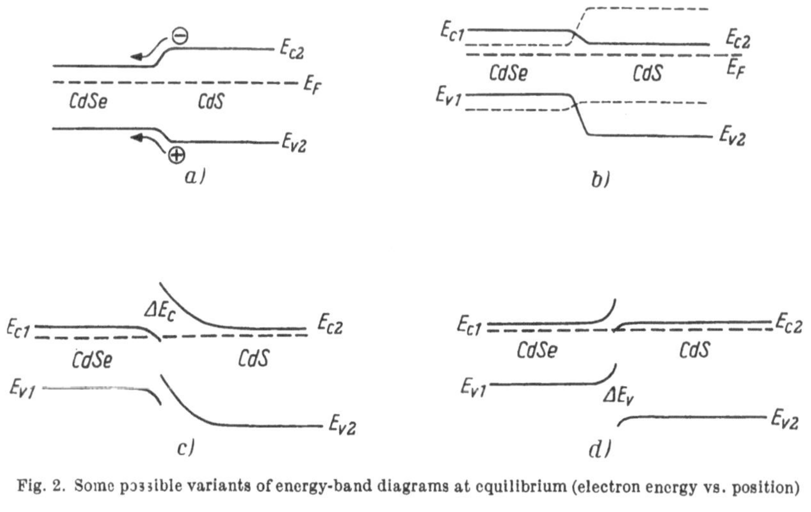

Fig. 2. Some possible variants of energy-band diagrams at equilibrium (electron energy vs. position)

between two layers. (Recently Muller and Zuleeg have given a successful interpretation of the rectification in two-layer heterojunctions using analogous diagrams [16].

In general a photovoltaic effect at a heterojunction may occur in two variants: (i) The electron-hole pairs, optically generated in the vicinity of the junction, are separated by the heterojunction; and (ii) the interface barriers, which are different for electrons and holes do not separate them, but the difference in the mobilities results in a kind of photodiffusion effect.

From the energy-band diagram it is easy to determine the sign and magnitude of the photovoltage which should be expected in each of the cases under consideration.

In our experiments a photoelement, prepared by vacuum evaporation as mentioned above, shows the best results when the CdSe is deposited on a CdS substrate and each of the layers is about 2 to 3 μm thick. The photovoltage obtained from such a photoelement reaches values up to 300 mV, and its sign is independent of the side of illumination (CdS or CdSe). For samples prepared under given conditions the cadmium sulfide electrode remains positive and the only difference is that the magnitude of the photovoltage measured in the second case is reduced.

All this excludes the possibility to classify the observed photoeffect as a Dember-effect. Taking into account that both CdS and CdSe are n-type semiconductors, the only explanation of the photoeffect under consideration is to classify it as a n–n heterojunction photo-effect.

Consider a heterojunction between pure CdSe and highly doped CdS. One might expect that the photovoltage reverses sign as the heterojunction changes

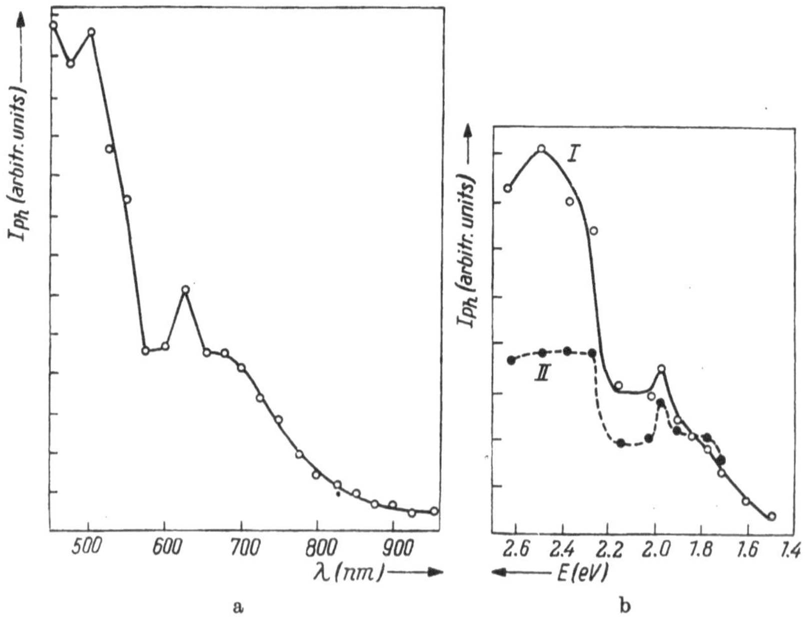

Fig. 3. Spectral dependence of short-circuit photocurrent (per incident energy) for two samples. On part (b) curves I and II present the data obtained by illumination from the CdS and CdSe side, respectively

from n^+–n type to n–n^+ type. In order to give a further check of the picture previously mentioned, we have prepared several photoelements using suitably doped materials. In all these cases the electrode on the cadmium selenide was positive and the magnitude of the photovoltage obtained was reduced.

It was of some interest to examine the spectral sensitivity of the photocurrent. The spectral sensitivity curves are shown on Fig. 3. It should be noticed that the short-circuit photocurrent curve has a shape similar to that established for some other heterojunctions. Two maxima, one at the CdS absorption edge and the other in the vicinity of the CdSe absorption edge, are clearly shown.

A more detailed study of the spectral dependence of the photocurrent in various samples, produced under different conditions, has shown a drastic dependence of the magnitude of the short-wave length maximum on the conditions under which the CdS substrate is deposited on the glass (substrate temperature as well as evaporation rate). Both the magnitude and the shape of the short-wavelength maximum depend on the structure of the glass substrate, too.

A paper concerning a more detailed study of the influence of the structure of the glass surface as well as that of the interface on the photovoltaic effect is in preparation.

Acknowledgement

The authors are greatly indebted to Mrs. J. Stanislavova for her helpful assistance in the sample preparation as well as in some of the measurements.

References

[1] R. P. Ruth, J. C. Marinace, and W. C. Dunlap, J. appl. Phys. **31**, 995 (1960).
[2] R. L. Anderson, IBM J. Res. Developm. **4**, 283 (1960).
[3] R. L. Anderson, Solid State Electronics **5**, 341 (1962).
[4] V. I. Ivanov-Omsky, N. K. Kiseleva, and B. T. Kolomiets, Fiz. tverd. Tela **3**, 1621 (1961).
[5] W. G. Oldham and A. G. Milnes, Solid State Electronics **6**, 121 (1963).
[6] L. J. Van Ruyven and W. Dekker, Physica **28**, 307 (1962).
[7] B. W. Sloope and C. O. Tiller, J. appl. Phys. **33**, 3458 (1962).
[8] M. Aven and W. Garwacki, J. Electrochem. Soc. **110**, 401 (1963).
[9] D. A. Cusano, Solid State Electronics **6**, 217 (1963).
[10] L. J. Van Ruyven, phys. stat. sol. **5**, K109 (1964).
[11] A. Lopez and R. L. Anderson, Solid State Electronics **7**, 695 (1964).
[12] Zh. Alferov, N. S. Zimogorova, A. M. Samolyanov, and M. K. Trukan, Ukr. fiz. Zh. **9**, 659 (1964).
[13] J. S. Preston, Proc. Roy. Soc. **202**, 449 (1950).
[14] H.-G. Sanner, Ann. Phys. (Germany) **7**, 416 (1950).
[15] T. K. Lakshmanan, Proc. IRE **48**, 1646 (1960).
[16] R. S. Muller and R. Zuleeg, J. appl. Phys. **35**, 1550 (1964).
[17] B. Kandilarov, Report Int. Symposium on Phys. Prop. of Semic. Comp., Warsaw, June 1963.
[18] B. Kandilarov, J. Stanislavova and R. Andreytchin, Rep. Third Nation. Conf. Phys., Sofia, May 1964 (p. 12).
[19] R. Kaischew, Bull. Acad. Bulg. Sci.-Série Phys. **1**, 100 (1951).
[20] W. G. Oldham and A. G. Milnes, Solid State Electronics **7**, 153 (1964).

(Received January 7, 1965)

phys. stat. sol. 8, 903 (1965)

IV. Physikalisches Institut der Humboldt-Universität zu Berlin
und Physikalisch-Technisches Institut der Deutschen Akademie der Wissenschaften zu Berlin, Bereich elektrischer Durchschlag

Zur Intensitäts- und Temperaturabhängigkeit des Photostromes von CdS-Einkristallen

Von
L. Elstner[1]) und J. Voigt

Die Photoleitfähigkeit von Kadmiumsulfideinkristallen wird als Funktion der Anregungsintensität und Temperatur experimentell untersucht. Die Meßergebnisse werden mit reaktionskinetischen Modellrechnungen verglichen. Es zeigt sich, daß selbst sehr allgemeine Modelle mit direktem Anregungsprozeß [1] keine vollständige Beschreibung der Meßergebnisse liefern. Gute Übereinstimmung zwischen Experiment und Theorie erhält man mit einem Modell mit sowohl direktem als auch indirektem [2] Anregungsprozeß der Leitungselektronen.

The photoconductivity of CdS single crystals is investigated experimentally as a function of light intensity and temperature. The results are compared with calculations based on reaction kinetic models. It is shown that even very general models involving a direct generation process [1] cannot give a complete explanation of the results. Good agreement between theory and experiment is obtained with a model involving direct and indirect [2] generation processes of conduction electrons.

1. Einleitung

In zwei vorangegangenen Arbeiten [1, 2] wurde von einem der Autoren die Intensitäts- und Temperaturabhängigkeit der Leitungselektronenkonzentration von photoleitenden Phosphoren in zwei sehr unterschiedlichen reaktionskinetischen Modellen ausführlich untersucht. Bei dem in [1] diskutierten Modell, das eine Erweiterung des Schönschen Minimalmodells [3] auf beliebig viele diskrete Hafttermgruppen bzw. auf Hafttermverteilungen (Zentren der Klasse 1 im Roseschen Modell [4]) darstellt, wurde angenommen, daß die Photoleitfähigkeitseffekte allein durch den Rekombinationsprozeß der freien Ladungsträger, speziell die Majoritätsträgerlebensdauer bestimmt sind. Als Anregungsprozeß wurde eine direkte Anregung von Elektronen ins Leitungsband mit konstanter Quantenausbeute betrachtet. Im Gegensatz hierzu wurde bei dem in [2] diskutierten Modell vorausgesetzt, daß die Photoleitfähigkeitseffekte bei konstanter Majoritätsträgerlebensdauer durch einen indirekten Anregungsprozeß der Ladungsträger, d. h. durch eine variable Quantenausbeute bestimmt werden.

Obwohl schon — im Zusammenhang mit anderen Untersuchungen — Messungen zur Intensitäts- und Temperaturabhängigkeit des Photostromes von Kadmiumsulfid-Einkristallen vorgenommen wurden, war es — im Zusammenhang mit den oben erwähnten Untersuchungen — wichtig, diese Abhängigkeiten nochmals systematisch an dieser „Modellsubstanz" für photoleitende Phosphore zu messen,

[1]) Jetzt VEB Halbleiterwerk Frankfurt/O, Werkteil Stansdorf.

um durch Vergleich zwischen Theorie und Experiment Aussagen über die Anwendbarkeit beider Modelle, im besonderen über die Rolle indirekter Anregungsprozesse im CdS, zu erhalten. Beide Modelle sind für einen quantitativen Vergleich zwischen Theorie und Experiment gut geeignet, da beide den im Experiment im allgemeinen vorliegenden Fall von Störtermverteilungen zu beschreiben gestatten.

2. Experimenteller Aufbau

Zur Messung wurden aus der Dampfphase gezüchtete Einkristalle in Plättchenform ohne bewußte Dotierung sowie solche mit Zusätzen von Cu, Co und Cl benutzt. Die Dotierung erfolgte durch Zugabe von $CuCl_2$ bzw. CoS zum hochgereinigten [5] CdS-Pulver. Der Dotierungsnachweis wurde mit Hilfe der Spektralanalyse durchgeführt. Eine Abschätzung ergab Fremdatomkonzentrationen der Größenordnung 10^{-4} Gew.-%. Mit Sicherheit war die Dotierungskonzentration aber nicht größer als 10^{-3} Gew.-%.

Die Kontaktierung der Proben erfolgte nach einer Vorbehandlung an den Kontaktflächen in einer Glimmentladung durch Vakuumaufdampfung von Indium oder Gold in Form von Spaltzellen. Der Elektrodenabstand betrug 1 bis 2 mm.

Das Schema der Meßanordnung ist in Fig. 1 dargestellt. Während der Messungen befanden sich die Proben in einem Glasrezipienten im Hochvakuum unter einem Druck $p < 10^{-5}$ Torr. Sie wurden mit Bronzefedern auf eine Kupferplatte aufgedrückt, wobei sie durch dünne Glimmerscheibchen gegen die Unterlage isoliert waren. Die Bronzefedern, an deren unterem Ende Goldspitzen angelötet waren, dienten gleichzeitig als Stromzuführungen.

Die Temperatur der Kupferplatte konnte zwischen der Temperatur der flüssigen Luft und 300 °C variiert werden. Die Temperaturmessung erfolgte mit einem dicht neben dem Kristall auf die Glimmerplatte aufgedrücktem Kupfer-Konstantan-

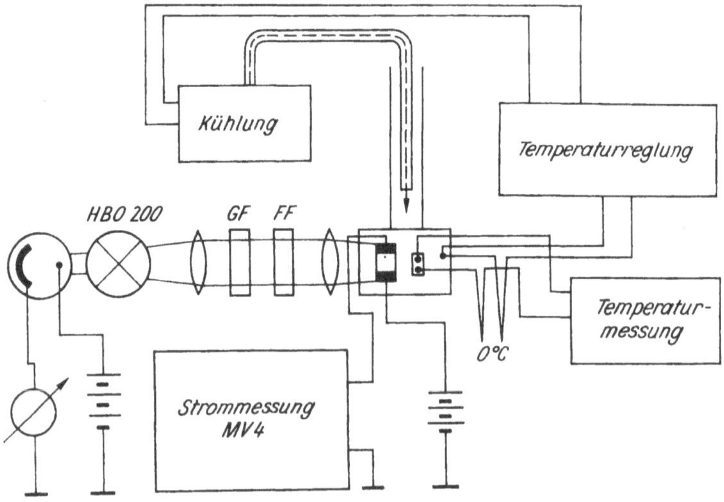

Fig. 1. Schema der Meßanordnung

Thermoelement. Ein zweites, an der Kupferplatte angelötetes Thermoelement befand sich in einem Regelkreis zur Einstellung einer konstanten Temperatur unterhalb von Zimmertemperatur. Am Nullinstrument dieses Thermoelements war neben der Nullmarke ein CdS-Photowiderstand angebracht. Überstrich der Lichtzeiger des Instruments infolge Erwärmung der Unterlage diesen Photowiderstand, so wurde mit dem fließenden Photostrom ein Thyratron gezündet, das seinerseits ein Relais betätigte, durch welches die Spannung für eine in einem Dewargefäß mit flüssiger Luft angebrachte Heizpatrone angeschaltet wurde. Damit strömte kalte Luft gegen den Kristallhalter und verhinderte seine weitere Erwärmung. Beim damit verbundenen Wegwandern des Lichtzeigers vom Photowiderstand wurde die Spannung an der Heizpatrone wieder abgeschaltet. Mit dieser Vorrichtung konnten die Temperaturschwankungen des am Kupferblock angelöteten Thermoelements in den Grenzen von $\pm 1°$ gehalten werden. Die Temperaturschwankungen des aufgedrückten Thermoelements, also auch des Kristalls, waren geringer. Die Temperaturwerte oberhalb Zimmertemperatur wurden durch direktes Heizen des Kupferblocks mit einer Heizpatrone erreicht.

Zur Messung wurde an die Proben eine Saugspannung von etwa 6 V angelegt. Der Photostrom konnte wahlweise mit einem Gleichstrommeßverstärker oder mit einem Mikroamperemeter gemessen werden. Der erfaßte Strombereich reichte von 10^{-12} bis 10^{-4} A.

Die Belichtung der Proben erfolgte mit einer HBO 200 im Gebiet des Photoleitungsmaximums. Die Intensität der Lampe wurde mit einer Photozelle kontrolliert. Die verschiedenen Belichtungsintensitäten wurden mit einem Satz geeichter Graufilter eingestellt.

3. Durchführung der Messungen und Meßergebnisse

Vor Beginn der eigentlichen Messungen wurden die Kontakte der Proben durch Strom-Spannungsmessungen (bei sehr geringer und sehr hoher Anregungsintensität) auf Sperrschichtfreiheit geprüft. Bei ohmscher Charakteristik wurden sodann bei etwa 5 Temperaturen unterhalb Zimmertemperatur Photostrom-Beleuchtungscharakteristiken aufgenommen. Oberhalb Zimmertemperatur wurde die Reproduzierbarkeit der Meßkurven besonders geprüft, um sicher zu sein, daß die in diesem Temperaturbereich am CdS häufig beobachteten photochemischen Reaktionen [6] die Meßergebnisse nicht verfälschen. Die Meßtemperatur wurde bis zur Einsatztemperatur photochemischer Reaktionen schrittweise erhöht.

Im Temperaturbereich unterhalb Zimmertemperatur war bei den sechzehn untersuchten Proben unabhängig von ihrer Dotierung (Cu, Co, Cl) der Photostrom nahezu unabhängig von der Temperatur. Er nahm entweder überhaupt nicht oder nur schwach mit wachsender Temperatur ab.

Im Temperaturbereich oberhalb Zimmertemperatur traten bei sieben der sechzehn Proben unterhalb +130 °C keine photochemischen Reaktionen auf. Bei allen übrigen lag deren Einsatztemperatur bereits in diesem Bereich.

Drei der sieben Proben zeigten oberhalb +70 °C eine temperaturabhängige Superlinearität. In Fig. 2 sind die Photostrom-Beleuchtungscharakteristiken bei verschiedenen Temperaturen für eine von diesen Proben (V 523 (1)) angegeben. Die Diskussion der vorliegenden Arbeit wird sich im wesentlichen auf Proben mit derartigen Charakteristiken beziehen.

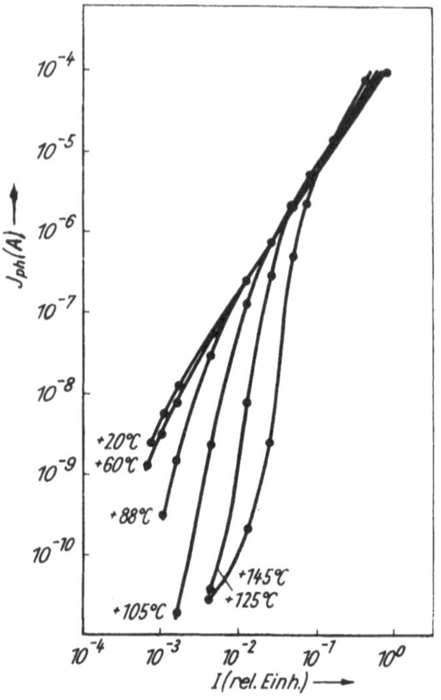

Fig. 2. Intensitätsabhängigkeit des Photostromes bei verschiedenen Temperaturen für die Probe V 523 (1)

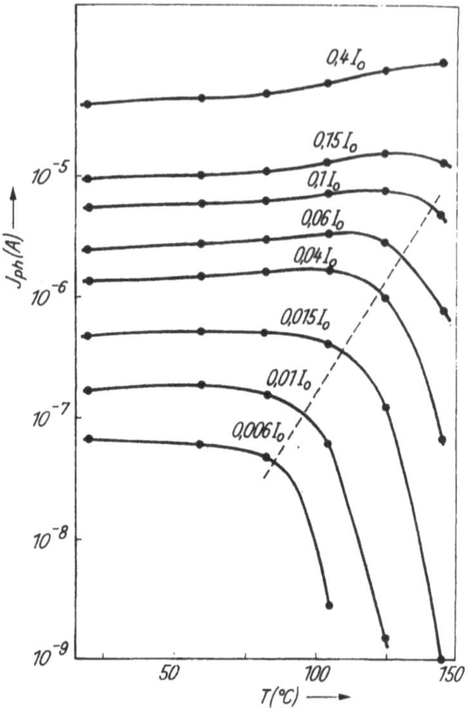

Fig. 3. Temperaturabhängigkeit des Photostromes bei verschiedenen Intensitäten für die Probe V 523 (1)

Fig. 3 zeigt die in Fig. 2 dargestellten Kurven in der Darstellung $J_{ph} = f(T)$ mit der Anregungsintensität als Scharparameter. Aus beiden Darstellungen läßt sich der energetische Abstand des Zentrums der Klasse 2 (Aktivators) vom Valenzband ΔE_A sowie das Verhältnis der Einfangquerschnitte dieses Terms für Löcher und Elektronen S_p/S_n ermitteln. Einmal wird dazu die in Fig. 2 ersichtliche Verschiebung des superlinearen Bereiches bei höheren Temperaturen zu höheren

Tabelle 1

	V 523 (1)	V 523 (3)	Sp 39: Co (3)
ΔE_A nach Methode 1	1,16 eV	0,86 eV	0,86 eV
ΔE_A nach Methode 2	1,11 eV	0,99 eV	—
S_p/S_n nach Methode 1	$5{,}6 \cdot 10^8$	$8{,}7 \cdot 10^5$	$1 \cdot 10^6$
S_p/S_n nach Methode 2	$6{,}3 \cdot 10^8$	$1 \cdot 10^7$	—
q_m	8,6	5,3	4,4

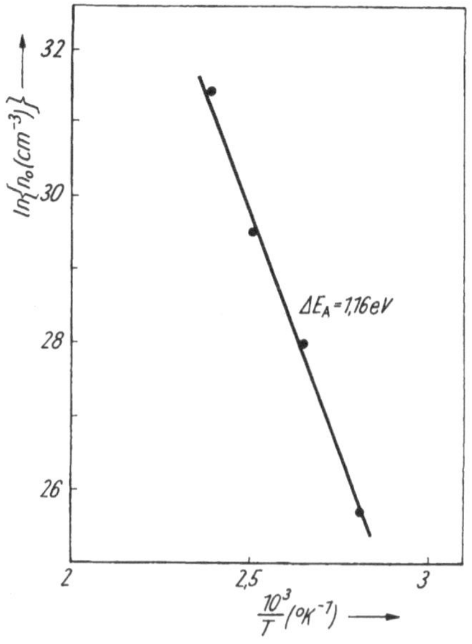

Fig. 4. Abhängigkeit der Elektronenkonzentration am „Endpunkt" des superlinearen Bereiches von der reziproken Temperatur für die Probe V 523 (1)

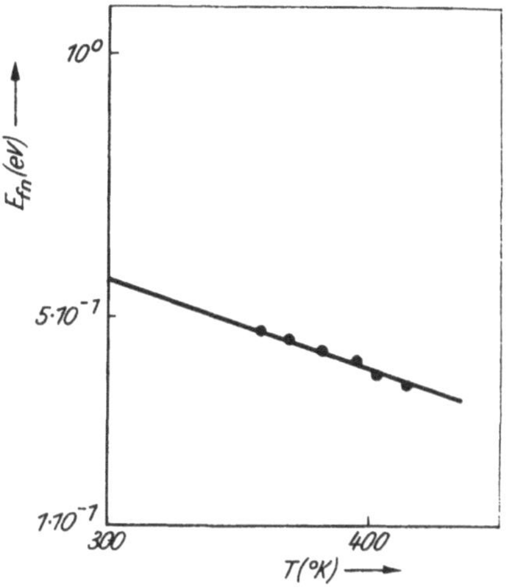

Fig. 5. Abhängigkeit der Quasifermigrenze für Elektronen von der Tilgungstemperatur für die Probe V 523 (1)

Anregungsintensitäten ausgenutzt [1, 7] (Methode 1). Der Logarithmus der Elektronenkonzentration am Endpunkt des temperaturabhängigen superlinearen Bereiches n_0 wird über der reziproken Temperatur aufgetragen (s. Fig. 4). Der Anstieg der sich ergebenden Geraden liefert ΔE_A, aus dem Ordinatenabschnitt $(1/T = 0)$ läßt sich S_p/S_n bestimmen.

Die andere Methode (Methode 2) [4] besteht darin, in Fig. 3 Tilgungstemperaturen festzulegen (gestrichelte Linie) und den Abstand der Quasifermigrenze für Elektronen vom Leitungsband bei den verschiedenen Tilgungstemperaturen über diesen aufzutragen (s. Fig. 5). Aus dem Anstieg der sich ergebenden Geraden läßt sich S_p/S_n bestimmen, der Ordinatenabschnitt ergibt $(T = 0)$ ΔE_A.

Die nach beiden Verfahren bestimmten Werte für ΔE_A sowie S_p/S_n für die Probe V 523 (1) sowie die beiden restlichen Proben sind in Tabelle 1 zusammengestellt. In der letzten Reihe sind zusätzlich die maximalen superlinearen Anstiege q_m angegeben.

In Fig. 6 sind die Photostrom-Beleuchtungscharakteristiken für die Probe V 523 (1) bei -60 und $+20$ °C dargestellt. Die Charakteristiken für Zwischentemperaturen lagen zwischen beiden Kurven; sie wurden aus Gründen der Übersichtlichkeit nicht in Fig. 6 eingezeichnet.

In Fig. 7 sind schließlich die Photostrom-Beleuchtungscharakteristiken angegeben (V 524), die für Proben charakteristisch sind, die bei höheren Temperaturen keine ausgeprägte temperaturabhängige Superlinearität zeigten (4 der oben genannten 7 Proben, bei denen bis $+130$ °C keine photochemischen Reaktionen auftraten). Die Dunkelleitfähigkeit dieser Proben war gegenüber der Dunkelleitfähigkeit der Proben mit ausgeprägter temperaturabhängiger Superlinearität um einige Größenordnungen höher (etwa $10^{-8}\ \Omega^{-1}\ cm^{-1}$ gegenüber $10^{-13}\ \Omega^{-1}\ cm^{-1}$).

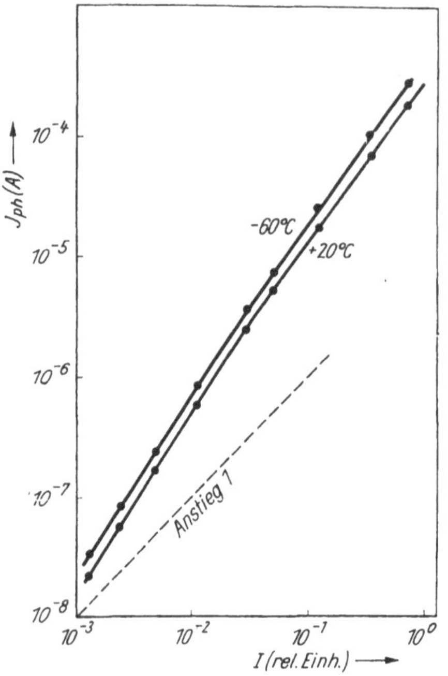

Fig. 6. Intensitätsabhängigkeit des Photostromes für die Probe V 523 (1) bei −60 und +20 °C

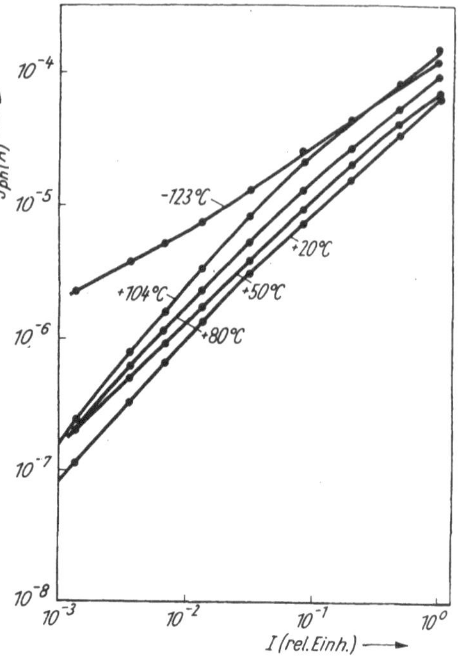

Fig. 7. Intensitätsabhängigkeit des Photostromes bei verschiedenen Temperaturen für die Probe V 524 Cu (7)

4. Diskussion der Ergebnisse

Die in Fig. 2 angegebenen Meßkurven (von V 523 (1)) zeigen deutlich zwei sehr unterschiedliche superlineare Bereiche des Photostromes. Im ersten Bereich (Bereich I), der nur auf einem relativ kleinen Intensitätsbereich beschränkt ist, sind die superlinearen Anstiege sehr groß (s. Tabelle 1). Als Funktion der Anregungsintensität besitzen sie einen glockenkurvenförmigen Verlauf (siehe besonders die Kurve bei $T = 145$ °C). Mit wachsender Temperatur findet eine Verschiebung des superlinearen Bereiches zu höheren Anregungsintensitäten statt. Im zweiten superlinearen Bereich (Bereich II), der sich bei hohen Elektronenkonzentrationen an den ersten anschließt und der über einen relativ großen Intensitätsbereich ausgedehnt ist, ist der superlineare Anstieg nahezu konstant und gegenüber den Anstiegen im Bereich I gering. Von der Temperatur ist der Bereich II nahezu unabhängig, so daß die zu verschiedenen Temperaturen gehörenden Kurven aus dem Bereich I in eine einheitliche, von der Temperatur unabhängige Grenzkurve einmünden. Besonders deutlich werden die genannten Eigenschaften des Bereiches II an den in Fig. 6 dargestellten Photostrom-Beleuchtungscharakteristiken (ebenfalls von V 523 (1)) bei tieferen Temperaturen, bei denen Bereich I im untersuchten Intensitätsbereich nicht mehr auftritt.

Da sich die in der Einleitung angegebenen reaktionskinetischen Untersuchungen auf die Intensitäts- und Temperaturabhängigkeit der Elektronenkonzentration beziehen, die angeführten Messungen aber auf die Intensitäts- und Temperaturabhängigkeit des Photostromes, sind die Aussagen der folgenden Diskussion nur bis auf die Intensitäts- und Temperaturabhängigkeit der Beweglichkeit der Leitungselektronen bestimmt.

Vergleicht man jedoch die in Bereich I auftretenden großen Änderungen des Photostromes in Abhängigkeit von der Temperatur bzw. Anregungsintensität mit den entsprechenden Änderungen der Beweglichkeit, wie sie an CdS bisher gefunden wurden [8, 9], so erkennt man, daß in diesem Bereich der Beweglichkeitseinfluß vernachlässigbar ist.

. Das trifft nicht mehr vollständig zu auf den Bereich II der Photostrom–Beleuchtungscharakteristik, da in diesem Bereich die Photostromänderungen mit der Temperatur bzw. Anregungsintensität wesentlich geringer sind. Bei einer quantitativen Diskussion der Meßkurven müßte daher der Beweglichkeitseinfluß berücksichtigt werden. Wichtig für die folgende Diskussion ist jedoch die Tatsache, daß der im Bereich II auftretende superlineare Anstieg von etwa 1,4 bei Zugrundelegung bisher bekannter experimenteller Werte für die Beweglichkeitsänderung mit der Anregungsintensität [8] nicht als Beweglichkeitseffekt erklärt werden kann. Für den günstigsten Fall der monomolekularen Rekombination ($n \sim I$) müßte in dem gemessenen Intensitätsbereich von drei Größenordnungen (Fig. 6) zur Deutung des superlinearen Anstiegs von 1,4 die Beweglichkeit um mehr als eine Größenordnung zunehmen. Die gemessenen Beweglichkeitsänderungen in einem entsprechenden Intensitätsbereich betrugen jedoch nur etwa einen Faktor 2.

Vergleicht man die in Fig. 2 angegebenen Meßkurven im ersten superlinearen Bereich mit Kurven, die auf Grundlage reaktionskinetischer Modelle mit direktem Anregungsprozeß berechnet wurden [1], so erkennt man eine gute Übereinstimmung zwischen Theorie und Experiment, wenn man den Rechnungen ein Modell mit einer diskreten Hafttermgruppe (Zentrum der Klasse 1) bzw. ein Modell mit einer Hafttermgruppe endlicher Breite, die jedoch nicht mehr mit dem Leitungsband in einem Quasigleichgewicht steht, zugrunde legt. Aus der Elektronenkonzentration am Anfangspunkt des superlinearen Bereiches (Fig. 2, Kurve bei $T = 145$ °C) läßt sich der energetische Abstand dieser Termgruppe vom Leitungsband ΔE_H abschätzen. Es ergibt sich ein Wert $\Delta E_H \approx 1{,}0$ eV.

Die nach zwei verschiedenen Verfahren (Fig. 4, Fig. 5) ermittelten Werte für ΔE_A stimmen für die Probe V 623 (1) gut untereinander und auch gut mit den Werten überein, die Bube [10] aus der Temperaturtilgung des Photostromes für CdS bestimmte. Dasselbe gilt für das Verhältnis der Einfangquerschnitte S_p/S_n. Die Übereinstimmung mit Bube besteht im letzten Fall auch darin, daß S_p/S_n bei undotierten Proben besonders groß ist.

Die gute Übereinstimmung zwischen Theorie und Experiment im ersten superlinearen Bereich läßt den Schluß zu, daß in diesem Bereich die Superlinearität des Photostromes mit der Anregungsintensität ein Effekt der Majoritätsträgerlebensdauer ist, die mit wachsender Anregungsintensität stark zunimmt. Diese Zunahme läßt sich gut im Roseschen bzw. Schön-Klasenschen Zweitermmodell verstehen.

Besonders übersichtliche Verhältnisse bekommt man bei Proben, bei denen die Zentren der Klasse 1 sehr tief (bzw. in einem sehr schmalen Energiebereich) liegen und die wenige sehr flache zusätzliche Hafttermgruppen besitzen, wie dies offensichtlich für die Probe V 523 (1) zutrifft. Die größeren Abweichungen der in der Tabelle 1 angegebenen Werte für ΔE_A und S_p/S_n für die Probe V 523 (3) sowie das Versagen der Methode 1 bei der Probe Sp 39: Co (3) sind wahrscheinlich auf das Wirksamwerden von Termverteilungen zurückzuführen. Hierfür spricht, daß die maximalen superlinearen Anstiege von der Probe V 523 (1) bis zur Probe Sp 39: Co (3) um die Hälfte abnehmen (siehe hierzu [1]).

Wollte man auch den zweiten superlinearen Bereich der in Fig. 2 dargestellten Meßkurvenschar durch ein Modell mit direktem Anregungsprozeß erklären, so würde sich nach [1] eine Erweiterung des schon diskutierten Modells mit einem diskreten Term der Klasse 1 um eine zusätzliche Termverteilung der Klasse 1 anbieten. Aus der Lage der Quasifermigrenze am Anfang und Ende des Bereiches II

(Fig. 2, Fig. 6, eine Beweglichkeit $\mu = 200$ cm²/Vs vorausgesetzt) könnte man entnehmen, daß diese Verteilung dann bei etwa 0,5 eV (unterhalb des Leitungsbandes) wirksam werden muß und bis etwa 0,3 eV das Rekombinationsgeschehen bestimmt. Aus dem superlinearen Anstieg ergäbe sich eine Abfallbreite der Verteilung zu $\delta E_{H_2} = 3\,k\,T$. Beide notwendigen Annahmen über die Termverteilung wären durchaus mit experimentellen Aussagen über das Hafttermspektrum von CdS verträglich, wie sie besonders aus Wechsellichtmessungen her bekannt sind [11].

Um zu prüfen, inwieweit das durch eine Termverteilung erweiterte Modell auch den Bereich II der in Fig. 2 dargestellten Meßkurvenschar zu beschreiben gestattet, wurde unter Zugrundelgung der aus der Meßkurvenschar bestimmten Parameter

$$\Delta E_A = 1{,}1 \text{ eV}, \qquad \Delta E_{H_1} = 1{,}0 \text{ eV}, \qquad \Delta E_{H_2\,\text{min}} = 0{,}3 \text{ eV},$$

$$\delta E_{H_2} = 0{,}08 \text{ eV}, \quad S_p/S_n = 10^8$$

sowie mit den weiteren Parametern [2])

$$H_1 = 10^{14} \text{ cm}^{-3}, \quad H_2 = 10^{16} \text{ cm}^{-3}, \quad \tilde{\beta} = \beta = 10^{-7} \text{ cm}^3 \text{ s}^{-1},$$

$$K \equiv \frac{\beta\,(H_1 + H_2)}{\tilde{\beta}\,A} = 10^{-1}$$

die Intensitätsabhängigkeit der Leitungselektronenkonzentration bei verschiedenen Temperaturen nach der in [1] angegebenen Gleichung (12) berechnet. Die Kurvenschar ist in der Fig. 8 dargestellt.

Man erkennt, daß im superlinearen Bereich I die berechnete Kurvenschar mit der gemessenen qualitativ gut übereinstimmt. Die Überschneidung der zu verschiedenen Temperaturen gehörenden Kurven der berechneten Schar bei kleinen Anregungsdichten ist allerdings aus der gemessenen Kurvenschar nur andeutungsweise zu entnehmen.

Hiergegen zeigt die berechnete Kurvenschar im superlinearen Bereich II ein vom Experiment völlig abweichendes Verhalten. Der Anstieg der Kurven ist sublinear. Grund hierfür ist, daß in diesem Bereich (obwohl – wie aus dem Experiment bestimmt – $S_p/S_n = 10^8$ gesetzt wurde!) die Rekombination nicht mehr über die Termverteilung der Klasse 1 H_2 verläuft, sondern über die Termgruppe der Klasse 2 A. Der sublineare konstante Anstieg wird durch das Auffüllen der nur noch als Hafttermgruppe wirkenden Termverteilung H_2 mit Elektronen verursacht.

Dieses Verhalten ist im wesentlichen durch den experimentell bestimmbaren Wert für S_p/S_n bestimmt. Aber selbst im ungünstigsten Grenzfall $S_p/S_n \to \infty$ erlaubt das betrachtete Modell keine Beschreibung des superlinearen Bereiches II. Das zeigt die Betrachtung der Temperaturabhängigkeit der Meßkurven in diesem Bereich. In Fig. 9 sind die in Fig. 8 gezeigten Kurven für den Grenzfall $S_p/S_n = \infty$ dargestellt. Man erkennt, daß sich in diesem Fall tatsächlich an den ersten superlinearen Bereich ein zweiter mit dem gewünschten konstanten superlinearen Anstieg anschließt. Jedoch verschiebt sich dieser superlineare Bereich,

[2]) β bedeutet den Übergangskoeffizienten für einen Elektronenübergang aus dem Leitungsband in die Termgruppen der Klasse 1 mit den Konzentrationen H_1 und H_2, $\tilde{\beta} = v_{\text{th}} \cdot S_p$ den Übergangskoeffizienten für einen Löcherübergang aus dem Valenzband in die Termgruppe der Klasse 2 mit der Konzentration A. Die Parameterwerte entsprechen den am CdS bei gleichartigen Proben im allgemeinen auftretenden Werten [11]. ($H_2 \gg H_1$ muß angenommen werden, damit die Termverteilung H_2 überhaupt wirksam wird.)

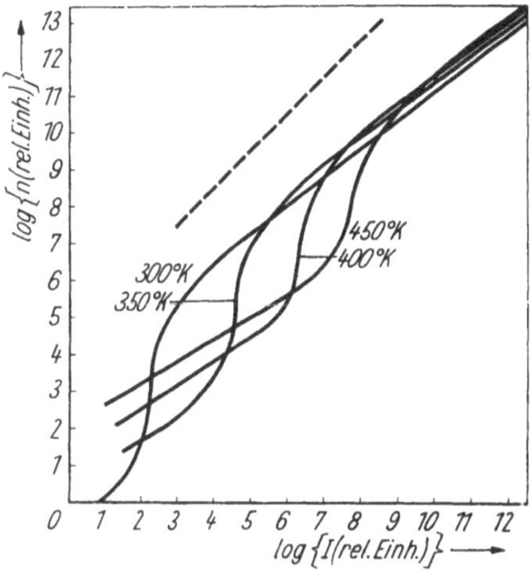

Fig. 8. Intensitätsabhängigkeit der Elektronenkonzentration bei verschiedenen Temperaturen, berechnet auf der Grundlage eines Modells mit direktem Anregungsprozeß. $S_p/S_n = 10^8$

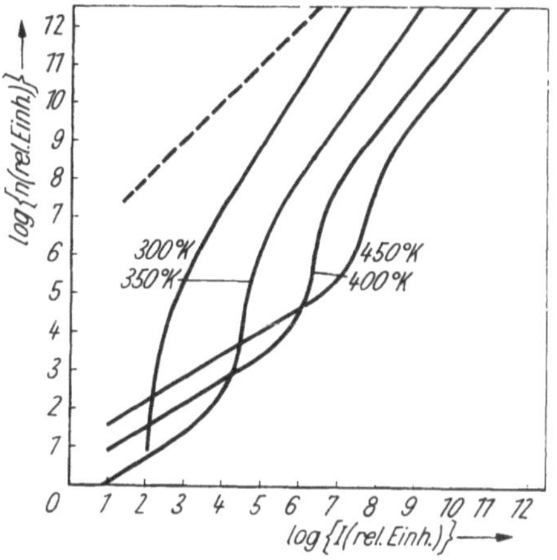

Fig. 9. Wie bei Fig. 8. $S_p/S_n = \infty$

ebenso wie der erste Bereich mit ausgeprägter Superlinearität, mit wachsender Temperatur zu höheren Anregungsdichten, im Gegensatz zu den experimentellen Kurven, die im zweiten superlinearen Bereich nahezu temperaturunabhängig sind.

Ein entsprechendes Verhalten ist zu erwarten, da der Mechanismus der Superlinearität des Photostromes in 2-Zentrenmodellen wesentlich von der Konzentration freier Löcher bestimmt wird, die mit wachsender Temperatur zunimmt. Das kommt auch in der zur Berechnung der Kurven benutzten Gleichung (12) [1] zum Ausdruck: Der dritte Summand, der die Rekombinationsprozesse über die Zentren der Klasse 1 (H-Terme) beschreibt, ist proportional $\exp(-\Delta E_A/k\,T)$. Somit kann das Modell mit direktem Anregungsprozeß, unabhängig von der speziellen Wahl der Parameter, niemals gleichzeitig die im Experiment gefundene Intensitäts- *und* Temperaturabhängigkeit des Photostromes im zweiten superlinearen Bereich erklären.

Eine andere Möglichkeit zur Deutung des beobachteten superlinearen Bereiches II in einem Modell mit direktem Anregungsprozeß wäre die von Hooge und Polder [12] in einem 2-Termmodell untersuchte sogenannte quadratische Superlinearität. Die Autoren betrachteten den in [1] nicht diskutierten Fall der Sättigung der Terme der Klasse 2 mit Löchern und erhielten einen superlinearen Zusammenhang zwischen der Elektronenkonzentration und Anregungsintensität mit dem Exponenten 2.

Abgesehen davon, daß der beobachtete superlineare Anstieg im superlinearen Bereich II (Fig. 2, Fig. 6) nicht 2, sondern nur etwa 1,4 beträgt und die Abweichungen beider Werte voneinander durch zusätzliche Effekte erklärt werden müßten, versagt die „quadratische Superlinearität" zur Deutung des beobachteten superlinearen Bereiches II aus folgenden beiden Gründen:

1. Die Bedingungen für das Auftreten des superlinearen Bereiches I („extreme Superlinearität" vgl. [12]) schließen die Möglichkeit aus, daß das Zentrum der Klasse 2 ein Akzeptor und das Zentrum der Klasse 1 ein Donator ist (soweit nicht zusätzliche Terme vorhanden sind). Diese Bedingungen sind erfüllt, wenn umgekehrt das Zentrum der Klasse 2 ein Donator und das Zentrum der Klasse 1 ein Akzeptor ist und zusätzlich die Konzentration der Zentren der Klasse 2 groß gegenüber der Konzentration der Zentren der Klasse 1 ist.

Das widerspricht aber den Bedingungen für das Auftreten des superlinearen Bereiches II („quadratische Superlinearität“ [12]), die bei der zuletzt vorausgesetzten Natur der Zentren nur dann erfüllt werden können, wenn die Konzentrationen beider Termgruppen nahezu gleich sind.

2. Die Elektronenkonzentration im Bereich der „quadratischen Superlinearität“ ist proportional exp $(-\Delta E_A/2\,k\,T)$. Der im Experiment beobachtete superlineare Bereich II ist dagegen nahezu temperaturunabhängig.

Da die betrachteten recht allgemeinen Modelle mit direktem Anregungsprozeß bei der Deutung des superlinearen Bereiches II offenbar versagen, und da man auch mit der Intensitätsabhängigkeit der Beweglichkeit nicht den beobachteten superlinearen Anstieg erklären kann, ist es naheliegend, das in [2] diskutierte Modell mit indirektem Anregungsprozeß zur Erklärung der Meßergebnisse heranzuziehen.

In Fig. 10 sind Kurven der Intensitätsabhängigkeit der Leitungselektronenkonzentration bei verschiedenen Temperaturen, berechnet auf Grundlage des Modells mit indirektem Anregungsprozeß ([2], Gleichung (10)), angegeben. Den Rechnungen wurde ein Modell mit einer zum Leitungsband hin exponentiell ansteigenden Störtermverteilung in der schon bei den obigen Rechnungen benutzten Form zugrunde gelegt. Über diese Verteilung sollen die indirekten Anregungsprozesse vor sich gehen, in der Form, daß die Zahl der ins Leitungsband angeregten Elektronen proportional dem Besetzungsgrad dieser Verteilung (sowie der Anregungsintensität) ist (s. [2]).

Als physikalische Interpretation für diesen Anregungsmechanismus wäre einmal ein Zweistufen-Anregungsprozeß denkbar [13]. Wahrscheinlicher ist jedoch — besonders nach den Untersuchungen von Gross und Mitarbeitern [14] über den Zusammenhang zwischen Exzitonen und Photoleitfähigkeit — ein photoleitungsaktiver Zerfall von Exzitonen an der mit Elektronen besetzten Störtermverteilung.

Von der aus den Meßkurven bestimmten Hafttermgruppe H_1 wurde angenommen, daß sie eine konstante Lebensdauer der Leitungselektronen τ_c verursacht.

Man erkennt, daß die berechneten Kurven das Verhalten der gemessenen Kurven im superlinearen Bereich II qualitativ richtig wiedergeben. Der superlineare Anstieg ist über den gesamten Intensitätsbereich konstant, die Kurven sind im Bereich höherer Anregungsintensitäten nur relativ schwach (verglichen mit der in Fig. 9 angegebenen Kurvenschar) von der Temperatur abhängig. Die auf-

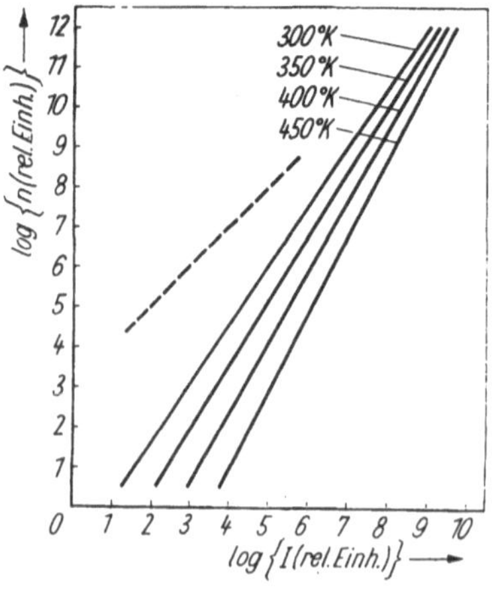

Fig. 10. Intensitätsabhängigkeit der Elektronenkonzentration bei verschiedenen Temperaturen, berechnet auf der Grundlage eines Modells mit indirektem Anregungsprozeß

tretende Temperaturverschiebung ist bei diesen Kurven nicht durch die Temperaturabhängigkeit der Löcherkonzentration, sondern durch die Temperaturabhängigkeit der effektiven Zustandsdichte N_c sowie des Quotienten $k\,T/\delta E$ bedingt.

Die Meßergebnisse in beiden superlinearen Bereichen lassen sich also am besten in einem Modell erklären, in dem sowohl direkte als auch indirekte Anregungsprozesse wirksam sind. Ein Mangel der bisherigen Überlegungen besteht jedoch noch darin, daß beide Anregungsprozesse getrennt voneinander behandelt wurden. Im folgenden soll daher ein einheitliches Modell mit sowohl direktem als auch indirektem Anregungsprozeß betrachtet werden, das ohne Zusatzannahmen, allein bei einer Intensitäts- bzw. Temperaturvariation, beide in Fig. 2 angegebenen superlinearen Bereiche zu beschreiben gestattet.

Die Anregungsrate a der insgesamt pro s und cm³ ins Leitungsband angeregten Elektronen wird additiv aus einer Rate a_1 der auf indirektem Wege [2] und aus einer Rate a_2 der auf direktem Wege angeregten Elektronen zusammengesetzt. Für die Anregungsraten a_1 und a_2 gilt, wenn man mit n_1 bzw. n_2 die durch a_1 bzw. a_2 angeregte Leitungselektronenkonzentration bezeichnet [1, 2]

$$\left.\begin{aligned} \frac{n_1}{\tau_c(n)} &= a_1 \equiv I \int_{E_c}^{E_v} q(E)\, f(E,\zeta)\, \overline{H}(E)\, dE\,, \\ \frac{n_2}{\tau_c(n)} &= a_2 \equiv \mu\, Q\, I\,, \qquad n = n_1 + n_2 \end{aligned}\right\} \tag{1}$$

(I Anregungsintensität; $q(E)$ Wirkungsquerschnitt einer in der energetischen Tiefe E gelegenen Störtermgruppe mit der Dichte $\overline{H}(E)$ für indirekte Anregungsprozesse; $f(E,\zeta)$ Fermifunktion mit der Quasifermigrenze ζ; E_c, E_v Energieniveaus der Leitungs- bzw. Valenzbandkante; μ Absorptionskonstante der Probe für die die direkten Anregungsprozesse erzeugende Strahlung; Q Quantenausbeute für die direkten Anregungsprozesse).

Die Lebensdauer der Leutungselektronen τ_c hängt, ebenso wie die Quasifermigrenze ζ, nur von der Gesamtelektronenkonzentration ab. $\tau_c(n)$ ist durch den gewählten Rekombinationsmechanismus der Leitungselektronen bestimmt.

Aus (1) erhält man

$$\left.\begin{aligned} n_1 &= \frac{n\, g(n)}{1 + g(n)}\,, \qquad n_2 = \frac{n}{1 + g(n)} \\ \text{mit} \qquad g(n) &= \frac{\int_{E_c}^{E_v} q(E)\, f(E,\zeta)\, \overline{H}(E)\, dE}{\mu\, Q}\,. \end{aligned}\right\} \tag{2}$$

$g(n)$ ist der Quotient von Quantenausbeute × Absorptionskonstante des indirekten und direkten Anregungsprozesses. Bei kleinen Anregungsdichten (kleinen Elektronenkonzentrationen) und somit kleinem Besetzungsgrad der Termgruppe $\overline{H}(E)$ wird $g(n) \ll 1$ sein und somit $n_2 \approx n$ gelten. Der indirekte Anregungsprozeß ist zu vernachlässigen. Mit wachsender Anregungsintensität und somit wachsendem Füllungsgrad der Termgruppe $\overline{H}(E)$ wird, soweit $\int_{E_c}^{E_v} q(E)\, \overline{H}(E)\, dE \gg \mu\, Q$ erfüllt ist, $n_1 \approx n$ gelten, d. h. der gesamte Anregungsprozeß im wesentlichen im indirekten Prozeß bestehen.

In Fig. 11 sind Kurven der Intensitätsabhängigkeit der Leitungselektronenkonzentration bei verschiedenen Temperaturen angegeben, die auf Grundlage der Beziehungen (1) berechnet wurden. Das Störtermspektrum ist identisch mit dem, das den in Fig. 8 bzw. Fig. 9 dargestellten Kurven zugrunde liegt und das zum Teil aus den in Fig. 2 angegebenen Meßkurven bestimmt wurde. Die Rekombinationsprozesse sollen über die Zentren der Klasse 1 (die diskrete Termgruppe H_1 und die exponentielle Verteilung H_2) sowie die Zentren der Klasse 2 (die diskrete Termgruppe A), die indirekten Anregungsprozesse nur über die exponentielle Verteilung H_2 verlaufen. $\tau_c(n)$ wurde, bei Zugrundelegung des angegebenen Rekombinationsmechanismus, wieder nach der Beziehung (12) [1] berechnet. Für die in (1) noch unbestimmte Größe $q/\mu Q$ $\left(q(E) = \text{const vorausgesetzt}\right)$ wurde ein Wert von 10^{-13} cm^3 gewählt.

Wie man Fig. 11 entnimmt, ist die Übereinstimmung der auf Grundlage des Modells mit direktem *und* indirektem Anregungsprozeß berechneten Kurvenschar mit der in Fig. 2 angegebenen gemessenen Kurvenschar qualitativ sehr gut. Bei kleinen Anregungsdichten erhält man den Bereich der ausgeprägten, stark von der Temperatur abhängigen Superlinearität; bei höheren Anregungsdichten münden alle Kurven in den Bereich der schwachen, nahezu temperaturunabhängigen Superlinearität.

Diese gute Übereinstimmung zwischen Theorie und Experiment läßt den Schluß zu, daß die am CdS beobachtete Superlinearität des Photostromes mit der Anregungsintensität verschiedene Ursachen besitzen kann. Die bei höheren Temperaturen (bei etwa +150 °C) auftretende stark temperaturabhängige Superlinearität ist gut als Effekt der Majoritätsträgerlebensdauer zu verstehen; die bei tieferen Temperaturen (zwischen etwa −150 °C und Zimmertemperatur) beobachtbare im allgemeinen nur wenig von der Temperatur abhängige Superlinearität läßt sich gut als Effekt der Quantenausbeute deuten. Weiterhin kann man schließen, daß die indirekten Anregungsprozesse im CdS mit im Energiebereich von etwa 0,5 bis 0,3 eV unterhalb des Leitungsbandes gelegenen Störtermverteilungen verbunden sind.

Diese Aussage ist in guter Übereinstimmung mit der Aussage der Arbeit von Dittfeld und einem der Autoren [15]. Durch Glowkurvenmessungen bei ver-

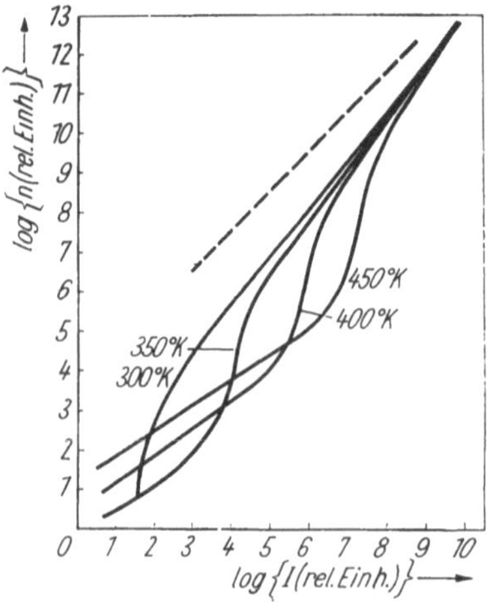

Fig. 11. Intensitätsabhängigkeit der Elektronenkonzentration bei verschiedenen Temperaturen, berechnet auf der Grundlage eines Modells mit sowohl direktem als auch indirektem Anregungsprozeß

schiedenen Heizgeschwindigkeiten zusammen mit Messungen der Intensitätsabhängigkeit des Photostromes konnte in [15] gezeigt werden, daß bei Variation der Quasifermigrenze in einem Energiebereich etwa 0,4 eV unterhalb des Leitungsbandes die Quantenausbeute von der Anregungsintensität abhängt.

Die in Fig. 7 angegebenen Photostrom-Beleuchtungscharakteristiken, die für Proben mit höherer Dunkelleitfähigkeit charakteristisch sind, lassen sich auch mit der oben diskutierten Modellvorstellung verstehen. Daß bei höheren Temperaturen kein Bereich mit ausgeprägter Superlinearität auftritt, erklärt sich leicht dadurch, daß bei diesen Proben, verursacht durch die hohe Donatorenkonzentration, auch bei höheren Temperaturen und geringen Lichtintensitäten die Zentren der Klasse 1 noch voll mit Elektronen besetzt sind. Daß kein Bereich entsprechend dem superlinearen Bereich II zu beobachten ist, läßt sich dadurch verstehen, daß zusätzliche flache Hafttermgruppen vorhanden sind, die nicht mit Exzitonen in Wechselwirkung stehen und die eine Abnahme der Lebensdauer der Leitungselektronen mit wachsender Anregungsintensität verursachen.

Die letzte Aussage läßt sich direkt durch die in Fig. 7 angegebenen Meßkurven bestätigen. Die Kurve für $T = -123$ °C zeigt bei kleinen Anregungsdichten zuerst einen Anstieg 0,5, bei höheren einen konstantenAnstieg von 0,7. Aus der Elektronenkonzentration am ,,Knickpunkt" errechnet man eine Lage der Quasifermigrenze zu ungefähr 0,15 eV unterhalb des Leitungsbandes. Von dieser Energie wird also eine zum Leitungsband hin exponentiell ansteigende Hafttermverteilung mit einer Abfallbreite von 2,3 $k\,T$ merklich.

Die Lage dieser aus Photoleitungsmessungen bestimmten Termgruppe stimmt recht gut mit der von Niekisch [11] aus Wechsellichtmessungen am CdS bestimmten Termgruppe bei 0,12 eV überein. Die Übereinstimmung mit Niekisch besteht auch darin, daß diese Termgruppe bei Proben mit Cd-Überschuß, also höherer Dunkelleitfähigkeit, besonders ausgeprägt war.

Herrn Dr. W. Borchardt und Herrn Dr. E. Gutsche danken wir für anregende Diskussionen und wertvolle Hinweise.

Literatur

[1] J. Voigt, phys. stat. sol. **5**, 123 (1964).
[2] J. Voigt, phys. stat. sol. **7**, 643 (1964).
[3] M. Schön, Abh. Dtsch. Akad. Wiss. Berlin **7**, 2 (1960); Halbleiterprobleme IV, Vieweg, 1958 (S. 302).
[4] A. Rose, Proc. IRE **43**, 1850 (1955); Phys. Rev. **97**, 322 (1955).
[5] K. W. Böer und K. Zimmermann, Monatsber. Dtsch. Akad. Wiss. **1**, 1 (1959).
[6] W. Borchardt, phys. stat. sol. **2**, 1575 (1962).
[7] R. H. Bube, J. Phys. Chem. Solids **1**, 234 (1957).
[8] R. H. Bube, H. E. MacDonald und J. Blanc, J. Phys. Chem. Solids **22**, 173 (1961).
[9] F. A. Kröger, H. J. Vink und J. Volger, Philips Res. Rep. **10**, 39 (1955).
W. W. Piper und R. E. Halsted, Proc. Internat. Conf. Semiconductor Physics, Prague 1960 (S. 1046).
M. Itakura und H. Toyoda, J. Phys. Soc. Japan **18**, 150 (1963).
K. Kobayashi, H. Fujita und T. Kawai, J. Phys. Soc. Japan **19**, 135 (1964).
[10] R. H. Bube, J. appl. Phys. **32**, 9 (1961).
[11] E. A. Niekisch, Z. phys. Chem. **217**, 110 (1961).
[12] F. N. Hooge und D. Polder, J. Phys. Chem. Solids **25**, 977 (1964).
[13] V. E. Lashkarev und G. A. Fedorus, Isv. Akad. Nauk SSSR, Ser. Fiz. **16**, 81 (1952).
[14] E. F. Gross und B. V. Novikov, J. Phys. Chem. Solids, **22**, 87 (1961).
[15] H.-J. Dittfeld und J. Voigt, phys. stat. sol. **3**, 1941 (1963).

(Received January 8, 1965)

phys. stat. sol. 8, 917 (1965)

Institut für experimentelle Physik der Universität Halle

Bildung und Ausheilung von Punktdefekten in KCl-Kristallen, gemessen mit Hilfe der Verfärbbarkeit durch F-Zentren

Von

F. FRÖHLICH und P. GRAU [1])

Es wird die Bildung von punktförmigen Gitterdefekten in KCl-Einkristallen nach Abschrecken von 530 °C, nach plastischer Deformation und nach intensiver Röntgenvorbestrahlung mit Hilfe der Röntgenverfärbbarkeit durch F-Zentren untersucht. Die Vorbehandlungen führen zu einer Erhöhung der Verfärbbarkeit, die durch die charakteristischen Parameter der F-Zentren-Bildungskurve erfaßt werden kann. Dabei wird eine bevorzugte Zunahme der Konzentration von Anionenleerstellen gefunden, die in der Umgebung von Gitterstörungen lokalisiert sind.

Die thermische Ausheilung der für die Verfärbbarkeitserhöhung verantwortlichen Defekte wird im Anfangsteil der F-Zentren-Bildungskurve verfolgt. Die Defekte nach Abschrecken heilen mit einer diskreten Aktivierungsenergie von 1,6 eV aus und sind offenbar verbunden mit der Art des Einbaues der als Verunreinigungen vorliegenden 2-wertigen Kationen. Bei deformierten und röntgenvorbestrahlten Proben ergeben sich diskrete Werte zwischen 1,6 eV und 2,0 eV. Daraus wird geschlossen, daß die für die Verfärbbarkeitserhöhung verantwortlichen zusätzlichen Anionenleerstellen sicher nicht isoliert, sondern wahrscheinlich in Form von relativ kleinen Leerstellen-Clustern ziemlich einheitlicher Größe nach diesen Vorbehandlungen im Kristall zurückbleiben. Ihr Abbau erfolgt über die Diffusion von isolierten Anionenleerstellen zu geeigneten Senken im Kristall.

The formation of point defects in KCl single crystals has been investigated by means of X-ray colourability by F-centres, after quenching from 530 °C, plastic deformation, and intense pre-X-ray irradiation, respectively. The pre-treatments enhance the colourability which can be measured by the characteristic parameters of the F-centre growth curve. Thereby a preferred increase in the concentration of anion vacancies localized in the neighbourhood of lattice perturbations has been found.

Thermal annealing of defects, responsible for the enhancement of colourability, is observed in the initial part of the F-centre growth curve. The defects craeted by quenching anneal with a discrete activation energy of 1.6 eV and obviously they are related to the mode of incorporation of bivalent cations which are present as impurities. In the case of defects, induced by deformation or X-ray pre-treatment, discrete values between 1.6 eV and 2.0 eV are obtained. One can, therefore, conclude that the additional anion vacancies, responsible for the enhancement of colourability, are certainly not isolated but remain in the crystal after these pre-treatments probably in form of relatively small vacancy clusters of rather uniform size. Their annealing takes place by diffusion of single anion vacancies to appropriate sinks in the crystal.

1. Einleitung

Die F-Zentren-Bildung in Alkalihalogenidkristallen unter der Einwirkung von ionisierender Strahlung ist bekanntlich in der Umgebung der Zimmertemperatur eine strukturempfindliche Kristalleigenschaft. Durch Vorbestrahlung mit ionisierender Strahlung, durch plastische Deformation, durch Abschrecken von höherer Temperatur und durch Fremdionenzusätze werden Gitterdefekte erzeugt, die die

[1]) Jetzt Institut für Grubensicherheit, Leipzig.

F-Zentren-Bildung wesentlich erhöhen, vgl. z. B. die zusammenfassenden Darstellungen von Pohl, Seitz, Pick, Stasiw, Schulman-Compton und Amelinckx [1a bis f]. Es wird heute allgemein angenommen, daß die F-Zentren-Bildung unter der Einwirkung ionisierender Strahlung in zwei voneinander unabhängigen Prozessen vor sich geht [2, 3]. Im ersten Prozeß sollen mit hoher Bildungsrate F-Zentren aus Anionenleerstellen gebildet werden, die in irgendeiner Form von vornherein im Kristall vorhanden sind. Der zweite Prozeß ist charakterisiert durch die Bildung zusätzlicher Anionenleerstellen mit relativ kleiner Bildungsrate für die F-Zentren. Beim zeitlichen Anstieg der F-Zentren-Absorption (Bildungskurve) werden den beiden Bildungsprozessen entsprechend die Bereiche I und II unterschieden. Die Bildungskurve läßt sich bei homogener Verfärbung durch die folgende Gleichung beschreiben [4, 5]:

$$f(t) = n_0 (1 - e^{-bt}) + \left(n_1 - \frac{a}{c}\right) (1 - e^{-ct}) + a\,t\,. \tag{1}$$

Dabei bedeuten

- n_0 Konzentration der im unvorbehandelten Kristall statistisch verteilten Anionenleerstellen
- n_1 Konzentration der im Kristall in der Umgebung von Gitterstörungen (z. B. Versetzungen) lokalisierten Anionenleerstellen
- b bzw. c Bildungsraten der F-Zentren aus n_0- bzw. n_1-Leerstellen
- a Bildungsrate für die während der Bestrahlung zusätzlich erzeugten Anionenleerstellen.

Der Einfluß der verschiedenartigen Vorbehandlungen eines Kristalles auf die Bildung von F-Zentren kann durch die Parameter der Bildungskurve erfaßt werden. Weitere Aussagen über die physikalische Natur der durch die Vorbehandlung erzeugten Defekte sollte man aus der quantitativen Untersuchung der thermischen Ausheilung dieser Defekte gewinnen. Insbesondere interessiert dabei, ob der Ausheilvorgang mit diskreten Aktivierungsenergien verbunden ist. Bei deuteronen- und röntgenvorbestrahlten KCl-Kristallen hat Thommen [6] für die Ausheilung der zusätzlichen F-, V- und M-Zentren-Verfärbbarkeit oberhalb von 200 °C definierte Abbaustufen gefunden und insbesondere für die Ausheilung der durch Deuteronenbeschuß erhöhten F-Zentren-Bildung eine diskrete Aktivierungsenergie ermitteln können.

In der vorliegenden Arbeit wird die Bildung und thermische Ausheilung von Gitterdefekten in KCl-Kristallen untersucht, die durch Röntgenvorbestrahlung, Deformation und unterschiedliche thermische Vorbehandlung im Kristall erzeugt werden können und für eine erhöhte F-Zentren-Bildung im Bereich I der Bildungskurve verantwortlich sind.

2. Verfahren zur Ermittlung der Aktivierungsenergie für die Ausheilung

Zur Bestimmung der Aktivierungsenergie hat man die Ausheilung der Gitterdefekte durch eine physikalische Eigenschaft p zu untersuchen, die in definierter Weise von der Konzentration der auszuheilenden Defekte abhängig ist.

Für die zeitliche Änderung der defektgebundenen Eigenschaft p während der thermischen Ausheilung gilt unter der Voraussetzung, daß die Aktivierungsenergie E_A konstant ist

$$\frac{dp}{dt} = F(p; q_1, \ldots, q_n)\, e^{-E_A/kT}\,. \tag{2}$$

$F(p; q_1, \ldots, q_n)$ ist die charakteristische Ausheilfunktion; die q_ν sind zeit- und temperaturunabhängige Parameter, die das Verhalten des Kristalles in bezug auf die auszuheilenden Defekte bestimmen, k die Boltzmannkonstante und T die absolute Temperatur. In (2) sind $\frac{dp}{dt}$ und T meßbare Größen, während die zu bestimmende Aktivierungsenergie E_A und im allgemeinen auch die Ausheilfunktion F unbekannt sind. Meechan und Brinkman [7] haben jedoch für die Ausheilung von elektronenbestrahltem Kupfer ein Verfahren angegeben, bei dem sich die Aktivierungsenergie ohne Kenntnis der Ausheilfunktion ermitteln läßt. Man benötigt dazu zwei Proben gleicher Vorgeschichte, die zwei verschiedenen Temperprozessen unterworfen werden.

Mit der ersten Probe wird eine isochronale pulsierte Temperung durchgeführt. Das ist eine in Intervalle unterteilte isotherme Wärmebehandlung mit der Zeitdauer Δt_ν und der Temperatur T_ν. Für die Ausheilung im ν^{ten} Intervall gilt

$$\int_{p_{\nu-1}}^{p_\nu} \frac{dp}{F(p; q_1, \ldots, q_n)} = \int_{t_{\nu-1}}^{t_\nu} e^{-E_A/k T_\nu}\, dt = e^{-E_A/k T_\nu}\, \Delta t_\nu \,. \tag{3}$$

Die zweite Probe wird isotherm bei der Temperatur T_i getempert und ergibt in der Zeit $\tau_\nu - \tau_{\nu-1}$ eine Ausheilung von p zwischen den Werten $p_{\nu-1}$ und p_ν

$$\int_{p_{\nu-1}}^{p_\nu} \frac{dp}{F(p; q_1, \ldots, q_n)} = \int_{\tau_{\nu-1}}^{\tau_\nu} e^{-E_A/k T_i}\, dt = e^{-E_A/k T_i}\, \Delta \tau_\nu \,. \tag{4}$$

Durch Vergleich von (3) und (4) erhält man

$$e^{-E_A/k T_\nu}\, \Delta t_\nu = e^{-E_A/k T_i}\, \Delta \tau_\nu \,, \tag{5}$$

also einen linearen Zusammenhang zwischen $\ln \Delta\tau_\nu$ und $\frac{1}{T_\nu}$

$$\ln \Delta\tau_\nu = -\frac{E_A}{k T_\nu} + \frac{E_A}{k T_i} + \ln \Delta t_\nu \,. \tag{6}$$

Der Anstieg $m = -E_A/k$ der entsprechenden Geraden liefert die Aktivierungsenergie E_A, die weiterhin auch aus dem Ordinatenabschnitt b

$$b = \frac{E_A}{k T_i} + \ln \Delta t_\nu \tag{7}$$

zu ermitteln ist. Die Übereinstimmung der beiden Werte für die Aktivierungsenergie aus dem Anstieg m und dem Ordinatenabschnitt b ist ein empfindliches Kriterium für die Genauigkeit des Ergebnisses. In Kapitel 4 werden deshalb die beiden aus b und m berechneten Werte $E_{A,b}$ und $E_{A,m}$ gesondert mitgeteilt. Bei verschiedenen diskreten Aktivierungsenergien ist die Kurve $\ln \Delta\tau_\nu, \frac{1}{T_\nu}$ aus mehreren Geraden zusammengesetzt.

In den vorliegenden Untersuchungen wird als physikalische Eigenschaft p, die eine eindeutige Funktion der wirksamen Gitterdefekte sein soll, der Absorptionskoeffizient im Maximum der F-Bande nach 15 min Röntgenbestrahlung mit Zimmertemperatur als Meßtemperatur T_M verwendet. Dieser Absorptionskoeffizient wird im folgenden immer als k^* bezeichnet. Die Daten der Röntgenstrahlung sind so gewählt, daß die $\alpha\, t$-Leerstellen-Produktion vernachlässigbar klein ist und die erreichte Verfärbung optisch leicht gebleicht werden kann. Eine durch diese „Meßbestrahlung" bewirkte Verfärbbarkeitserhöhung („Erinnerungseffekt"[2])) fällt bei den thermischen Ausheilexperimenten wegen der geringen Aktivierungsenergien der für den Erinnerungseffekt verantwortlichen Defekte nicht ins Gewicht.

Der Absorptionskoeffizient k^* ist bei homogener Verfärbung proportional zur F-Zentren-Konzentration f aus (1) mit $a \approx 0$ und $t = \frac{1}{4} h$

$$f_{t=\frac{1}{4}h} = n_0 (1 - e^{-b/4}) + n_1 (1 - e^{-c/4}) . \tag{8}$$

Der k^*-Wert ist in Näherung auch proportional dem Anstieg der Verfärbungskurve zur Zeit $t = 0$

$$\left.\frac{df}{dt}\right|_{t=0} = n_0 b + n_1 c , \tag{9}$$

aus dem sich wie aus (8) die lineare Beziehung zwischen k^* und den Leerstellenkonzentrationen n_0 und n_1 unmittelbar ablesen läßt.

3. Experimentelles

Als Kristallmaterial wurden nach dem Kyropoulos-Verfahren aus analysenreinen Ausgangssubstanzen an Luft hergestellte Kristalle von unserem schon länger zurückliegendem Züchtungsversuch Nr. 12 (KCl–12) und unserem Züch-

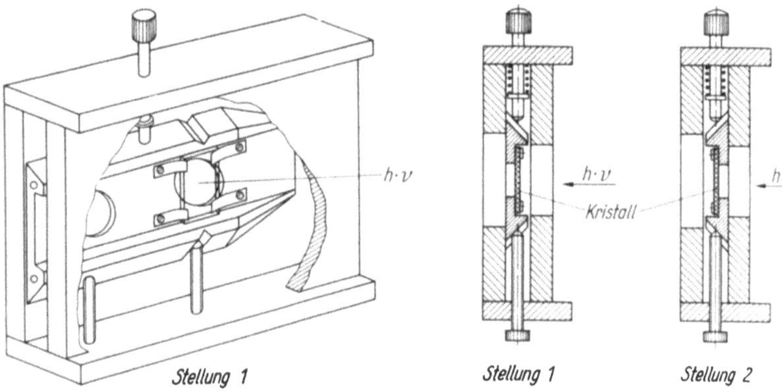

Fig. 1.

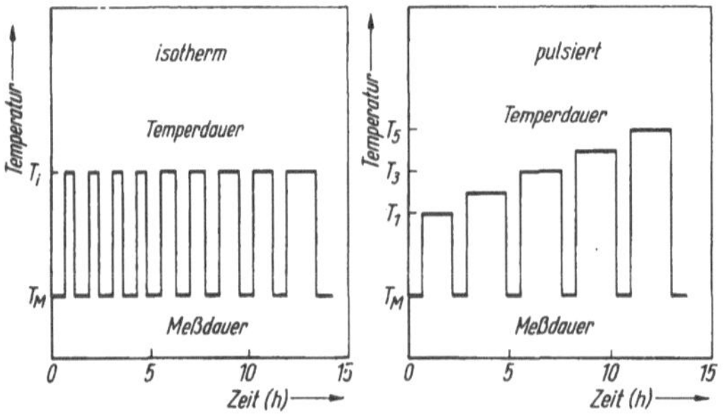

Fig. 1. Darstellung der Vorrichtung zur reproduzierbaren Anordnung des Probenhalters. Der Kristall wird mit Federn auf dem Probenhalter befestigt

◄ Fig. 2. Schematische Darstellung des Temperatur-Zeit-Programmes bei isothermer und pulsierter Temperung

[2]) Auf den Erinnerungseffekt wird in einer nachfolgenden Arbeit eingegangen [8].

tungsversuch Nr. 97 (KCl–97) verwendet. Aus benachbarten Bereichen wurden Meßproben von 1 mm Dicke abgespalten und auf einem geeigneten Probenhalter montiert. Der Probenhalter konnte in reproduzierbarer Lage an der Röntgenröhre, am Spektralphotometer und im Temperofen eingesetzt werden. Die Meßbestrahlung zur Bestimmung von k^* erfolgte mit 65 kV/2 mA Röntgenstrahlung (Wolfram-Anode) bei Zimmertemperatur. Um Homogenität der Verfärbung zu gewährleisten, wurde die Strahlung mit 1 mm Aluminium gefiltert und ein Kristall jeweils 7,5 min von beiden Seiten bestrahlt, siehe Fig. 1. Die durch die Meßbestrahlung bewirkte Verfärbung wurde bei Zimmertemperatur mit dem weißen Licht einer Mikroskopierleuchte gebleicht. Die Temperung der Proben erfolgte an getrockneter Luft mit einer Temperaturkonstanz von $\pm 0{,}5$ °C. Das Temperaturprogramm für die isotherme und pulsierte Temperung ist in Fig. 2 dargestellt. Der Betrag der F-Zentren-Konzentration wurde aus der Smakula-Formel [9] mit der Oszillatorenstärke 0,81 ermittelt.

4. Ergebnisse

4.1 Defekte in „as cleaved“ Kristallen [3])

Bei unseren Untersuchungen zur Ausheilung von Röntgenstrahldefekten in KCl–97 ergab sich, daß die am Ende einer Temperkurve gemessene F-Zentren-Verfärbung k^* bis zu 20% unter dem jungfräulichen Wert[4]) lag. In KCl–12 dagegen wurde diese Eigenschaft nicht beobachtet. Der jungfräuliche Wert für KCl–97 nahm im Laufe eines Jahres von 0,45 mm^{-1} auf 0,41 mm^{-1} ab. Für KCl–12 ließ sich im gleichen Zeitraum keine Änderung des jungfräulichen Wertes feststellen. Er lag im Mittel bei 0,34 mm^{-1}. Diese Ergebnisse lassen den Schluß zu, daß KCl–97 schon im as cleaved Zustand Defekte enthält, die bei Röntgenbestrahlung F-Zentren liefern und thermisch ausheilbar sind. Da das Herstellungsdatum von KCl–12 mehr als $3^1/_2$ Jahre gegen das von KCl–97 zurückliegt, sollte man weiter schließen können, daß diese as cleaved Defekte schon bei Zimmertemperatur über längere Zeiträume (Jahre) abgebaut werden können.

Der thermische Abbau solcher Defekte ist z. B. aus Fig. 3 zu ersehen. Sie zeigt die pulsierte und die isothermen Temperaturkurven eines as cleaved Kristalles aus KCl-97 mit einer durch den Erinnerungseffekt bewirkten Verfärbbarkeitserhöhung. In den ersten beiden Stufen zwischen 80 °C und 180 °C wird der Erinnerungseffekt mit Aktivierungsenergien von 0,68 eV und 0,94 eV ausgeheilt. Mit dem Erreichen des jungfräulichen Wertes schließt sich zwischen 180 °C und 300 °C der Abbau der as cleaved Defekte an. Mit einer zweiten Probe wurde bei 250 °C eine Isotherme aufgenommen, in der vorher mit einer Temperung bei 180 °C der Erinnerungseffekt beseitigt worden war. Bei der Auswertung ergab sich eine Aktivierungsenergie von 1,70 eV. Dieser Wert kann jedoch wegen der Unsicherheit des Anfangsverlaufes der Isothermen durch einen systematischen Fehler verfälscht sein. In gesonderten Experimenten wurden mit Kristallen, in denen der Erinnerungseffekt jeweils durch isotherme Temperungen bei 180 °C ausgeheilt worden war, Werte bei 1,60 eV gefunden, Versuche Nr. 10 und 13 in Tabelle 3.

---

[3]) Kristalle, die von größeren Blöcken abgespalten und danach nicht zusätzlich vorbehandelt wurden, werden nachfolgend als „as cleaved“ Proben bezeichnet.

[4]) Unter jungfräulichem Wert wird der Absorptionskoeffizient k^* verstanden, der vor der jeweiligen Erzeugung der zusätzlichen Defekte gemessen wurde.

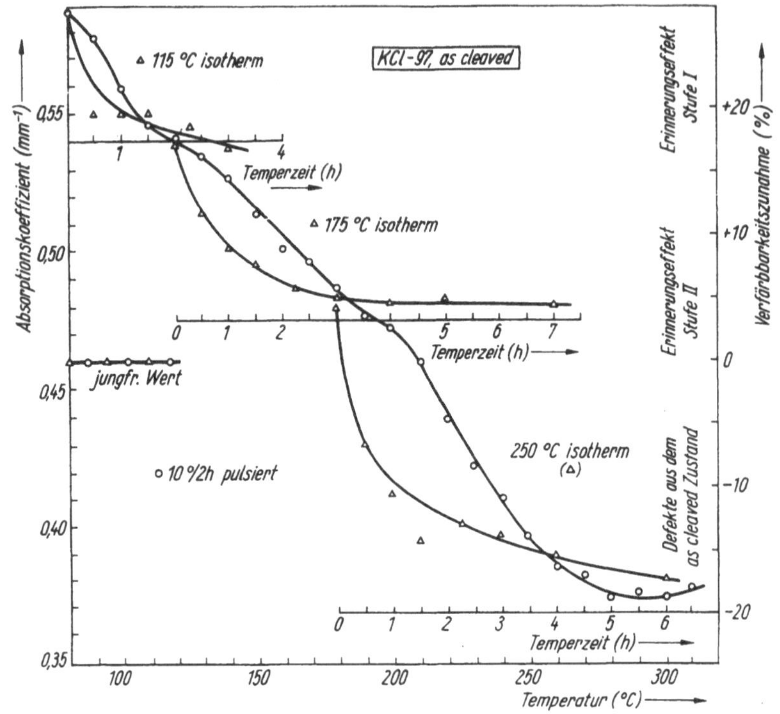

Fig. 3. Ausheilung der Verfärbbarkeitszunahme bei pulsierter (○) und isothermer (△) Temperung von as cleaved KCl-97 mit Erinnerungseffekt

Bei Versuchen mit nicht zusätzlich vorbehandelten KCl-Kristallen ist demnach immer das Auftreten der as cleaved Defekte zu berücksichtigen. Diese können sich bei der Ausheilung den durch eine andere Vorbehandlung erzeugten Defekten störend überlagern. Da aber die Ausheilung der as cleaved Defekte mit einer festen und bekannten Aktivierungsenergie verbunden ist, sollten sie sich von anderen Ausheilprozessen (durch einen Knick in der Aktivierungsenergie-Geraden) trennen lassen, wenn die Aktivierungsenergie für die Ausheilung der zusätzlich erzeugten Defekte nach der hier angewendeten Bestimmungsmethode um mehr als 0,20 eV von 1,60 eV abweicht. In vielen Fällen empfiehlt es sich jedoch, die KCl-Kristalle von ihren as cleaved Defekten zu „reinigen“. Hierzu ist nicht unbedingt der beschriebene pulsierte Temperprozeß notwendig, sondern es genügt auch eine etwa 12-stündige Temperung bei 320 °C. Für KCl–97 wurde nach einer solchen Wärmebehandlung im Mittel ein k^* von 0,34 mm^{-1} gemessen. Eine Verfärbungskurve für KCl–97 nach 320 °C-Temperung ist in Fig. 4 (untere Kurve) dargestellt. Man erkennt, daß sich der Kristall durch diese Temperung weitgehend dem Zustand des gealterten KCl–12 angenähert hat.

4.2 Defekte in abgeschreckten Kristallen

As cleaved Proben wurden etwa 10 h bei 530 °C getempert und danach durch schnelles Herausnehmen aus dem Temperofen auf Zimmertemperatur abgeschreckt. Da der Temper- und Abschreckvorgang in Luftatmosphäre vorgenommen wurde,

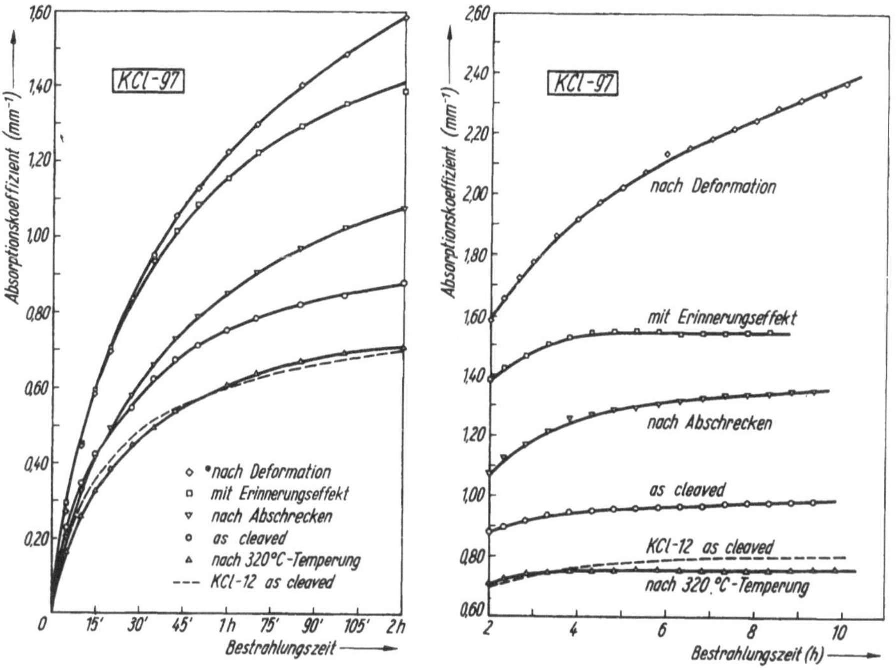

Fig. 4

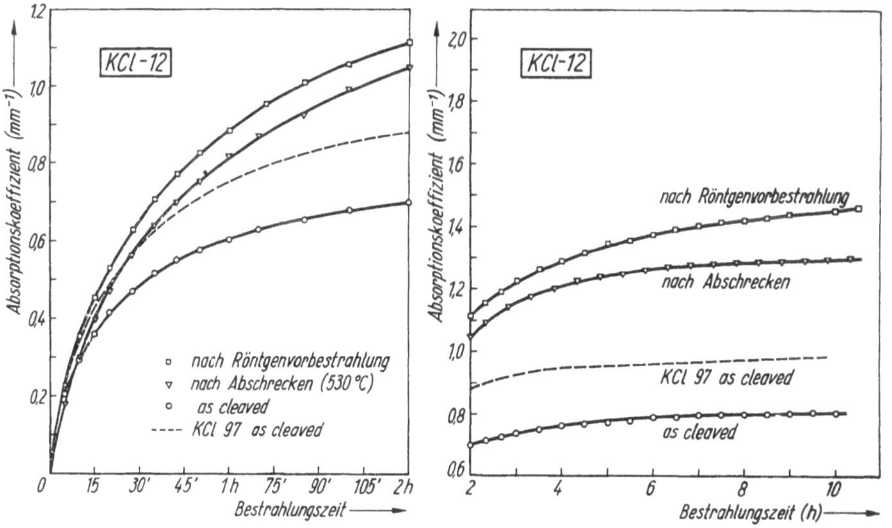

Fig. 5

Fig. 4 und Fig. 5. Zeitlicher Verlauf der F-Zentren-Bildung für KCl-97 (Fig. 4) und KCl-12 (Fig. 5) nach unterschiedlicher Vorbehandlung

Tabelle 1

Zusammenstellung der charakteristischen Parameter der Verfärbungskurven

Material	Vorbehandlung	a	n_0		n_1		$n_0 + n_1$
		(10^{16}cm^{-3}h^{-1})	(10^{16} cm^{-3})				
97	as cleaved	0,033	2,09		2,32		4,41
97	abgeschreckt	0,050	2,13	2%*)	3,83	53%*)	5,96
97	deformiert	0,283	3,27	56%	5,19	125%	8,46
97	mit Erinnerungseffekt	0	2,52	20%	4,89	110%	7,41
12	as cleaved	0,020	2,39		1,28		3,67
12	abgeschreckt	0,034	1,91	-20%	4,08	240%	6,00
12	röntgenvorbestrahlt	0,056	2,64	10%	3,44	165%	6,08

*) Die %-Zahlen geben die Zunahme der Leerstellenkonzentration in bezug auf die as cleaved Proben an.

Tabelle 2

	KCl-97, as cleaved Beteiligung an k^*		KCl-97, abgeschreckt Beteiligung an k^*		Zunahme absolut	Zunahme relativ
	(10^{16} cm^{-3})	(%)	(10^{16} cm^{-3})	(%)	(10^{16} cm^{-3})	(%)
n_0	1,42	74,0	1,37	69,0	−0,05	− 3,5
n_1	0,51	25,5	0,60	30,5	+0,09	+18

Tabelle 3

Zusammenstellung der ermittelten Aktivierungsenergien für die thermische Ausheilung der im as cleaved Zustand vorhandenen und der durch Abschrecken, Deformation und intensive Röntgenbestrahlung erzeugten Defekte

Versuch Nr.	Material	Vorbehandlung	Aktivierungsenergie (eV) $E_{A, m}$	$E_{A, b}$
10	97	keine (as cleaved)	1,53 ± 0,09	1,52 ± 0,09
13	97	keine (as cleaved)	1,57 ± 0,07	1,64 ± 0,09
18	97	abgeschreckt	1,62 ± 0,08	1,63 ± 0,08
30	97	10% deformiert	2,00 ± 0,04	2,03 ± 0,06
33	97	15% deformiert	1,63 ± 0,08	1,71 ± 0,09
34	97	10% deformiert	1,84 ± 0,11	1,80 ± 0,12
35	97	9% deformiert	1,81 ± 0,03	1,88 ± 0,04
36	97	15% deformiert	1,73 ± 0,05	1,73 ± 0,07
38	97	13% deformiert	1,80 ± 0,21	1,81 ± 0,22
52	97	10% deformiert	1,76 ± 0,08	1,77 ± 0,10
70	12	röntgenvorbestrahlt	1,71 ± 0,08	1,72 ± 0,09
71	12	röntgenvorbestrahlt	1,56 ± 0,09	1,52 ± 0,10
72	12	röntgenvorbestrahlt	1,99 ± 0,10	1,97 ± 0,11
85	12	röntgenvorbestrahlt	1,67 ± 0,06	1,67 ± 0,08
86	12	röntgenvorbestrahlt	1,66 ± 0,09	1.58 ± 0,10

sind die Proben in Übereinstimmung mit Mitchell, Wiegand und Smoluchowski [5] als „air quenched“ zu bezeichnen. Es kann dabei eintreten, daß so abgeschreckte Proben von as cleaved Material aus KCl–97 in k^* keine Verfärbbarkeitsänderung gegenüber dem jungfräulichen Wert erkennen lassen (vgl. z. B. die 15-min-Werte in Fig. 4).

Dieses Ergebnis ist noch kein hinreichendes Kriterium dafür, daß Abschrecken die Anionenleerstellenkonzentration von KCl-97 nicht verändert. Das kann erst durch die Aufnahme einer Verfärbungskurve entschieden werden: In Fig. 4 übertreffen die Absorptionswerte eines abgeschreckten Kristalles die eines as cleaved Kristalles bei längeren Bestrahlungszeiten beträchtlich. Aus den graphisch ermittelten Parametern der F-Zentren-Bildungskurve in Tabelle 1 ist eine Zunahme der n_1-Leerstellenkonzentration um 53% zu ersehen. Diese Zunahme wird von der Meßbestrahlung nur ungenügend erfaßt. Dies ist aus Tabelle 2 zu ersehen, die angibt, welcher Teil der n_0- und n_1-Leerstellenkonzentrationen als F-Zentren zu k^* beitragen, berechnet nach Gl. (1) mit den zugehörigen n_0- und n_1-Werten der Tabelle 1.

Auf KCl-12 hat der Abschreckvorgang – wie man aus Fig. 5 und Tabelle 1 erkennt – die gleiche Wirkung: n_1 wird vergrößert, und zwar auf den Wert des KCl-97 nach Abschrecken. Die n_0-Werte dagegen bleiben in KCl-12 und KCl-97 praktisch ungeändert.

Bei KCl–12 jedoch wird die Verfärbbarkeitszunahme bereits durch k^* erfaßt. Das gleiche sollte man auch mit KCl–97 erreichen, wenn man vorher die as cleaved Defekte ausheilt. In einem Versuch wurde darum KCl–97 zunächst bei 320 °C von den as cleaved Defekten gereinigt. Es ergab sich dabei ein k^*-Wert von 0,35 mm^{-1}. Durch Abschrecken von 530 °C wurde k^* auf 0,46 mm^{-1} erhöht. Damit wurde eine Aktivierungsenergiebestimmung mit dem k^*-Wert möglich. Eine pulsierte und isotherme Wärmebehandlung – dargestellt in Fig. 6a – läßt einen definierten Ausheilprozeß mit einer Aktivierungsenergie von 1,62 eV erkennen (Fig. 6b). Dieses Ergebnis ist in Tabelle 3, Versuch Nr. 18 aufgeführt.

4.3 *Defekte in deformierten Kristallen*

Der Einfluß der Deformation auf die F-Zentren-Bildung wurde durch Meßbestrahlungen vor und nach der Verformung ermittelt. Zwischen der Zunahme von k^* und der prozentualen Dickenänderung beim Druckversuch konnte zunächst keine eindeutige Abhängigkeit gefunden werden. Als Ursache dafür ist der Erinnerungseffekt anzusehen, der die mit der Meßbestrahlung erfaßbare Verfärbbarkeitszunahme durch Deformation vermindert.

Nach Beseitigung des Erinnerungseffektes ergab sich in zehn Experimenten mit as cleaved Kristallen aus KCl-97 nach einer Deformation von etwa 10% eine Erhöhung von k^* um etwa 11%. Bei einem Kristall, der bis 15% deformiert werden konnte, wurde eine Zunahme des k^*-Wertes von 32% gefunden. Die Verfärbungskurve für diesen besonders stark deformierten Kristall ist in Fig. 4 (obere Kurve) dargestellt. Die zugehörigen Parameter in Tabelle 1 ergeben,

Tabelle 4

	KCl-97, as cleaved Beteiligung an k^*		KCl-97, deformiert Beteiligung an k^*		Zunahme	
	(10^{16} cm^{-3})	(%)	(10^{16} cm^{-3})	(%)	absolut (10^{16} cm^{-3})	relativ (%)
n_0	1,42	74,0	1,86	68	0,44	31
n_1	0,51	22,5	0,81	29,5	0,30	59
$a \cdot t$	0,008	0,5	0,07	2,5	–	–

daß durch die Deformation beide Leerstellenarten vermehrt worden sind. Die n_1-Zunahme ist dabei wesentlich größer als die von n_0. Während der Meßbestrahlung kommen die n_0-Leerstellen aber weit mehr zur Geltung (vgl. dazu Tabelle 4).

Die Deformation bewirkt weiterhin eine starke Erhöhung des Ratenparameters a im Bereich II der Verfärbungskurve (siehe Fig. 4). Die Ausheilung der dafür verantwortlichen Defekte wird jedoch mit dem hier angewendeten Verfahren durch k^* nicht erfaßt (vgl. dazu auch Tabelle 4).

Als Beispiel für die Aktivierungsenergiebestimmungen werden die Ergebnisse von zwei Proben aus KCl–97 angegeben, die zunächst bei 320 °C bis auf ein k^* von 0,31 mm^{-1} gereinigt und nach Beseitigung des Erinnerungseffektes 10% deformiert worden waren. Dies führte zu einer Verfärbbarkeitserhöhung von 43%. Bei der anschließenden Ausheilung erfolgte die pulsierte Temperung in 10 °C/2 h Intervallen, die isotherme Temperung bei 275 °C (vgl. Fig. 7a). Es ergab sich eine Aktivierungsenergie von 1,76 eV (vgl. Fig. 7b). Dieser Wert ist in Tabelle 3

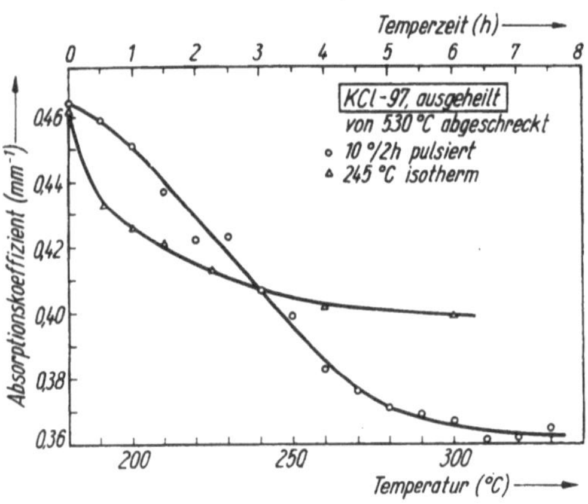

Fig. 6a. Ausheilung der Verfärbbarkeitszunahme bei pulsierter (○) und isothermer (△) Temperung für KCl-97 nach Abschrecken von 530 °C

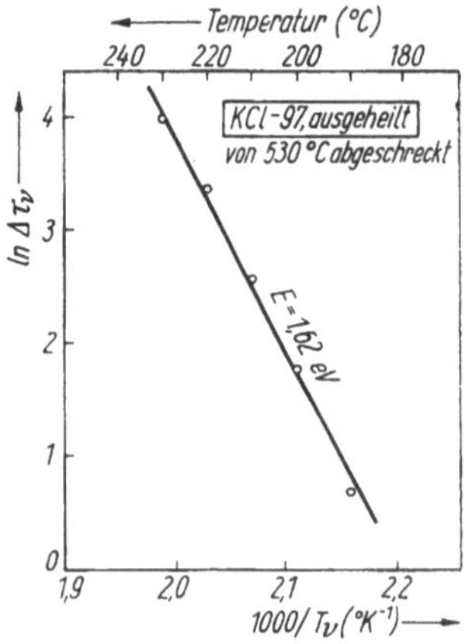

Fig. 6b. Ermittlung der zugehörigen Aktivierungsenergie aus dem Anstieg

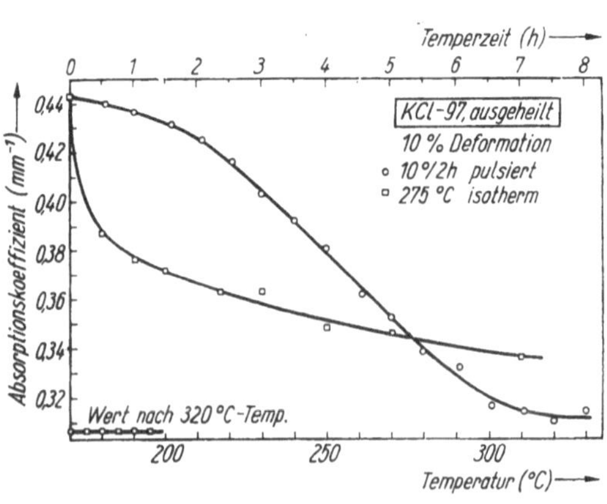

Fig. 7a. Ausheilung der Verfärbbarkeitszunahme bei pulsierter (○) und isothermer (□) Temperung für ausgeheiltes KCl-97 nach plastischer Deformation

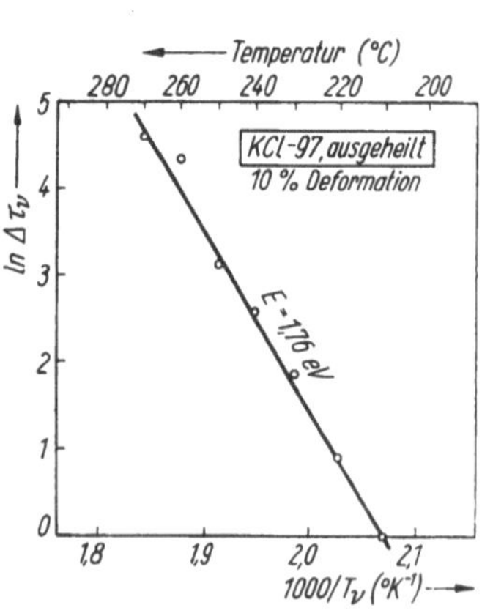

Fig. 7b. Ermittlung der zugehörigen Aktivierungsenergie aus dem Anstieg

(Versuch Nr. 52) aufgeführt zusammen mit den Aktivierungsenergien aus sechs weiteren Versuchen, bei denen die Kristalle im as cleaved Zustand deformiert wurden. In as cleaved Kristallen ist die Zunahme der Verfärbbarkeit nach plastischer Deformation geringer als in ausgeheilten Proben.

4.4 Defekte in röntgenvorbestrahlten Kristallen

Ehe die thermische Ausheilung der Röntgendefekte behandelt wird, muß auf die Röntgenvorbestrahlung und die nachfolgende Bleichung der bestrahlten Proben eingegangen werden.

4.4.1 Röntgenvorbestrahlung

Zur Erzeugung einer ausreichenden Konzentration von $a\,t$-Leerstellen durch Röntgenbestrahlung ist ein möglichst großer Röhrenstrom notwendig, da der Parameter a näherungsweise proportional zu diesem ist [4, 5, 10]. Die verwendete Röntgenanlage lieferte bei 60 kV maximal 15 mA. Auf eine Abfilterung des energiearmen Anteils des Röntgenspektrums wurde verzichtet, um die gesamte Röntgenintensität wirksam werden zu lassen. Die Forderung nach Homogenität der Verfärbung in bezug auf die Kristalldicke wurde durch beidseitige Bestrahlung der Proben berücksichtigt (Fig. 2). Für diese Röntgenvorbestrahlung ergab sich bei KCl–12, das allein zu dieser Versuchsreihe verwendet wurde, ein a-Wert zwischen 1,7 und $2{,}5 \cdot 10^{16}$ cm^{-3} h^{-1}. In Fig. 8 ist eine entsprechende Verfärbungskurve angegeben. Sie hat nach 1 bis 2 Stunden Bestrahlungszeit einen Wendepunkt, der charakteristisch ist für große Röntgenintensitäten [11, 5, 4]. Die Streuung der Meßpunkte erklärt sich aus der Unsicherheit der Meßwerte bei hohen Absorptionen. Für eine quantitative Untersuchung des Ausheilvorganges ist die Erzeugung von mindestens $2 \cdot 10^{17}$ Leerstellen (cm^{-3}) erforderlich. Diese Konzentration erreicht man mit den verwendeten Röntgenintensitäten nach 8 bis 12 h Bestrahlung. Die KCl-Kristalle enthalten dann neben einer dominierenden F-Bande eine breite V-Bande (215 nm) und in geringem Maße Komplexzentren (R, M, N) [12, 13].

4.4.2 Bleichung

Die Entfärbung der röntgenvorbestrahlten Kristalle bereitete Schwierigkeiten. Es wurden verschiedene Experimente mit dem Ziel durchgeführt, eine wirksame thermische oder optische Bleichung zu finden, die aber die $a\,t$-Röntgendefekte nur wenig angreift.

Zur optischen Bleichung wurde das weiße Licht einer 50 W-Mikroskoplampe benutzt. Die Rückbildung der F-Bande in Abhängigkeit von der Bleichzeit ist aus Fig. 8 zu ersehen. Einer anfangs schnellen Bleichung von etwa 10^{17} F-Zentren (cm^{-3}) schließt sich nach 15 min eine langsame Entfärbung an. Nach 30 h (Meßwert in Fig. 8 nicht eingetragen) wird immer noch eine Abnahme der F-Zentren-Konzentration festgestellt, so daß eine vollständige optische Bleichung nach beliebig langer Bleichzeit möglich scheint [12]. Die Farbe der teilweise optisch entfärbten Proben schlug von tief violett nach blau um. Im Absorptionsspektrum ist die F-Bande nach 30-stündiger Einstrahlung von weißem Licht zwar noch bestimmend, auf ihrer langwelligen Seite machen sich aber deutlich die R-Banden [14, 15, 16] bemerkbar, denen sich ab 720 nm langwelligere Komplex- oder Kolloidbanden anschließen.

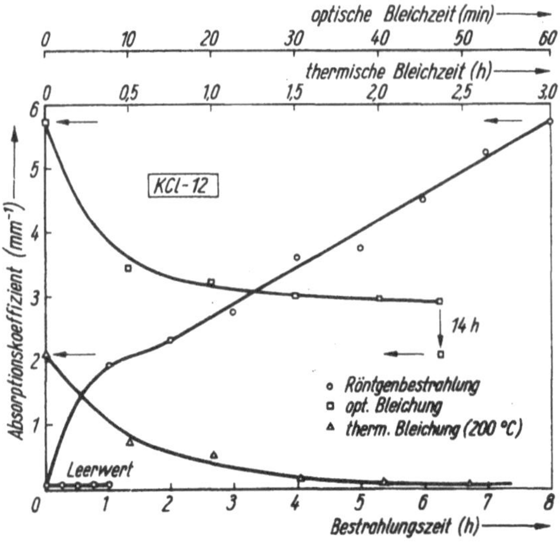

Fig. 8. Zeitlicher Verlauf der F-Zentren-Bildung für KCl-12 bei intensiver Röntgenbestrahlung (○), der optischen Bleichung der Verfärbung (□) und der anschließenden thermischen Bleichung der Verfärbung bei 200 °C (△)

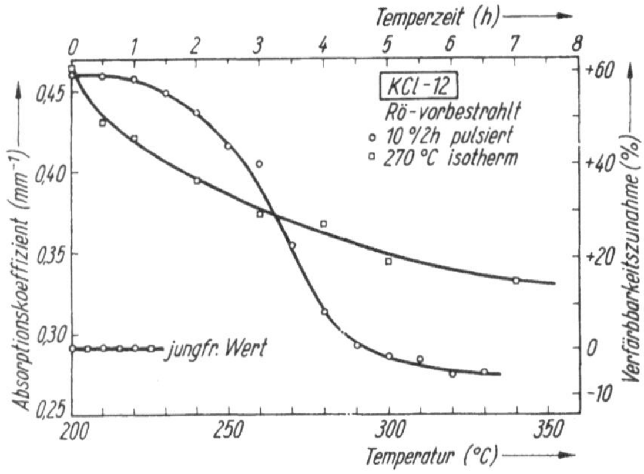

Fig. 9a. Ausheilung der Verfärbbarkeitszunahme bei pulsierter (○) und isothermer (□) Temperung für KCl-12 nach Röntgenvorbestrahlung

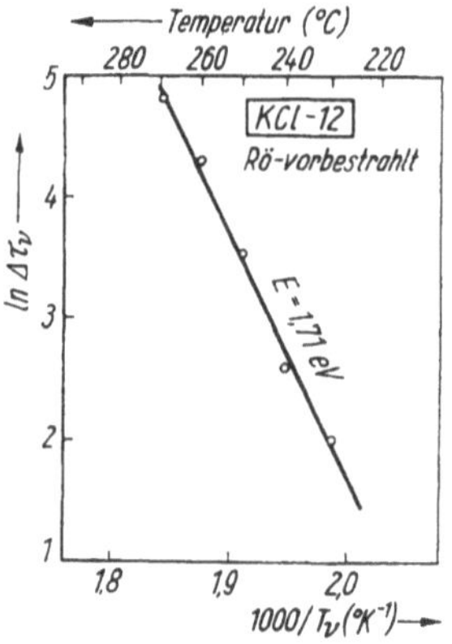

Fig. 9b. Ermittlung der zugehörigen Aktivierungsenergie aus dem Anstieg

Durch Temperaturerhöhung auf 135 °C wurde versucht, die Wirksamkeit der optischen Bleichung zu steigern. Dabei verschiebt sich das Maximum im Absorptionsspektrum nach 670 nm. Das erklärt sich aus einem rapiden Abbau der F-Bande, während die R-Banden noch erhalten bleiben. Diese werden erst nach mehrstündigen Bleichzeiten beseitigt. Auch nach 30 h bleibt im gesamten Spektralbereich eine Restabsorption zurück, die oberhalb 720 nm stärker ausgeprägt ist.

Neben der kombiniert thermisch-optischen Bleichung wurde eine rein thermische Bleichung bei 175 °C erprobt. Dabei nimmt die F-Bande stetig ab, wobei ihr Maximum um wenige nm ins Langwellige wandert. Es entsteht wieder eine R-Bande, deren Absorptionsmaximum aber immer kleiner als das der F-Bande bleibt. F- und R-Zentren sind nach 20 h restlos vernichtet. Dagegen tritt mit zunehmender Bleichzeit die Absorption ab 720 nm verstärkt auf. Zur Beseitigung dieser langwelligen Restabsorption waren wie bei Thommen [6] längere Temperungen bei etwa 200 °C notwendig. Die Wirksamkeit einer solchen Temperung ist aus Fig. 8 zu ersehen.

4.4.3 Ausheilung der Röntgendefekte

Um zu erfahren, wieviel von den gebildeten $a\ t$-Leerstellen bei erneuter Röntgenbestrahlung für die F-Zentren-Bildung verfügbar sind, wurden Verfärbungskurven von as cleaved und röntgenvorbestrahlten KCl–12-Kristallen verglichen. Ein Beispiel dafür ist in Fig. 5 angegeben. Ihre Parameter in Tabelle 1 zeigen an, daß durch Röntgenvorbestrahlung vorwiegend n_1-Leerstellen neu gebildet werden. Von $1{,}1 \cdot 10^{17}$ $a\ t$-Leerstellen (cm^{-3}) bleiben nach dem Bleichvorgang $2{,}4 \cdot 10^{16}$ Leerstellen (cm^{-3}) erhalten. Diese zusätzlichen Defekte bewirken eine Zunahme des k^*-Wertes um 25%.

Für die Ausheilexperimente wurden Proben verwendet, in denen jeweils mindestens $2 \cdot 10^{17}$ $a\ t$-Leerstellen (cm^{-3}) erzeugt worden waren. Diese Proben ergaben nach erfolgter Bleichung eine Zunahme des k^*-Wertes von 60%. Danach darf angenommen werden, daß in diesen Kristallen nach der Bleichung 5 bis $7 \cdot 10^{16}$ Röntgendefekte (cm^{-3}) zurückbleiben. Für diese Experimente wurden wieder nur Proben mit gleicher Vorgeschichte, insbesondere auch mit gleich großem a-Wert benutzt. Die pulsierte Temperung erfolgte in 10 °C/2 h Intervallen, die isotherme Temperung bei 270 °C (vgl. Fig. 9a). Bei dem in Fig. 9a dargestellten Beispiel wurde zwischen 230 °C und 270 °C eine Aktivierungsenergie von 1,71 eV ermittelt (vgl. Fig. 9b). In Tabelle 3 sind dazu noch die aus fünf weiteren Experimenten mit röntgenvorbestrahlten Kristallen bestimmten Aktivierungsenergien angegeben.

5. Diskussion

Die in der Tabelle 1 angegebenen Parameter der Verfärbungskurven und die erhaltenen Aktivierungsenergien weisen daraufhin, daß die verschiedenartige Vorbehandlung in den Proben im wesentlichen zu zwei voneinander verschiedenen Gruppen von Defekten führt. Zu einer Gruppe gehören die durch Abschrecken von 530 °C erzeugten Defekte und auch die im Ausgangszustand vorhandenen as cleaved Defekte, die beim Abkühlen des aus der Schmelze gezogenen Kristalles entstehen. In der zweiten Gruppe werden die Deformations- und Röntgenstrahldefekte zusammengefaßt. Ihr gemeinsames Merkmal ist die Wirksamkeit von Versetzungen. In der Diskussion wird jeweils zuerst auf die Bildung und anschließend auf die Ausheilung der Defekte eingegangen.

5.1 Defekte durch thermische Vorbehandlung

5.1.1 Bildung

Die Ergebnisse über die Bildung der Defekte durch die verschiedenen angewendeten thermischen Vorbehandlungen lassen sich wie folgt zusammenfassen:

1. Im as cleaved Zustand ist die F-Zentren-Konzentration im Bereich I der Bildungskurve bei KCl–97 wesentlich größer als bei KCl–12 auf Grund einer größeren n_1-Leerstellenkonzentration. Dieser Überschuß nimmt über größere Zeiträume (Jahre) hin ab.

2. Eine Temperung bei 320 °C führt bei KCl–97 zu einer Abnahme der erreichbaren F-Zentren-Konzentration im Bereich I und zu einer weitgehenden Annäherung an die Verfärbungseigenschaften von KCl–12 im as cleaved Zustand.

3. Temperung bei 530 °C und Abschrecken haben auf KCl–12 und KCl–97 dieselbe Wirkung: Zunahme der nachweisbaren n_1-Leerstellenkonzentration auf vergleichbare absolute Werte.

Zur Erklärung der drei Eigenschaften hat man vor allem zu berücksichtigen, daß die F-Zentren-Bildung sehr stark von Verunreinigungen, insbesondere von 2-wertigen Kationen beeinflußt wird [1b, 1e, 1f, 17, 18, 42], die bei den hier verwendeten Ausgangssubstanzen sicher in merklicher Konzentration im Kristall vorhanden sind. Der überwiegende Teil entfällt dabei auf Ca-Ionen.[5]) Die Wirksamkeit auf die Verfärbung hängt ab von der Art des Einbaues der Ca-Ionen. Dieser kann in drei verschiedenen Arten erfolgen: 1. atomdispers gelöst und von der zur Ladungskompensation notwendigen Kationenleerstelle dissoziiert, 2. atomdispers verteilt aber mit der zugehörigen Kationenleerstelle assoziiert,[6]) 3. ausgeschieden zu größeren Komplexen (bevorzugt an Versetzungen und möglicherweise als $CaCl_2$ [19, 20]). Die gemäß 2) eingebauten Ca-Ionen (im folgenden als „assoziierte Ca-Ionen" bezeichnet) bewirken nach Crawford und Nelson [21] die starke Erhöhung der F-Zentren-Bildung.

Zwischen den verschiedenartig eingebauten Ca-Ionen existiert ein temperaturabhängiger Gleichgewichtszustand (vgl. z. B. die zugehörigen Leitfähigkeitsbereiche [22]). Bei den in den vorliegenden Untersuchungen angewendeten Temperaturen wird die Wirksamkeit der Ca-Ionen durch das Gleichgewicht zwischen assoziierten und ausgeschiedenen Ca-Ionen bestimmt: Bei hoher Temperatur ist das Gleichgewicht zu Gunsten des assoziierten, bei niedriger Temperatur zu Gunsten des ausgeschiedenen Materials verschoben. Der Einfluß der thermischen Vorgeschichte auf den für die F-Zentren-Bildung wirksamen Ca-Gehalt kann mithin folgendermaßen verstanden werden: Bei langsamem Abkühlen der Kristalle nach der Herstellung scheiden sich die ursprünglich atomdispers gelöst vorliegenden Verunreinigungen in größeren Komplexen aus und gehen dadurch für eine zusätzliche F-Zentren-Bildung verloren. Durch Tempern bei hinreichend hoher Temperatur können die ausgeschiedenen Verunreinigungen wieder gelöst und durch plötzliches Abschrecken atomdispers und assoziiert eingefroren werden. Die Richtigkeit dieser Vorstellungen kann durch Leitfähigkeitsmessungen im Störleitungsbereich nachgewiesen werden [23, 24]: In langsam abgekühlten Kristallen ergibt sich eine geringere Leitfähigkeit als in abgeschreckten. Weiterhin werden in abgeschreckten NaCl-Proben verstärkte Z-Banden gefunden [25], zu deren Deutung isoliert eingebaute Ca-Ionen notwendig sind [26, 27].

Bei den in den vorliegenden Untersuchungen verwendeten Kristallen KCl–97 und KCl–12 wird durch die Kyropoulos-Züchtung jeweils ein Gleichgewichtszustand von höherer Temperatur eingefroren. Das Gleichgewicht verschiebt sich jedoch durch extrem lange „Temperung" bei Zimmertemperatur allmählich zu Gunsten der Ausscheidungen. Dieser Prozeß ist bei den um mehrere Jahre

---

[5]) Die im Kristall als Verunreinigungen vorhandenen Ca-Ionen wurden komplexometrisch zu etwa $3 \cdot 10^{-6}$ Ca-Ionen/K-Ion bestimmt. Dies entspricht $5 \cdot 10^{16}$ Ca-Ionen/cm³. Dieser Wert liegt in der gleichen Größenordnung wie die Änderung der n_1-Konzentration nach thermischer Vorbehandlung (vgl. Tabelle 1).

[6]) In Gegenwart von OH-Ionen möglicherweise auch als Komplex $Ca(OH)_2$ + Doppellücke [62].

älteren KCl–12 wesentlich weiter fortgeschritten als bei KCl–97 (Eigenschaft 1). An KCl–97 ist ferner auch noch der allmählich weiter fortschreitende Abbau zu erkennen (Eigenschaft 1). Die Abnahme in einem Zeitraum von einem Jahr kann bereits durch k^* erfaßt werden (vgl. Kapitel 4.1).

Das ist in Übereinstimmung mit Ergebnissen an γ-bestrahltem NaCl, bei dem Proben aus extrem langsam abgekühlten Kristallblöcken weniger verfärbbar waren als solche mit schneller Abkühlgeschwindigkeit [28]. Mit dieser Alterungserscheinung verbunden ist auch eine langsamere F-Zentren-Bildung, die möglicherweise von gleicher Natur wie die unterschiedliche Bildungsgeschwindigkeit für F-Zentren in natürlichem und synthetischem NaCl ist [29].

Die allmähliche Ausscheidung des atomdispers eingebauten Ca kann in KCl–97 durch Temperung bei einer relativ niedrigen Temperatur — z. B. 320 °C — wesentlich beschleunigt werden: Die Verfärbbarkeit nimmt auf Werte des KCl–12 ab (Eigenschaft 2). Das bedeutet, daß sich in KCl–12 durch die 4-jährige Lagerung bei Zimmertemperatur ein Verteilungszustand der Ca-Verunreinigungen eingestellt hat, der etwa dem nach Temperung bei 320 °C entspricht.

Tempern bei 530 °C bringt die Ausscheidungen praktisch vollständig in Lösung. Nachfolgendes Abschrecken friert diesen Zustand ein. Bei gleicher Gesamtverunreinigung an Ca — wie sie für KCl–12 und KCl–97 bei gleichen Ausgangssubstanzen und gleichen Züchtungsbedingungen vorliegen dürfte — sollte sich mithin die gleiche Wirkung auf sonst unterschiedlich vorbehandelte Kristallproben ergeben: Zunahme der Verfärbbarkeit auf gleiche absolute Werte. Die relative Zunahme ist bei KCl–12 erwartungsgemäß größer, die erreichten Absolutwerte aber nach Abschrecken für KCl–12 und KCl–97 von etwa gleicher Größe (Eigenschaft 3), vergleiche auch Tabelle 1.

Die eben besprochenen Prozesse über die Verschiebung des Gleichgewichtes der Ca-Verunreinigungen in KCl-Kristallen durch thermische Behandlung sind mit einer charakteristischen Veränderung der n_1-F-Zentren-Konzentration verbunden (die zugehörigen Werte sind in Tabelle 1 aufgeführt). Dies ist in Übereinstimmung mit unserer früheren Feststellung, daß zwischen der n_1-Konzentration und dem Ca-Ionengehalt in KCl eine eindeutige Beziehung besteht [30].

Tempern bei 530 °C und nachfolgendes Abschrecken bringt nicht nur das Ca wieder in Lösung, sondern führt auch durch die auftretenden mechanischen Spannungen zu einer merklichen plastischen Deformation [31], wodurch ebenfalls die n_1-Konzentration erhöht wird. Auf den Zusammenhang zwischen n_1 und der plastischen Deformation wird im zweiten Teil der Diskussion eingegangen.

5.1.2 Ausheilung

Die Ausheilung der durch thermische Vorbehandlung erzeugten Defekte ist durch zwei Eigenschaften charakterisiert:

1. Der Ausheilvorgang erfolgt mit Aktivierungsenergien von etwa 1,60 eV.
2. Die Ausheilung führt zu minimaler Verfärbbarkeit durch F-Zentren, wie sie in lange bei Zimmertemperatur gelagerten Kristallen gemessen wird.

Nach der obigen Diskussion ist die Ausheilung der für die zusätzliche F-Zentren-Bildung verantwortlichen Defekte offenbar mit der Ausscheidung von atomdispers vorliegenden Ca-Ionen verbunden. Außerdem müssen bei der Ausheilung die nach der optischen Bleichung noch vorhandenen zusätzlichen Anionenleerstellen beseitigt werden. Die Ausscheidung der Ca-Ionen erfordert deren Transport über endliche Entfernungen im Gitter. Das ist nur möglich über die Diffusion des neutralen Assoziates (Ca-Ion + Kationenleerstelle) [32]. In NaCl-Kristallen ist dafür eine Aktivierungsenergie von 0,96 eV [33] bzw. 1,0 eV für Sr-Ionen [34] gefunden worden. Wenn man annimmt, daß für KCl ähnliche Werte gültig sind, so liegen diese um 0,6 eV bis 0,7 eV unter den hier bei der Ausheilung ermittelten E_A-Werten. Man muß jedoch dabei berücksichtigen, daß das Assoziat (Ca-Ion + Kationenleerstelle) sicher nicht frei im Kristall vorliegt, sondern an weitere Störstellen gebunden ist, von denen es erst mit Energieaufwand getrennt werden muß, und daß — falls der Crawford-Nelson-Mechanismus gültig ist — auch Energie verbraucht wird, um das Cl-Atom an einen normalen Cl-Ionen-Platz zurückzubringen. Werte von 0,6 eV bis 0,7 eV erscheinen dafür durchaus plausibel.

Es ist weiter möglich, daß beim Abschrecken auch ein zu höherer Temperatur gehörendes thermisches Gleichgewicht in bezug auf die Anionenleerstellen im Kristall eingefroren wird. Da sich die Konzentration der statistisch verteilten n_0-Leerstellen aber bei unseren Versuchen im Gegensatz zu den Ergebnissen in [5] praktisch nicht ändert, muß angenommen werden, daß ein großer Teil der überschüssigen thermischen Leerstellen relativ schnell an Gitterinhomogenitäten (z. B. Versetzungen) abgebaut wird, möglicherweise aber zum Teil auch zu Leerstellen-Clustern koaguliert. Dieser durch die Diffusion der Leerstellen bestimmte Prozeß sollte bei hoher Versetzungsdichte und großer Konzentration innerer Senken (internal sinks [22]) besonders schnell ablaufen. Ein höherer Störgrad unserer Kristalle könnte mithin den Unterschied im n_0-Verhalten im Vergleich zu [5] erklären. Die nach Abschrecken in der Umgebung von Versetzungen und anderen Störstellen vorliegenden Leerstellen-Cluster können ebenfalls als zusätzliche Quellen für n_1-Leerstellen bei der Röntgenverfärbung dienen. Die Ausheilung solcher Cluster mit Aktivierungsenergien von etwa 1,6 eV wird auch in deformierten und röntgenvorbestrahlten Kristallen gefunden und im nachfolgenden Kapitel 5.2.2 diskutiert.

5.2 *Defekte durch Deformation und Röntgenvorbestrahlung*

5.2.1 *Bildung*

Die Ergebnisse über die Bildung der Defekte durch Deformation und Bestrahlung lassen sich wie folgt zusammenfassen:

1. Die plastische Deformation bewirkt eine erhöhte Verfärbbarkeit im Bercich I der F-Zentren-Bildungskurve durch Zunahme der n_0- und n_1-Leerstellenkonzentration. Die n_1-Konzentration nimmt dabei wesentlich mehr als die n_0-Konzentration zu.
2. Die Wirksamkeit der plastischen Deformation wird in ausgeheilten Kristallen vergrößert, in Kristallen mit Erinnerungseffekt dagegen verkleinert.
3. Intensive Röntgenbestrahlung und nachfolgende Bleichung führt bei erneuter Bestrahlung zu einer erhöhten Verfärbbarkeit im Bereich I der F-Zentren-Bildungskurve, die durch eine Zunahme der n_1-Konzentration bewirkt wird.

Der Einfluß der Deformation auf den Bereich I der F-Zentren-Bildungskurve ist schon mehrfach nachgewiesen worden [5, 28, 35]. Einen besonders eindrucksvollen Beweis dafür erbrachte Sibley [36]: ein bereits bis zum Bereich II verfärbter Kristall liefert nach Deformation bei Weiterbestrahlung erneut einen für den Bereich I charakteristischen Kurvenverlauf. Die Deformation muß demnach

Defekte erzeugt haben, die sich wie n_0- und n_1-Leerstellen verhalten. In der vorliegenden Arbeit wird in Übereinstimmung mit den Ergebnissen von Mitchell, Wiegand und Smoluchowski [5] aus den Verfärbungskurven deformierter KCl-Kristalle eine dominierende Zunahme der n_1-Leerstellenkonzentration gefunden.

Der Einfluß der plastischen Deformation ist zu erklären durch die Wechselwirkung zwischen Versetzungen und Leerstellen im Kristall. Für die Bildung von Leerstellen sind nichtkonservative Bewegungen von Versetzungen („climbing") verantwortlich [37, 38]. Climbing-Prozesse können angeregt werden durch Störungen in der Gleitebene (z. B. Verunreinigungen) oder durch Schneiden von Versetzungen aus verschiedenen Gleitebenen. Leerstellen, die danach im Bereich der Versetzungen liegen, bilden wegen der hohen räumlichen Dichte in diesem Gebiet größere Leerstellen-Cluster [35, 39, 40], die bei Zimmertemperatur zusätzliche Quellen für n_1-Leerstellen darstellen (Eigenschaft 1). Leerstellen, die in nach dem Durchlaufen der Versetzungen ungestörten Kristallbereichen zurückbleiben, werden zu Leerstellenpaaren oder anderen kleinen Clustern zusammendiffundieren [41, 42] und sollten als Quellen für isolierte n_0-Leerstellen wirken (Eigenschaft 1).

Die Ergiebigkeit der Versetzungen in Zusammenhang mit dem Climbing-Prozeß richtet sich auch nach dem Leerstellenangebot, welches sie im Kristall vorfinden, da sie nicht nur als Quellen, sondern auch als Senken für Leerstellen wirken können [42 bis 45]. Das vergrößerte Leerstellenangebot in Kristallen mit Erinnerungseffekt (vergleiche den steilen Anfangsverlauf der entsprechenden Bildungskurve in Fig. 4 und den hohen n_1-Wert in Tabelle 1) dürfte deshalb hemmend wirken auf die Produktion weiterer Leerstellen bei Deformation (Eigenschaft 2).

Die bevorzugte Wirksamkeit der Deformation in ausgeheilten Proben aus KCl–97 (Eigenschaft 2) kann durch den Gehalt an 2-wertigen Verunreinigungen erklärt werden. In ausgeheilten Proben sind — wie in 5.1.1 dargelegt — die Verunreinigungen zu Komplexen ausgeschieden, die nicht mehr als n_1-F-Zentren-Quellen wirksam sind. Diese Ausscheidungen können auch durch Deformation wieder „gelöst" werden [23, 24, 28]. Die Deformation erzeugt mithin in ausgeheiltem KCl–97 auch noch zusätzliche F-Zentren-Quellen durch Verteilung der 2-wertigen Verunreinigungen.

Für die Bildung der *a t*-Leerstellen durch Röntgenquanten haben Seitz [1 b], Varley [41] und Mitchell, Wiegand und Smoluchowski [5] Modelle angegeben, die die Mitwirkung von Versetzungen voraussetzen. Die primär entstehenden Anionenleerstellen laufen im allgemeinen zu Clustern zusammen, wenn sie nicht durch Einfang eines Elektrons zu einem F-Zentrum stabilisiert werden. Diese *a t*-F-Zentren können lichtoptisch und thermisch gebleicht werden (siehe Kapitel 4.4.2). Beide Bleichprozesse führen über Komplexzentren ebenfalls zu Leerstellen-Clustern [46 bis 50]. Diese Cluster sollten als zusätzliche Quellen für n_1-Leerstellen verantwortlich sein für die Zunahme der F-Zentren-Bildung im Bereich I der Bildungskurve bei erneuter Verfärbung (Eigenschaft 3).

5.2.2 Ausheilung

Für die Ausheilung der durch Deformation und Röntgenvorbestrahlung erzeugten Defekte ergaben sich die folgenden zwei Eigenschaften:

1. Die Ausheilung erfolgt mit einer diskreten Aktivierungsenergie.
2. Von Versuch zu Versuch ergeben sich unterschiedliche Werte für die Aktivierungsenergie, bei plastisch deformierten Proben zwischen 1,65 eV und 2,0 eV,

bei röntgenvorbestrahlten Proben zwischen 1,55 eV und 2,0 eV (vergleiche Tabelle 3). Die Schwankungen liegen außerhalb der Fehlerbreite des Verfahrens zur Bestimmung der Aktivierungsenergie.

Nach der voranstehenden Diskussion sind für die Verfärbbarkeitszunahme sowohl nach plastischer Deformation als auch nach Röntgenvorbestrahlung Leerstellen-Cluster verantwortlich. Ein weiterer Hinweis dafür, daß Cluster und nicht isolierte Anionenleerstellen die Verfärbbarkeitserhöhung bewirken, ergibt sich hier nun unmittelbar aus der Streubreite der gefundenen Aktivierungsenergien. Die Ausheilung isolierter Leerstellen müßte nämlich innerhalb der Fehlerbreite des Verfahrens in allen Versuchen mit konstanter Aktivierungsenergie — der Wanderungsenergie der Anionenleerstellen — erfolgen.

Bei der Ausheilung der Verfärbbarkeitserhöhung hat man mithin den Abbau der Cluster zu betrachten. Zu einer analogen Schlußfolgerung gelangt Thommen [6] bei der Ausheilung deuteronen- und röntgenvorbestrahlter KCl-Kristalle. Für deuteronenbestrahlte Proben ergab sich in Übereinstimmung mit den in der vorliegenden Arbeit gefundenen Werten eine Aktivierungsenergie von 1,97 eV; für röntgenvorbestrahlte Proben soll der Wert etwas niedriger liegen. Die Ausheilung der Cluster wird im Zusammenhang mit der Wanderung von Doppellücken diskutiert. Auch die von Davidge und Whitworth [51] an deformierten NaCl-Kristallen untersuchten „debris" sollten den hier beschriebenen Clustern entsprechen, da die debris etwa in dem gleichen Temperaturgebiet wie die Cluster ausheilen.

Am einfachsten lassen sich diese Cluster abbauen durch die Diffusion isolierter Anionenleerstellen zu geeigneten Senken im Gitter. Die wirksame Aktivierungsenergie setzt sich dann aus der Energie für die Wanderung (migration) der Anionenleerstellen E_M und der Energie für deren Bildung (formation) E_F zusammen: $E_A = E_M + E_F$. Für E_M werden in der Literatur Werte zwischen 0,89 eV und 1,86 eV angegeben [52 bis 56]. Die experimentellen Bedingungen von Damm und Kowalczyk [54] kommen denen der vorliegenden Arbeit am nächsten. Mit ihrem Wert von 1,3 eV für E_M ergeben sich aus unseren E_A-Werten etwa 0,3 bis 0,7 eV für die Bildungsenergie E_F der Anionenleerstellen. Diese Werte sind wesentlich kleiner als die Hälfte der Bildungsenergie für ein Schottky-Paar, wofür Werte zwischen 2,1 und 2,6 eV angegeben wurden [55, 57 bis 61]. Die Cluster haben mithin nicht die Eigenschaften einer „freien" inneren Oberfläche, sondern dürften nur relativ kleine, aus wenig Leerstellen aufgebaute Aggregate darstellen, für die E_F stark vermindert ist. Der genaue Wert von E_F ist sicher abhängig von der Form und Größe des Clusters, die Größe des Clusters wiederum von den speziellen Bedingungen der Vorbehandlung, insbesondere von der Konzentration der erzeugten Defekte. Dies erklärt die Streubreite der gefundenen Werte für die Aktivierungsenergie (Eigenschaft 2). Da sich bei den einzelnen Experimenten diskrete Aktivierungsenergien ergeben (Eigenschaft 1), dürften sich die in einer vorgegebenen Probe vorhandenen Cluster in ihrer Größe nur sehr wenig unterscheiden.

Herrn Prof. Dr. W. Messerschmidt danken wir für sein der Arbeit entgegengebrachtes förderndes Interesse, Herrn Prof. Dr. H. Bethge für vielerlei Anregung und Unterstützung während der Arbeit und für die kritische Durchsicht des Manuskriptes.

Literatur

[1a] R. W. POHL, Phys. Z. **39**, 36 (1938).
[1b] F. SEITZ, Rev. mod. Phys. **18**, 384 (1946); **26**, 7 (1954).
[1c] H. PICK, Nuovo Cimento Suppl. **7**, 498 (1958).

[1d] O. Stasiw, Elektronen- und Ionenprozesse in Ionenkristallen, Springer 1959.
[1e] J. H. Schulman und W. D. Compton, Color Centers in Solids, Pergamon Press, 1962.
[1f] S. Amelinckx, Radiation Damage in Solids, Academic Press, New York/London 1962.
[2] R. B. Gordon und A. S. Nowick, Phys. Rev. **101**, 977 (1956).
[3] H. Rabin, Phys. Rev. **116**, 1381 (1959).
[4] F. Fröhlich, Z. Naturf. **16a**, 211 (1961).
[5] P. V. Mitchell, D. A. Wiegand und R. Smoluchowski, Phys. Rev. **121**, 484 (1961).
[6] K. Thommen, Z. Naturf. **16a**, 992 (1961).
[7] C. J. Meechan und J. A. Brinkman, Phys. Rev. **103**, 1193 (1956).
[8] F. Fröhlich und P. Grau, phys. stat. sol., erscheint demnächst.
[9] A. Smakula, Z. Phys. **59**, 603 (1930).
[10] F. Abramson und W. E. Caspari, Phys. Rev. **129**, 536 (1963).
[11] B. Mozer und P. Levy, zitiert in [5].
[12] R. W. Christy und D. H. Phelps, Phys. Rev. **124**, 1053 (1961).
[13] R. W. Christy und E. Fukushima, Phys. Rev. **118**, 1222 (1960).
[14] C. Z. van Doorn und Y. Haven, Philips Res. Rep. **11**, 479 (1956).
[15] C. Z. van Doorn und Y. Haven, Philips Res. Rep. **12**, 309 (1957).
[16] H. Pick, Z. Phys. **159**, 69 (1960).
[17] H. Hummel, Dissertation, Göttingen 1950.
[18] F. Okamoto und A. B. Scott, J. Phys. Soc. Japan **18**, 243 (1963).
[19] S. Miyake und K. Suzuki, J. Phys. Soc. Japan **9**, 802 (1954).
[20] S. Amelinckx, W. van der Vorst, R. Gevers und W. Dekeyser, Phil. Mag. **46**, 450 (1955).
[21] J. H. Crawford jr. und C. M. Nelson, Phys. Rev. Letters **5**, 314 (1960).
[22] A. B. Lidiard, Hdbch. der Physik, Bd. XX, Springer 1957.
[23] D. B. Fischbach und A. S. Nowick, J. Phys. Chem. Solids **2**, 226 (1957).
[24] F. Fröhlich und D. Suisky, phys. stat. sol. **4**, 151 (1964).
[25] J. B. Caffyn und B. K. Ridley, Proc. Phys. Soc. **77**, 153 (1961).
[26] H. Pick, Ann. Phys. (Germany) **35**, 73 (1939); Z. Phys. **114**, 127 (1939).
[27] F. Seitz, Phys. Rev. **83**, 134 (1951).
[28] A. S. Nowick, Phys. Rev. **111**, 16 (1958).
[29] H. Adler, Acta phys. Austriaca **13**, 369 (1960).
[30] F. Fröhlich, Z. Naturf. **17a**, 327 (1962).
[31] B. H. Kear und P. L. Pratt, Phil. Mag. **37**, 56 (1959).
[32] F. A. Kröger, Chemistry of Imperfect Solids, North-Holland, Amsterdam 1964 (S. 220ff.).
[33] A. H. Murin, C. H. Banasewitsch und J. C. Gruschko, Fiz. tverd. Tela **3**, 2427 (1961).
[34] M. Chemla, zit. bei Lidiard [22] (S. 335).
[35] R. W. Davidge und P. L. Pratt, phys. stat. sol. **3**, 665 (1963).
[36] W. A. Sibley, Phys. Rev. **133**, 1176 (1964).
[37] F. Seitz, Adv. Phys. **1**, 43 (1952).
[38] A. Seeger, Hdbch. der Physik, Bd. VII, Springer 1955 (S. 476ff.).
[39] M. Ueta und W. Känzig, Phys. Rev. **97**, 1591 (1955).
[40] F. Seitz, Phys. Rev. **80**, 239 (1950).
[41] J. H. O. Varley, J. Phys. Chem. Solids **23**, 985 (1962).
[42] J. H. Crawford jr., J. Phys. Soc. Japan **18**, 329 (1963).
[43] F. Bassani und R. Thompson, Phys. Rev. **102**, 1264 (1956).
[44] Vgl. z. B. auch H. G. van Bueren, Imperfections in Crystals, North-Holland, Amsterdam 1961.
[45] H. Bethge, Sixth Congress of the International Union of Crystallography, Rom 1963, Acta cryst. **16**, A 108 (1963).
[46] W. E. Bron und A. S. Nowick, Phys. Rev. **119**, 114 (1960).

[47] N. Itoh und T. Suita, J. Phys. Soc. Japan **17**, 345 (1962).
[48] H. Rabin, Phys. Rev. **129**, 129 (1963).
[49] H. W. Etzel und J. B. Allard, Phys. Rev. **116**, 885 (1959).
[50] H. N. Hersh, J. chem. Phys. **30**, 790 (1959).
[51] R. W. Davidge und R. W. Whitworth, Phil. Mag. **6**, 217 (1961).
[52] J. D. Laurent und J. Benard, J. Phys. Chem. Solids **3**, 7 (1957).
[53] R. Guccione, M. P. Tosi und M. Asdente, J. Phys. Chem. Solids **10**, 162 (1959).
[54] J. Z. Damm und J. Kowalczyk, phys. stat. sol. **6**, 693 (1964).
[55] R. A. Allnatt und P. W. Jacobs, Trans. Faraday Soc. **58**, 116 (1962).
[56] J. Ewles und S. C. Jain, Proc. Roy. Soc. **A 243**, 353 (1958).
[57] C. Wagner und P. Hantelmann, J. chem. Phys. **18**, 72 (1950).
[58] T. E. Phipps und E. G. Partridge, J. Amer. Chem. Soc. **51**, 1331 (1929).
[59] F. G. Fumi und M. P. Tosi, Disc. Faraday Soc. **23**, 92 (1957).
[60] H. Kelting und H. Witt, Z. Phys. **126**, 697 (1949).
[61] A. Scholz, phys. stat. sol. **7**, 973 (1964).
[62] B. Fritz, F. Lüty und J. Anger, Z. Phys. **174**, 240 (1963).

(Received December 14, 1964)

See Short Note by L. DELAEY and H. WARLIMONT

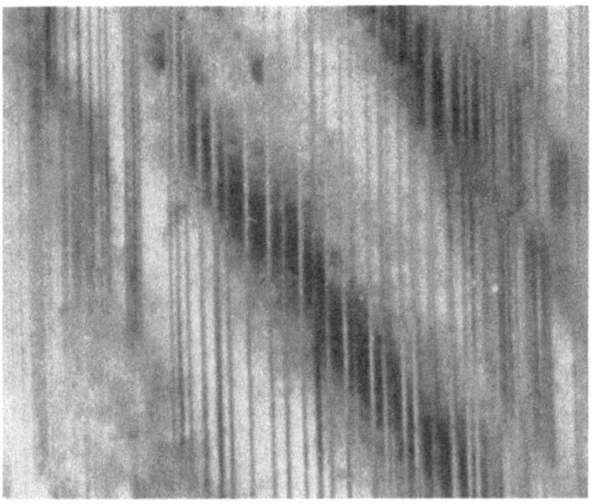

Fig. 1. Gefügebild des β_1''–Cu–Zn–Ga-Martensits. Innere Lamellierung parallel zum Elektronenstrahl (105000 ×)

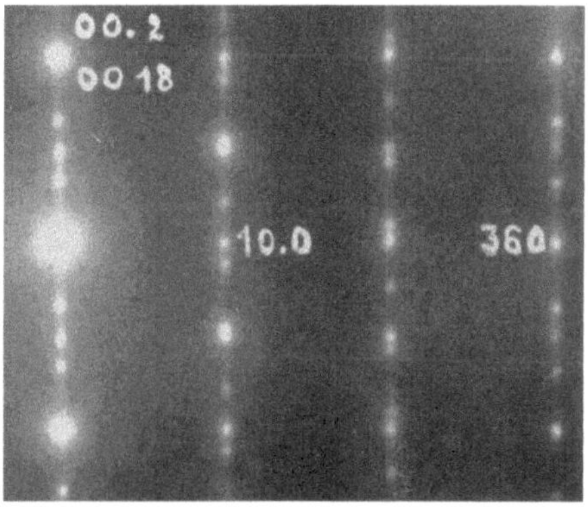

Fig. 2. Elektronenbeugungsdiagramm des β_1''–Cu–Zn–Ga-Martensits. Zone $[3\bar{2}0]_0$ und $[01\bar{1}0]_h$

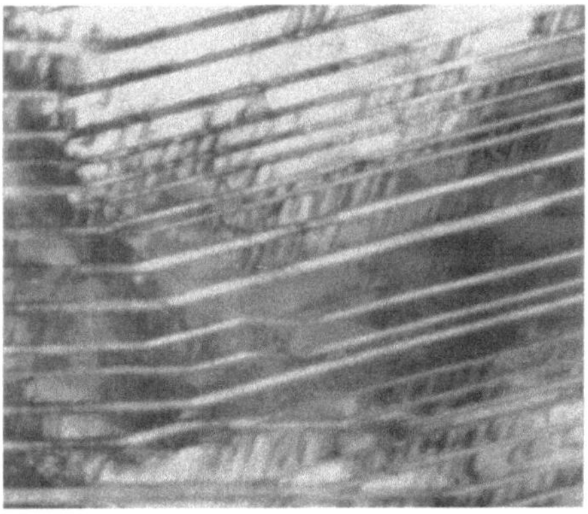

Fig. 3. Innere Verzwillingung des γ_1'–Cu–Ga-Martensits parallel zum Elektronenstrahl, mit sekundärer Verzwillingung (50000 ×)

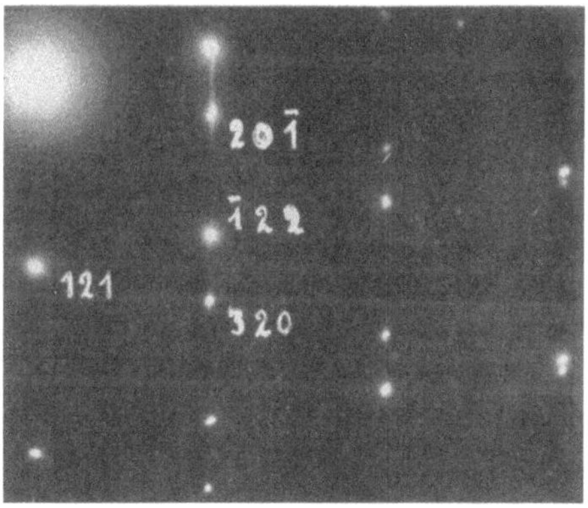

Fig. 4. Elektronenbeugungsdiagrmm des γ_1'–Cu–Ga-Martensits. Zone $[2\bar{3}4]$

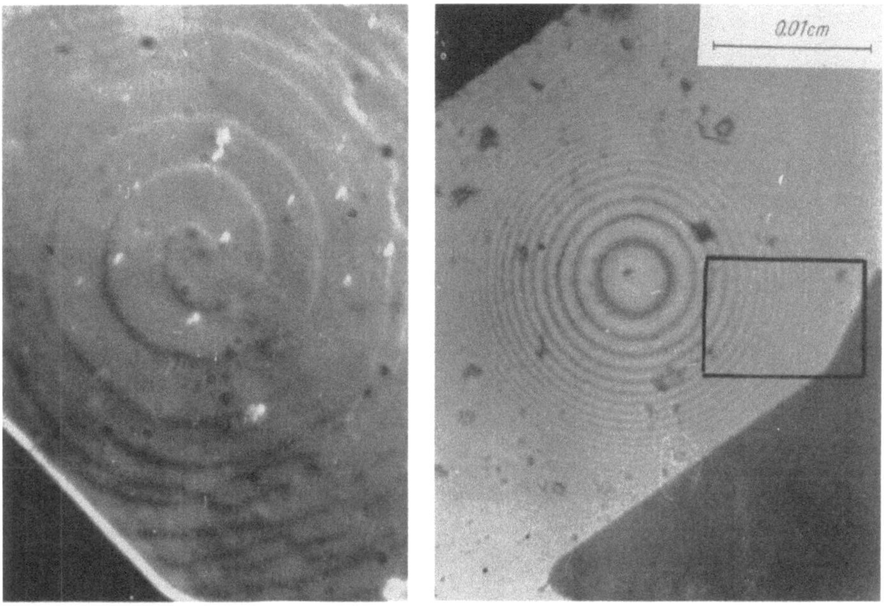

a b

Fig. 1.

a) Round growth spiral on the basal plane of a ZnS needle. The microphotograph was made by means of Nomarski's interference-contrast arrangement at maximal contrast

b) Microphotograph of the basal plane of a crystal containing a microspiral. This microphotograph was made with the interference tester in Na light. The marked region corresponds to the area of Fig. 1a with the microspiral

Fig. 2. Polygonal macrospiral on the basal plane of a ZnS needle. The microphotograph was made by means of Nomarsk'is interference-contrast arrangement

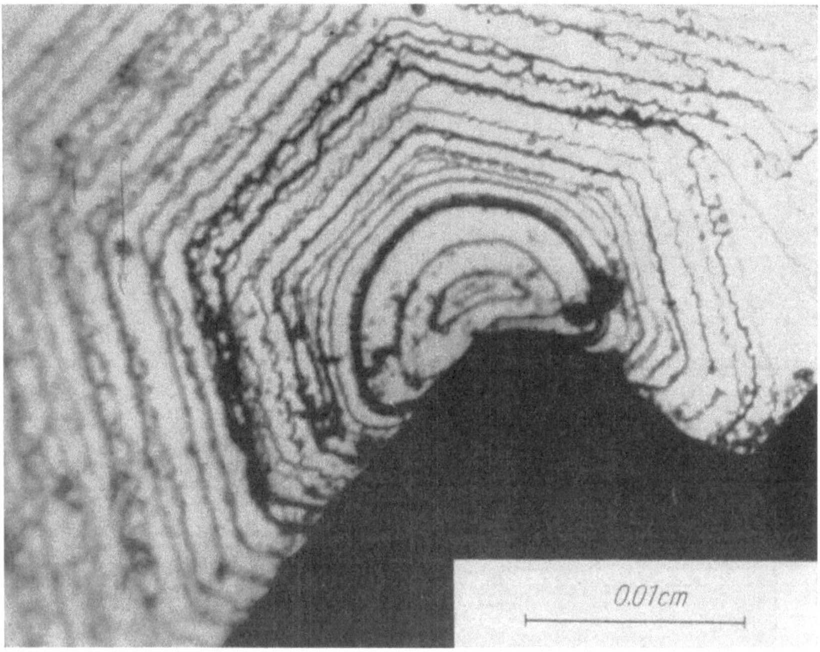

Fig. 3. Growth pattern on the basal plane of a ZnS needle produced by screw dislocations of opposite sign, which emerge near to each other. The microphotograph was made by means of Nomarski's interference-contrast arrangement

Short Notes

phys. stat. sol. 8, K121 (1965)
Max-Planck-Institut für Metallforschung, Stuttgart

Diffusionslos gebildete Phasen in Cu-Ga-, Cu-Zn-Ga- und Cu-Zn-Legierungen

Von

L. DELAEY und H. WARLIMONT

Das Gefüge und die Struktur der aus der kubisch-raumzentrierten β-Phase durch diffusionslose Umwandlung entstehenden Martensitphasen wurden an verschiedenen Cu-Ga- und Cu-Zn-Ga-Legierungen elektronenmikroskopisch untersucht und für Cu-Zn-Legierungen auf Grund dieser und anderer Untersuchungen berechnet.

Nach U. JOLLEY und D. HULL (1) soll der Cu-Zn-Martensit eine verzwillingte orthorhombische Struktur aufweisen. Nach Röntgenuntersuchungen von G. KUNZE (2, 3, 4) soll sich in β-Cu-Zn-Legierungen bei mäßiger Unterkühlung eine monoklin flächenzentrierte und bei stärkerer Unterkühlung eine triklin basiszentrierte martensitische Struktur einstellen. Durch D.B. MASSON und R.K. GOVILA (5) war bisker bekannt, daß sich Cu-Zn-Ga- und Cu-Zn-Martensit röntgenographisch nur wenig unterscheiden.

Während der Durchführung der vorliegenden Untersuchungen wurde die Bestimmung von 2 Cu-Ga-Martensitstrukturen durch T. SABURI und C.M. WAYMAN (6) aus einer Vortragszusammenfassung bekannt. Dabei soll es sich um eine kubisch-flächenzentriert ungeordnete und eine kubisch-flächenzentriert geordnete Struktur mit Stapelfehlern analog zu β' - und β'_1-Cu-Al-Martensit (7, 8) handeln.

Bei unseren elektronenmikroskopischen Untersuchungen an den bei Raumtemperatur stabilen martensitischen Zuständen von Cu- 19,2 At.% Zn- 12,6 At.% Ga- und Cu- 23,7 At.% Ga-Legierungen wurden 2 Martensitphasen β''_1 und γ'_1 beobachtet. Das Gefüge des β''_1-Martensits besteht aus Platten, die in Lamellen von 20 bis 400 Å Stärke unterteilt sind (Fig. 1). Die sehr komplexen Elektronenbeugungsdiagramme, wie z.B. Fig. 2, dieser Platten lassen bei leichtem Verkippen des Präparates (um ca. $1/2^{\circ}$) erkennen, daß sie zwei einander überlagerte Strukturen wiedergeben. Eine dieser Strukturen ist den Martensitstrukturen β_1'-Cu-Al (7, 8) und β_1'-Cu-Zn (9) sehr ähnlich. Sie kann als Stapelfolge $AB_1'CB'C_1A'CA_1'BA'B_1C'BC_1'AC'A_1B'$ nahezu dichtest gepackter Netzebenen mit der folgenden Atomordnung beschrieben werden:

$$(A) \;:\; \mathrm{Cu}\ (0,0;\ \tfrac{1}{2},\ \tfrac{1}{4};\ \tfrac{1}{2},\ \tfrac{3}{4})\ ;\ (\mathrm{Zn},\ \mathrm{Ga})\ (0,\ \tfrac{1}{2})$$

$$(A_1) \;:\; \mathrm{Cu}\ (0,0;\ \tfrac{1}{2},\ \tfrac{3}{4})\ ;\ (\mathrm{Zn},\ \mathrm{Ga})\ (0,\ \tfrac{1}{2};\ \tfrac{1}{2},\ \tfrac{1}{4})$$

$$(B) = (A) + \overrightarrow{(0,\ \tfrac{2}{3})}\ ;\ (C) = (A) + \overrightarrow{(0,\ \tfrac{1}{3})}$$

$$(B_1) = (A_1) + \overrightarrow{(0,\ \tfrac{2}{3})}\ ;\ (C_1) = (A_1) + \overrightarrow{(0,\ \tfrac{1}{3})}$$

$$(A') = (A) + \overrightarrow{(0,\ \tfrac{1}{2})}\ ;\ (B') = (B) + \overrightarrow{(0,\ \tfrac{1}{2})}\ ;\ (C') = (C) + \overrightarrow{(0,\ \tfrac{1}{2})}$$

$$(A_1') = (A_1) + \overrightarrow{(0,\ \tfrac{1}{2})};\ (B_1') = (B_1) + \overrightarrow{(0,\ \tfrac{1}{2})};\ (C_1') = (C_1) + \overrightarrow{(0,\ \tfrac{1}{2})}$$

$(A) + \overrightarrow{(0,\ \tfrac{1}{2})}$ bedeutet, daß alle Atome in der Ebene A um $\overrightarrow{(0,\ \tfrac{1}{2})}$ verschoben sind.

Die zweite dieser Strukturen hat ebenfalls orthorhombische Symmetrie und kann als eine nahezu hexagonal dichteste Kugelpackung mit geordneter Atomverteilung beschrieben werden. Sie ist damit der Martensitstruktur γ'-Cu-Al (7) sehr ähnlich. Aus Dunkelfeldaufnahmen des Gefüges konnte man nachweisen, daß der β_1''-Cu-Zn-Ga-Martensit aus einander abwechselnden Lamellen dieser beiden Strukturen besteht. Die Röntgenaufnahmen von D.B. MASSON und R.K. GOVILA (5) stehen mit dieser Strukturbestimmung im Einklang.

Der γ_1'-Martensit, der hauptsächlich in Zn-armen Legierungen und in binären Cu-Ga-Legierungen auftritt, besteht aus fein verzwillingten Platten (Zwillingsdicken etwa 175 und 375 Å) (Fig. 3). Seine Struktur ist orthorhombisch und besteht aus einer geordneten, nahezu hexagonal dichtesten Kugelpackung wie der eine Strukturanteil des β_1''-Martensits (Fig. 4). Gefüge und Struktur entsprechen damit denen des γ'-Cu-Al-Martensits (7). Aus der Überstruktur dieses Martensits konnte auf eine bisher unbekannte Überstruktur der Hochtemperatur-β-Phase mit dem Strukturtyp Fe_3Si ($F^{3,\ 1}$) (10) geschlossen werden.

Die Struktur des Cu-Zn-Martensits ist bei Raumtemperatur nicht stabil und konnte deshalb auch nicht elektronenmikroskopisch untersucht werden. Aus der Kenntnis der Struktur des β_1''-Cu-Zn-Ga-Martensits, der Gitterparameter der α- und β-Phase und der Überstruktur vom Typ CsCl ($C^{1,\ 1}$) (10) der reinen Cu-Zn-Legierungen konnte jedoch auf Grund der Röntgendaten von D.B. MASSON und R.K. GOVILA (5) und mit Hilfe von Berechnungen nach der phänomenologischen Martensittheorie (11) darauf geschlossen werden, daß der Cu-Zn-Martensit eine orthorhombische Struktur besitzt, deren Atomverteilung aus der Überstruktur der β-Phase und deren Stapelfolge in der c-Richtung aus der martensitischen inhomogenen Scherung eines kubisch-flächenzentrierten Gitters abzuleiten ist.

Die so berechnete Struktur des Cu-Zn-Martensits, den wir mit β_2' bezeichnen, besteht (bei 39,5 At.% Zn) aus Netzebenen mit der Atomverteilung

$$(A)\ :\ Cu\ (0,\ 0)\ ;\ Zn\ (\tfrac{1}{2},\ \tfrac{1}{2})$$

$$(B)\ =\ (A) + \overrightarrow{(0,\ \tfrac{2}{3})}$$

$$(C)\ =\ (A) + \overrightarrow{(0,\ \tfrac{1}{3})}$$

etwa in der Stapelfolge ABCBCACABABCBCACA und nimmt damit eine orthorhombische Elementarzelle mit den Abmessungen a = 4,417 Å; b = 2,665 Å und c = 36,29 Å an.

Diese Ergebnisse stehen im Einklang mit Beobachtungen und allgemeinen Zusammenhängen der Martensitphasen in Kupferlegierungen (12). Sie werden in ausführlicher Form an anderer Stelle veröffentlicht (13).

Wir danken Herrn Dr. M. WILKENS für wertvolle Hinweise. Der Deutschen Forschungsgemeinschaft sind wir für Unterstützung der Versuche zu Dank verpflichtet. L. DELAEY hat diese Arbeit teilweise als Stipendiat des DAAD und teilweise als Aspirant des belgischen NFWO ausgeführt.

Literatur

(1) W. JOLLEY und D. HULL, J. Inst. Metals 92, 129 (1963/64).
(2) G. KUNZE, Z. Metallk. 53, 329 (1962).
(3) G. KUNZE, Z. Metallk. 53, 396 (1962).
(4) G. KUNZE, Z. Metallk. 53, 565 (1962).
(5) D.B. MASSON und R.K. GOVILA, Z. Metallk. 54, 293 (1963).
(6) T. SABURI und C.M. WAYMAN, Vortrag beim Fall Meeting 1964 of Met. Soc. of AIME, Philadelphia.
(7) P.R. SWANN und H. WARLIMONT, Acta metall. 11, 511 (1963).
(8) H. WARLIMONT und M. WILKENS, Z. Metallk. 55, 382 (1964).
(9) H. WARLIMONT und D. HÄRTER, Z. Metallk. (demnächst).
(10) K. SCHUBERT, Kristallstrukturen zweikomponentiger Phasen, Springer-Verlag, Berlin-Göttingen-Heidelberg 1964.
(11) M.S. WECHSLER, D.S. LIEBERMAN und T.A. READ, Trans. AIME 197, 1503 (1953).
(12) H. WARLIMONT, Proceedings of "Physical Metallurgy of Bainite and Martensite", British Iron and Steel Research Association and Iron and Steel Institute, 1965.
(13) L. DELAEY und H. WARLIMONT, Z. Metallk. 56 (1965).

(Received December 7, 1964)

phys. stat. sol. 8, K125 (1965)

Research Institute for Technical Physics of the Hungarian Academy of Sciences, Budapest

Growth Spirals on ZnS Crystals

By

E. LENDVAY and P. KOVÁCS

1. Introduction

Frank's theory of the dislocation mechanism of crystal growth has been proved by topological studies for numerous substances (1 to 10). In the case of many hexagonal crystals, screw dislocation produce spiral growth-steps on the basal plane, and these may join into dislocation-terraces. For CdS such growth patterns were observed by several authors (3, 11, 12).

So far there are only a small number of investigations dealing with ZnS. Evidence for a growth of ZnS by screw dislocations has been found by GREENE et al. (13). SAMELSON and BROPHY support the dislocation mechanism of growth for ZnS, but growth spirals could not be observed by them (14). Similar results are described in (15, 16).

2. Growth Spirals on ZnS

According to the theory of crystal growth the rate of advance of a straight step is

$$U = 2(\sigma - 1)X_s \nu \exp(-\frac{W}{kT}), \qquad (1)$$

where $(\sigma - 1)$ is the supersaturation, X_s the mean diffusion length of an atom or molecule on the crystal surface, ν an atomic vibrational frequency, and W the evaporation energy. It is expected that during growth from vapor the surface diffusion length exceeds the mean distance between kinks in a growth step (X_o):

$$X_s \gg X_o , \qquad (2)$$

therefore the step will advance independently of its orientation. If this step originates from a screw-dislocation during growth, a circularly symmetric spiral forms. As the step height of this type of growth spiral is often not large enough to be resolved optically, it was seldom possible to observe such microspirals.

Their optical observation is only possible by means of the interference-contrast method. Using a sensitive reflection microscope, it is possible to observe growth layers as thin as a few Å (17). Nomarski's interferometric contrast arrangement which is a double-beam interferometer was used in our experiments. In this arrangement spiral steps were often found on the basal planes of ZnS needles, if the basal or top planes were intact. Fig. 1a shows a microphotograph of such a spiral. The whole basal plane of the investigated crystal is covered by spiral centre, too. Fig. 1b shows, for comparison, the interferogram of the examined basal plane. It is shown that the step height of the rounded spiral cannot be well measured with the interference tester, i.e., the step height is about a few unit cells.

If $X_s \ll X_o$ the rate of advance of a step has its minimum when the step lies parallel to the close-packed direction. Directional growth results, the edges of the steps becoming straight and perpendicular to the slow growth direction. Generally, this is the case in which macrospirals form due to the presence of screw dislocations. Such macrospirals were observed on the basal planes of ZnS needles. The step height of these spirals is of the same order as in the case of CdJ_2, SiC, and CdS. Fig. 2 shows a characteristic polygonal growth-spiral on ZnS. In this Figure it is shown that certain polygonal edges are parallel to the habit faces. The spiral steps show a strictly six-fold symmetry corresponding to the symmetry of the crystal.

In addition to the mentioned growth-spirals on the basal planes of ZnS we observed patterns which originate from emerging screw-dislocations of opposite sign which are close to each other. The resulting growth-steps form closed loops similar to the distribution around a Frank-Read source. Fig. 3 shows a microphotograph of such a pattern. It is very remarka that after some curved steps the growth layers have a polygonal character with a high kink density at the edges. The value of the step height is between those of the mentioned two types of spirals.

The latter growth-pattern is very rare. It can be only observed at very low density of screw dislocations, i.e., practically, in cases when only two screw dislocations of opposite sign emerge at points sufficiently near to each other. According to the theory of FRANK (6) curved edges are formed at high densities of the emerging screw-dislocations, and the advance of the randomly distributed curved steps makes it impossible to observe the mentioned phenomena.

3. Acknowledgements

The authors' thanks are due to Prof. G. SZIGETI for his valuable interest, and to Mrs. T. SELLEI and P. LÖRIK for their assistance.

References

(1) A.R. VERMA, Nature 167, 939 (1951); 168, 430, 783 (1951).
(2) S. AMELINCKX, Nature 167, 938 (1951); 168, 431 (1951).
(3) J. WOODS, Brit. J. appl. Phys. 10, 529 (1959).
(4) A.H. FORTHY, Phil. Mag. 42, 670 (1951); 43, 377 (1952).
(5) J.B. NEWKINK, Acta metall. 3, 121 (1955).
(6) F.C. FRANK, Disc. Farad. Soc. 5, 48 (1949).
(7) L.M. DAWSON and I. VAUD, Nature 167, 476 (1951).
(8) L.J. GRIFFIN, Phil. Mag. 41, 196 (1950).
(9) R.M. TICHANE and L.B. WILSON, J. appl. Phys. 35, 2537 (1964).
(10) E.N. SICKAFUS and D.R. WINDER, J. appl. Phys. 35, 2541 (1964).
(11) D.C. REYNOLDS and L.C. GREENE, J. appl. Phys. 29, 559 (1958).
(12) D.C. REYNOLDS and S.J. CZYZAK, Phys. Rev. 79, 543 (1960).
(13) L.C. GREENE, D.C. REYNOLDS. S.J. CZYZAK, and N.M. BAKER, J. chem. Phys. 29, 6 (1957).
(14) H. SAMELSON and V.A. BROPHY, J. Electrochem. Soc. 108, 150 (1961).

(15) K. HAUPTMANOVÁ and K. PÁTEK, phys. stat. sol. 3, 383 (1963
(16) H. HARTMANN, phys. stat. sol. 2, 585 (1962).
(17) I. SUNAGAWA, Miner. Soc. Am. Special Paper 1, 258 (1963).

(Received December 28, 1964)

phys. stat. sol. 8, K129 (1965)

Institute of Physics, Academy of Sciences of the Lithuanian SSR, Riga (a), and Institute of Solid State Physics, Czechoslovak Academy of Sciences, Prague (b)

Influence of γ-Irradiation on the Time Dependence of the Hysteresis Loop in Mn-Cu Ferrites

By

E.I. TRINKLER (a) and K. ZÁVĚTA (b)

The recently observed time variation of a rectangular hysteresis loop in a Mn-Cu ferrite (1) was tentatively explained on the basis of the magnetic after-effect connected with the induced anisotropy. Because in some experiments performed earlier by one of the authors it was found that γ-irradiation leads to changes of the magnetic after-effect in ferrites at low temperatures (2), we tried, in the present work, to study the effect of γ-rays on the time dependence of the hysteresis curve, in order to check the explanation adopted in (1).

The ring-shaped sample used for our work had the same composition as that studied in (1), viz. $Cu_{0.47}Mn_{0.53}Fe_2O_4$. The hysteresis loops obtained for an a.c. magnetic field of frequency 50 Hz were displayed on the screen of an oscilloscope (Ferrotester "Orion"), photographed and their parameters determined.

As a source of γ-radiation we use the "In-Ga-Sn radiation loop" working at the nuclear pile IRT-2000 of the Latvian Academy of Sciences, Riga. The liquid alloy In-Ga-Sn is circulated through the neutron irradiated zone, where the In atoms are excited to γ-radioactivity, which is concentrated in the spectral region 0.4 to 2.1 MeV, with its strongest line at 1.49 MeV (3). Our experiments with γ-irradiation were carried out at a total intensity of 700 R/s.

The sample was demagnetized at room temperature and then cooled down in liquid nitrogen. After maintaining the sample for 20 minutes at this temperature, an a.c. field of constant amplitude was applied. The time variation of the hysteresis

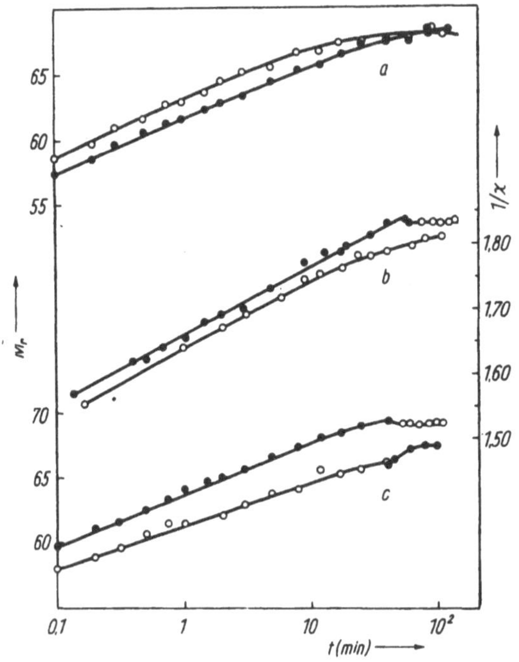

Fig. 1.

a) Increase of remanent magnetization in the absence (o) and presence (●) of irradiation

b) Time dependence susceptibility after demagnetization in the absence (o) and presence (●) of irradiation (irradiation cut off at 60 minutes)

c) Effect of applying (o → ●) and cutting off (● → o) irradiation on the increase of remanent magnetization

loop was recorded, between 0.1 min and about 2 h after field application, keeping all the time the sample at liquid nitrogen temperature. All the described operations at this temperature were performed both in the presence and absence of γ-irradiation. The curves measured at H_{max} = 7.7 Oe are shown for both these cases in Fig. 1a, where the remanent magnetization in arbitrary units is plotted against the logarithm of time. Obviously the difference between the non-irradiated and irradiated case consists only in the character of the saturation of magnetization which occurs at longer times under irradiation. The effect of irradiation obtained does not seem, however, convincing enough, in spite of earlier results concerning the influence of γ-irradiation on permeability disaccommodation (2). Therefore we measured directly the influence of irradiation upon the decay of the permeability of our sample after demagnetization, at the same temperature. The results are presented in a $1/\chi$ vs log t plot in Fig. 1b. From this curve we can see that the change caused by irradiation is analogous to that obtained on Ni-Zn ferrites (2), but the magnitude of the effect is much smaller in our case (2.5 % after 40 min). Besides, cutting off the irradiation leads to a clear kink in this time dependence and a quasi-saturation of permeability. These data suggested a more pronounced demonstration of the effect of irradiation: In one series of experiments the irradiation was applied at 40 min and the measurements extended to 100 min; the reciprocal experiments consisted in cutting off the irradiation at the same time.

These two series of measurements are illustrated in Fig. 1c. It should be noted that, because of the smallness of the effect to be demonstrated and the insufficient stability of our sources, only the qualitative changes of the time dependences were reproducible, the absolute values were reproducible to a smaller extent. In both curves of Fig. 1c the changes in the slopes of our time dependences which occur when irradiation is switched on and off are quite clearly pronounced, and their character is in agreement with the effect of irradiation on the decay of permeability. This result supports the idea that both processes, i.e., the time dependence of the hysteresis curve and the decay of permeability after demagnetization, are based on the same mechanism of magnetic aftereffect.

The authors want to acknowledge the support received by the Committee for Peaceful Utilization of Atomic Energy of the USSR and the Czechoslovak Commission for Atomic Energy which enabled this work to be done. Our thanks are also due to G.KONOVALOVA for her help during the measurements.

References

(1) E. KRATOCHVÍLOVÁ, S. KRUPIČKA, J. ŠTERNBERK, and B. ZÍTKA, phys. stat. sol. 6, K23 (1964).

(2) E.I. TRINKLER, Izv. Akad. Nauk SSSR, ser. fiz. 2, 95 (1961); E.I. TRINKLER, V.A. GOLDIN, and V.B. OSINOV, Izv. Akad. Nauk SSSR, ser. fiz. 25, 1411 (1961);

E.I. TRINKLER, E.A. RAJTMAN, and J.V. VOROPAJEV, to be published in the Proceedings of the Conference on the Physics and Chemistry of Ferrites, Minsk 1963.

(3) J.S. RJABUCHIN, Industrial Uses of Large Radiation Sources International Atomic Energy Agency, Vol. II, Vienna 1963 (p. 175).

(Received January 4, 1965)

phys. stat. sol. 8, K133 (1965)

Institute of Physics, Kiev

Some Remarks on the Review Article of M. BERNDT and P. GÖRLICH

By

P. BORZJAK

In the article of M. BERNDT and P. GÖRLICH (1) it was written that the infrared sensitivity of silver-oxygen-cesium cathodes "is attributed by BORZJAK et al. (2) to a partially polarized surface layer of Cs, by SPICER, however, to colloidal particles of silver embedded in a cesium oxide matrix" (p. 975). This wrong statement was taken from the article of G.A. MORTON (3) in which our conception was attributed to SPICER, while an alien conception was attributed to us.

In our article (2) published in 1956 a discussion was given of whether the silver-oxygen-cesium cathode is an impurity semiconductor (cesium oxide film with cesium and silver impurities). From the experimental results given in our article we drew the conclusion that the sensitivity of this cathode can be explained as follows: "It seems to us that it is more correct to think of a silver-oxygen-cesium photocathode as of a compound which is additively composed of two different cathodes: a semiconducting cesium-oxygen cathode which determines the sensitivity of the compound cathode in the ultraviolet range, and a metallic cathode of dispersed Ag-particles coated with cesium oxide, which determines the sensitivity of the compound cathode in the long-wavelength range" (p. 1047). To our knowledge, SPICER had not published such a statement previously.

References

(1) M. BERNDT and P. GÖRLICH, phys. stat. sol. 3, 963 (1963).

(2) P.G. BORZJAK, V.F. BIBIK, and G.S. KRAMARENKO, Izv. Akad. Nauk SSSR, Ser. fiz. 20, 1039 (1956).

(3) G.A. MORTON, Proc. IRE 47, 1467 (1959).

(Received January 7, 1965)

phys. stat. sol. 8, K135 (1965)

Institute of Physics, Polish Academy of Sciences, Warsaw

Band Structure of HgTe

By

R. PIOTRZKOWSKI, S. POROWSKI, Z. DZIUBA, J. GINTER, W. GIRIAT, and L. SOSNOWSKI

The properties of HgTe have been investigated by a number of authors. None of the band-structure models proposed so far gives a consistent explanation of all experimental data. Recently, measurements of the electrical properties of high-purity material were carried out by one of us (1).

The product $R_H T^{3/2}$ was found to be constant in the wide temperature-range from 20 °K to 400 °K. This fact indicates that the thermal energy gap is approximately zero, whereas the deviation of the conduction band from the parabolic form is not too strong.

S. GROVES and W. PAUL suggested a new model for gray tin (2, 3) which can explain these two features. The essential point of this model is the location of the Γ_8 level above the Γ_6 level, contrary to the configuration observed in all others semiconductors with diamond and zincblende structures. The $\vec{k}\vec{p}$ interaction between these levels leads to the structure shown in Fig. 1. At k = 0 the conduction band is degenerate with the band of heavy holes, and this corresponds to zero energy gap at $k = 0 (E_g^o = 0)$.

The thermal energy gap E_g may be even negative, if one takes into account the possibility of maxima in the valence band occurring at $k \neq 0$ due to terms linear in k in Kane's

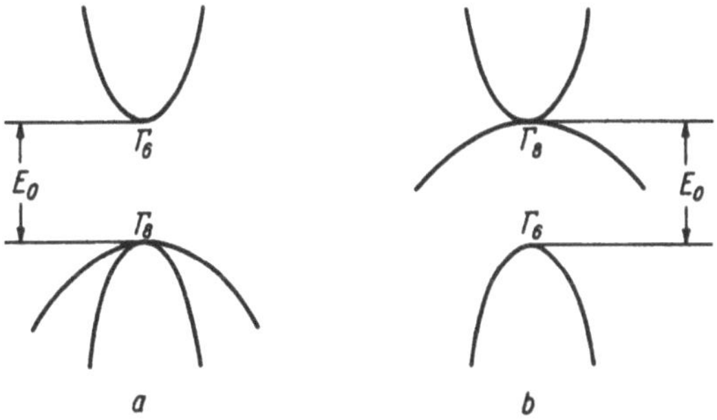

Fig. 1

formula for the heavy-hole band (4). From the evidence given in (1) it appears that at least at low temperatures $E_g^o \sim 0$.

We denote by E_o the energy difference between the Γ_6 and Γ_8 levels.

E_o is positive for all known substances of the zincblende type, but negative for the new proposed model. The shape of the new conduction band is given by Kane's formula for light holes:

$$E_c = \frac{\hbar^2 k^2}{2m_o} + \frac{1}{2}\left[\left(E_o^2 + \frac{2\hbar^2}{m_n}|E_o|\right)^{1/2} - |E_o|\right] . \qquad (1)$$

It is known (5) that in all diamond and zincblende type compounds examined so far the level Γ_6 shifts upwards with pressure faster than Γ_8. Thus $\frac{d}{dp}\left[E_{\Gamma_6} - E_{\Gamma_8}\right]$ is positive and of the order 10^{-5} eV/atm. It is easy to notice that for the model with $E_o > 0$ $\frac{d|E_o|}{dp} > 0$ and for the new model with $E_o < 0$ $\frac{d|E_o|}{dp} < 0$. Hence hydrostatic-pressure experiments can provide data which enable a choice between these two models. For both models pressure changes the shape of the conduction band due to the change in E_o. According to EHRENREICH (6) one has approximately $\frac{m_n}{m_o} = \frac{|E_o|}{P}$. Thus for $E_o > 0$ $\frac{dm_n}{dp} > 0$ and for $E_o < 0$ $\frac{dm_n}{dp} < 0$.

Still more important for $E_o > 0$, $E_o = E_g^o$ which shows that $\frac{dE_g^o}{dp} \neq 0$, whereas for the new model $E_g^o = 0$ independent of pressure.

Fig. 2 gives the experimental dependence of the thermoelectric power on pressure ($\alpha(p)$) for an n-type HgTe sample with an electron concentration $n = 1.1 \times 10^{18}$ cm^{-3}. Measurements of the Hall coefficient $R_H(p)$ of this sample show that the electron concentration is independent of pressure.

Fig. 3 gives the dependence of the Hall coefficient on pressure for an intrinsic sample.

Fig. 4 gives the theoretical dependence $\alpha(E_o)$ for constant n calculated for a Kane-type conduction band according to (7).

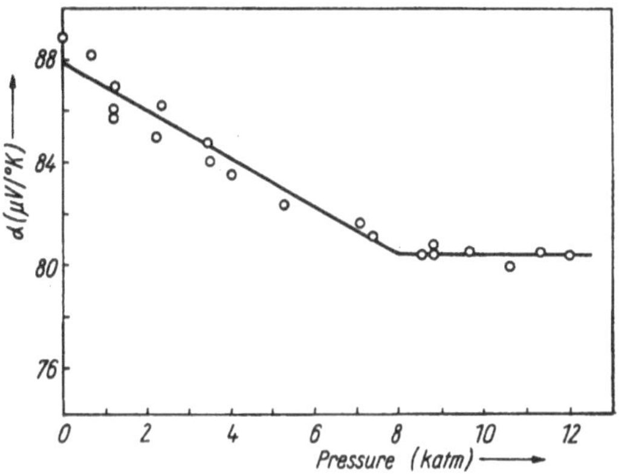

Fig. 2

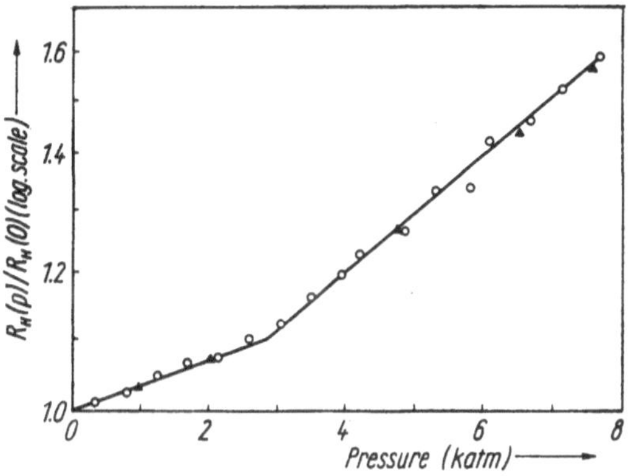

Fig. 3

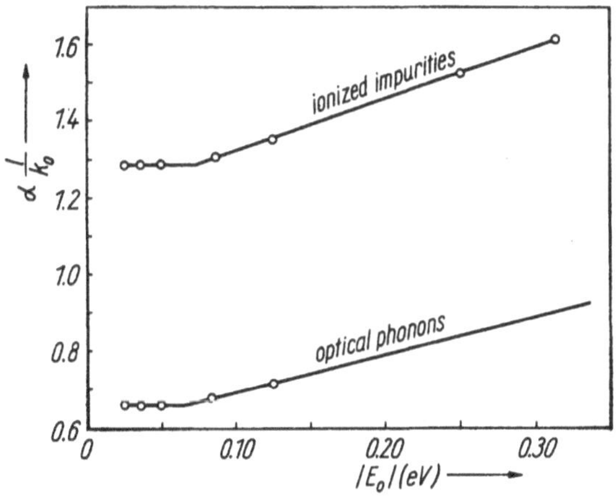

Fig. 4

$$n = \frac{4}{3}\left(\frac{2\pi m_n k_o T}{h^2}\right)^{3/2} \times {}^oL_o^{3/2}(\beta,\eta), \quad (2$$

$$\alpha = -\frac{k_o}{e}\left[\frac{{}^1L_{-2}^{\frac{r+5}{2}}(\beta,\eta)}{{}^oL_{-2}^{\frac{r+5}{2}}(\beta,\eta)} - \eta\right], \quad (3$$

where r = 1 and -1 for scattering by ionized impurities and optical phonons at $T > \Theta_D$, respectively.

${}^mL_k^n(\beta,\eta)$ are the integrals introduced in (7), where $\beta = k_oT/$ and $\eta = \zeta/k_oT$. The shape of the curves is almost identical for scattering by ionized impurities and optical phonons. By comparing the theoretical dependence with the experimental curve (Fig. 2 and 4) it is easily seen that in HgTe $d|E_o|/dp < 0$, and this indicates that we have to deal with band structure as described by the model of GROVES and PAUL

This model explains also very well the curve $R_H(p)$ (Fig. 3) of the intrinsic sample.

The triangles in Fig. 3 are the theoretical points calculated from $R_H = r/ne$ with $r = \frac{{}^oL_o^{3/2}\ {}^oL_{-4}^{5/2}}{({}^oL_{-2}^2)^2}$ (8), taking n from equation (2).

Usually optical phonon scattering is assumed to be dominant. The Fermi level is determined from the condition $n = p$, $E_g = 0$, assuming a parabolic valence band with a independent on pressure.

From the best fit with the experimental data one obtain

$$\frac{d|E_o|}{dp} = 14.5\text{x}10^{-6} \pm 2.5\text{x}10^{-6} \text{ eV/atm},$$

$$E_o = -0.19 \pm 0.04 \text{ eV},$$

$$P = 18 \pm 3 \text{ eV}.$$

Investigations which aim at a more precise determination of the band parameters, particularly effective masses of electrons and holes, are in progress.

References

1) Z. DZIUBA and T. ZAKRZEWSKI, phys. stat. sol. 7, 1019 (1964).
2) S. GROVES and W. PAUL, Phys. Rev. Letters 11, 194 (1963).
3) S. GROVES and W. PAUL, Proceedings of the 7th International Conference on the Physics of Semiconductors, Paris 1964 (p. 41).
4) J. KOŁODZIEJCZAK, Proceedings of the 7th International Conference on the Physics of Semiconductors, Paris 1964 (p. 1147).
5) W. PAUL, Progress in Semiconductors 7, 135 (1963).
6) H. EHRENREICH, J. appl. Phys. 32, 2155 (1961).
7) J. KOŁODZIEJCZAK and L. SOSNOWSKI, Acta phys. Polon. 21, 399 (1962).
8) J. KOŁODZIEJCZAK, Acta phys. Polon. 20, 289 (1961); 20, 379 (1961).

(Received January 15, 1965)

phys. stat. sol. 8, K141 (1965)

Institute of Experimental Physics, University of Warsaw

The Anisotropy of the Conduction Band in CdSe

By

J. GINTER

As CdSe crystallizes in the hexagonal wurtzite structure it is reasonable to expect an influence of this fact on the band structure. An anisotropy of the valence band has been reported by WHEELER and DIMMOCK (1) and by GUTSCHE and LANGE (2). However, their measurements of the effective mass of electrons have not revealed an anisotropy; the estimated value of the mass (0.13 m_0) was independent of direction within the limits of experimental error.

In this work an anisotropy of the effective mass of electrons has been established from the positions of the minima of plasma reflectivity for linearly polarised light. The measurements have been made at room temperature. Fig. 1 shows the reflectivity as a function of wavelength (in a logarithmic plot) for an indium-doped sample with a carrier concentration of 8.1×10^{18} cm^{-3}. Similar results have been obtained for a sample with a carrier concentration of 1.45×10^{19} cm^{-3}.

The reflectivity at the minimum is lower than 1 %, and therefore we can neglect the influence of the extinction coefficient. At the minimum we have n = 1 and

$$1 = \varepsilon_{\parallel} - \frac{4\pi N e^2}{m_{\parallel}(\omega_{\min \parallel})^2} , \qquad 1 = \varepsilon_{\perp} - \frac{4\pi N e^2}{m_{\perp}(\omega_{\min \perp})^2} .$$

The ratio of the masses is given by

$$\frac{m_{\perp}}{m_{\parallel}} = \left(\frac{\omega_{\min \parallel}}{\omega_{\min \perp}}\right)^2 \left(\frac{\varepsilon_{\parallel} - 1}{\varepsilon_{\perp} - 1}\right)$$

($\parallel$, $\perp$ electric vector parallel and perpendicular to the c-axis, respectively).

The values of ε at infra-red frequencies were given in WARDZYSNKI's paper (3); $\varepsilon_{\parallel} - 1 / \varepsilon_{\perp} - 1 = 1/1.018$

From the present measurements we obtain $\omega_{min\parallel}/\omega_{min\perp}$ = = 1.026 and $m_\perp/m_\parallel$ = 1.035. Therefore the surfaces of constant energy in k-space are oblate spheroids, like in CdS.

The average value of the effective mass can be evaluated from the position of the reflectivity minima and is found to be

$$\bar{m}_n = 0.125 \pm 0.01\ m_o\ .$$

This value agrees quite well with the result of WHEELER and DIMMOCK.

The anisotropy of the conduction band is also confirmed by the anisotropy of the free-carrier absorption in CdSe, which is shown in Fig. 2 for the same sample as in Fig. 1. The anisotropy of the absorption coefficient is found to be about 16 %. Similar results (20 %) have been obtained for samples with carrier concentrations of 1.2×10^{18} and 4.8×10^{16} cm^{-3}. These results agree with those of CZAN CZUN-DA, RAUŁUSZKIEWICZ, and WARDZYŃSKI (4).

The author wishes to thank Prof. Dr. L. SOSNOWSKI for his constant interest in the present investigation. The author is especially indebted to L. ŚNIADOWER for his help in the measurements, and to Doc. W. WARDZYŃSKI for helpful discussions.

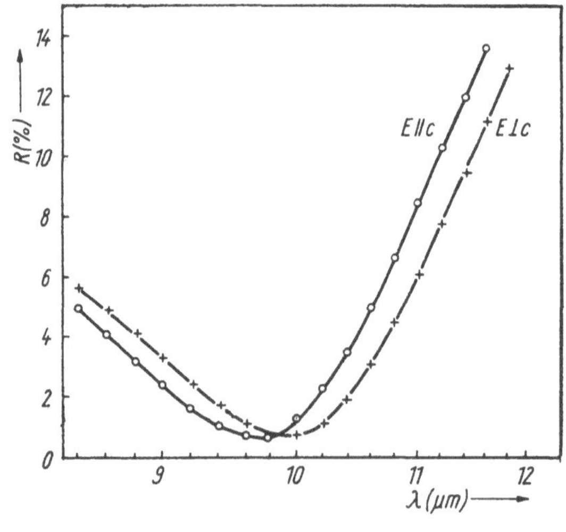

Fig. 1

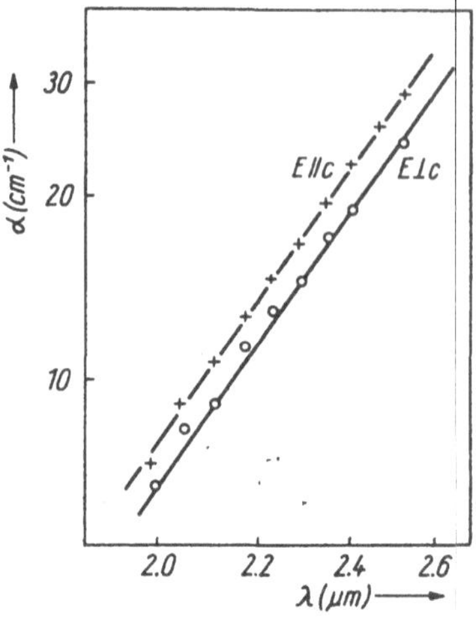

Fig. 2

References

1) R.G. WHEELER and J.O. DIMMOCK, Phys. Rev. 125, 1805 (1962).
2) E. GUTSCHE and H. LANGE, Proceedings of the 7th International Conference on the Physics of Semiconductors, Dunod, Paris 1964 (p. 129).
3) W. WARDZYŃSKI, Proc. Roy. Soc. A 260, 370 (1961).
4) CZAN CZUN DA, J. RAUŁUSZKIEWICZ, and W. WARDZYŃSKI, phys. stat. sol. 2, 1254 (1962).

(Received January 18, 1965)

phys. stat. sol. 8, K145 (1965)

Institut für experimentelle Physik der
Martin-Luther-Universität Halle

Untersuchungen über den reziproken Piezoeffekt des Eises

Von
I. TEICHMANN und G. SCHMIDT

Durch Strukturuntersuchungen läßt sich nicht eindeutig feststellen, ob Eis der Kristallklasse D_{6h} oder C_{6v} zuzurechnen ist (1). Die Zugehörigkeit zu C_{6v} würde bedeuten, daß die sechszählige c-Achse des Kristalls polar ist und die Piezomoduln d_{33}, d_{31} und d_{15} von Null verschieden sind. Entsprechende Untersuchungen lieferten teils positive (+), teils negative (-) Resultate. Es wurden alle bekannten Meßverfahren angewendet: Die Methode nach GIEBE und SCHEIBE (2) (+), (3) (-), nach BERGMANN (4) (+), (5) (+), (6) (-), die Resonanzmethode (6) (-), Messung der piezoelektrisch erzeugten Wechselspannung mechanisch angeregter, schwingender Eisstäbe (7) (-), sowie Messung der Dehnung im elektrischen Feld (7) (-).

Für die zuletzt genannte Methode stand eine Apparatur zur Verfügung, mit der man im niederfrequenten Wechselfeld noch sehr kleine piezoelektrische und elektrostriktive Dehnungen (ca. 10^{-9} cm) messen kann (8, 9). Ein Piezoeffekt mit einem Modul der Größenordnung 10^{-14} mV^{-1} - das ist 1/200 von d_{11} des Quarzes - müßte sich damit noch nachweisen lassen.

Gemessen wurde an Eiseinkristallen von einigen zehntel Millimeter Dicke und mehreren Millimeter Kantenlänge. Zum Meßergebnis trägt bei dem vorliegenden Verfahren allerdings nur ein sehr kleiner Teil des gesamten Kristalls bei (9). Die Eisplatten waren frei oder nach Impfen an der Wasseroberfläche gewachsen. Ihre Orientierung - c-Achse senkrecht zur Plattenoberfläche - wurde polarisationsmikroskopisch überprüft. Aufgefrorene Zinnfolien bildeten die Elektroden. Die Untersuchungen erfolgten zwischen -10 °C und -80 °C an Proben unterschiedlichen Alters (eine Stunde bis drei Monate).

Bei den 25 untersuchten Kristallen trat nur in drei Fällen ein linearer Effekt auf, der nicht reproduzierbar war.

Insgesamt sprechen die Ergebnisse gegen eine Piezoelektrizität des Eises. Dagegen wurde stets ein quadratischer Effekt beobachtet:

$$S_3 = R_{333} \cdot E_3^2$$

(S_3 und E_3 ist die Dehnung bzw. die elektrische Feldstärke in Richtung der c-Achse des Kristalls).

Die gemessene elektrostriktive Konstante hat die Größenordnung -10^{-20} m^{2}V^{-2}. Dieser Wert liegt an der unteren Grenze des Meßbereichs für die Elektrostriktion.

Der quadratische Effekt könnte eine Folge elektrostatischer Kräfte sein. Das elektrische Feld zwischen den Elektroden müßte die Dehnung

$$S_3 = -\frac{1}{2} s_{33} \varepsilon_3 E_3^2$$

erzeugen. Setzt man für die elastische Nachgiebigkeit s_{33} bzw. die Dielektrizitätskonstante ε_3 die Werte

$$s_{33} = 0{,}85 \cdot 10^{-10} \text{ m}^2\text{N}^{-1}$$

bzw. $$\varepsilon_3 = 81 \cdot \varepsilon_0 \text{ As V}^{-1}\text{m}^{-1}$$

ein (10, 11), so ergibt sich die scheinbare Elektrostriktionskonstante

$$R_{333} \approx -3 \cdot 10^{-20} \text{ m}^2\text{V}^{-2}.$$

Das entspricht der beobachteten Größenordnung.

War der NF-Wechselspannung am Kristall ein Gleichfeld E_0 überlagert, so ergab sich ein linearer Effekt. Deutet man diesen als linearisierte (scheinbare) Elektrostriktion

$$S_3 = R_{333} (E_0 + E_3)^2 = 2R_{333} E_0 E_3 + \dots ,$$

so erhält man für R_{333} dieselbe Größenordnung wie oben.

Eine mögliche Abhängigkeit der Resultate von der Temperatur und dem Alter der Proben konnte nicht beobachtet werden.

Die Verfasser danken Herrn Professor Dr. W. MESSERSCHMIDT für die stete Förderung dieser Arbeit und für die Bereitstellung von Institutsmitteln.

Herr Professor Dr. K. MOTHES und Herr Professor Dr. H. BETHGE stellten in dankenswerter Weise einen Arbeitsplatz im Institut für Biochemie bzw. in der Arbeitsstelle für Elektronenmikroskopie der DAW zu Berlin in Halle zur Verfügung.

Literatur

(1) M. BLACKMANN und N.D. LISGARTEN, Adv. Phys. 7, 26, 189 (1958).
(2) E. KUMM, Arch. Meteorol. Geophys. A3, 382 (1951).
(3) A. HETTICH, Z. Phys. 76, 700 (1932).
(4) F. ROSSMANN, Experientia 6, 182 (1950).
(5) A. DEUBNER, R. HEISE und K. WENZEL, Naturwissenschaften 47, 600 (1960).
(6) S. STEINEMANN, Experientia 9, 135 (1953).
(7) D. ROSSBERG und S. MAGUN, Naturwissenschaften 44, 59 (1957).
(8) G. SCHMIDT, Exper. Tech. Phys. 6, 250 (1958).
(9) G. SCHMIDT, Z. Phys. 161, 579 (1961).
(10) F. JONA und P. SCHERRER, Helv. Phys. Acta 25, 35 (1952).
(11) R. BASS, D. ROSSBERG und G. ZIEGLER, Z. Phys. 149, 199 (1957).

(Received January 22, 1965)

phys. stat. sol. 8, K149 (1965)

Institute of Physics of the Slovak Academy of Sciences,
Bratislava

On Some New Dielectric-Loss Maxima in Alkali Halides Occurring between 100 and 400 °C at 1 kHz

By

A. KESSLER and E. MARIANI

Some pronounced dielectric-loss maxima exhibiting unusual properties have recently been observed by KADERKA (1), TRNKA, KADERKA, and BOHUN (2) and the present authors (3, 4, 5, 6). They were observed in NaCl, KCl, and LiF crystals which were either "pure" undoped (3, 6, 7) or doped with different divalent cation admixtures (2, 3, 4), Na_2SO_4, Na_2CO_3, and NaOH (1, 5). The maxima lie at 1 kHz at temperatures from 140 to 390 °C. A short report on these findings is given in the following paper.

The properties of these maxima can be summarized as follows:

1. The temperature of the maximum, T_{max}, depends on the concentrations of impurities added to the crystal. With increasing concentration of cation impurities T_{max} shifts to

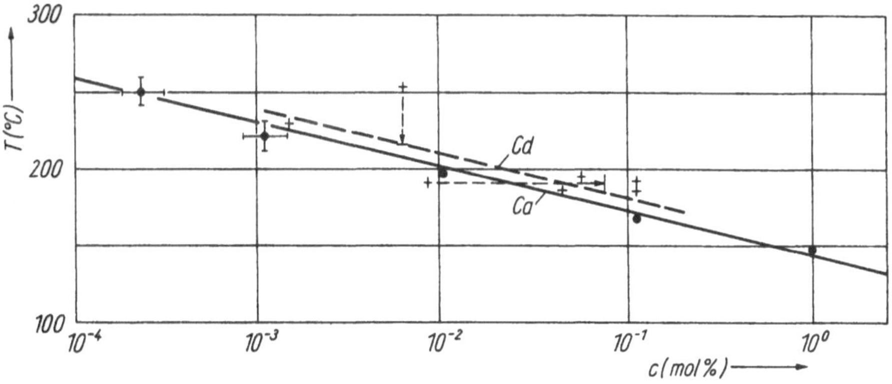

Fig. 1. T_{max} = f(c) curves of calcium (•) and cadmium (+) doped crystals (4). The crystal doped with 9×10^{-3} molar % $CaCl_2$ contained about 8×10^{-2} molar % $CaCl_2$. The crystal doped with about 8×10^{-3} molar % $CdCl_2$ was found to be considerably alkalines; therefore T_{max} is higher

lower temperatures (Fig. 1) (2, 3, 4); with increasing concentration of anion impurities it goes to higher temperatures (Fig. 2) (1, 5). In extremly "pure" crystals T_{max} is eventually as high as 390 ^{o}C (7).

2. On the other hand, the magnitude of the relaxational component of the loss angle, tan $\delta_{rel\ max}$, is nearly independent of impurity concentrations (3), but proves to be proportional to the thickness of the samples (Fig. 3a, 3b). (It shou be noted that, in full accord with theory, the value tan δ_{rel} : of the calcium-complex maximum at about 67 ^{o}C, measured simultaneously, was found to be independent of thickness.) The same dependence on thickness is observed for the real part, ε', and the imaginary part, ε'', of the complex dielectric constant, as follows from Fig. 3a. The loss tangent of a one millimeter thick sample is nearly 1. From Fig. 3a it is obvious that the dependences of ε' and ε'' on frequency are described by Debye relations.

3. TRNKA, KADERKA, and BOHUN (2) reported also that the maxima occurred eventually only after thermal treatment of the crystals.

It should be emphasized that the experiments under dis-

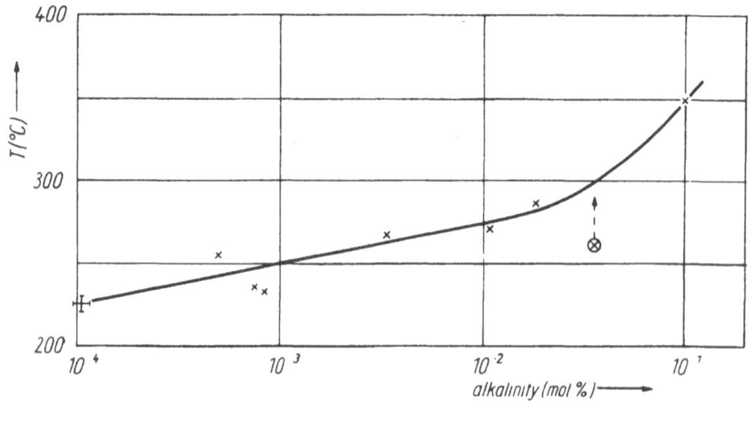

Fig. 2. $T_{max} = f(\sigma)$ curve of alkaline crystals doped with NaOH or grown in humid air (5). (⊕) crystal doped additionally with $CaCl_2$; therefore T_{max} is lower

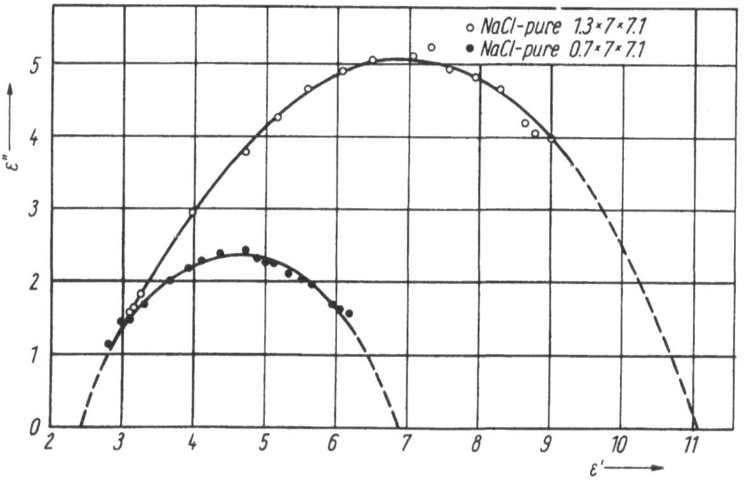

Fig. 3a) Cole-Cole diagrams of "pure" NaCl crystals of thickness 0.7 mm (●) and 1.3 mm (o) (6)

cussion were carried out with samples without any conducting layer applied to the contact surface (3); but experiments have also been performed with samples whose contact surfaces were coated with colloidal graphite, and these yielded practically the same result. The same maxima occurred; however, it is not sure whether the magnitudes of the relaxational loss maxima were exactly the same, because conductivity losses are considerably increased under those conditions, and exceed the value of $\tan \delta_{rel\ max}$ several times.

From the observed facts it can be concluded primarily that the loss maxima are most probably not caused by any kind of dipoles distributed over the crystal, i.e. complexes etc. It is quite obvious that the above findings are not consistent with any theory of relaxational losses caused by the reorientation of dipoles, e.g., the theory of the losses caused by the reorientation of impurity-vacancy complexes as given by LIDIARD (8). For instance, if we estimate, on the basis of this theory the number of dipoles responsible for the loss maxima we get unreasonably high values for all impurity concentrations, even in the case of very pure crystals.

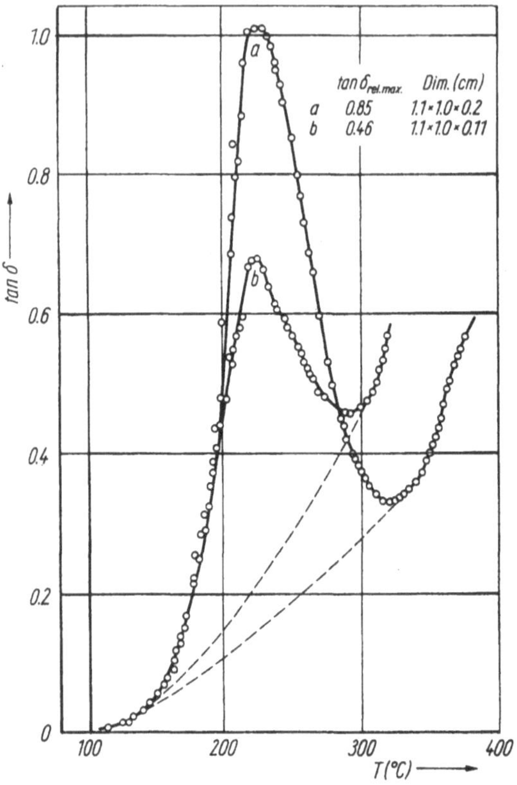

Fig. 3b) Loss angle vs temperature for "pure" NaCl crystals of thickness 2.0 mm (a) and 1.1 mm (b) (6)

Nevertheless, the same measurements revealed a fairly good agreement between this theory and the mentioned maxima at about 67 °C (3). Furthermore, according to Debye's theory (which obviously holds in our case), the relation $\omega\tau = 1$ (ω frequency, τ relaxation time) must hold for every value of T_{max}, if a maximum occurs in the tan δ_{rel} = f(T) curve for ω = const. One should therefore expect that τ has always the same value in the case of the discussed maxima, in spite of the change of T_{max}, because ω is always the same. But for a complex the jumping probability varies with temperature exponentially, i.e. very markedly, and so would τ, of course.

A change in impurity concentration need not necessarily cause the observed change of T_{max} directly, but perhaps by changing the conductivity of the crystals. An investigation of the latter in fact supports such a view (5). It was found that the conductivities σ of all the investigated crystals at their respective T_{max} were nearly the same, about 3×10^{-9} to 7×10^{-9} (Ω cm $^{-1}$). Let us assume now that τ depends on σ. Whatever the dependence of τ on σ, the constancy of σ leads to that of τ. This would provide an explanation of the behaviour of the maxima, since for different crystals value of τ

which satisfies $\omega\tau = 1$ will be reached at different temperatures.

The shift of T_{max} in opposite directions which occurs for divalent cation and anion admixtures would then be due to their opposite effects on conductivity (5,9) . Then the observation of a T_{max} as high as 390 °C for the purest (natural) crystal could be explained by the very small conductivity of this crystal. It is also clear that 1) we should get different $T_{max} = f(c)$ curves for cadmium and calcium as indicated by Fig. 1; 2) that the effect of two different divalent admixtures should be additive; and 3) that divalent anion and cation admixtures should compensate each other (to a certain degree), as shown in (3, 5) and Fig. 1 and 2.

It is interesting to note in connection with the given facts that in the case of interfacial polarisation, as treated theoretically e.g. by MAXWELL (10), we expect 1) the occurrence of a relaxational maximum with a relaxation time depending on the conductivity of the dielectric, 2) a variation of ε' and $\tan\delta$ with the thickness of the dielectric layers, and 3) a frequency response according to Debye relations. This suggests another direction in which to look for the cause of the observed phenomena. This could perhaps be a blocking effect of the electrodes resulting in the formation of surface charge, or some similar effect.

But there are arguments against such an idea. Considerable research, theoretical as well as experimental, has been done on the a.c. effects of polarisation, namely by FRIAUF (11), MCDONALD (12), ALLNATT and JACOBS (13), JACOBS and MAYCOCK (14), and others, but none of the results indicates the presence of maxima of the discussed type. Only the equivalent circuit used by MCDONALD (Fig. 5 in (12)) to represent his theoretical results seems to suggest a loss maximum. Furthermore, to our knowledge, no polarisation has so far been found in the temperature range of about $10^3/T = 1.6$ to $2.2\ (^{\circ}K)^{-1}$, i.e. the range where the discussed maxima were observed. Therefore it is felt that, until more suitable experimental evidence is available, there is only little chance for positive conclusions as to the origin of the loss maxima. Further investigations are therefore in progress.

References

(1) M. KADERKA, Czech. J. Phys. B 13, 378 (1963).
(2) J. TRNKA, M. KADERKA, and A. BOHUN, Czech. J. Phys. B 14, 63 (1964).
(3) A. KESSLER and E. MARIANI, Czech. J. Phys. B 14, 757 (1964).
(4) A. JOHNOVÁ, A. KESSLER, and E. MARIANI, Mat. fyz. čas. (in print).
(5) M. HARTMANNOVÁ, A. KESSLER, and E. MARIANI (in preparation).
(6) A. KESSLER and E. MARIANI, Czech. J. Phys. B 15,(1965) (in print).
(7) A. KESSLER and E. MARIANI, Mat. fyz. čas. 14, 317 (1964).
(8) A.B. LIDIARD: Ionic conductivity, in: Encyclopedia of Physics, Vol. XX, 2, Springer-Verlag, Berlin 1957 (p. 316).
(9) B. FRITZ, B. LÜTY, and J. ANGER, Z. Phys. 147, 240 (1963).
(10) J.C. MAXWELL, Treatise on Electricity and Magnetism, 3rd Ed., Vol. 1, Chap. X, Oxford University Press 1904.
(11) R.J. FRIAUF, J. chem. Phys. 22, 1329 (1954).
(12) J.R. MCDONALD, Phys. Rev. 92, 4 (1953).
(13) A.R. ALLNATT and P.W.M. JACOBS, J. Phys. Chem. Solids 19, 281 (1961).
(14) P.W.M. JACOBS and J.N. MAYCOCK, J. chem. Phys. 39, 757 (1963).

(Received January 25, 1965)

phys. stat. sol. 8, K155 (1965)

A.F. IOFFE Physical Technical Institute, Academy of Sciences of the USSR, Leningrad, and Institute of Physics, Academy of Sciences of the Azerbaidzhan SSR, Baku

Dependence of the Thermal E.M.F. on the Hole Concentration in Gallium Arsenide Crystals

By

O.V. EMELYANENKO, F.P. KESAMANLY, D.N. NASLEDOV, V.G. SIDOROV, and G.N. TALALAKIN

Up to the present p-type GaAs crystals are very little investigated. This is due to the facts that a) GaAs crystals normally, i.e. without impurity incorporation, exhibit n-type conductivity and b) the technique of impurity incorporation into this substance is only little developed. In this note the results about doping of GaAs with Zn, Cd, Cu, Mn, and Ti impurities are reported and the experimental data of thermal e.m.f. measurements made with these crystals discussed.

Single crystals have been prepared by the method of horizontal zone melting. Doped samples with hole concentrations of 10^{17} to 10^{19} cm^{-3} have been obtained from the gas phase at a temperature of 610 ^{o}C at the cold end of the ampule, whereas doped crystals with hole concentrations $p > 10^{19}$ cm^{-3} have been prepared else from the gas phase at an elevated temperature of 800 to 900 ^{o}C at the cold end of the ampule or by a preliminary solution of the impurities in Ga with subsequent synthesis and growing. The last mentioned method allowed samples to be prepared with hole concentrations as high as $2x10^{20}$ cm^{-3}. Single crystals with higher concentrations of current carriers could not be obtained. All crystals investigated were electrically homogeneous, the mobility was close to the theoretical value of 50 to 300 cm^2/Vs (1).

The differential thermal e.m.f. of these crystals was measured in the temperature range 25 to 100 ^{o}C with an experimental arrangement described in (2). The thermal e.m.f. depends linearly on temperature for all crystals investigated. This behaviour should be observed for semiconductors with impurity depletion. The value of the thermal e.m.f. at room tem-

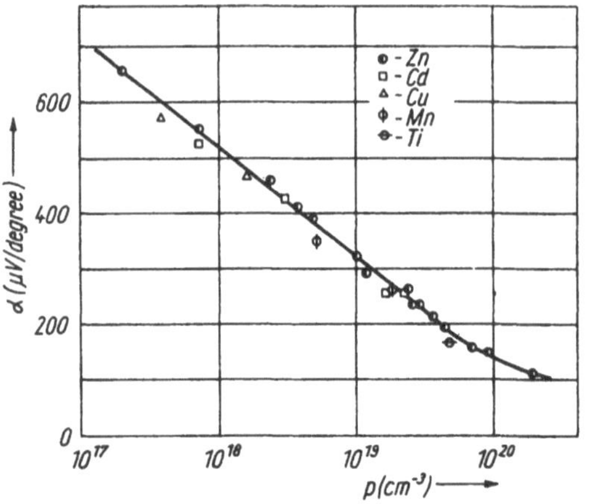

Fig. 1. Thermal e.m.f. as a function of hole concentration at 300 °K

perature decreases with increasing hole concentration (Fig. 1). The experimental method applied allowed the differential thermal e.m.f. at room temperature to be measured with an accuracy of ± (2 to 3) %. The experimental error for the determination of the hole concentration from the Hall coefficient was also ± (2 to 3) %. The total experimental error for $\alpha(p)$ was therefore about ± 5 %. Within these limits the values of the thermal e.m.f. can be represented by a curve $\alpha(p)$, shown in Fig. 1, which can be taken as a standard for the determination of the hole concentration in GaAs crystals from measurements of the thermal e.m.f. (3). Within the limits mentioned above the magnitude of the differential thermal e.m.f. does not depend on the nature of the acceptors incorporated. Only with Zn impurities the values of the thermal e.m.f. are somewhat higher than for the other crystals.

The valence bands of GaAs is similar to that of Ge and Si. By assuming equal scattering mechanisms for light and heavy holes (4) which can be described by $l \sim \varepsilon^r$ (l mean free path, ε kinetic energy), the thermal e.m.f. is given by the simple formula

$$\alpha = \frac{k}{T}\left[\frac{r+2}{r+1} \cdot \frac{F_{r+1}(\tilde{\mu})}{F_r(\tilde{\mu})} - \tilde{\mu}\right]. \qquad (1)$$

The scattering parameter r in equation (1) can be determined from measurements of other transport effects (transverse Nernst Ettingshausen effect, temperature dependence of the mobility (1). It can be shown that with this experimentally determined value of r, equation (1) is also approximately valid for mixed scattering or different scattering mechanisms for light and heavy holes. The quantity $\tilde{\mu}$ is thereby determined with an accuracy of ± 0,2. The degeneracy of the hole gas in p-type GaAs crystals begins at room temperature for hole concentrations higher than $2x10^{19}$ cm^{-3}.

With the $\tilde{\mu}$-values determined from the thermal e.m.f. measurements, the density of states effective mass m_p^* which is equal to $(m_1^{3/2} + m_2^{3/2})^{2/3}$ (m_1 and m_2 are the effective masses of light and heavy holes, respectively) has been calculated for T = 300 °K from equation (2)

$$p = \left(\frac{m_p^*}{m_o}\right)^{3/2} \frac{4}{\sqrt{\pi}} \left(\frac{2\pi m_o kT}{h^2}\right)^{3/2} F_{1/2}(\tilde{\mu}) \qquad (2)$$

The coefficient a_2 in the relation $p = a_2/eR$ (R Hall coefficient) has been taken from the $a_2(\tilde{\mu})$ curves given in (5). The value of m_p^* is equal to 1,1 ± 0,3 m_o in the GaAs crystals investigated. The error is due to the approximations made for the application of equation (1). This value of the effective mass is considerably higher than the value reported in (6) (m_p = 0,5m_o) which has been also obtained from thermal e.m.f. and Hall measurements. This lower value in (6) is probably due to the fact that the influence of the scattering mechanism on the results could not be evaluated exactly, and thus the values of the thermal e.m.f. were underestimated. This follows that the effective electron mass was also underestimated by the authors. It shall be noticed that the density of states effective mass for the valence band has been calculated in (7) and (8) and found to be 0,72 and 1m_o, respectively.

References

(1) O.V. EMELYANENKO, T.C. LAGUNOVA, and D.N. NASLEDOV, Fiz. tverd. Tela 2, 192 (1960).

(2) O.V. EMELYANENKO and F.P. KESAMANLY, Fiz. tverd. Tela 2, 1494 (1960).

(3) M. SHOJI, J. Phys. Soc. Japan 16, 2590 (1961).

(4) G.L. BIR, E. NORMANTAS, and G.E. PIKUS, Fiz. tverd. Tela 4, 1180 (1962).
(5) V. VZDOPNOV and I.M. TSIDILKOVSKII, Fiz. Metallov i Metallovedenie 7, 801 (1959).
(6) J.T. EDMOND, R.E. BROOM, and F.A. CUNNEL, Report of the Meeting on Semiconductors at Rugby, London 1956 (p. 109).
(7) H. EHRENREICH, Phys. Rev. 120, 1951 (1960).
(8) O. MADELUNG, Physics of III-V Compounds, New York-London-Sydney 1964 (p. 353).

(Received January 26, 1965)

phys. stat. sol. 8, K159 (1965)

A.F. IOFFE Physical Technical Institute, Academy of Sciences of the USSR, Leningrad, and Institute of Physics, Academy of Sciences of the Azerbaidzhan SSR, Baku

Electrical Properties of p-Type $ZnSnAs_2$ Crystals at Low Temperatures

By

F.P. KESAMANLY, D.N. NASLEDOV, and YU.V. RUD

The preparation of semiconducting diamond-like $A^{II}B^{IV}C_2^V$ compounds and the investigation of its properties is the subject of a considerable experimental literature. However, little is known about the electrical behaviour of these crystals at low temperatures. The investigations published in (1, 2) have been continued and the results of experiments on the temperature dependence of the electrical properties of p-type $ZnSnAs_2$ crystals, in the temperature range 2 to 600 ^oK, are reported in this note. The experimental method is described in (3). By a half-automatic recording of the results on the recorder chart of a potentiometer the reliability of the results could be increased and the measurements more quickly performed. The samples investigated had at room temperature a hole concentration of 10^{18} to 10^{20} cm^{-3} which was determined from $p = 1/eR$. By the use of the highly sensitive photo-compensated amplifiers F-116 the experimental error of the Hall coefficient did not exceed $\pm$ 5 % for all crystals (including the samples with hole concentrations of $\sim 10^{20}$ cm^{-3}).

The Hall coefficient R as a function of temperature T exhibits for all samples investigated a strongly pronounced maximum at a temperature denoted by T_{max}. The maximum of R decreases with increasing hole concentration and thereby shifts towards higher temperatures.

According to the experimental results on the temperature dependences of the conductivity $\sigma(T)$ and the Hall coefficient R(T) the crystals can be divided into two groups. The first group includes all crystals with hole concentrations lower than $6x10^{18}$ cm^{-3} and the second one all crystals with hole concentrations higher than $6x10^{18}$ cm^{-3}. By lowering the tem-

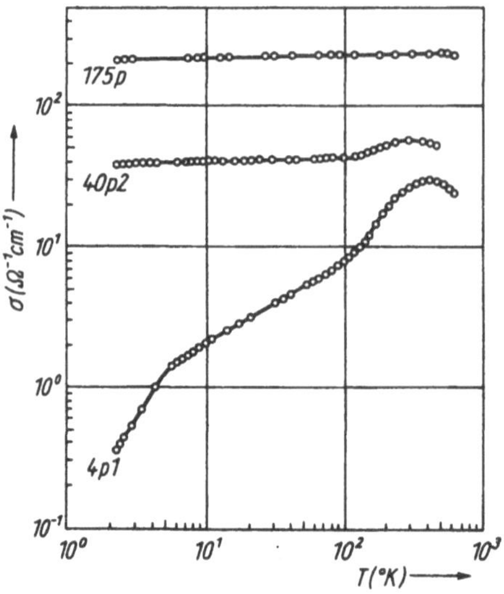

Fig. 1. The conductivity as a function of temperature in p-type $ZnSnAs_2$ crystals

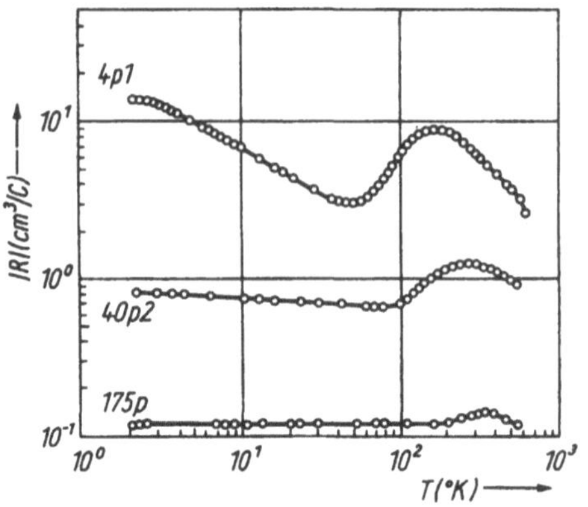

Fig. 2. The Hall coefficient as a function of temperature in p-type $ZnSnAs_2$ crystals

perature below T_{max} the Hall coefficient R for the first group of crystals reaches a constant value, but increases again for the crystals of the second group. In Fig. 1 and 2 the temperature dependences of σ and R are shown for three typical crystals (altogether six crystals have been investigated at low temperatures). For the samples 40p2 and 175p the conductivity changes only slightly with decreasing temperature T (for the temperature range $T < T_{max}$), but decreases strongly for the sample 4p1 at temperatures below room temperature, and in a double logarithmic plot vs temperature, $\ln \sigma$ changes twice the slope. A similar behaviour of $\sigma(T)$ is found with other crystals.

The maximum in the R(T) dependence at lower temperatures is also found with Ge, InSb, and GaAs crystals (4 to 6) and explained by a conduction process due to current carriers in the impurity and valence bands. The maximum of R(T) found with p-type $ZnSnAs_2$ crystals cannot be explained by an impurity conductivity. According to the theory, an impurity conductivity in these crystals should be observed, in view of the small effective hole mass[1] ($m^* = 0,6m_o$), at much lower temperatures and hole concentrations. It can be seen from the Table that the maximum of R(T) in the analogous compounds p-type Ge, p-type InSb, and p-type GaAs, which is due to an impurity conductivity, is observed at much lower temperatures and hole concentrations ($< 10^{19}$ cm^{-3}). In this connection it shall be noted that the effective hole mass, which determines the properties of the impurity band, is higher in comparable semiconductors than in p-type $ZnSnAs_2$.

semiconductor	effective hole mass	p_{max} (cm^{-3})	T (^oK)
p-GaAs	~ 1 m_o (9)	$4,5 \times 10^{18}$	200 (6)
p-Ge	$0,3m_o$ (8)	$1,35 \times 10^{18}$	50 (4)
p-InSb	$0,2m_o$ (8)	$8,2 \times 10^{16}$	50 (5)
p-$ZnSnAs_2$	$0,6m_o$	$1,3 \times 10^{19}$	250

The notation of the Table are: m_o the free electron mass, p_{max} the maximum hole concentration at 300 ^oK, giving still a maximum in the R(T) curve.

The assumption that the scattering mechanism of holes changes at higher temperatures from scattering by ionized impurities to scattering by lattice vibrations is also doubtful.

According to our opinion the temperature dependence of R is best explained by the assumption that the conductivity is due to two types of current carriers of the same sign. The results of (7) support this hypothesis. According to this paper the valence band of $A^{II}B^{IV}C_2^V$ crystals with chalcopyrite structure has the following form: The maximum of the heavy hole band is in the centre of the Brillouin zone. The light hole band is not in contact with the heavy hole band, as for the valence bands in the $A^{III}B^V$ compounds, but split off by about 10^{-2} eV. The spin-orbit band of the heavy holes lies below these two bands. Because of the small energetic distance bet-

---

1) Through an oversight of the authors a value of $m^* = 0,13m_o$ has been published in (2).

ween the light and the heavy hole bands, the temperature dependence of the Hall coefficient above 80 °K can be qualitatively explained, but for a quantitative analysis the exact value of this distance must be known.

References

(1) N.A. GORYUNOVA, F.P. KESAMANLY, D.N. NASLEDOV, and YU.V. RUD, Fiz. tverd. Tela 6, 113 (1964).
(2) F.P. KESAMANLY, D.N. NASLEDOV, and YU.V. RUD, Fiz. tverd. Tela 6, 2187 (1964).
(3) O.V. EMELYANENKO and D.N. NASLEDOV, Zh. teor. Fiz. 28, 1177 (1958).
(4) H. FRITZSCHE and M. GUEVAS, Phys. Rev. 119, 1238 (1960).
(5) H. FRITZSCHE and K. LARK-HOROWITZ, Phys. Rev. 99, 400 (1955).
(6) O.V. EMELYANENKO, T.S. LAGUNOVA, and D.N. NASLEDOV, Fiz. tverd. Tela 3, 198 (1961).
(7) F.M. GASHIMZADE, Fiz. tverd. Tela 5, 1199 (1963); A.A. VAIPOLIN et al., Izv. Akad. Nauk SSSR, ser. fiz. 28, 1085 (1964).
(8) D. LONG, Uspekhi fiz. Nauk 80, 639 (1963).
(9) O.V. EMELYANENKO, F.P. KESAMANLY, D.N. NASLEDOV, V.G. SIDOROV, and G.N. TALALAKIN, phys. stat. sol. 8, K155 (1965).

(Received January 26, 1965)

phys. stat. sol. 8, K163 (1965)

Lehrstuhl für Festkörperphysik der Mathematisch-Physikalischen-Fakultät der Karlsuniversität, Prag

Geschwindigkeitswechselversuche an Zinkkristallen

Von

P. LUKÁČ

Bekanntlich kann die Fließspannung $\tau(a)$, die sich im dynamischen Zugversuch nach einer Abgleitung a am Kristall einstellt, aufgefaßt werden als die Summe zweier Spannungsanteile $\tau_G(a)$ und $\tau_S(a, \dot{a}, T)$ (1). Dieser Sachverhalt wurde in letzter Zeit bei Untersuchungen über die Temperaturabhängigkeit der kritischen Schubspannung verschiedener Metalle (Zn (2, 3, 4), Cd (5), Mg (6), Al (7)) und bei Temperaturwechselversuchen am Al und Cu (8) wiederholt bestätigt. Legt man diesen Befunden das Seeger'sche Modell (1) zugrunde, so bedeutet τ_G den zur Überwindung der weitreichenden Spannungsfelder notwendigen temperaturunabhängigen Spannungsanteil und τ_S den temperatur- und geschwindigkeitsabhängigen Spannungsbetrag, der bei Vorgängen aufzubringen ist, denen ein kleines Aktivierungsvolumen zukommt (kurzreichende Spannungsfelder). Kennt man die Größe beider Anteile, bzw. ihr Verhältnis, und, was noch bedeutsamer ist, die Abhängigkeit dieser Spannungsbeträge von der Verformung, so ist es möglich, zusammen mit der Bestimmung des Aktivierungsvolumens, Aussagen über die Art der Verfestigungsmechanismen zu machen.

In einigen älteren Arbeiten, hauptsächlich an Al, Cu und Ni (9, 10, 11) wurde gefunden, daß das Verhältnis τ_G/τ_S unabhängig ist von der Verformung; dieses Verhalten wurde später als das Cottrell-Stokesche-Gesetz bezeichnet. Neuerdings wurden verschiedentlich Abweichungen von diesem Gesetz beobachtet, die darauf hinauslaufen, daß dieses nicht für die ganze Verfestigungskurve Gültigkeit hat (12, 13).

Im Zusammenhang mit der Tatsache, daß an Zinkeinkristallen (4), ähnlich wie an einer Reihe anderer Metalle, ein von der Theorie abweichender Verlauf der Temperaturabhängigkeit der kritischen Schubspannung bei höheren Temperaturen beobachtet wurde, der in einem starken Spannungsabfall zum Aus-

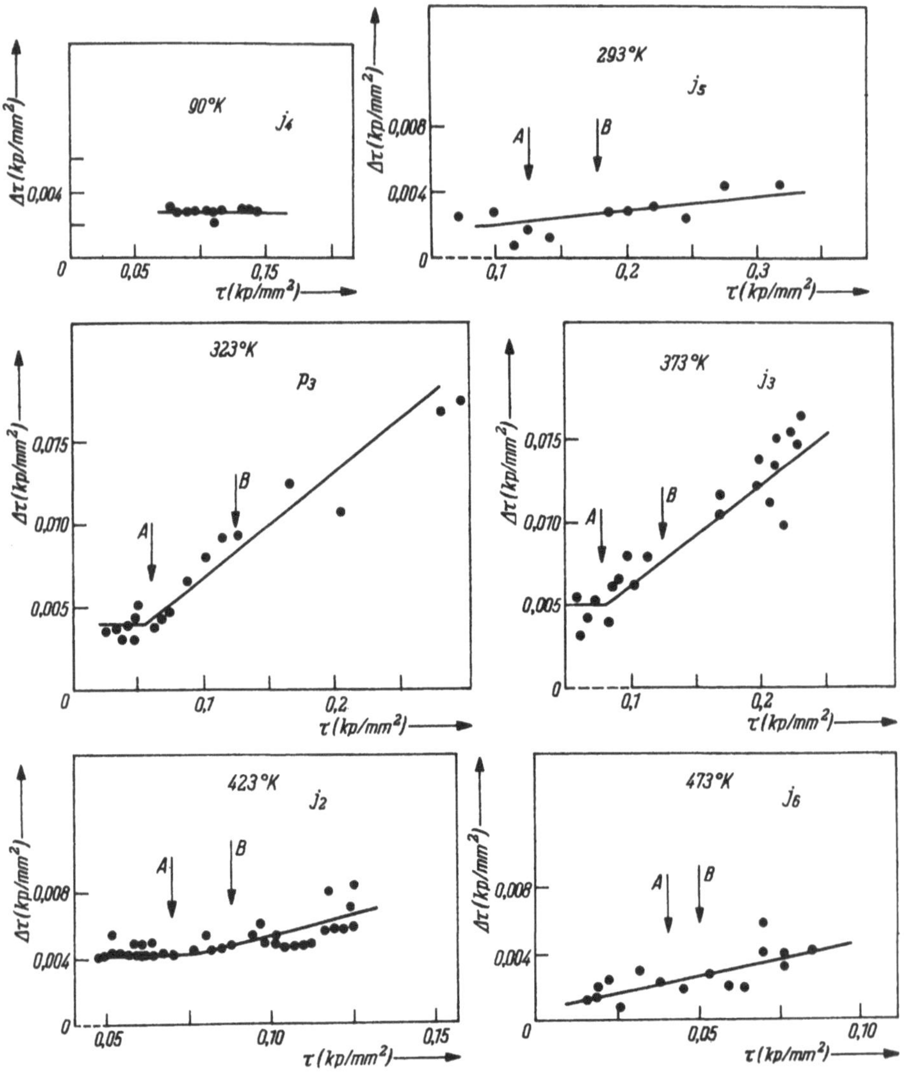

Fig. 1. Abhängigkeit der durch einen Geschwindigkeitswechsel hervorgerufenen Spannungsänderung $\Delta\tau$ von der ursprünglichen Spannung (Die Orientierung der Kristalle der Serie j ist $\chi_o = 39^o$, $\lambda_o = 44^o$, für p ist $\chi_o = 42^o$, $\lambda_o = 46^o$)

druck kommt, wurden an Zinkkristallen, zur Bestimmung von τ_G/τ_S, Geschwindigkeitswechselversuche in einem breiten Temperaturbereich unternommen. Über die wesentlichen Tatsachen wird hier kurz berichtet; die quantitative Auswertung der Ergebnisse und ihre Diskussion folgt in einem weiteren Bericht.

Die nach einem Bridgman-Verfahren hergestellten Zinkeinkristalle (mit Pb, Cu, Cd, Fe < $2\cdot10^{-4}$ At.% als Verunreinigungen) hatten annähernd gleiche Orientierung (s. Fig. 1) und wurden in einem Polanyi-Apparat im Temperaturbereich von 90 bis 473 ^{o}K verformt. Die Verformungsgeschwindigkeit wurde alternierend von $1{,}5\cdot10^{-3}$ auf $1{,}5\cdot10^{-2}$ mm/s gewechselt (wobei der Wechsel innerhalb von weniger als 0,1 s durchgeführt wurde). Die Spannungsänderungen wurden (für den Bereich A und B der Verfestigungskurve), bei einer Empfindlichkeit der Lastregistrierung von 56 ± 1,2 p/mm, beim Übergang von kleinerer zu höherer Verformungsgeschwindigkeit aus der Verfestigungskurve entnommen. Diese Spannungsänderungen sind gegenüber der Schubspannung τ (für die niedrigere Verformungsgeschwindigkeit) für verschiedene Verformungstemperaturen in Fig. 1 aufgetragen. Die Ergebnisse, die man daraus entnehmen kann, können wie folgt zusammengestellt werden:

1. Die Spannungsänderungen $\Delta\tau$ sind bei 90 ^{o}K unabhängig von der Spannung τ. Für alle anderen Temperaturen wird $\Delta\tau$ mit zunehmender Spannung größer.

2. Die Neigungen der linearen Teile der ($\Delta\tau$ - τ)-Abhängigkeit werden zunächst mit wachsender Temperatur (bis etwa 373 ^{o}K) größer und fallen daraufhin wieder ab.

3. Die Spannungsänderung $\Delta\tau$, hervorgerufen durch einen Geschwindigkeitswechsel, ist bei Raumtemperatur größer als bei 90 ^{o}K. (Eine Übersicht über die Temperaturabhängigkeit von $\Delta\tau$ für τ = 0,15 kp/mm^2 gibt Tabelle 1; die Werte bei 423 ^{o}K und 473 ^{o}K wurden durch Extrapolation ermittelt.)

Tabelle 1

Verformungstemperatur (^{o}K)	90	293	323	373	423	473
$\Delta\tau_{0,15}$ (kp/mm^2)	0,003	0,003	0,01	0,01	0,007	0,006

4. In einem Temperaturbereich, in dem bei Zink eine 3-Bereichs-Verfestigungskurve sehr ausgeprägt ist (293 bis

393 °K) kann die ($\Delta\tau$ - τ)-Abhängigkeit für die Bereiche A und B durch zwei Äste mit unterschiedlichen Neigungen dargestellt werden, was darauf schließen läßt, daß das Cottrell-Stokesche-Gesetz nicht für die ganze Verfestigungskurve gilt. Zu ähnlichen Ergebnissen führten auch Untersuchungen an Cd- (14) und Mg- (15) Einkristallen.

5. Eine lineare Abhängigkeit zwischen $\Delta\tau$ und τ tritt besonders im Bereich B der Verfestigungskurve zutage.

Literatur

(1) A. SEEGER, Z. Naturf. 9a, 758 (1954).

(2) M. BOČEK, Czech. J. Phys. B 10, 830 (1960).

(3) A. SEEGER und H. TRÄUBLE, Z. Metallk. 51, 435 (1960).

(4) M. BOČEK und G. HÖTZSCH, phys. stat. sol. 6, 777 (1964).

(5) M. BOČEK, P. LUKÁČ, B. SMOLA und M. ŠVÁBOVÁ, phys. stat. sol. 7, 173 (1964).

(6) M. BOČEK, G. HÖTZSCH und B. SIMMEN, phys. stat. sol. 7, 833 (1964).

(7) F. BENZE, S.E. BÜHLER und K. LÜCKE, Acta metall. 11, 1179 (1963).

(8) P.B. HIRSCH und D.H. WARRINGTON, Phil. Mag. 6, 735 (1961).

(9) A.H. COTTRELL und R.J. STOKES, Proc. Roy. Soc. A 233, 17 (1955).

(10) M.A. ADAMS und A.H. COTTRELL, Phil. Mag. 46, 1187 (1955).

(11) P. HAASEN, Phil. Mag. 3, 284 (1958).

(12) P.R. THORTON, T.E. MITCHELL und P.B. HIRSCH, Phil. Mag. 7, 337 (1962).

(13) J. DIEHL und R. BERNER, Z. Metallk. 51, 522 (1960).

(14) K.G. DAVIS, Canad. J. Phys. 41, 1454 (1963).

(15) Z.S. BASINSKI, Austral. J. Phys. 13, 284 (1960).

(Received February 8, 1965)

phys. stat. sol. 8, K167 (1965)

Institute of Nuclear Techniques, Cracow

Magnetic Structure of MnHg

By

A. OLEŚ

The magnetic properties of MnHg having an equiatomic composition have already been studied by Y. NAKAGAWA and T. HORI (1), as well as P. DE V. DU PLESSIS and L. ALBERTS (2). Measurements of the magnetic susceptibility in the temperature range 90 to 360 °K have shown that the susceptibility has a maximum value at 198 °K. According to these authors, MnHg is antiferromagnetic at low temperatures, with the Néel point at 198 °K. However, neutron diffraction patterns obtained by DE WET (6) revealed no magnetic ordering in MnHg.

In view of the results of all these authors the study of the antiferromagnetism of this compound seemed to be interesting.

The utilized method of preparing the MnHg compound was as follows: pure Hg and electrolytic Mn powders were mixed in the desired proportion, sealed in an evacuated quartz tube,

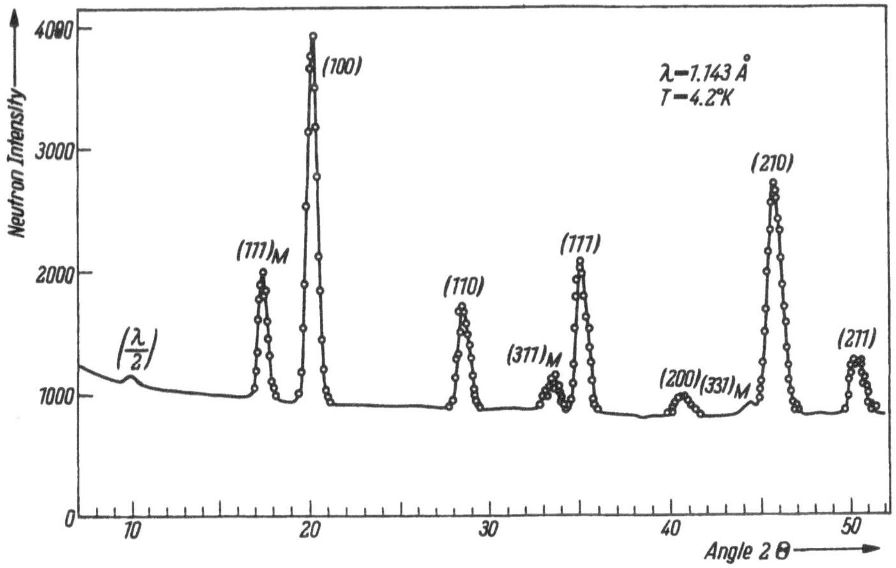

Fig. 1. Neutron diffraction pattern of MnHg (Monitor: 360 000 counts per point)

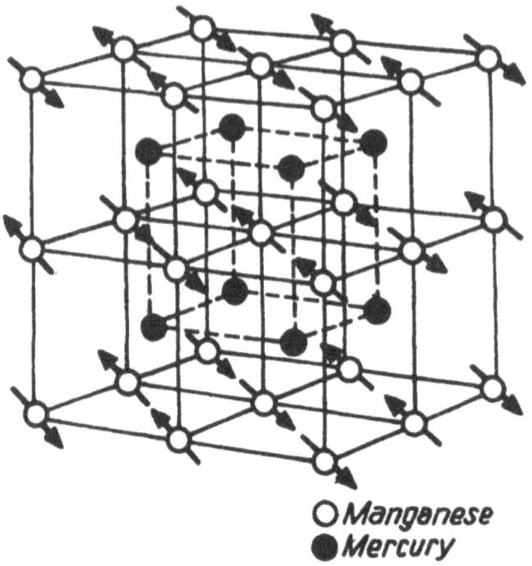

Fig. 2. The magnetic structure of MnHg

and heated to 400 °C. At this temperature the specimen was kept for 36 h. Next the obtained compound was powdered and carefully annealed at 150 °C for 30 h in an evacuated tube, in order to obtain sharp diffraction lines.

It was confirmed by the X-ray diffraction patterns that the products consisted only of MnHg (the simple ordered B2 type of structure). These investigations yielded a = (3.318 ± 0.002)Å at 24 °C. By chemical analysis the atomic composition was found to be 52.4 % Mn.

The neutron diffraction study was carried out using the reactor EL 3 at Saclay. The measurements were performed at room temperature and liquid helium temperature, at scattering angles 2Θ from 4° to 54°. The results obtained at 4.2 °K are shown in Fig. 1.

The integrated intensities were obtained by measuring the areas of the "peaks" in the intensity vs angle plots. It is interesting to note that the integrated intensities are insensitive to the change in the spin arrangement which occurs at 198 °K. The "peaks" $(111)_M$, $(311)_M$, $(331)_M$ indexed on a doubled cell exist also at room temperature. An analysis of the data shows that MnHg is antiferromagnetic already at room temperature. The magnetic structure of MnHg proposed in the previous paper (3) is illustrated in Fig. 2. Table 1 gives the calculated and observed integrated intensities.

The magnetic intensities were calculated using the magnetic form-factor found by SHULL and WOLLAN (4) for Mn^{++}. The obtained magnetic moment of Mn atoms is $\mu = (3.7 \pm 0.2)\mu_B$.

The antiferromagnetic structure of MnHg and the obtained magnetic moment of the manganese atom as published in (3) are

Table 1

Calculated and observed intensities (barns)		
hkl	$J_{calc.}$	$J_{obs.}$
$(111)_M$	81.0	82.2
100	259.5	257.2
110	79.0	80.8
$(311)_M$	28.8	28.1
111	119.0	122.5
200	20.4	20.0
$(331)_M$	7.8	6.0
210	221.6	221.2
211	56.3	54.0

in very good agreement with the results of Y. NAKAGAVA et al. (5).

In order to determine the Néel temperature, the intensity of the $(111)_M$ reflection was measured as a function of temperature. The results of these measurements are shown in Fig. 3.

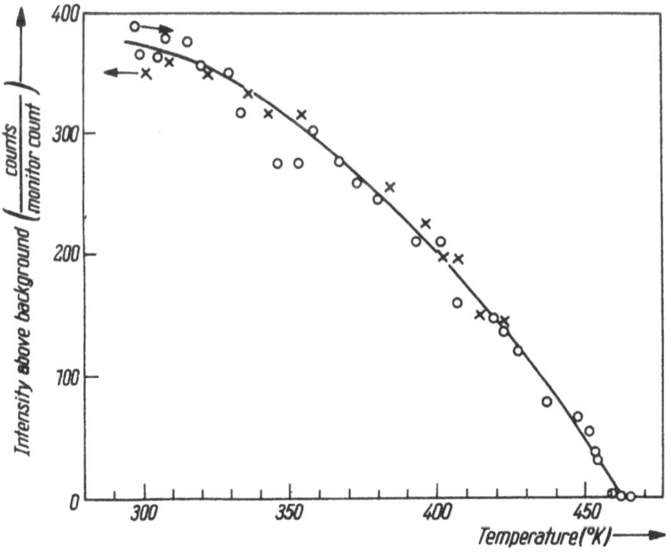

Fig. 3. Variation with temperature of the antiferromagnetic $(111)_M$ reflection of MnHg

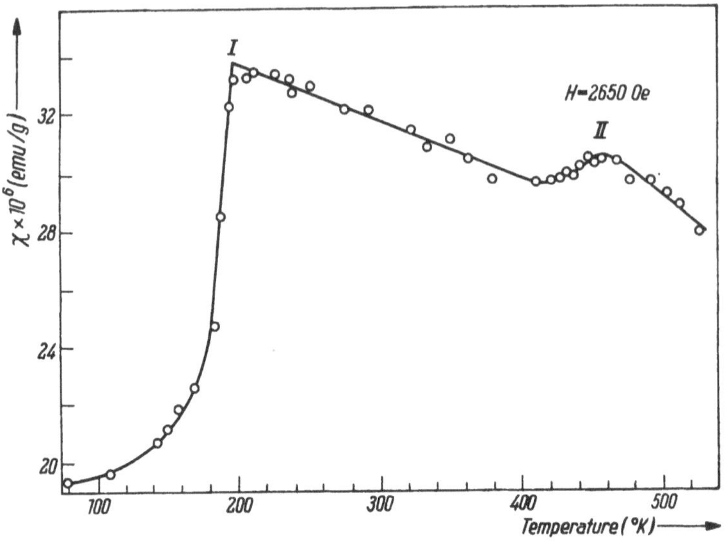

Fig. 4. Magnetic susceptibility of MnHg

The $(111)_M$ reflection disappears at 460 °K. Since there are no data about the magnetic susceptibility of MnHg at high temperatures, corresponding measurements were also carried out by means of the pendulum magnetometer constructed by Y. ALLAIN (Saclay). Fig. 4 shows the temperature dependence of the magnetic susceptibility. At 460 °K the second maximum of susceptibility was observed.

The results of neutron diffraction and the susceptibility vs temperature curve show that the Néel point is at 460 °K.

It is found from the neutron investigations that at 198 °K, which corresponds to the greatest value of magnetic susceptibility, a change occurs in the easy-magnetisation direction (3).

The author would like to express his sincere thanks to Dr. P. MERIÈL, Dr. R. PLUMIER, and Mr. B. BOUCHER for their interest and discussions. Thanks are due Mr. JEANAU for his cooperation in the X-ray investigations.

References

(1) Y. NAKAGAWA and T. HORI, J. Phys. Soc. Japan 16, 1470 (1961).

(2) P. DE V. DU PLESSIS and L. ALBERTS, Brit. J. appl. Phys. 15, 109 (1964).

(3) A. OLEŚ, Institute of Nuclear Research, Rep. Nr. 559/VI, Warsaw 1964.

(4) C.G. SHULL and E.O. WOLLAN, Solid State Physics, Vol. 2, 1956 (p. 201).

(5) Y. NAKAGAWA, H. WATANABE, and T. HORI, J. Phys. Soc. Japan 19, 2078 (1964).

(6) DE WET, Thesis, Pretoria 1959.

(Received February 8, 1965)

phys. stat. sol. 8, K173 (1965)

Institut für Magnetische Werkstoffe der Deutschen Akademie der Wissenschaften zu Berlin, Jena

Zur Anisotropiekonstanten K_1 bei Eisen, Kobalt, Ni-Ferrit und Mn-Ferrit[1]

Von

E. JÄGER

Das Problem der Kristallanisotropie bei ferromagnetischen Stoffen wurde in der Arbeit (1) mit Hilfe der Methode der Greenschen Funktionen behandelt. Ausgangspunkt war ein Hamiltonoperator, der als das für die kubische Anisotropie verantwortliche Glied eine Quadrupolkopplung enthält. Für die Anisotropiekonstante K_1 wurde der folgende Ausdruck erhalten (aus der Winkelabhängigkeit der Spinwellenenergie)

$$K_1 = \alpha Q N_v S^2 \sigma (1 - 2S\sigma)(S + 1 - 3S\sigma^2). \qquad (1)$$

Dabei bedeuten: α einen Gitterfaktor ($\alpha = -\frac{1}{2}, \frac{4}{9}, \frac{1}{4}$ für das einfache kubische, kubisch raumzentrierte bzw. kubisch flächenzentrierte Gitter), Q die Quadrupolkopplungskonstante, N_v die Teilchenzahl pro Volumen, S die Spinquantenzahl pro Gitteratom, σ die relative Magnetisierung (als Funktion des äußeren Feldes, der Temperatur und der Richtungskosinus der Gleichgewichtsmagnetisierung). In (2) wurde versucht, dieses Ergebnis auf dem direkten Weg aus der freien Energie abzuleiten; es ergab sich dabei

$$K_1 = \alpha Q N_v S^2 (S + 1 - 3S\sigma^2)^2. \qquad (2)$$

Wegen verschiedener in (1) und (2) gemachter Näherungen gelten beide Formeln nur für tiefe Temperaturen. Für T = 0 stimmen sie überein, es kann aber nicht exakt abgeschätzt werden, bis zu welcher Temperatur beide gültig sind bzw. beide, falls sie gültig sind, physikalisch das gleiche liefern. Als Prüfstein bleibt deshalb der Vergleich mit dem Experiment.

In den Fig. 1 bis 4 sind die berechneten relativen Werte der Anisotropiekonstanten $K_1(T)/K_1(T_0)$ über der Temperatur aufgetragen (Fig. 1 Eisen, Fig. 2 kubisches Kobalt,

---

1) Vorgetragen auf dem Internationalen Kolloquium über Ferromagnetismus in Poznan vom 3. - 6. 12. 1964.

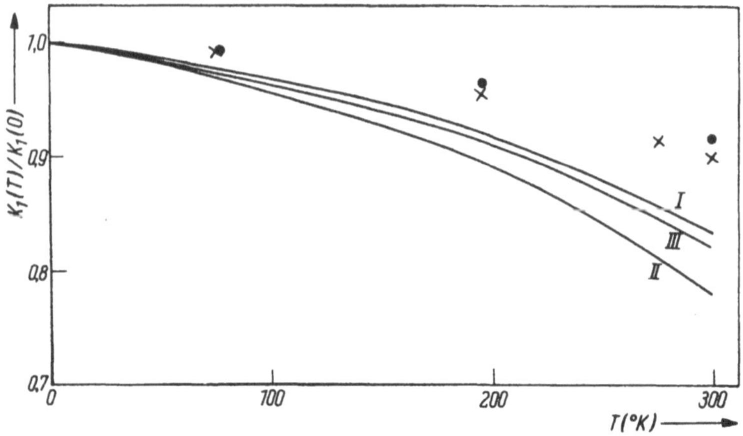

Fig. 1. $K_1(T)/K_1(0)$ von Eisen als Funktion der Temperatur
● Meßwerte von GRAHAM (4),
× Meßwerte von BOZORTH (bei GRAHAM angegeben)

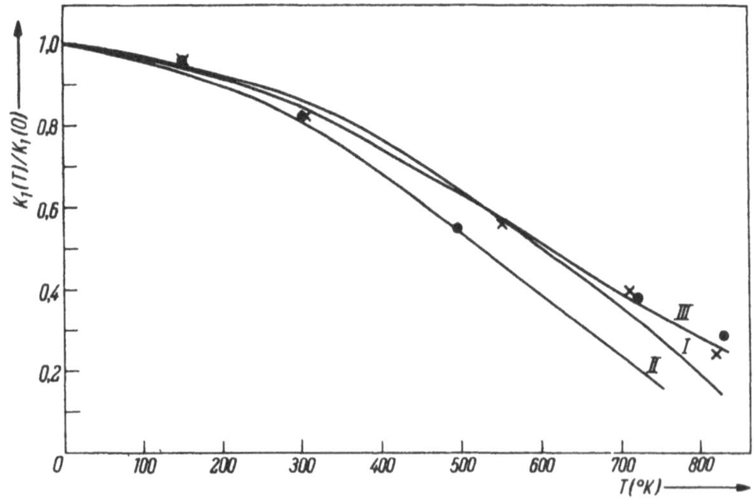

Fig. 2. $K_1(T)/K_1(0)$ von kubischem Kobalt als Funktion der Temperatur
● Meßwerte von RODBELL (5) an kleinen Teilchen,
× Meßwerte an dünnen Schichten

Fig. 3. $K_1(T)/K_1(T_o = 88\ ^oK)$ von Ni-Ferrit als Funktion von T/T_c
× Meßwerte von ELBINGER (7)

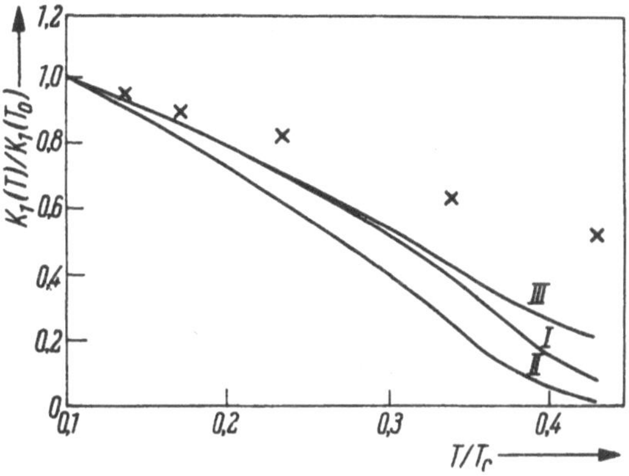

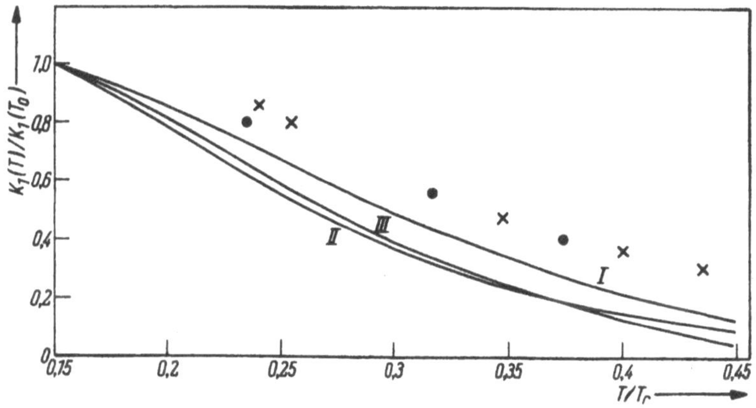

Fig. 4. $K_1(T)/K_1(T_o = 88\ ^oK)$ von Mn-Ferrit als Funktion von T/T_c
● Meßwerte von DILLON, GESCHWIND und JACCARINO (8),
× Meßwerte von ELBINGER (7)

Fig. 3 Ni-Ferrit, Fig. 4 Mn-Ferrit) Die Kurven I gehören zu Formel (1), II zu Formel (2). Als Vergleich ist noch das bekannte M^{10}-Gesetz aufgetragen (Kurven III). Dabei wurde so verfahren, daß die experimentellen Werte für die Magnetisierung mit der Temperatur T als Parameter in (1), (2) eingesetzt bzw. σ^{10} gebildet wurde. Die Meßwerte von K_1 sind durch Punkte bzw. Kreuze gekennzeichnet. Für Eisen wurden σ-Werte von GENGNAGEL (3) verwendet, die K_1-Werte, die durch Punkte dargestellt sind, wurden von GRAHAM (4), die durch Kreuze gekennzeichnet sind, von BOZORTH (bei GRAHAM zitiert) übernommen. Bei Co wurden sowohl die σ - als auch die K_1-Werte von RODBELL (5) genommen. Die Punkte stellen Meßwerte von kleinen Co-Teilchen, die Kreuze Meßwerte von dünnen Schichten dar. Die Magnetisierungswerte von Ni- und Mn-Ferrit wurden aus Arbeiten von PAUTHENET (6) entnommen. K_1-Messungen lagen von ELBINGER (7) (Kreuze), bzw. DILLON, GESCHWIND und JACCARINO (8) bei Mn-Ferrit vor (Punkte). Bei den Ferriten ist $K_1(T)/K_1(T_0 = 88\ ^oK)$ aufgetragen ($T \geq 88\ ^oK$), da die experimentellen K_1-Werte von ELBINGER nur bis zu dieser Temperatur reichten.

Aus den Figuren ist zu entnehmen, daß die berechneten Werte durchweg zu niedrig sind. (Vor allem die Kurve II liegt sehr tief, das ist aber wegen gewisser Näherungen in (2) verständlich.) Man kann dafür verschiedene Gründe anführen: Einerseits spielt die thermische Ausdehnung eine Rolle (9, 10), andererseits ist bei den Metallen Fe und Co das Modell mit lokalisierten Spins vielleicht nicht richtig, und bei den Ferriten sind vermutlich die Untergittereffekte von Bedeutung, so daß man die Anisotropie bei Ferriten wahrscheinlich anders behandeln muß (z.B. Kristallfeldtheorie). Man bemerkt in den Figuren weiterhin, daß die Kurven I und III meistens sehr dicht zusammen liegen. Innerhalb der Meßgenauigkeit stimmen sowohl I als auch III bei Co mit den Meßwerten gut überein. Ein Modell mit Quadrupolkopplung für die Anisotropiekonstante K_1 gibt trotz einer anderen funktionalen Abhängigkeit praktisch das gleiche wie das Zener-Akulowsche M^{10}-Gesetz.

Zum Schluß sei bemerkt, daß Ni mit diesem Modell nicht behandelt werden kann, weil für Ni $S \approx \frac{1}{2}$ ist, für $S = \frac{1}{2}$ verschwindet aber die Quadrupolkopplung.

Herrn Dr. W. HAUBENREISSER danke ich für wertvolle Diskussionen und den Herren Dr. H. GENGNAGEL und Dr. G. ELBINGER für die Bereitstellung von Meßdaten.

Literatur

(1) W. HAUBENREISSER und E. JÄGER, phys. stat. sol. 5, 33 (1964).

(2) W. BRODKORB, W. HAUBENREISSER und E. JÄGER, phys. stat. sol. 8, 551 (1965).

(3) H. GENGNAGEL, private Mitteilung.
(4) C.D. GRAHAM, Phys. Rev. 112, 1117 (1958).
(5) D.S. RODBELL, J. Phys. Soc. Japan, Suppl. B-I 17, 313 (1962).
(6) R. PAUTHENET, Ann. Phys. France, 7, 710 (1952); C.R. Acad. Sci. (France) 230, 1842 (1950).
(7) G. ELBINGER, Dissertation Jena 1961.
(8) J.F. DILLON, S. GESCHWIND und V. JACCARINO, Phys. Rev. 100, 750 (1955).
(9) W.J. CARR, Phys. Rev. 109, 1971 (1958); J. appl. Phys. 31, 69 (1960).
(10) A.I. MIZEK, Fiz. tverd. Tela 5, 1800 (1963).

(Received February 10, 1965)

phys. stat. sol. 8, K179 (1965)

Physics Department, University of Delaware, Newark

Franz-Keldysh-Effect for Appreciable Deflection of Light Beams[1)]

By

K.W. BÖER

As shown by FRANZ (1) and KELDYSH (2) high electric fields cause the absorption edge to shift towards longer wavelength. This effect is accompanied by a change in the index of refraction near the band edge. As computed by SERAFIN and BOTTKA (3) using the Kramers-Kronig dispersion relation the change in the refraction index at the band edge in a range in which the absorption constant is less than 10^2 cm^{-1} can amount to $\Delta n \simeq 0.02$ in germanium at field strength of about 10^5 V/cm.

In pn-junctions of semiconductors field strengths of this order are easy to achieve. These fields usually extend over a small region only, causing an appreciable field gradient, and therefore a gradient in the index of refraction. A light beam passing through a dielectric medium with changing index of refraction becomes deflected. This effect is most pronounced if the light beam passes perpendicular to the direction of highest gradient of refractive index.

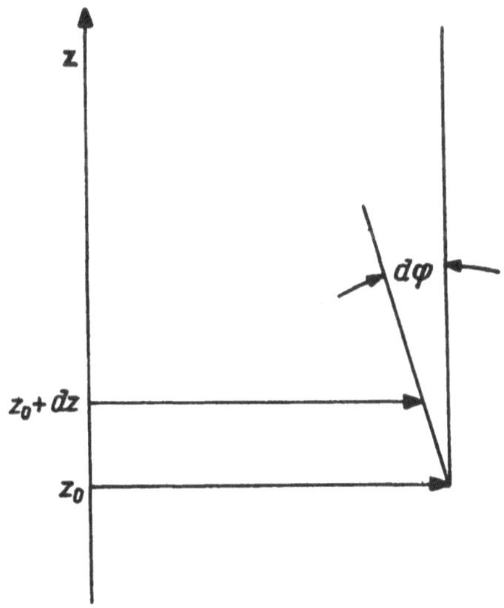

Fig. 1

Assuming the light beam enters a crystal parallel to a pn-junction at the portion near the highest field strength (z_o in Fig. 1) it will travel the distance $c\Delta t = nc_o\Delta t$ in a time interval Δt, (c light velocity, c_o vacuum light velocity, n index of refraction). A neighboring beam at z_o + dz travels $(c + dc)\,\Delta t = (n + dn)\, c_o\Delta t$,

1) Work supported by the U.S. Army Research Office Durham.

thus a deflection of the beam by $d\varphi = \frac{dn}{dz}$ will result. If the crystal is long enough so that the beam can be deflected completely out of the pn-junction, the total deflection in the crystal will be $\varphi = \Delta n$. If the beam enters the crystal perpendicular to its surface, the total angle of deflection after leaving the crystal will be

$$\alpha = n\Delta n.$$

In order to leave the high field region (width d) completely the crystal must have a thickness L of at least

$$L \geq \frac{2d}{\Delta n}$$

as can easily be seen by simple geometric discussion assuming small deflection angles and constant gradient of the refraction index.

Using the computation of SERAFIN and BOTTKA, one obtains deflection angles of the order of 10 degrees by a crystal of a thickness[2)] of about 0.5 mm. The laser technique, e.g. injection laser, promises to make this experimentally feasable in supplying a narrow collimated monochromatic beam. If it is possible to select the wavelength of the laser beam at a position where a large change of index of refraction and still high enough transparency is observed (4) e.g. close to the absorption edge, fast scanning over appreciable angles should be possible.

References

(1) W. FRANZ, Z. Naturf. 13a, 484 (1958).

(2) L.V. KELDYSH, Zh. eksper. teor. Fiz. 34, 1138 (1958).

(3) B.O. SERAFIN and N. BOTTKA, Bull. Amer. Phys. Soc. 10, 106 (1965).

(4) B.O. SERAFIN, R.B. HESS, and N. BOTTKA, Bull. Amer. Phys. Soc. 9, 714 (1964).

(Received February 11, 1965)

---

2) Assuming relatively wide pn-junctions of 5×10^{-4} cm in order to accomodate a narrow laser beam.

Pre-printed Titles and Abstracts

Papers to be published in "physica status solidi"
Vol. 9, No. 1

Review Article

H. STOLZ, Physikalisch-Technisches Institut der Deutschen Akademie der Wissenschaften zu Berlin, Theorie der magnetoakustischen Effekte in Metallen (IV)

Original Papers

F. KROUPA, Institute of Physics, Czechoslovak Academy of Sciences, Prague, Long-Range Elastic Field of Semi-Infinite Dislocation Dipole and of Dislocation Jog Expressions are given for the displacement field and the dilatation stress of a semi-infinite dipole of infinitesimal width. This corresponds to a long-range approximation for a dipole of finite width terminated within a crystal. The expressions also apply to a jog in a dislocation. Apart from the angular dependence, the displacement field decreases with distance as r^{-1} and the stress field as r^{-2}.

V.N. KASCHTSCHEEV, Institute of Physics, Academy of Sciences of the Lithuanian SSR, Riga, Spin Heat Capacity of a Ferromagnetic at High Temperatures (in Russian) On the basis of level splittings for the Green's spin functions an integral equation is obtained for the correlation $\langle S^z(q)S^z(-q)\rangle$, by using the method of TYABLIKOV (1). With this equation the energy and heat capacity of the spin system on both sides of the Curie point are calculated. At the Curie point T_c the energy of the spin system is discontinuous and the ferromagnetic heat capacity has a singularity of the form $(T_c - T)^{-1/2}$. The origin of this behaviour and the limits of applicability of the TYABLIKOV splitting are discussed in detail.

(1) S.V. TYABLIKOV, Ukrain. mathem. J. 11, 287 (1959) (in Russian).

E. GRAGERT und K. MEYER, Institut für Physikalische Chemie der Deutschen Akademie der Wissenschaften zu Berlin, Einfluß der Temperatur auf die Deformationsstrukturen in NaCl-Kristallen bei Druck- und Stoßbeanspruchung (100)-NaCl-Spaltflächen werden im Temperaturbereich von -150 bis 300 °C mit einer Stahlkugel statisch belastet sowie durch mechanische Impulsübertragung plastisch verformt. Die beobachteten Verformungsstrukturen an der Oberfläche und im Kristallinnern werden im untersuchten Temperaturbereich beschrieben. Bei den statischen Belastungsversuchen ist das Verformungsvolumen stark temperaturabhängig, während bei den Stoßversuchen die Bearbeitungstemperatur eine untergeordnete Rolle spielt.

M. ASCHE, Physikalisch-Technisches Institut der Deutschen Akademie der Wissenschaften zu Berlin, B.L. BOITSCHENKO und O.G. SARBEJ, Institut für Physik der Akademie der Wissenschaften der Ukrainischen SSR, Kiew, Abhängigkeit der Anisotropie der elektrischen Leitfähigkeit des Si vom elektrischen Feld Die elektrische Leitfähigkeit in Abhängigkeit von der Feldstärke wird experimentell und theoretisch an n-Si bei 77 °K untersucht. Die longitudinale und transversale Anisotropie werden an jeweils ein und denselben Proben als Funktionen der Orientierung in der (110)- und (100)-Ebene gemessen. Zur Analyse werden mit der Bilanzmethode Verteilungsfunktionen vom Maxwelltyp als Näherungen berechnet, dabei sind für die Zwischentalstreuung verschiedene mögliche f- und g-Übergänge herangezogen worden. In allen Fällen ergibt sich, daß die größten Unterschiede der Elektronenkonzentrationen in den verschiedenen Tälern sowie die Maxima der longitudinalen und transversalen Anisotropie bei kleineren Feldstärken als die maximalen "Temperatur"-Differenzen liegen. Der qualitative Verlauf der berechneten Anisotropie entspricht den experimentellen Kurven, für eine quantitative Übereinstimmung müßten die Besetzungszahlen der einzelnen Tälergruppen sich noch stärker unterscheiden.

V. TOPA, L. MATEI, and T. SERBAN, Institute of Physics, Bucharest, New Exciton Bands in KJ:Na Crystals with A-Centers

S.V. VONSOVSKII and M.S. SVIRSKII, Institute of Metal Physics, Academy of Sciences of the USSR, Sverdlovsk, and Tchelyabinsk Pedagogical State Institute, On the Theory of Superconductivity of an Electron System with Singlet and Triplet Pairs The general case of a superconductor with singlet and triplet pairs is investigated. The method uses the B-operators for creation and annihilation of singlet and triplet pairs, and A-operators for interchange of multiplicity of these pairs. The theory utilises the properties resulting from the symmetry type of the electron-pair wave-function which do not involve the conventional assumptions regarding the dependence of the matrix element $V_{\kappa\kappa'}$ on the vectors κ and κ. A more general form of electron-pair Hamiltonian is proposed which includes both the previously discussed (1) exchange terms and those due to the alternation of electron-pair multiplicity. The variational problem for the coexistence of singlet and triplet pairs is solved. The trial wave-functions are established both from the properties of the B-operator and the generalized unitary Yosida operator (2) for singlet and triplet pairs. The physical significance of the limitations imposed in (3, 4) on the solutions of superconductive problem with triplet pairs is clarified. The distribution functions are established for the different types of pair and it is shown that the solutions for an isotropic gap may correspond to an anisotropic distribution of these pairs. The properties of a superconductor at non-zero temperature are investigated using both the method of double-time Green's functions and the method of variation of the thermodynamic potential with the interaction parameter. Some modifications are considered which allow an analysis of the problem by quantum-field techniques.

(1) H. KOPPE and B. MÜHLSCHLEGEL, Z. Phys. 151, 613 (1958).
(2) K. YOSIDA, Phys. Rev. 111, 1255 (1958).
(3) R. BALIAN and N.R. WERTHAMER, Phys. Rev. 131, 1553 (1963).
(4) A.V. SVIDZINSKY, Dokl. Akad. Nauk SSSR 153, 1044 (1963).

B. LALEVIC, The Franklin Institute Laboratories, Philadelphia, Unstable Domain Structure in the Intermediate State of Superconductors Measurements of resistance fluctuations, metastable resistance levels, and related parameters have been extended to the case of the intermediate state of a specimen carrying a variable current in a constant external magnetic field. The origin of peculiar peaks in the plots of relative changes of resistance $\frac{\Delta R}{R}$ is now clarified and related to metastable resistance levels. The influence of specimen current on the shape of the resistance vs external magnetic field plot is explained on the basis of the present measurements. From a comparison of these and previously obtained data with the theory of domain boundary propagation, the conditions for the occurrence of domain boundary instabilities are derived and a mechanism responsible for the appearance of resistance fluctuations is discussed.

G. MAYER, Compagnie Générale de TSF, Orsay, Effets de stimulation multiple dans l'interaction de plusieurs ondes de bosons On considère des interactions élémentaires entre lumière et matière, où le nombre de bosons produits peut être supérieur de plusieurs unités au nombre de bosons incidents. De telles interactions, souvent trop faibles pour être décelées expérimentalement dans des conditions normales peuvent produire des phénomènes considérables dès que l'excitation est assez intense pour mettre en jeu les effets de stimulation. Le cas d'un photon donnant naissance par interaction avec des molécules à plusieurs phonons identiques et à un photon est particulièrement discuté à cause de ses incidences possibles sur les effets Raman stimulés et sur la réalisation de systèmes lasers à gain croissant.

E. IGRAS and T. WARMINSKI, Institute of Physics, Polish Academy of Sciences, Warsaw, Electron Mirror Observations of Semiconductor Surfaces at Low Temperatures A new type of electron mirror microscope is described schematically. The microscope allows the observation of electrical micro-fields on semiconductor surfaces in the temperature range 77 to 1300 $^{\circ}$K. Preliminary results of observations on silicon surfaces at low temperatures are presented. It is suggested that the observed pictures can be interpreted in terms of surface fields resulting from the trapping of minority carriers. Results concerning impurity precipitation in silicon and germanium are also given.

R.P. GUPTA and B. DAYAL, Department of Physics, Banaras Hindu University, Varanasi, Theoretical Specific Heats of Beryllium The specific heat of beryllium has been calculated from the electron gas model proposed by us in an earlier paper (1). The theoretical values of C_v computed from this model are in good agreement with the experimental values.

(1) R.P. GUPTA and B. DAYAL, phys. stat. sol. 8, 115 (1965).

J. BARANOWSKI, University of Warsaw, and J. MYCIELSKI, Institute of Physics, Polish Academy of Sciences, Warsaw, PEM and Nernst Effects in Modulated Light The investigation of the PEM effect involves always the sum of the light-induced Nernst and PEM effects. The relative value of these effects depends on the properties of a given material and frequency of light modulation. The theory of the Nernst effect, induced by modulated light, is developed by taking into account that the temperature gradient is not constant in the sample. The expression for the sum of the two effects is given. Results of measurements on mixed $Cd_{0.1}Hg_{0.9}Te$ crystals are reported and found to be in satisfactory agreement with the theory.

J. MATHON and D. FRAITOVÁ, Institute of Physics, Czechoslovak Academy of Sciences, Prague The Spin Dependent Effective Mass of the Conduction Electrons in Ferromagnetic Metals The energy correction of conduction electrons in ferromagnetic metals due to the s-d exchange is calculated by a perturbation method. It is found that the ordered d-system affects the energy of one s-electron in a manner to give allowance for the introduction of a spin dependent effective mass. The energy of one conduction electron is not changed if its spin and that of d-electrons are parallel, the energy correction becomes important if the spins are antiparallel. With this result the resonance frequency and the line width of the spin resonance of conduction electrons are evaluated according to IZJUMOV and SKROCKIJ (1).

(1) J.A. IZJUMOV and G.V. SKROCKIJ, Fiz. tverd. Tela 2, 1766 (1960).

A. BOHUN, J. SAK und J. DOLEJŠI, Institut für Festkörperphysik der Tschechoslowakischen Akademie der Wissenschaften, Prag, Optische und elektrische Erscheinungen in bleiaktivierten Alkalihalogeniden Die Absorptions- und Emissionsspektren von bleiaktivierten Alkalihalogenidkristallen (und orientierend auch von KCl-Tl) wurden bei 20 und -150 °C und verschiedener Anregung gemessen und mit den Absorptionsspektren von entsprechenden gesättigten Lösungen verglichen. Es wurde festgestellt, daß die Absorptionsbanden A bei den Alkalichloriden und -bromiden sowie bei KCl-Tl aus einer ganzen Reihe von Subbanden bestehen. Bei den Lösungen ist diese Feinstruktur weniger ausgeprägt. Die Absorptionskurven von Alkalijodiden sind glatt (ohne Feinstruktur). Die Erklärung der Feinstruktur basiert auf der Existenz von Bleikomplexen und deren Vibration, wird also als Elektronen-Vibrationsspektrum der Bleikomplexe gedeutet. Bei der thermisch stimulierten Elektronenemission und Lumineszenz werden die Herkunft der Elektronen und der Elementarmechanismus diskutiert.

M.S.R. CHARI, National Physical Laboratory of India, New Delhi, Exchange Interactions in Dilute Ag-Mn Alloys at Helium Temperature The expression for the drop in the electrical resistivity below the temperature of magnetic ordering, derived on the basis of the Kasuya-Yosida model (1 to 3) for dilute alloys containing a transitional metal solute, has been modified recently by COLLINGS, HEDGCOCK, and MUTO (4), in order to take the resistivity minimum into account. This has been used here to estimate the exchange interaction parameter in polycrystalline rods of silver containing 0.14, 0.32, or 0.55 at.-% of manganese in random solid solution. The predictions of the model as to the behaviour of the electronic Lorentz parameter in these alloys at liquid helium temperatures, with regard to the applied magnetic field and the temperature are found to agree qualitatively with the behaviour inferred from experimental results.

(1) T. KASUYA, Progr. theor. Phys. 16, 58 (1956).
(2) K. YOSIDA, Phys. Rev. 107, 396 (1957).
(3) T. KASUYA, Progr. theor. Phys. 22, 227 (1959).

(4) E.W. COLLINGS, F.T. HEDGCOCK, and Y. MUTO, Phys. Rev. 134, A 1521 (1964).

M.J. MAKIN and J.V. SHARP, Metallurgy Division, A.E.R.E., Harwell, A Model of "Lattice" Hardening in Irradiated Copper Crystals with the External Characteristics of "Source" Hardening A model is proposed to explain the formation of slip lines in irradiated copper crystals during the propagation of the Lüders front. In this model it is shown that, due to the sweeping up of the irradiation produced defect clusters by deformation, the critical stress to produce a slip line is that required to operate the source in producing the first dislocation loop in the environment of the defect clusters. As the slipped area expands a pile-up forms and the stress required to expand this pile-up falls. Since on this scale a normal tensile machine is soft, i.e. the load is constant, the pile-up rapidly increases in size and velocity and hence slip lines should form very quickly. This has been verified by high-speed photography. The sweeping up of the damage by deformation has also been verified experimentally. The model has important consequences for "source" and "lattice" hardening. It is shown that the lattice hardening of the dispersed defects produces all the characteristics of the deformation which have formerly been used as evidence of "source" hardening.

H. LAMATSCH, Institut für Angewandte Physik an der Universität Erlangen-Nürnberg, Elektronische Zustände an reinen Siliziumoberflächen Nach theoretischen Vorstellungen treten an reinen Kristalloberflächen mit Elektronen besetzbare Quantenzustände auf (TAMM, SHOCKLEY). Folgende Experimente wurden zur Prüfung der Existenz solcher Zustände unternommen. Die Oberflächenleitfähigkeit und ihre Änderung durch ein transversales elektrisches Feld (Feldeffekt) wurden an Siliziumkristallen beobachtet (1000 Ω cm, p-Typ). Die Proben hatten die Form zylindrischer Stäbchen, die auf einen Durchmesser von 0,3 mm geschliffen waren. Die Zylindersymmetrie der Meßröhre ermöglichte einen gleichmäßigen Beschuß mit Argonionen zur Reinigung und eine empfindliche Feldeffektmessung. Dem Ionenbeschuß folgte stets eine Glühbehandlung. Es gelang, das Minimum der

Oberflächenleitfähigkeit zu beobachten und damit Absolutwerte dieser Größe zu erhalten. Die gereinigte Oberfläche ist schwach p-leitend. Das Ferminiveau liegt 0,25 eV unterhalb der Mitte der Bandlücke. Die Zustandsdichte in der Umgebung des Ferminiveaus beträgt $1,5 \cdot 10^{15}$ $cm^{-2}eV^{-1}$. Für die integrierte Dichte der Oberflächenzustände in der ganzen verbotenen Zone kann nur eine untere Grenze mit $2 \cdot 10^{14}$ cm^{-2} angegeben werden. Durch vorübergehende Einwirkung von Sauerstoff sinkt die Zustandsdichte um etwa eine Größenordnung.

J. VAN LANDUYT and R. GEVERS, Solid State Physics Department, S.C.K. - C.E.N., Mol, and S. AMELINCKX, Carnegie Institute of Technology, Pittsburgh, Dynamical Theory of the Images of Micro-Twins as Observed in the Electron Microscope (I) The diffraction contrast at micro-twins, observed by transmission electron microscopy, is discussed on the basis of the dynamic theory and including anomalous absorption. The diffraction conditions and geometrical properties are analysed, and their effect on the image deduced. The contrast at overlapping twins, is discussed for the reflecting position. It is shown that a fringe pattern having similar properties to stacking-fault patterns may be observed. The analysis does, however, establish criteria which distinguish this image from one due to a stacking fault. The properties are illustrated by observations on (101) micro-twins in rutile, and the analytical conclusions are compared with computed profiles for thick foils and those of intermediate thickness. The conditions existing at twinning dislocations and at the intersection of anti-phase boundaries with twin boundaries in rutile are discussed.

R.R. HASIGUTI, Department of Metallurgy, University of Tokyo, Relaxation Peaks in Cold-Worked Metals Internal friction peaks, which appear in cold-worked metals above the Bordoni peak temperature, were considered theoretically on the basis of point-defect-associated dislocation models. The present theory as well as the previous theory of the same author give a relaxation type loss. Calculated activation energies, pre-exponential factors and relaxation strengths were discussed in the light of experimental data of ten metals and an alloy, and even of an ionic crystal. It is of importance in the present theory that activation energies give experimental values of binding energies between dislocations and point defects concerned (e.g. vacancies, interstitial atoms, impurity atoms, etc.).

P. HÖSCHL and C. KOŇÁK, Department of Solid State Physics, Faculty of Mathematics and Physics, Charles University, Prague, Sublimation of Cadmium Telluride and Cadmium Selenide under a Vapour Pressure of one of their Components and the Equilibrium Form of Crystal Growth The dependence of the sublimation velocity V_{AB}^{dif} $(g \cdot s^{-1} \cdot cm^{-2})$ of binary $A^{II}B^{VI}$ compounds on the vapour pressure of one of their components p_A' has been investigated. For this purpose the kinetics of elements sublimation and the mechanism of diffusion transport have been extended. A relationship $V_{AB}^{dif} \sim (p_A')^{-3}$ has been derived. A method for growing single crystals by heating a mixture of small (r_1)- and larger crystals (r_2) of binary AB compounds under a vapour pressure of the component A is described and established. The relationship $\alpha = \exp(4v_0\sigma_i/kT_s r_1)$ has been derived for the supersaturation (α) of the vapour phase at the crystal growth site by a generalization of the Thompson-Gibbs equation. By equilibrium growth of single crystals the G_I planes of CdTe and CdSe could be determined.

S. OLSZEWSKI and M. DUBEJKO, Institute of Physical Chemistry, Polish Academy of Sciences, Warsaw, Electron Correlations in Statistical Model and Cohesion of Solids A method is presented for calculating the cohesive energy of solids by means of a statistical model including both the exchange and the Coulomb correlations. The model takes into account the local lowering of the electron density due to the finite volume of the exchange and Coulomb correlation holes produced in the electron gas of the statistical atomic system. The cohesive energy and the volume occupied by one crystal atom have been estimated numerically for elements with atomic numbers Z varying from 1 to about 30, and the data of the theory are compared with the corresponding average experimental dependences on Z. A rather good agreement between theory and experiment is obtained for the cohesive energy if only the exchange correlations are considered. The introduction of the Coulomb correlations increases the cohesive energy by about one third of the value obtained by the exchange correlations alone.

E. HEGENBARTH, Arbeitsstelle für Tieftemperaturphysik der Deutschen Akademie der Wissenschaften zu Berlin, Dresden, Dielektrische Untersuchung an festen $Ba_xSr_{1-x}TiO_3$-Lösungen im Konzentrationsbereich von x = 0 bis 0,1 bei tiefen Temperaturen

An polykristallinen festen Lösungen von $(Ba_xSr_{1-x})TiO_3$ im Konzentrationsbereich von x = 0,01 bis 0,1 wurden die Temperatur- und Feldstärkeabhängigkeit der Dielektrizitätskonstanten im Temperaturbereich von 10 °K bis 80 °K gemessen. Es zeigte sich, daß die Abhängigkeit der Übergangstemperaturen von der Konzentration im Bereich x = 0,01 bis 0,1 viel größer ist als im Bereich x > 0,1. Die Ergebnisse werden diskutiert.

J. DANIEL-SZABO, V. HAJKO, Physikalisches Institut der Naturwissenschaftlichen Fakultät der Safárik-Universität, Kosice und H. GENGNAGEL, Institut für magnetische Werkstoffe der Deutschen Akademie der Wissenschaften zu Berlin, Jena, Studium des Reptations- und Bascule-Effektes verschiedener Ferromagnetika bei unterschiedlichen Magnetisierungszuständen

An drei Fe-Proben mit verschiedenen Kohlenstoffgehalten und an einer Probe aus Mo-Permalloy wurde der Reptation-Effekt für verschiedene Magnetisierungszustände untersucht. Die Messungen wurden nur bei geringen Schaltzahlen (n_{max} = 20) vorgenommen, dabei wurde während des Schaltens jede Magnetisierungsänderung gemessen (ballistisch). Sämtliche Messungen wurden für $H_B = 0$ und $H_A \approx H_c$ (Koerzitivfeldstärke) durchgeführt. Am Mo-Permalloy konnte kein Reptation-Effekt gemessen werden. Die Untersuchungen an den drei verschiedenen Fe-Proben ergaben: a) Die Größe des Reptation-Effektes hängt sehr wesentlich von dem gewählten Magnetisierungszustand ab. Es wurden sowohl für $H = H_B = 0$ als auch für $H = H_A$ Magnetisierungszustände gefunden, in denen die Magnetisierung sich in Abhängigkeit von der Schaltzahl nicht änderte. b) Die Größe des Reptation-Effektes hängt von der Reinheit des verwendeten Materials (Eisen) ab und ist bei sehr reinem Eisen mit einer Koerzitivfeldstärke von $H_c \approx 0{,}7$ Oe wesentlich kleiner als bei verunreinigtem Eisen mit H_c-Werten zwischen 1,5 und 8 Oe.

H. WAGNER und H. GENGNAGEL, Institut für magnetische Werkstoffe der Deutschen Akademie der Wissenschaften zu Berlin, Jena, Magnetische Untersuchungen über die Bildung von Nahordnungsbereichen von Fe-Al-Legierungen in der Nähe der stöchiometrischen Zusammensetzung Fe_3Al bei höheren Temperaturen

Messungen der während einer Magnetfeldglühung induzierten Anisotropie an Fe-Al-Legierungen ergeben, daß bei höheren Temperaturen in der Nähe der stöchiometrischen Zusammensetzung Fe_3Al nach langem isothermen Tempern relativ große einachsige Anisotropien auftreten. Die Größe dieser Anisotropie (etwa $4 \cdot 10^4$ erg cm^{-3}) und die Größe der Relaxationszeit (etwa 80 h bei 500 °C) zeigen, daß ein anderer Prozeß als der der Pärchenbildung und -ausrichtung dominieren muß. Messungen der Koerzitivfeldstärke im gleichen Temperaturbereich ergeben einen ähnlichen zeitlichen Verlauf; die Koerzitivfeldstärke einer polykristallinen Probe mit 23,5 At.-% Al erreicht nach etwa 40stündiger Temperung bei 550 °C (ohne Magnetfeld) einen Sättigungswert von ca. 170 Oe. Die Meßergebnisse können unter der Annahme von Nahordnungsinseln mit einem gewissen Magnetisierungsunterschied gegenüber der Matrix gedeutet werden. Das Auftreten dieser Nahordnungsinseln läßt wahrscheinlich einerseits eine Formanisotropie entstehen und andererseits die Koerzitivkraft auf den hundertfachen Wert ansteigen.

J. HESSE, Institut für Metallphysik der Universität Göttingen, Die plastische Verformung von Natriumchlorid

Die plastische Verformung von NaCl wird im dynamischen Druckversuch für verschiedene Kristallorientierungen, Verformungstemperaturen T und -geschwindigkeiten a untersucht. Die Verfestigungskurven (VK) zeigen eine Dreiteilung, wie sie von den kfz. Metallen her bekannt ist. Die Änderung der VK von NaCl mit T und $\dot{a}$ entspricht der der Metalle, dagegen ist die Orientierungsabhängigkeit in beiden Kristallsystemen völlig verschieden. Es wird versucht, die Verfestigung von NaCl versetzungstheoretisch zu interpretieren. Dabei werden Versetzungszählungen mittels Ätzgruben benutzt, die für den Bereich I der VK einen Zusammenhang zwischen der Ätzgrübchendichte N_P und Fließspannung τ von der Form $\tau = A\sqrt{N_P}$ ergeben. A ist unerwartet groß und stark von T, $\dot{a}$ und der Orientierung abhängig. Für [001]-orientierte Kristalle wird für den Bereich I geschlossen, daß die Fließspannung durch schwer bewegliche Sprünge bestimmt ist (und nicht allein durch elastische Versetzungswechselwirkung oder Schneidprozesse), und daß die Orientierungsabhängigkeit auf dem Mitwirken sekundärer Gleitsysteme vom Typ {100} ⟨110⟩ beruht. Der Verfestigungsmechanismus im Bereich II von NaCl, in dem τ stärker als $\sim\sqrt{N_P}$ anwächst, ist ungeklärt. Der Bereich III von [001] -Kristallen läßt sich formal auf thermisch aktivierte Quergleitung von Schraubenversetzungen zurückführen. Sowohl Sekundär- wie Quergleitung müßten sich elektronenmikroskopisch nachweisen lassen auf Kristallen, die unter Luftabschluß verformt wurden, wie Vorversuche zeigen.

G. SCHOTTKY, A. SEEGER und V. SPEIDEL, Institut für theoretische und angewandte Physik der Technischen Hochschule Stuttgart und Max-Planck-Institut für Metallforschung, Stuttgart, Thermisch aktivierte Quergleitung in Zink und Cadmium Die Deutung der Experimente von GILMAN (1) zur Prismengleitung von Zn und Cd durch thermisch aktivierte Quergleitung von aufgespaltenen Schraubenversetzungen aus der Basisebene heraus wird quantitativ behandelt. Die Rechnung geht von flächenhaften Versetzungsverteilungen aus, die kontinuumstheoretisch behandelt werden. Die theoretische Spannungsabhängigkeit der Aktivierungsenergie stimmt gut mit den Experimenten überein; als Stapelfehlerenergien ergeben sich 300 erg/cm^2(Zn) und 170 erg/cm^2(Cd).

(1) J.J. GILMAN, Trans. AIME 206, 1326 (1956).

M. VERANI-BORGUCCI and L. VERDINI, Istituto Nazionale di Ultracustica "O.M. Corbino", Roma, Electrical Resistance Measurements on Hydrogen-Charged Tantalum and Niobium By measuring the electrical resistance of tantalum and niobium wires, it has been possible to detect the effective introduction of hydrogen in these b.c.c. metals by electrolytic charging. In both metals this process gives an increase of the electrical resistivity which, for low concentrations, at room temperature is $\Delta\rho_H = 0.80$ µΩcm/at.-% of hydrogen in tantalum and $\Delta\rho_H = 0.75$ µΩcm/at.-% of hydrogen in niobium. By means of subsequent isochronal annealings in vacuum (10^{-4} mm Hg) it has been found that the hydrogen introduced by electrolytic charging is completely removed at temperatures above 800 °C. This degassing process becomes important only for temperatures above 500 °C for tantalum and 400 °C for niobium.

H. COHEN and W.S. DIETHORN, Nuclear Engineering Department, The Pennsylvania State University, Diffusion of Tritium from Neutron-Irradiated Lithium Fluoride Release of tritium from LiF is well described by volume diffusion kinetics below 500 °C and an activation energy of 51 kcal/mole. The tritium is introduced into the LiF by the $Li^6(n,\alpha)H^3$ reaction. Tritium release from massive LiF is complex; important factors include temperature history, initial tritium concentration (neutron exposure) and trapping of tritium in the LiF structure.

H. WILKENS, Max-Planck-Institut für Metallforschung, Stuttgart, Streuung von Blochwellen schneller Elektronen in ferromagnetischen Kristallen Die Beugung schneller Elektronen in Kristallen mit magnetischem Feld wird theoretisch untersucht, wobei das Vektorpotential des magnetischen Feldes als schwache Störung betrachtet wird. Man erhält Differentialgleichungen für die Ortsabhängigkeit der Anregungsstärken der Blochwellen, wie sie in ähnlicher Form für die Streuung von Blochwellen in Kristallen mit Gitterbaufehlern gefunden wurden (1). In der Säulennäherung ist die Wirkung des magnetischen Feldes einer Krümmung des Gitters gleichwertig. Die Differentialgleichungen werden für den Zweistrahlfall und ein homogenes Magnetfeld näherungsweise gelöst und der elektronenmikroskopische Hellfeld- bzw. Dunkelfeld-Flächenkontrast von ferromagnetischen Domänen diskutiert. Es wird gezeigt, daß dieser Flächenkontrast als Folge der Wanderung der Anregungspunkte sogenannter "modifizierter Blochwellen" im Sinne von PENNING und POLDER (2) bzw. KATO (3) verstanden werden kann. Die bisher vorliegenden experimentellen Untersuchungen, über die teilweise schon berichtet wurde (4), sind mit der Theorie im Einklang.

(1) M. WILKENS, phys. stat. sol. 6, 939 (1964).
(2) H.-M. THIERINGER und M. WILKENS, phys. stat. sol. 7, K5 (1964).
(3) P. PENNING und D. POLDER, Philips Res. Rep. 16, 419 (1961).
(4) N. KATO, J. Phys. Soc. Japan 18, 1785 (1963) und 19, 67 (1964).

Original Papers published in "Fizika Tverdogo Tela"
Vol. 7, No. 1

V.YA. KRAVCHENKO and V.L. VINETSKII, Kiev, Double-Phonon Processes in the Spin-Lattice Relaxation of F-Centres

The probability for the double-phonon relaxation of an F-centre due to hyperfine and spin-orbital interactions is calculated by a method developed in (1, 2). The results are applied to KCl.

(1) V.YA. KRAVCHENKO and V.L. VINETSKII, Fiz. tverd. Tela 6, 2075 (1964).
(2) V.L. VINETSKII, and V. YA. KRAVCHENKO, Zh. eksper. teor. Fiz. 47, 902 (1964).

V.A. TYAGAI and YU.YA. GUREVICH, Moscow, Calculation of the Dynamic Charging Curve of a Semiconductor Surface

The formation of an inversion layer in a doped semiconductor due to the passage of a square current-pulse is treated taking into account the generation of minority carriers in the quasi-neutral region. Expressions are given for the time dependence of the concentration and current of minority carriers at the boundary of the diffusion region. The contribution of the minority carriers to the total current is found to be negligible in an initial stage whose length depends upon the properties of the semiconductor. It is shown that the non-stationary processes which result from the flow of minority carriers from the bulk to the surface, lead to the appearance of an additional capacity in series with the capacity of the bulk charge.

V.L. BONCH-BRUEVICH and SH.M. KOGAN, Moscow, Formation of Domains in Semiconductors Possessing Negative Resistance

Conditions are given for the stability of a spatially homogeneous system of electrons possessing negative resistivity. It is shown that such a system can be stabilised by diffusion and recombination processes if its dimensions are sufficiently small.

M.K. SHEINKMAN, Kiev, Possibility of Auger Recombination at Multiply Charged Centres in Germanium and Silicon

An attempt is made to interpret available experimental data

on recombination at multiply charged centres in Ge and Si in terms of Auger processes (1, 2). Experiments are proposed which may give further evidence of the recombination mechanisms.

(1) L. BESS, Phys. Rev. 111, 129 (1958).
(2) M.K. SHEINKMAN, Ukrainskii fiz. Zh. 7, 1364 (1962) and Fiz. tverd. Tela 5, 2780 (1963).

A.M. KOSEVICH and V.D. NATSIK, Kharkov, Equation of Motion of Dislocations in Anisotropic Media The equation of motion of a dislocation in an anisotropic medium is discussed and a general expression derived for its mass-density tensor. The case of an arbitrary dislocation loop in a weakly anisotropic cubic crystal is studied in detail. The stress field produced by a system of moving dislocations in an infinite anisotropic medium is calculated to first order in the velocities.

V.N. VATULEV and A.F. PRIKHOTKO, Kiev, Martensite-Type Polymorphic Transformation in Octahydroanthracene Crystals Measurements of double refraction and intrinsic absorption spectra show that a martensite-type polymorphic transformation occurs in octahydroanthracene crystals.

L.I. DOMANSKAYA, G.I. KHARUS, and I.M. TSIDILKOVSKII, Sverdlovsk, Nernst-Ettingshausen Effect in Doped n-Type Silicon The transverse Nernst-Ettingshausen coefficient ($Q^{\perp}$) of Si single crystals containing 10^{16} to 10^{20} As atoms per cm^3 is measured between 100 and 400 ^{o}K. It is found that $Q^{\perp}$ changes sign at high As concentrations, and this is not understood on the basis of the Brooks-Herring theory. An order-of-magnitude agreement with the experimental results is obtained by including in this theory the deviation from the screened Coulomb potential due to the "cores" of the impurities.

L.S. KUKUSHKIN, Kharkov, Inverse Problems for Processes Connected with Many-Phonon Transitions A method is proposed which enables parameters of centres and of the lattice to be deduced from the shape of absorption and emission bands. The method also allows the density of states in the vibrational spectrum to be deduced from the shape of Mössbauer flanks and the incoherent scattering cross-section for slow neutrons.

The theory is based on an integral equation which includes many-phonon transitions to arbitrary order. This equation is solved for the case of low temperatures. It is found that this method requires only a moderate degree of experimental accuracy.

I.M. LIFSHITS and YA.E. GEGUZIN, Kharkov, Surface Phenomena in Ionic Crystals The equilibrium distribution of anion and cation vacancies near the surface of an ionic crystal is studied theoretically. It is shown that an electric double layer may appear near the surface due to the accumulation of one type of vacancies or charged impurities. An expression is given for the stationary distribution of vacancies established when different stresses are applied to two surfaces of a crystal. The charge appearing in the vicinity of an isolated spherical pore due to the inhomogeneous distribution of vacancies is calculated taking into account the Laplace pressure at the pore surface. Conditions are obtained for the compensation of the space charges due to an isolated pore and a system of pores.

A.A. VOLFSON, S.M. GORODETSKII, and V.K. SUBASHIEV, Leningrad, Photoconductivity in Heavily Doped p-Type Silicon The diffusion length of electrons and the lifetimes of electrons (τ_n) and holes (τ_p) in B-doped p-type Si are determined from measurements of photoconductivity. It is found that τ_n and τ_p differ by several orders of magnitude, and this is thought to be due to trapping. The concentration and positions of the trapping levels are deduced from photocurrent vs light intensity curves.

A.A. GUTKIN, D.N. NASLEDOV, and V.E. SEDOV, Leningrad, Spectral Characteristics of GaAs Photoelements The spectral distribution of the photovoltage of GaAs p-n junctions is measured between 78 and 430 °K. Recombination parameters of the investigated junctions are obtained by comparing the measured and theoretical spectral distribution between 1.6 and 2.7 eV. Deviations from the theoretical dependence are found and explained qualitatively.

R. TYAGARADZHAN and A.A. URUSOVSKAYA, Moscow, Motion and Multiplication of Dislocations in CsI The motion and multiplication of dislocations in CsI under compressional stress is investigated by means of an etching technique. The minimum stresses for motion and multiplication are found to be 1 and 3 g/mm^2, respectively. It is found that the dislocations in this substance can move by jumps between imperfections, (analogous to the cases of NaCl and LiF), as well as in a uniform manner. For the latter type of motion the velocity increases exponentially with applied stress. It is believed that the multiplication of dislocations at stresses above 20 g/mm^2 is due to double cross-slip.

P.A. BAZHULIN and A.A. RAKHIMOV, Moscow, Low-Frequency Raman Spectrum of Crystalline Stilbene and Tolane at Various Temperatures The temperature dependence of the position, integral intensity, and width of "low-frequency" Raman peaks of polycrystalline stilbene and crystalline tolane is measured between 123 and 373 $^{\circ}$K. The results are discussed in detail.

M.M. ZARIPOV and G.K. CHIRKIN, Kazan, Electron Paramagnetic Resonance of Iron-Group Ions in NH_4Cl Single Crystals An EPR study is made of interstitial Cu^{2}, Ni^{2+}, Co^{2+}, Fe^{2+}, and Fe^{3+} ions in NH_4Cl crystals. It is found that a compensation of the excess charge of interstitial Cu^{2+} and Fe^{3+} may occur along second-order crystal-axes, and Ni^{2+}, Co^{2+}, Fe^{2+}, and Fe^{3+} are compensated along fourth-order axes. The coefficients of the spin Hamiltonian are given for Cu^{2+}, Ni^{2+}, and Co^{2+}.

O.V. KOVALEV, Kharkov, Spiral Structures in MnO_2 The occurrence of spiral structures in MnO_2 is discussed on the basis of the theory of second-order phase transitions. It is found that the magnetic moments may be arranged in elliptic spiral configuration.

V.M. AGRANOVICH and YU.V. KONOBEEV, Theory of Long-Wavelength Surface-Excitons in Dielectrics A macroscopic method is developed which enables the polarisation and frequencies of long-wavelength surface-excitons to be determined. The method is based on an analytic extension of the dielectric tensor $\varepsilon_{ij}(\omega, k)$ to complex k-values and additional boundary-conditions arising from the presence of spatial dispersion. It is found that two types of surface excitons exist in the case of an isotropic medium surrounded by vaccum. The first type is characterised by non-zero electric field strength ($\vec{E}$), a frequency given by the condition $\varepsilon(\omega) = -1$, and negligible spatial dispersion. However for the second type $\vec{E} = 0$, $\vec{D} \neq 0$, and spatial dispersion is essential.

A.A. DOROZHKIN and N.N. PETROV, Leningrad, Potential Extraction of Electrons from Tungsten and Molybdenum by Mercury Ions The extraction of electrons from W and Mo by 500 to 2800 eV Hg^+ ions is studied using a modulated ionic beam. It is shown that potential extraction is dominant for pure targets, and this process is strongly temperature-dependent. In contrast to this, the potential as well as kinetic extraction of electrons from W by 0.5 to 8 keV Ar^+ ions is independent of temperature between 300 and 2100 ^oK. The results are discussed qualitatively.

YA.A. KRAFTMAKHER and E.B. LANINA, Novosibirsk, Formation Energy and Concentration of Vacancies in Platinum The formation of vacancies in Pt is studied by measuring the electric resistivity and specific heat at high temperatures. It is found that the relative concentration of vacancies is given by 90 exp(-1.6 eV/kT) and reaches 1 % at the melting point. The electric resistivity for 1 % vacancies is found to be 2.4 $\mu\Omega$cm.

S.N. KOMNIK, V.Z. BENGUS, and V.I. STARTSEV, Kharkov, Effect of the Formation of Dislocations on Twinning in Calcite The effect of the deformation rate and crystal imperfections on the broadening of stable twin-layers in calcite is investigated. It is shown that the broadening rate is determined by the formation of twinning dislocations.

N.N. KRAINIK, N.P. KHUCHUA, A.A. BEREZHNOI, and A.G. TUTOV, Leningrad, Nature of Phase Transitions in $BiFeO_3$-$PbFe_{1/2}Nb_{1/2}O_3$ Solid Solutions The phase transitions in $BiFeO_3$-$PbFe_{1/2}Nb_{1/2}O_3$ which cause or change the spontaneous polarisation are studied by X-ray diffraction at room temperature. ϵ and $\tan\delta$ are measured in the temperature range -150 to 800 °C.

V.V. ZHDANOVA, Leningrad, Dilatometric Investigation of $BiFeO_3-PbFe_{1/2}Nb_{1/2}O_3$ Solid Solutions The coefficient of thermal expansion of $xBiFeO_3+(1-x)PbFe_{1/2}Nb_{1/2}O_3$ is measured for 15 values of x, between 20 and 800 °C. The results indicate the presence of phase transitions and an extended morphotropic boundary in the range $0.65 < x < 0.75$.

S.L. GINZBURG, Leningrad, Damping Spin-Complexes The existence of both stable and unstable complexes of two spin waves is established theoretically.

S.A. ABASOV, B.B. KULIEV, and KH.M. KHALILOV, Baku, Time and Temperature Dependence of the Tensile Strength of Amorphous Selenium The destruction time (τ) of Se whiskers is measured at 12, 20, and 30 °C. It is found that, as in many other substances, τ is proportional to $\exp\left(\frac{U_0 - \gamma\sigma}{kT}\right)$ (σ tensile strength, U_0, γ constants).

A.I. STEKHANOV, A.A. KARAMYAN, and N.I. ASTAFEV, Leningrad, Infra-Red Absorption Spectra of Perovskite Type Ferroelectric Crystals Infra-red absorption spectra of $BaTiO_3$, $PbTiO_3$, and $CaTiO_3$ are measured between 2 and 300 μm. For all these substances two absorption peaks possessing a triplet structure are found at 550 and 360 cm^{-1}, respectively, and an additional peak between 180 and 190 cm^{-1}. The results are treated group-theoretically.

R.I. GARBER and E.I. STEPINA, Kharkov, Annihilation Velocity of Elastic Twins in Calcite The velocity (v) of the annihilation of elastic twins in calcite which occurs after the removal of the applied load is measured by a kinematographic technique. v is found to be about 10 m/s and to increase with temperature. This is considerably larger than the values obtained in (1, 2). The results indicate that the velocity of twinning dislocations exceeds 100 m/s.

(1) F.P. BOWDEN and R.E. COOPER, Nature 195, 1091 (1962).
(2) R.E. COOPER, Proc. Royal Soc. 270A, 89 (1962).

R.I. GARBER and L.M. POLYAKOV, Kharkov, Density Change of Ionic Crystals during Plastic Compression Measurements are made of the stress, deformation, and density changes in NaCl and KCl crystals under uniaxial compression. At small deformations the density at first increases, and then decreases to a constant value. This behaviour is explained by a splitting of the crystal into micro-blocks and the formation of ultra-microscopic imperfections. Further deformation leads to a sharp decrease of density, and this is followed by the destruction of the crystal. This decrease is related to the formation and growth of macroscopic imperfections. It is found that the average relative changes of density in a sample are of the order 10^{-5}, whereas local changes of the order 10^{-4} occur in the vicinity of glide bands. It is also shown that plastic deformation leads to remanent changes of density.

G.A. ANDREEV and B.P. ALEKSANDROV, Leningrad, Flotation Measurement of the Distribution of Univalent Impurities in NaCl Crystals during Growth The distribution coefficient of alkali and halogen ions between a NaCl crystal and its melt is obtained by measuring the density of the crystal by a flotation method. The effective distribution-coefficient is given as a function of impurity concentration and growth velocity. It is found that the equilibrium distribution of impurities is established at growth velocities below 0.5 mm per minute. The density of the crystal varies linearly with the impurity concentration in the melt. A correlation is found between the distribution coefficient and the difference between the radii of the impurity and substituted ion.

G.A. ANDREEV, Leningrad, Effect of Admixtures of Molecular Anions Containing Oxygen on the Density of NaCl and KCl Single Crystals The density changes of NaCl and KCl crystals due to the admixture of molecular ions to the melt are measured by a flotation technique. It is shown that admixtures of SO_4^{2-} and BO_2^- do not affect the density whereas considerable changes are observed for NO_3^-, NO_2^-, CO_3^{2-} (only in NaCl), OH^-, and O_2^{2-}. The magnitude of these changes depends upon the atmosphere surrounding the melt. The admixture of CO_3^{2-} in a CO_2 atmosphere does not affect the density of the crystals. This behaviour is attributed to thermal dissociation of the molecular ions and the interaction of the decay products with the atmosphere. The distribution coefficients of NO_3^-, NO_2^-, and OH^- in NaCl, and NO_3^- and OH^- in KCl are determined by the method of double crystallisation.

I.N. PENKOV and I.A. SAFIN, Kazan, Effect of Impurities on the NQR Spectrum of α - and γ-Bi_2O_3 An investigation is made of the effect of impurities on the NQR spectrum of Bi_2O_3. In the α-modification the impurity atoms are located near one of the two types of Bi lattice-sites. Al atoms are incorporated in the lattice of the γ-modification only if the crystal is pre-doped with Si.

S.V. TSIVINSKII and A.V. STEPANOV, Leningrad, Preparation of Germanium Single Crystals Having a Controlled Shape A method is described for the preparation of Ge single crystals having a controlled shape and crystallographic orientation.

S.A. ABAGYAN, S.M. GORODETSKII, T.B. ZHUKOVA, A.I. ZASLAVSKII, A.V. LISHINA, and V.K. SUBASHIEV, Leningrad, X-Ray and Optical Investigation of $GaAs_xP_{1-x}$ Crystals The lattice constants, density, refractive index, and infra-red transmission spectrum of GaAs-GaP crystals are measured at various compositions. The validity of Vegard's law is discussed. The observed absorption peaks are attributed to lattice vibrations.

V.P. PISAREV, Leningrad, Absorption Spectrum of Fe^{3+} Ions in $PbFe_{0.5}Nb_{0.5}O_3$ and $Ga_{2-x}Fe_xO_3$ Optical absorption spectra of antiferromagnetic $PbFe_{0.5}Nb_{0.5}O_3$ and ferromagnetic $Ga_{2-x}Fe_xO_3$ are measured at 77 and 295 ^{o}K. The observed peaks are attributed to transitions in Fe^{3+} ions occupying octahedral lattice-sites. For $PbFe_{0.5}Nb_{0.5}O_3$ a shift of these peaks is observed at 143 ^{o}K, and this is explained by the occurrence of magnetic ordering. It is found that the absorption spectrum of Fe^{3+} ions is only weakly affected by the crystal field in $Ga_{2-x}Fe_xO_3$, and this disagrees with conclusions drawn from magnetic data.

N.G. NAKHODKIN, A.A. OSTROUKHOV, and V.A. ROMANOVSKII, Kiev, Inelastic Scattering of Electrons in Thin Layers The inelastic scattering of electrons in a thin layer is treated, on the basis of (1), by solving the transport equation for normal and oblique incidence of the electronic beam, taking account of the exit conditions. For normal incidence the transmission coefficient is given as a function of thickness. An expression is derived for the dependence of the reflection coefficient on the atomic number (Z) of the target substance and found to agree with experimental data for small as well as large Z.

(1) N.G. NAKHODKIN, A.A. OSTROUKHOV, and V.A. ROMANOVSKII, Fiz. tverd. Tela 4, 1514 (1962) and 5, 41 (1963).

E.A. TUROV and V.G. SHAROV, Sverdlovsk, Energy Gap of Spin Waves in Ferro- and Antiferromagnetics Due to Magnetoelastic Energy It is shown that an energy gap appears in the spectrum of spin waves in ferro- and antiferromagnetics. This is partly due to the spontaneous deformations associated with the magnetoelastic energy, and partly due to the magnetic anisotropy energy and energy of the demagnetising fields. This magnetoelastic gap is independent, to first order, of the magnetisation vector and antiferromagnetic vector in ferro- and antiferromagnetics, respectively, and is most likely to appear in the following cases: 1) in uniaxial antiferromagnetics possessing a natural antiferromagnetic axis parallel to a basal plane whose crystallographic anisotropy is small, and 2) in uniaxial ferromagnetics possessing a negative anisotropy-constant of large magnitude. It is shown that energy gaps observed for hematite and dysprosium are of magnetoelastic origin. The effect of mechanical stress and piezomagnetism on the magnetoelastic gap is also considered.

YA.P. ZINGERMAN and V.A. ISHCHUK, Kiev, Investigation of the Interaction of Oxygen with Tungsten by Means of Electron-Stimulated Desorption The absorption of oxygen at W surfaces and its penetration into the interstitial space are studied by measuring, during these processes, the desorption due to irradiation with slow electrons. It is found that below 1040 °K the number of oxygen atoms incorporated on interstitial sites near the surfaces only increases as long as adsorption occurs. It is concluded that the incorporation energy is obtained from the released heat of adsorption.

L.V. KROTOVA, V.V. OSIKO, and V.T. UDOVENCHIK, Moscow, Effect of Oxygen Impurities on the Optical Properties of CaF_2-Sm Crystals A study is made of the effect of oxygen impurities on the valency transitions of Sm ions in fluorite crystals.

N.N. KUZIN and A.A. SEMERCHAN, Moscow, Temperature Dependence of the Resistivity of Germanium at Pressures up to 90 kbar The temperature dependence of the resistivity (ϱ) of n-type Ge (ϱ = 39.4 Ωcm at room temperature) is measured between 20 and 90 kbar, and 20 and 122 °C. It is found that the band gap has a maximum value of 0.84 eV at (40 $\pm$ 2.5) kbar. The application of high pressure causes insignificant mechanical damage of the crystals.

Z.N. GRIBNIKOV, Kiev, Theory of the Injection of Carriers in Long Diodes The current-voltage characteristics of a long diode (d/L > 4) is calculated. A relation of the form $\varrho \sim i^{1/2}[\delta - \delta_1(i)]^{3/2}$ with a logarithmic i-dependence of δ_1 is found in an intermediate range between the ohmic and saturation ranges. The special cases of p^+n diodes and p^+nn^+ diodes in which the current through the p^+n and nn^+ junction is carried by holes and electrons, respectively, are treated in detail. The limits of applicability of the theory are discussed.

V.P. NABEREZHNYKH and V.L. MELNIK, Kharkov, Resonance Effects in Zin in an Oblique Magnetic Field Cyclotron resonance occurs in zinc at 3.6×10^{10} Hz and field strengths up to 10 kG, over a certain range of angle between the magnetic field and the normal to the surface. The harmonic resonance peaks, and an additional peak of unknown origin, are observed.

M.E. LEVINSHTEIN, Leningrad, Effect of Local Surface-Inhomogeneities on the Width of Ferromagnetic Resonance Lines A discussion is given of the anisotropy of the width of the ferromagnetic resonance line which is observed for spherical samples in the presence of a local surface-inhomogeneity. It is shown that this effect can be explained by the occurrence of inhomogeneous precession rather than the propagation of plane spin-waves from the inhomogeneity to the centre of the sphere.

YU.K. VORONKO, V.V. OSIKO, V.T. UDOVENCHIK, and M.M. FURSIKOV, Moscow, Optical Properties of CaF_2 Crystals Containing Dy^{3+} Absorption and emission spectra, and excitation spectra of luminescence are measured on Dy^{3+} doped CaF_2 crystals possessing various concentrations of oxygen impurities.

Short Notes published in "Fizika Tverdogo Tela"
Vol. 7, No. 1

A.M. RODICHEV and R.G. KHLEBOPROS, Krasnoyarsk
Effect of Magnetoelastic Coupling on the Motion of the Magnetic Moment

I.F. TIGANE, Tartu
Electron-Microscopic Investigation of Conducting SnO_2 Layers

S.I. SHAKHOVTSOVA, I.D. KNOZENKO, and E.A. MUZALEVSKII, Kiev
Generation of Current Pulses by CdS Single Crystals

E.V. CHISLER, Leningrad
Temperature Dependence of the Intensity of First-Order Raman Spectra of Crystals

V.I. CHEREPANOV and A.N. MEN, Sverdlovsk
Calculation of the Parameters of the Spin Hamiltonian of Cr^{3+} in $MgAl_2O_4$

A.A. SAMOKHVALOV and A.N. MEN, Sverdlovsk
Relation between the Energy Spectrum of Fe^{2+} Ions and the Conductivity of Spinel-Type Ferrites

I.N. KURKIN and V.G. STEPANOV, Kazan
Electron Paramagnetic Resonance of Gd^{3+} in an Artificial $SrMoO_4$ Crystal

I.M. DYKMAN and P.M. TOMCHUK, Kiev
Anisotropy of the Conductivity due to Hot Electrons and Interelectronic Interactions

E.A. SALKOV, Kiev
Several Properties of p(SiC)-n(CdS) Junctions

E.F. GROSS, L.G. SUSLINA, and V.G. MOKEROV, Leningrad
Narrow Luminescence-Lines in Hexagonal ZnS Crystals

V.A. TOCHIN, Moscow
Photoconductivity in Irradiated Aromatic Compounds

V.I. SOKOLOV, Leningrad
Character of Cleavage Surfaces of Semiconducting Crystals

Containing a p-n Junction

I.G. ISMAILZADE and S.A. KIZHAEV, Leningrad
termination of the Curie Point of Ferroelectric $YMnO_3$ and $YbMnO_3$

V.E. ADAMYAN, A.V. GOLUBKOV, and G.M. LOGINOV, Leningrad
Magnetic Susceptibility of SmS

S.A. POLTINNIKOV and G.M. PRIKHODKINA, Leningrad
Temperature Dependence of the Initial Permeability of Co-Zn Ferrites

G.A. BOGACHEV, Moscow
Relaxation of Rare-Earth Ions

T.V. MASHOVETS and R.YU. KHANSEVAROV, Leningrad
Energy Spectrum of γ-Defects in Ge

S.A. ALTSHULER and R.I. MINEEVA, Kazan
Nuclear Magnetic Resonance Due to Nuclei of Paramagnetic Atoms

A.S. SHIBANOV, Moscow
Equilibrium Concentration of F-Centres in Additively Coloured Alkali Halide Crystals

A.V. SAVITSKII and Z.V. PANKEVICH, Chernovtsy
Piezoresistance Effect in p-Type ZnTe

I.M. TSIDILKOVSKII, M.M. AKSELROD, and V.I. SOKOLOV, Sverdlov
Oscillations of Magnetoresistance in Pure n-Type InSb

V.L. VINETSKII and C.YA. KRAVCHENKO, Kiev and Moscow
Single-Phonon Spin-Lattice Relaxation Due to Optical Vibrations

A.V. YUKHNEVICH, Leningrad
Emission Spectrum Due to the Capture of Holes by A-Centres in Si

A.I. MOROZOV, Moscow
Dependence of the Ultrasonic Absorption Coefficient of CdS Crystals on Light Intensity

V.V. PROKLOV, E.E. GODIK, and YA.E. POKROVSKII, Moscow
Generation-Recombination Noise in Si Doped with B and In

V.G. SHAVROV, Sverdlovsk
Effect of an Electric Field on the Resonance Frequency of Antiferromagnetics

B.V. KORNILOV
Generation of Oscillations by a Plate of Cu-Doped n-Type Ge

E.L. ZORINA, Moscow
Absorption of Light in Vitreous As_2S_3 and As_2S_5 near the Absorption Edge

F.A. MAKOVSKII, Kazan
Optical Properties of Fe Doped TiO_2

B.S. MURAVSKII, Krasnodar
Determination of Parameters of "Fast" States by Means of Electrical Excitation

N.G. BASOV, E.M. BELENOV, and V.S. LETOKHOV, Moscow
Synchronisation of the Oscillations in a Semiconductor Laser Containing Several p-n Junctions

List of Short Notes

List of Short Notes - Continued

Short Notes

(listed on the last pages of the issue)

Pre-printed Titles and Abstracts

of papers to be published in this or in the Soviet journal „Физика Твердого Тела“ (Fizika Tverdogo Tela).

Manuscripts and letters should be sent to

physica status solidi, 102 Berlin, Neue Schönhauser Str. 20

or Internationaler Buchversand GmbH, 1 Berlin 19, Eschenallee 21.

The object of this journal is to publish original work in the general field of solid state physics. Each issue will be published on the first day of the month. Items published will fall into the following categories:

1. Review articles on topical subjects in the field of solid state physics written by arrangement with the editors.
2. Original papers on new results of research, which have not been submitted for publication elsewhere, after being reviewed. (Contributions should be sent in by the 8th day of each month, 50 days before the delivery of the journal).
3. Short notes on topical results of research. These will be published in the photographically reproduced part of the journal and are the sole responsibility of the author. (Manuscripts for this part should be sent in by the 12th day of each month, 16 days before the delivery of the journal).
4. Pre-printed titles and abstracts of papers to be published in this or in the Soviet journal „Физика Твердого Тела“ (Fizika Tverdogo Tela), the latter translated into English.

Editorial Information

Manuscripts may be submitted in English, German, French, and Russian. Authors are asked to observe the following rules:

1. Manuscripts should be typewritten on one side only, with at least double spacing. For both review articles and original papers two copies should be sent. Manuscripts of original papers and short notes should not exceed 20 and 3 pages, respectively.
2. The author's name, initials and institution should be given.
3. An abstract of about ten lines, and, if possible, a translation of it into one of the above languages is required.
4. Drawings should be enclose don separate sheets of paper; photographs should be submitted, on white paper, only for review articles and original papers.
5. Captions for the figures should be submitted on separate sheets of paper, and special instructions with regard to the size of illustrations should be given on these.
6. References should be numbered in order of appearance and should be given, on a separate sheet, in the form used in this journal.
7. Units, symbols, etc. should be used in accordance with the recommendations given by the IUPAP.

The authors will be sent two galley proofs of review articles and original papers. Proof reading of short notes and pre-printed titles and abstracts will be done by the editorial staff.

For each original paper, and short note, a total of 75 reprints is available free of charge. The number of pre-prints free of charge per review article is 25. In addition review articles will be remunerated.

Kolibri 360 GmbH
Ferdinand-Jühlke-Straße 7
99095 Erfurt, Deutschland
produktsicherheit@kolibri360.de